FUNDAMENTALS OF ANALYTICAL TOXICOLOGY

Robert J Flanagan
Department of Clinical Biochemistry,
King's College Hospital NHS Foundation Trust,
London, UK

Andrew Taylor
Royal Surrey County Hospital, Guildford, Surrey, UK

Ian D Watson
Department of Clinical Biochemistry, University Hospital Aintree
Liverpool, UK

Robin Whelpton
School of Biological and Chemical Sciences, Queen Mary,
University of London, London, UK

BICENTENNIAL 1807 WILEY 2007 BICENTENNIAL

John Wiley & Sons, Ltd

Email (for orders and customer service enquiries): cs-books@wiley.co.uk
Visit our Home Page on www.wileyeurope.com or www.wiley.com

Other Wiley Editorial Offices

John Wiley & Sons Inc., 111 River Street, Hoboken, NJ 07030, USA
Jossey-Bass, 989 Market Street, San Francisco, CA 94103-1741, USA
Wiley-VCH Verlag GmbH, Boschstr. 12, D-69469 Weinheim, Germany
John Wiley & Sons Australia Ltd, 33 Park Road, Milton, Queensland 4064, Australia
John Wiley & Sons (Asia) Pte Ltd, 2 Clementi Loop #02-01, Jin Xing Distripark, Singapore 129809
John Wiley & Sons Canada Ltd, 6045 Freemont Blvd, Mississauga, Ontario, L5R 4J3, Canada

Wiley also publishes its books in a variety of electronic formats. Some content that appears in print may not be available in electronic books.

Anniversary Logo Design: Richard J. Pacifico

Library of Congress Cataloging-in-Publication Data

Fundamentals of analytical toxicology / Robert J. Flanagan ... [et al.].
p. ; cm.
Includes bibliographical references and index.
ISBN 978-0-470-31934-5 (hb : alk. paper)
ISBN 978-0-470-31935-2 (pbk. : alk. paper)
1. Analytical toxicology. I. Flanagan, Robert James.
[DNLM: 1. Toxicology—methods. 2. Chemistry, Clinical—methods. 3. Specimen Handling—methods.
QV 602 F9805 2007]
RA1221.F86 2007
615.9′07—dc22

2007013704

British Library Cataloguing in Publication Data

A catalogue record for this book is available from the British Library.

ISBN 9780470319345 (HB)
9780470319352 (PB)

Typeset by Aptara, New Delhi, India.
Printed and bound in Great Britain by Antony Rowe Ltd, Chippenham, Wiltshire.

This book is printed on acid-free paper responsibly manufactured from sustainable forestry in which at least two trees are planted for each one used for paper production.

Contents

Preface

The analytical toxicologist may be required to detect, identify, and in many cases measure a wide variety of compounds in samples from almost any component of the body or in related materials such as residues in syringes or in soil. Many difficulties may be encountered. The analytes may include gases such as carbon monoxide, drugs, solvents, pesticides, metal salts, and naturally occurring toxins. Some samples may be pure chemicals and others complex mixtures. New drugs, pesticides, and other substances continually present novel challenges in analysis and in the interpretation of results. The analyte might be an endogenous compound such as acetone, or an exogenous compound such as a drug and/or metabolite(s) of the drug, whilst the sample matrix may range from urine to bone.

Many biological samples contain muscle, connective tissue, and so forth, which may have to be separated or degraded prior to the analysis, as well as a multitude of small and large molecular weight compounds. The concentration of the analyte to be measured can range from g L^{-1} (parts per thousand) in the case of blood ethanol to μg L^{-1} (parts per thousand million) in the case of plasma digoxin. The stability of the analytes also varies considerably, ranging from a few minutes for protease sensitive peptides and esters such as aspirin and heroin, to several years for some other drugs and pesticides.

This book aims to give principles and practical information on the analysis of drugs, poisons and other relevant analytes in biological specimens, particularly clinical and forensic specimens. The material presented here is intended to cover part of the basic theoretical syllabus of the Association for Clinical Biochemistry (ACB) pre-registration training course in Clinical Chemistry, subspecialty Analytical Toxicology (http://acb.org.uk/training/documents/gradeA_tox.pdf, accessed 20 April 2006). The book should also prove useful to other analytical toxicology trainees and also to those undertaking preregistration training in Clinical Chemistry (http://acb.org.uk/training/documents/gradeA_biochem.pdf). As such, this volume extends the scope of the World Health Organization (WHO) basic analytical toxicology manual (Flanagan, et al. *Basic Analytical Toxicology*. WHO, Geneva: 1995), but is not intended to provide detailed coverage such as that given in *Clarke's Analysis of Drugs and Poisons* (Moffat, et al., eds. Edition 3. Pharmaceutical Press, London: 2004).

It has been assumed that the reader will be familiar with basic analytical laboratory operations. Many of the topics discussed here (use of clinical specimens, samples and standards, sample pretreatment, TLC, UV/visible spectrophotometry, GLC, HPLC, etc.) are the subject of monographs in the Analytical Chemistry by Open Learning (ACOL) series, and study of these volumes is recommended for those with a limited background in analytical chemistry. A distance learning course in analytical chemistry is available (http://www.rsc.org/education/careersandCPD/acol.asp, accessed 3 August 2007).

A detailed basic text (*Laboratory Skills Training Handbook*. Bailey and Barwick, 2007) has been produced under the Valid Analytical Measurement (VAM) programme. The VAM website (http://www.vam.org.uk/home.asp, accessed 3 August 2007) details a number of related publications, many of which can be downloaded free of charge.

The appendices to this book (incorporating a guide to interpreting analytical toxicology results, CAS registry numbers, a list of volatile compounds & conversion factors, and a glossary of commonly-used terms) are available online, and can be accessed at www.wiley.com/go/flanagantoxicology.

The authors and publishers would like to remind readers that web addresses cited in the text, including those in the preface, are current and accurate as we go to press. However, we cannot guarantee that cited resources or content will remain indefinitely at the exact addresses given.

Health and Safety

Care should be taken to ensure the safe handling of all chemical or biological materials, and particular attention should be given to the possible occurrence of allergy, infection, fire, explosion, or poisoning (including transdermal absorption or inhalation of toxic vapours). Readers are expected to consult local health and safety regulations and adhere to them.

Nomenclature, Symbols and Conventions

We have followed IUPAC nomenclature for chemical names except when Chemical Abstracts nomenclature or trivial names are more readily understood. With regard to symbols we have adopted the convention that variables and constants are italicized, but labels and mathematical operators are not (IUPAC Compendium of Chemical Terminology, Edition 2, 1997; http://www.iupac.org/publications/compendium/index.html, accessed 20 April 2006). Thus, for example, the acid dissociation constant is written K_a, a being a label to denote that it is an acid dissociation constant. The notation for the negative logarithm of K_a is $\mathrm{p}K_\mathrm{a}$; p is a mathematical symbol and is not italicized. Where the subscript is a variable then it is italicized, so the concentration at time t, is C_t, but the concentration at time 0 is C_0. Note especially that relative molecular mass (molecular weight, relative molar mass), ratio of the mass of an atom or molecule to the unified atomic mass unit (u), is referred to throughout as M_r. The unified atomic mass unit, sometimes referred to as the dalton (Da), is defined as $1/12$ of the mass of one atom of ^{12}C. The symbol amu for atomic mass unit can sometimes be found, particularly in older works. The unified atomic mass unit is not a Système International (SI) unit of mass, although it is (only by that name, and only with the symbol u) accepted for use with SI.

We have adopted the convention whereby square brackets [] used in figure legends, etc. denote analyte concentration.

As to drugs and pesticides, we have used recommended International Nonproprietary Name (rINN) or proposed International Nonproprietary Name (pINN) whenever possible. For drugs of abuse, the most common chemical names or abbreviations have been used. It is worth noting that for rINNs, it is now general policy to use 'f' for 'ph' (e.g. in sulfur), 't' for 'th' (e.g. chlortalidone not chlorthalidone) and 'i' for 'y' (mesilate not mesylate for methanesulfonate, for example). However, so many subtle changes have been introduced in recent years that it is difficult to ensure compliance with all such changes. Names that may be encountered include British Approved Name (BAN), British Pharmacopoeia (BP) name, United States Adopted Name (USAN), United States National Formulary (USNF) name, and United States Pharmacopoeia (USP) name. Where the rINN is markedly different from common US usage, e.g. acetaminophen rather than paracetamol, meperidine instead of pethidine, the alternative is given in parentheses at first use and in the index.

A useful source of information on drug and poison nomenclature is the *Merck Index* (Edition 14. O'Neil et al., eds. (2006) Merck & Co., Whitehouse Station, NJ); it should be noted that the reference number of a particular compound (Merck Index number) changes with edition. Chemical Abstracts Service (CAS) Registry Numbers (RN) provide a unique identifier for individual compounds, but it is again important to be clear which numbers refers to salts, hydrates, racemates, etc. Similarly, when discussing dosages we have tried to be clear when referring to salts, etc. and when to free bases or quaternary ammonium compounds.

We emphasize that cross-referral to an appropriate local or national formulary is mandatory before any patient treatment is initiated or altered. Proprietary names must be approached with caution – the same name is sometimes used for different products in different countries (see: http://www.fda.gov/oc/opacom/reports/confusingnames.html, accessed 20 April 2006).

The oxidation number of metals is given by, for example, iron(II), but older terminology such as ferrous and ferric iron for Fe^{2+} and Fe^{3+}, respectively, will be encountered, as discussed in Chapter 11.

Amount Concentration and Mass Concentration

In the UK and in other parts of Europe some laboratories report analytical toxicology data in 'amount concentration' using what have become known as SI molar units (μmol L^{-1}, etc.), while others, especially in the US, continue to use mass concentration [so-called 'traditional' units (mg L^{-1}, etc. or even mg dL^{-1})]. Most published analytical toxicology and pharmacokinetic data are presented in SI mass units per millilitre or per litre of the appropriate fluid [the preferred unit of volume is the litre (L)], or units which are numerically equivalent in the case of aqueous solutions:

$$\text{[parts per million]} = \mu\text{g g}^{-1} = \mu\text{g cm}^{-3} = \mu\text{g mL}^{-1} = \text{mg L}^{-1} = \text{mg dm}^{-3} = \text{g m}^{-3}$$

When preparing written statements for a court of law or other purpose outside the normal reporting channels it is advisable to write out the whole unit of measurement in full (milligrams per litre, for example). An exception to the above is carbon monoxide which is usually reported as % carboxyhaemoglobin (% COHb); the SI convention is that fractions of one should be used rather than percentages, but this is generally ignored.

We have followed the recommendations of the UK NPIS/ACB (National Poisons Information Service/Association for Clinical Biochemistry) for reporting analytical toxicology results and have used the litre as the unit of volume and SI mass units except for lithium, methotrexate, and thyroxine [NPIS/ACB (2002) Laboratory analyses for poisoned patients: joint position paper. *Ann Clin Biochem* **39**, 328–39]. More information on SI is available [see Flanagan R. J. (1995) SI units – Common sense not dogma is needed. *Br J Clin Pharmacol* **39**, 589–94].

Conversion from mass concentration (ρ) to amount concentration (c) ('molar units') and vice versa is simple if the molar mass (M) of the compound of interest is known. Thus, using in the example a compound with a molar mass of 151.2 g mol^{-1}:

$$c = \rho/M \quad \text{For example: } (1 \text{ mol L}^{-1}) = (151.2 \text{ g L}^{-1})/(151.2 \text{ g mol}^{-1})$$
$$\rho = cM \quad \text{For example: } (151.2 \text{ g L}^{-1}) = (1 \text{ mol L}^{-1}) \times (151.2 \text{ g mol}^{-1})$$

However, such conversions always carry a risk of error. Especial care is needed in choosing the correct M_r if the drug is supplied as a salt, hydrate, etc. This can cause great discrepancies especially if the contribution of the accompanying anion or cation is high. Most analytical measurements are reported in terms of the free acid or base and not the salt (see Section 14.1.2).

Acknowledgements

We thank Mr J Burrows (Huntingdon Life Sciences), Dr J Cocker (Health and Safety Laboratory), Ms B Earl (Guy's and St Thomas' Hospital), Dr J Haines (United Nations Institute for Training and Research), Mr M Hallworth (Royal Shrewsbury Hospital), Dr M Lennard (University of Sheffield), Prof H Maurer (Saarland University, Homburg), Dr G Mills (University of Portsmouth), Mr P Morgan (Guy's and St Thomas' Hospital), Prof P Myers (University of York), Dr I Ojanperä (University of Helsinki), Prof D Perrett (Queen Mary, University of London), Prof F Pragst (University Hospital Charité, Berlin), Miss J Smith (Selly Oak Hospital), Dr J Wilson (University of Cardiff), and Dr Peter Wyatt (Queen Mary, University of London) for help and advice.

List of Abbreviations

AAG	α_1-acid glycoprotein
AAFS	American Academy of Forensic Sciences
AAS	atomic absorption spectrometry
AASP	Advanced Automated Sample Processor
ABTS	2,2′-azino-bis(3-ethylbenzthiazoline-6-sulfonic acid)
ACB	Association for Clinical Biochemistry (UK)
ADH	alcohol dehydrogenase
ADME	absorption, distribution, metabolism, and excretion
AED	atomic emission detector
AES	atomic emission spectroscopy
AFID	alkali-flame ionization detector
AFS	atomic fluorescence
AGP	α_1-acid glycoprotein (AAG)
AIC	Akaike information criterion
ALDH	aldehyde dehydrogenase
ANOVA	analysis of variance
APC	apparent partition coefficient
APDC	ammonium pyrrolidine dithiocarbamate
APCI	atmospheric pressure chemical ionization
APPI	atmospheric pressure photoionization
AR	analytical reagent
ASE	accelerated solvent extraction
ASPEC	Automatic Sample Preparation with Extraction Columns
ASTED	Automated Sequential Trace Enrichment of Dialysates
ASV	anodic stripping voltametry
ATP	adenosine triphosphate
AUC	area under plasma concentration–time curve
BAC	blood alcohol concentration
BAN	British Approved Name
BBB	blood brain barrier
BE	benzoylecgonine
BEH	bridged ethylsiloxane/silica hybrid
BGE	background electrolyte
BDB	3,4-methylenedioxybutanamine
BMC	4-bromomethyl-7-methoxycoumarin
BMDMCS	bromomethyldimethylchlorosilane
BP	British Pharmacopoeia
BrAC	breath alcohol concentration
BSA	*N*,*O*-bis(trimethylsilyl)acetamide

BSTFA	*N,O*-bis(trimethylsilyltrifluoro)acetamide
CAD	charged aerosol detector
CAS	Chemical Abstracts Service
CBQCA	3-(4-carboxybenzoyl)quinoline-2-carboxaldehyde
CD	cyclodextrin (eluent additive); circular dichroism (chiral detection)
CDER	Center for Drug Evaluation and Research
CDMS	cyclohexanedimethanol succinate
C(Z)E	capillary (zone) electrophoresis
CEA	chiral eluent additive
CEC	capillary electrochromatography
CECF	1-chloroethyl chloroformate
CEDIA	cloned enzyme donor immunoassay
CHAPS	3-[(3-cholamidopropyl)dimethylammonio]-1-propanesulfonate
CHAPSO	3-([3-cholamidopropyl]dimethylammonio)-2-hydroxy-1-propanesulfonate
CI	chemical ionization (mass spectrometry), confidence interval (statistics)
CIEF	capillary isoelectric focusing
CITP	capillary isotachophoresis
CJD	Creutzfeldt–Jakob Disease
Cl	plasma (whole body) clearance
CLEC	chiral ligand exchange chromatography
CLD	chemiluminescence detector
CLIA	chemiluminescent immunoassay
CLND	chemiluminescent nitrogen detector
CMC	critical micellar concentration
CMDMCS	chloromethyldimethylchlorosilane
CMIA	chemiluminescent magnetic immunoassay
CNS	central nervous system
COMT	catechol *O*-methyl transferase
CPD	continuing professional development
CPHP	4-(4-chlorophenyl)-4-hydroxypiperidine
CPR	chlorophenol red
CR	controlled release
CT	computerized tomography/tomographic
CTAB	hexadecyltrimethylammonium bromide
CSF	cerebrospinal fluid
CSL	Commonwealth Serum Laboratories
CSP	chiral stationary phase
CSV	cathodic stripping voltammetry
CV	coefficient of variation (relative standard deviation, RSD)
CYP	cytochrome
2,4-D	2,4-dichlorophenoxyacetic acid
DAD	diode-array detector
DC	direct current
DFG	Deutsche Forschungsgemeinschaft

DFO deferoxamine (desferrioxamine)
DFSA drug facilitated sexual assault
DLIS digoxin-like immunoreactive substance(s)
DMCS dimethyldichlorosilane
DNOC dinitro-*ortho*-cresol (4,6-dinitro-2-methylphenol)
DNSCl dansyl chloride
DoA drugs of abuse
DOPA 3-(3,4-dihydroxyphenyl)alanine
DTAB dodecyltrimethylammonium bromide
DTAF 5-(4,6-dichlorotriazinyl)aminofluorescein
DTTO Drug Treatment and Testing Order
DUI driving under the influence
EA enzyme acceptor
EC electrochemical
ECD electron capture detection
ECF extracellular fluid
ED enzyme donor (biochemistry); emergency department (medicine)
EDDP 2-ethylidene-1,5-dimethyl-3,3-diphenylpyrrolidine (methadone metabolite I)
EDTA ethylenediaminetetra-acetic acid (or sodium salt)
EDXRF energy dispersive X-ray fluorescence
EI electron ionization
EIA enzyme immunoassay
ELISA enzyme-linked immunosorbent assay
ELSD evaporative light scattering detector
EM extensive metabolizer
EME ecgonine methyl ester
EMIT enzyme-multiplied immunoassay technique
EOF electro-osmotic flow
EPA Environmental Protection Agency (US)
EPC electroplanar chromatography
EQA external quality assessment
ESI electrospray ionization
ETAAS electrothermal atomization atomic absorption spectrometry
FAAS flame atomic absorption spectrometry
FAB fast atom bombardment
FFAP free fatty acid phase
FIA flow-injection analysis
FID flame ionization detector
FITC fluorescein isothiocyanate
FMO flavin-containing monooxygenase
FPD flame photometric detector
FPN ferric/perchloric/nitric (acids)
FPIA fluorescence polarisation immunoassay
FQ 3-(2-furoyl)quinoline-2-carboxaldehyde
FSCE free solution capillary electrophoresis

FSH	follicle stimulating hormone
FTICR-MS	Fourier-transform ion cyclotron resonance mass spectrometry
FTIR	Fourier-transform infrared
FRAT	free radical association technique
GC	gas chromatography
GCE	glassy carbon (working) electrode
GFR	glomerular filtration rate
GHB	γ-hydroxybutyrate
GI	gastrointestinal
GLC	gas–liquid chromatography
G-6-P	glucose-6-phosphate
GPC	gel permeation chromatography
G-6-PDH	glucose-6-phosphate dehydrogenase
GPR	general purpose reagent
GSC	gas–solid chromatography
GSH	reduced glutathione
GSSG	oxidized glutathione
GST	glutathione *S*-transferase
GTN	glyceryl nitrate (trinitrin)
HbA1c	glycated haemoglobin
HCG	human chorionic gonadotropin
HEMA	2-hydroxyethyl mercapturic acid [*N*-acetyl-*S*-(2-hydroxyethyl)-L-cysteine]
HETP	height equivalent to a theoretical plate
HFBA	heptafluorobutanoic acid
HFBAA	heptafluorobutyric anhydride
HFBI	heptafluorobutyrylimidazole
5-HIAA	5-hydroxyindoleactic acid
HIV	human immunodeficiency virus
HLoQ	higher limit of quantification
HMDS	hexamethyldisilazane
HPLC	high-performance liquid chromatography
HPTLC	high-performance thin-layer chromatography
HPT	hybrid particle technology
hR_f	retention (retardation) factor $\times$ 100
HRT	hormone replacement therapy
HS	headspace
HSA	human serum albumin
IC	ion chromatography
i.d.	internal diameter
IDE	insulin-degrading enzyme
IEMA	immunoenzymometric assay
IFMA	immunofluorometric assay
IgG	immunoglobulin
i.m.	intramuscular
IM	intermediary metabolizer

IP	ionization potential
IPC	inductively coupled plasma
ISO	International Organization for Standardization
IQC	internal quality control
IR	infrared
IRI	immunoreactive insulin
IRMA	immunoradiometric assay
ISE	ion selective electrodes
ISI	ionspray ionization
ISRP	Internal surface reverse-phase
i.v.	intravenous
IUPAC	International Union of Pure and Applied Chemistry
LAT	latex agglutination test
LC	liquid chromatography
LD	loading dose
LDV	low dead volume
LIF	laser-induced fluorescence
LLE	liquid–liquid extraction
LLoQ	lower limit of quantitation
LMPG	Laboratory Medicine Practice Guideline
LoD	limit of detection
LP	laboratory procedure
LPME	liquid phase microextraction
LSD	lysergic acid diethylamide
MALDI	matrix-assisted laser desorption ionization
MAOI	monoamine oxidase inhibitor
6-MAM	6-monoacetylmorphine
MBDB	3,4-methylenedioxy-*N*-methylbutanamine
MbOCA	2,2′-Dichloro-4,4′-methylene dianiline
MBTFA	*N*-Methyl-bis(trifluoroacetamide)
MCTA	microcrystalline triacetate
MDA	4,4′-methylenedianiline (occupational context); 3,4-methylenedioxyamfetamine (drug of abuse)
MDEA	3,4-methylenedioxyethylamfetamine
MD-GC	multidimensional gas chromatography
MDMA	3,4-methylenedioxymetamfetamine
MDR	multidrug resistance
MEIA	microparticle enzyme immunoassay
MEK	methyl ethyl ketone (butanone)
MEKC	micellar electrokinetic capillary chromatography (also MECC)
MFO	mixed function oxidase (oxygenase)
MHRA	Medicines and Healthcare Products Regulatory Agency (UK)
MIBK	methyl isobutyl ketone (4-methyl-2-pentanone)
MIP	molecularly imprinted polymer
M–M	Michaelis–Menten

MMDA	3-methoxy-4,5-methylenedioxyamfetamine
MMP	mixed-mode (stationary) phase
MOCA	4,4′-methylenebis(2-chloroaniline)
M_r	relative molecular (atomic) mass
MR	modified release
MS	mass spectrometry/spectrometric
MSTFA	*N*-methyl-*N*-trimetylsilyl-trifluoroacetamide
MTBE	methyl *tert*-butyl ether
NAA	neutron activation analysis
NACB	National Academy of Clinical Biochemistry (US)
NAC	*N*-acetyl-L-cysteine
NACE	nonaqueous capillary electrophoresis
NAD	nicotine adenine dinucleotide
NAPQI	*N*-acetyl-*p*-benzoquinoneimine
NDA	2,3-naphthalenedialdehyde
NADP	nicotine adenine dinucleotide phosphate
NED	*N*-naphthylethylenediamine
NBD	7-nitrobenz-2-oxa-1,3-diazole
NIH	National Institutes of Health (US)
NIMH	National Institute of Mental Health (US)
NIST	National Institute for Standards and Technology (US)
NLM	National Library of Medicine (US)
NMR	nuclear magnetic resonance
NPD	nitrogen–phosphorus detector
NPT	near patient testing
NSD	nitrogen-selective detector
OATP	organic anion-transporting protein
OCT	organic cation transporter
ODS	octadecylsilyl
OIA	optical immunoassay
OP	organophosphorus
OPA	*o*-phthaldialdehyde (1,2-phthalic dicarboxaldehyde)
OPLC	overpressured-layer chromatography
OPLC	Optimum Performance Laminar Chromatography
OR	optical rotation
OTC	over-the-counter
PAN	1-(2-pyridylazo)-2-naphthol
PBA	phenylboronic acid
PC	paper chromatography
PCA	perchloric acid
PCP	phencyclidine
PCT	proximal convoluted tubule
PDD	pulsed discharge detector
PDMS	polydimethylsiloxane
PDVB	polydivinylbenzene

PEEK	polyetheretherketone
PEG	polyethylene glycol
PFBB	pentafluorobenzyl bromide
PFBC	pentafluorobenzoyl chloride
PFBI	pentafluoropropanylimidazole
PFPA	pentafluoropropionic acid
PFPAA	pentafluoropropionic anhydride
PFK	perfluorokerosene
PFTBA	perfluorotributylamine
PGC	porous graphitic carbon
PGP	P-glycoprotein
P(I)CI	positive (ion) chemical ionization
PID	photoionization detector
pINN	proposed International Non-proprietary Name
PK	pharmacokinetics
PLOT	porous layer open tubular
PM	poor metabolizer
POC(L)T	point of care (laboratory) testing
PSDVB	polystyrene-divinylbenzene
PT	proficiency testing
PTFE	polytetrafluoroethylene
PVC	polyvinylchloride
QC	quality control
QED	quantitative ethanol detector
QIT	quadrupole ion trap
RAM	restricted access media
RCF	relative centrifugal force
RD	relative density (specific gravity)
RF	radiofrequency
R_f	retention (retardation) factor
RI	refractive index
RIA	radioimmunoassay
rINN	recommended International Non-proprietary Name
RLU	relative chemiluminescent unit
RN	(Chemical Abstracts) registry number
RPC	rotational planar chromatography
rpm	revolutions per minute
RSD	relative standard deviation
RTL	retention time locking
SAM	*S*-adenosylmethionine
SAMHSA	Substance Abuse and Mental Health Services Administration (US)
SAX	strong anion-exchange
s.c.	subcutaneous
SCOT	support coated open tubular
SCX	strong cation-exchange

SD	standard deviation
SDS	sodium dodecylsulfate
SDME	single-drop microextraction
SEC	size exclusion chromatography (gel permeation chromatography, GPC)
SELDI	surface enhanced laser desorption ionization
SER	smooth endoplasmic reticulum
SFC	supercritical fluid chromatography
SFE	supercritical fluid extraction
SG	specific gravity (relative density)
SI	Système Internationale
SIM	selected ion monitoring
SLD	soft laser desorption
S/N	signal-to-noise
SOFT	Society of Forensic Toxicologists (US)
SOP	standard operating procedure
SOS	sodium octyl sulfate
SPE	solid phase extraction
SPS	semipermeable surface
SR	sustained release
STA	systematic toxicological analysis
SPME	solid phase microextraction
SSRI	selective serotonin reuptake inhibitor
SVDK	snake venom detection kit
2,4,5-T	2,4,5-trichlorophenoxyacetic acid
TCA	trichloroacetic acid
TCAs	tricyclic antidepressants
TCD	thermal conductivity detector (katharometer)
TCPO	bis(2,4,6-trichlorophenyl)oxalate
TDM	therapeutic drug monitoring
TEC	trace-enrichment cartridge
TFAA	trifluoroacetic anhydride
TFAI	trifluoroacetylimidazole
THC	Δ^9-tetrahydrocannabinol
THF	tetrahydrofuran
TIA	turbidimetric immunoassay
TLC	thin-layer chromatography
TMAH	tetramethylammonium hydroxide
TMB	3,3′,5,5′-tetramethylbenzidine
TMCS	trimethylchlorosilane
TMS	trimethylsilyl
TMSDEA	trimethylsilyldiethylamine
ToF	time-of-flight
TPC	true partition coefficient
Tris	tris(hydroxymethyl)aminomethane
TSIM	*N*-trimethylsilylimidazole

TSP	thermospray
TXRF	total reflection X-ray fluorescence
TTCA	2-thiothiazolidine-4-carboxylic acid
UDP	uridine diphosphate
UDPGA	uridine diphosphate glucuronic acid
UKNQAS	UK National Quality Assessment Scheme for Drugs
ULoQ	upper (higher) limit of quantitation (HLoQ)
UM	ultrarapid metabolizer
USAN	United States Adopted Name
USNF	United States National Formulary
USP	United States Pharmacopoeia
UPLC	ultra-performance liquid chromatography
UV	ultraviolet
UV-Vis	ultraviolet-visible
V	volume of distribution
VSA	volatile substance abuse
WDXRF	wavelength dispersive X-ray fluorescence
WHO	World Health Organization
WCOT	wall coated open tubular
XRF	X-ray fluorescence

1 Analytical Toxicology: Overview

1.1 Introduction

Analytical toxicology is concerned with the detection, identification and measurement of drugs and other foreign compounds (xenobiotics) and their metabolites in biological and related specimens. The analytical toxicologist can play a useful role in the diagnosis, management and, in some cases, the prevention of poisoning, but to do so a basic knowledge of clinical and forensic toxicology is essential. Moreover the analyst must be able to communicate effectively with clinicians, pathologists, coroners, police and, possibly, others. In addition, a good understanding of clinical chemistry, pharmacology and pharmacokinetics is desirable.

1.1.1 Historical development

The use of physicochemical techniques in the analysis of drugs and other poisons in body fluids or tissues and related specimens has its origins in the development of forensic toxicology. Important contributions came later from work to improve food safety and from occupational toxicology.

The trial of Mary Blandy at Oxford in March 1752 for the murder of her father is the first reported use of chemical tests to detect a poison, in this case arsenic, in a criminal trial (Mitchell, 1938; Watson, 2004). Mathieu Joseph Bonaventure Orfila (1787–1853) in Paris, following the work of the physician François Emmanuel Fodéré (1764–1835), divided poisons into six classes: 'corrosives, astringents, acrids, stupifying or narcotics, narcotico-acrids and septics or putrefiants' (Orfila, 1821), and pioneered systematic study of the role of chemical analysis in the diagnosis of poisoning. This work was carried on by his pupils, notably Sir Robert Christison (1797–1882) in Edinburgh, Alfred Swaine Taylor (1806–1880) at Guy's Hospital in London (Coley, 1998), and Jöns Jacob Berzelius (1779–1848) in Sweden (Jones, 1998).

Plenck (1781) had observed that 'the only certain sign of poisoning is the botanical character of a vegetable poison or the chemical identification of a mineral poison found in the body'. However, this was not accepted by British toxicologists, notably Christison, who held that the medical probability, in conjunction with the general evidence, might be so strong that a diagnosis of poisoning was undoubted. Nevertheless, given this caveat the importance of chemical tests in confirming a suspicion of poisoning was clear and the partnership between clinical and analytical (chemical) toxicology has developed over the last 250 years.

Sensitive and selective chemical methods for the detection of metallic poisons such as arsenic (Marsh, 1836), antimony, bismuth and mercury (Reinsch, 1841) in biological specimens were introduced in the 1830s. Solvent extraction of alkaloids such as nicotine from biological specimens to facilitate their detection and identification using chemical or physiological tests was introduced by Stas in the 1850s, and developed into a systematic method for the extraction of nonvolatile organic compounds from body fluids by Otto. By the late nineteenth century, chemical and spectroscopic methods for the measurement of carbon monoxide and chemical methods for the measurement of ethanol in blood had also been developed (Niyogi, 1981).

Further relevant advances in chemical toxicology included the introduction of a colorimetric technique for the measurement of barbiturates (1933) and the use of ammonium sulfate to 'salt out' alkaloids (1937). An alternative approach to the purification of basic drugs from biological specimens, the precipitation of protein with trichloroacetic acid followed by kaolin adsorption and subsequent elution of the drugs of interest, was described in 1937, and sodium tungstate protein precipitation was introduced in 1946. Florisil, a synthetic magnesium silicate, was introduced in 1949 for the adsorption of basic drugs such as narcotics from biological specimens in place of kaolin. Microdiffusion methods (Conway, 1947) were developed for volatile analytes.

Major advances in analytical methodology followed the introduction and application of refined physicochemical techniques such as spectrophotometry and chromatography in the late 1940s. In particular, ultraviolet (UV) and infrared (IR) spectrophotometry, together with visible spectrophotometry (colorimetry), and paper and ion-exchange column chromatography were widely used. More recently, paper chromatography has been largely superseded by thin-layer chromatography (TLC) as this latter technique offers advantages of speed of analysis and lower detection limits. Improved instrumentation for UV spectrophotometry (including facilities for derivative spectroscopy), spectrophotofluorimetry, atomic absorption spectrometry, anodic stripping voltammetry, electrochemistry, X-ray diffraction, mass spectrometry, nuclear magnetic resonance and neutron activation analysis has led to these techniques being widely applied to particular problems.

1.2 Modern analytical toxicology

The last 25 years have seen many advances in methods for detecting, identifying and measuring drugs and other poisons in biological fluids with consequent improvement in the scope and reliability of analytical results. The value of certain emergency assays and their contribution to therapeutic intervention has been clarified. Some such assays are performed for clinical purposes, but have overt medicolegal implications and require a high degree of analytical reliability. Examples include 'brain death' and child abuse screening, and instances of suspected iatrogenic poisoning. In addition, demand for the measurement of plasma drug and sometimes metabolite concentrations to aid treatment [therapeutic drug monitoring (TDM)], for drugs of abuse and laxative/diuretic screening, and for laboratory analyses to monitor occupational exposure to certain chemicals, has increased.

Nowadays, a range of powerful analytical methods, typically chromatographic methods, ligand immunoassays and other techniques (Table 1.1) are available to the analytical toxicologist. However, it remains impossible to look for all poisons in all samples at the sensitivity required. It is vital therefore that the reason for any analysis is kept clearly in view. Although the underlying principles remain the same in the different branches of analytical toxicology, the nature and the amount of specimen available can vary widely, as may the time-scale over which the result is required and the purpose for which the result is to be used. All these factors may, in turn, influence the choice of method(s) for a particular analysis.

Gas–solid chromatography (GSC) and, more commonly, gas–liquid chromatography (GLC) have made a notable contribution to both the qualitative and quantitative analysis of drugs and other organic poisons, especially since the introduction of sensitive detectors such as the

Table 1.1 Some methods for the analysis of drugs and other organic poisons in biological samples

Principle	Technique
Chemical	Colour test
Electrochemical	Biosensors
	Differential pulse polarography (DPP)
Spectrometric	Mass spectrometry (MS), also known as mass fragmentography (MF)
	Nuclear magnetic resonance (NMR)
	Spectrophotofluorimetry (SPFM)
	Ultraviolet/visible absorption spectrophotometry (UV/Vis)
Chromatographic	Gas chromatography (GC), includes gas–solid chromatography (GSC) and gas–liquid chromatography (GLC)
	(High performance) liquid chromatography [(HP)LC]
	(High performance) thin-layer chromatography [(HP)TLC]
	Super(critical) fluid chromatography [S(C)FC]
Electrophoretic	Capillary (zone) electrophoresis [C(Z)E]
	Capillary electro-chromatography (CEC)
	Micellar electrokinetic capillary chromatography (MECC)
Immunoassay	Cloned enzyme donor immunoassay (CEDIA)
	Enzyme linked immunosorbent assay (ELISA)
	Enzyme multiplied immunoassay technique (EMIT)
	Fluorescence polarization immunoassay (FPIA)
	Latex agglutination tests (LAT)
	Microparticle enzyme immunoassay (MEIA)
	Radioimmunoassay (RIA)
Enzyme-based assay	Alcohol dehydrogenase – ethanol
	Aryl acylamide amidohydrolase – paracetamol

flame-ionization detector (FID) and the electron-capture detector (ECD). These detectors are often complementary because the ECD shows a selective response to certain compounds or derivatives of compounds, in practice those containing either a halogen or some other electronegative or 'electron-capturing' species such as a nitro moiety, while the FID responds to most organic compounds. More recently, the introduction of nitrogen-selective detectors (NSD), also known as alkali flame-ionization (AFID) and nitrogen–phosphorus (NPD) detectors, which show an enhanced and selective response to compounds containing C–N bonds or phosphorus, further extended the scope of GLC. However, the use of gas chromatography in combination with mass spectrometry (GC-MS), sometimes referred to as an example of a 'hyphenated technique', provides high sensitivity together with unparalleled selectivity and can identify unequivocally many compounds using only nanogram quantities of material, and has largely supplanted the use of the NPD, certainly as far as qualitative work is concerned.

High performance (originally high pressure) liquid chromatography (HPLC) has achieved wide application in analytical toxicology since the early 1970s. Gases and very volatile solvents excepted, most analytes are amenable to analysis by HPLC or a variant of the basic procedure, in contrast to GC, which is restricted to the analysis of compounds which are both stable and volatile at temperatures up to approximately 350 °C. However, the use of HPLC in the qualitative analysis of drugs is restricted to a certain extent by the lack of a sensitive universal detector analogous to the FID in GC, although a range of sensitive and reliable detectors [notably UV absorption, fluorimetric, electrochemical (ED) and MS] of varying sensitivities and selectivities are now available. In addition, fractions of column effluent corresponding to the chromatographic peaks of interest may be collected and analyzed off-line, for example, by immunoassay.

Immunoassays have also found wide application, whether radioimmunoassay (RIA) or more recent variants, for example, enzyme-multiplied immunoassay technique (EMIT) and cloned enzyme donor immunoassay (CEDIA), and are often highly sensitive. Enzyme-based assays, such as that for paracetamol (acetaminophen), have also been described. However, all of these assays have the disadvantage that antibodies, enzymes, or specific binding proteins have to be prepared for each analyte or group of analytes before an analysis is possible. On the other hand, these and similar assays may often be used directly in small volumes of aqueous media ('homogenous assay'), in contrast to chromatographic methods which often require some form of purification procedure, for example, solvent extraction, prior to the analysis. Although, immunoassays can be very sensitive, some may be poorly selective, that is the antibody may recognize several structurally similar molecules. Sometimes this cross-reactivity can be exploited, as in screening for classes of abused drugs such as opiates.

1.2.1 Drugs and pesticides

Capillary GC, often with MS detection, is widely used both in systematic toxicological analysis (STA) and in the assay of specific analytes, although packed column GC may still find a place for certain applications. HPLC, often nowadays in conjunction with MS, is used to analyze specific compounds or groups of compounds, although STA procedures based on diode-array detectors (DAD) and wavelength ratioing techniques are also used.

The problem in STA (poisons screening, drug screening, unknown screening) is simply to detect reliably as wide a range of compounds as possible in as little sample (plasma/serum/whole blood, urine, vitreous humour, stomach contents or vomit, or tissues) as possible at high sensitivity, but with no false positives. Ideally some sample should be left to permit confirmation of the results using another technique and also quantitation of any poison(s) present to aid clinical interpretation of the results.

When screening for unknown substances it is important to adopt a systematic approach in order to eliminate possible contenders and to 'home in' on the compound(s) present. STA can be divided into three key stages (Figure 1.1). The aim of the sample preparation step is to retain all the toxicologically important substances whilst removing potentially interfering sample matrix components. Thus, as wide a range as possible of analytes of interest, including lipophilic and

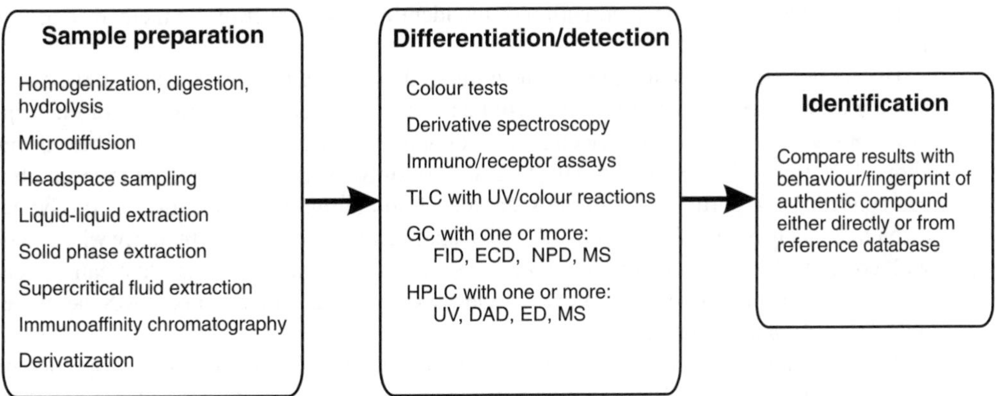

Figure 1.1 The three key steps in systematic toxicological analysis.

moderately polar, acidic, basic and neutral species, should be isolated. To increase the yield of analyte(s), the sample may be treated with β-glucuronidase/arylsulfatase to hydrolyze conjugated metabolites.

The aim of the differentiation/detection step is identify the relevant compounds in the minimum amount of time. This requires a combination of relatively nonspecific ('universal') assays with highly specific methods. Immunoassays, particularly if the antibody has wide cross-reactivity, are useful for identifying classes of drugs. TLC has the advantage that all the nonvolatile materials in the extract remain on the plate, whereas with GC and HPLC there is always the possibility that compounds have not been eluted from the column. Obviously, one analytical technique cannot separate and identify all the possible compounds of interest; for example, only a finite number of compounds can be resolved on a single TLC plate.

The ability of a given analytical method to identify a compound from a given set of test compounds is known as the identification power. One approach in quantifying identification power is the use of discriminating power (DP):

$$\mathrm{DP} = 1 - \frac{2M}{N(N-1)} \tag{1.1}$$

where M is the number of pairs of compounds which are not resolved and N is the number of compounds examined. The concept of discriminating power was introduced by Moffat *et al.* (1974) with the aim of quantifying the ability of paper chromatography, TLC and GLC to give unequivocal identification of unknowns. When this approach was applied to an investigation of the separation of 34 neutral compounds in 15 TLC systems, it was shown that one system had the greatest DP (0.75). However, by combining the results from two of the systems, the DP could be increased to 0.88 (Owen *et al.*, 1978). As the identification power increases so the DP increases towards 1.0.

A second approach to define the identification power is the mean list length (MLL). A list length is defined as the number of feasible candidates for a particular analytical parameter, for example, the retention index in a GC system (Section 5.3). The average of all list lengths gives the MLL for that set of compounds in that system. As in the example above, MLLs can be calculated for a combination of systems. MLLs are >1.0, but will approach 1.0 as identification power increases. In both cases (DP and MLL), examination of a low number of test compounds will give an overestimation of the identification power of the method (Boone *et al.*, 1999).

The greater the number and range of techniques that are available to the analyst, the greater the probability that unknown substance(s) will be identified correctly. Investigation of the responses of various analytes to different detectors, for example FID/ECD, can provide valuable information about the nature of a compound. HPLC-DAD not only provides spectral information, but also can confirm peak purity via multiple scanning of an eluting peak (Section 8.8.3). Hyphenated techniques such as GC-MS can provide robust analyte identification, particularly when combined with computerized libraries of electron ionization (EI) fragmentation data that can be searched rapidly to confirm compound identity. In addition, chemical ionization (CI) MS can be used to obtain the M_r of a substance.

Analytes may be chemically modified to improve their chromatographic properties or 'detectability', but derivatization can also give useful qualitative information. One old, but classic, example is the so-called 'acetone-shift' (reaction of acetone with a primary amine to give the corresponding Schiff's base). Amfetamine, for example, reacts with acetone to form *N*-(1-methyl-2-phenylethyl)propanimine (Figure 1.2).

(±)-Amfetamine —Acetone→ *N*-(1-Methyl-2-phenylethyl)propanimine

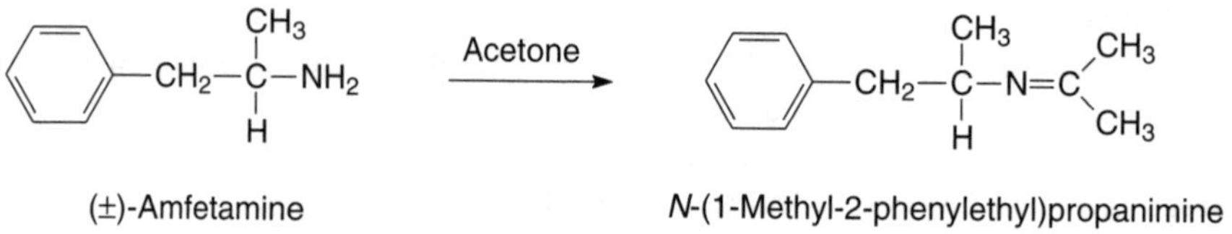

Figure 1.2 Reaction of amfetamine with acetone.

The third step in STA is to compare the observed data with validated database information. Clearly, databases used in compound identification need to be regularly updated, and must include information on not only patent compounds, but also metabolites, common interferences, and contaminants. It is important that the analytical techniques used in establishing such databases are reproducible, both within and between laboratories.

1.2.2 *Ethanol and other volatile substances*

Enzymatic methods for blood ethanol using alcohol dehydrogenase with spectrophotometric measurement of a coenzyme are available in kit form such as that available for the Abbott TDx/ADx. GC analysis of ethanol either by direct injection of blood or urine diluted with deionized water (Curry *et al.*, 1966), or by static headspace sampling (Machata, 1975), is also widely used, particularly in forensic work. GC is advantageous because methanol, 2-propanol and acetone may be separated and measured simultaneously. Methanol poisoning from ingestion of synthetic alcoholic drinks is one of the few causes of acute poisoning 'epidemics' and measurement of blood methanol is important in confirming the diagnosis and in monitoring treatment.

More than 20 additional volatile compounds may be encountered in acute poisoning cases arising, for example, from deliberate inhalation of vapour in order to become intoxicated ['glue sniffing', solvent abuse, inhalant abuse, volatile substance abuse (VSA)]. Some of these volatile compounds have metabolites that may be measured in urine in order to assess exposure, notably hippuric and methylhippuric (toluric) acids (from toluene and the xylenes, respectively) and trichloroacetic acid (from trichloroethylene). However, most volatile substances are excreted unchanged in exhaled air, and thus whole blood is the best sample in which to detect and identify these compounds (Flanagan *et al.*, 1990).

1.2.3 Trace elements and toxic metals

In order to help diagnose chronic poisoning, where elevations of only a few $\mu g\ L^{-1}$ (parts per billion, ppb, i.e. parts per thousand million) of blood or serum can be important, good accuracy and reproducibility are essential (Braithwaite and Brown, 1988). Sample contamination during collection (e.g. from sample tubes, or even from syringe needles in the case of chromium and manganese) and within the laboratory itself can be serious sources of error. This applies particularly to common elements such as lead and aluminium. Modern methods for measuring toxic metals in biological materials (Table 1.2) vary enormously in terms of complexity, cost, accuracy and sensitivity. Some techniques (isotope dilution MS, neutron activation analysis) are in reality reference methods. Atomic absorption spectrometry with either flame or electrothermal atomization using a graphite furnace has been employed widely, but is being superseded by inductively coupled plasma-mass spectrometry (ICP-MS). In the case of serum iron, however, reliable kits based on the formation of a coloured complex remain widely used in clinical chemistry.

Table 1.2 Methods for the analysis of toxic metals in biological materials

Technique	Mode	Variant
Electrochemical	Potentiometric	Ion selective electrodes
	Coulometric	(Differential pulse) polarography
		Anodic/cathodic stripping voltametry (A/CSV)[a]
Spectrophotometric	Atomic emission (AE)	Flame emission photometry (FEP)[b]
		DC plasma
		Inductively coupled plasma (ICP-AES)
	Atomic absorption (AA)	Flame
		Hydride generation
		Electrothermal
		Cold vapour
	X-Ray	Fluorescence
	Nuclear	Neutron activation
		Proton activation
Mass spectrometry		Inductively coupled plasma (ICP-MS)

[a]Also known as potentiometric stripping analysis (PSA)
[b]Normally refers to the use of filters to select the emission wavelength – used mainly for potassium, lithium and sodium assay

ICP-MS is a multi-element technique that can detect and measure elements with detection limits of $\mu g\ L^{-1}$ to $ng\ L^{-1}$. Different isotopes of an element can also be measured. For some elements, the relative abundance of the isotopes depends upon the source of the metal. Therefore, by measuring the isotope ratios of an element such as lead in a sample from a chronically poisoned patient with those found in material present in the patient's immediate environment it may be possible to localize the source of exposure (Delves and Campbell, 1988). Ethnic cosmetics such as surma may contain from 0 to 80 % elemental lead as either the oxide or sulfide and such products are important causes of lead poisoning. So-called 'traditional' medicines may also contain toxic doses of salts of lead or other toxic metals (Braithwaite and Brown, 1988).

1.3 Provision of analytical toxicology services

The stages in processing analytical work if poisoning is suspected can be divided into preanalytical, analytical and postanalytical phases (Table 1.3).

Table 1.3 Steps in undertaking an analytical toxicological investigation

Pre-analytical	Obtain details of current (suspected) poisoning episode, including any circumstantial evidence of poisoning, and the results of biochemical and haematological investigations, if any. Also obtain the patient's medical and occupational history, if available, and ensure access to the appropriate sample(s) Decide the priorities for the analysis
Analytical	Perform the agreed analysis
Postanalytical	Interpret the results in discussion with the physician looking after the patient or the pathologist. Perform additional analyses, if indicated, using either the original samples or further samples from the patient. Save any unused or residual samples in case they are required for further tests

1.3.1 Samples and sampling

In analytical toxicology, clinical chemistry and related fields, the words 'sample' and 'specimen' are used to denote a portion of a body fluid, tissue, incubation medium, and so on obtained under defined conditions. The samples encountered may range from relatively pure solutions of a drug to a putrefying piece of tissue. Liquids, such as blood, saliva, urine and cerebrospinal fluid (CSF), are generally easier to sample and to analyze than solids and semisolids, which require homogenization or digestion prior to analysis. Blood plasma is used in clinical work if quantitative measurements are needed in order to assess dosage or monitor treatment as in TDM. Urine is commonly used in qualitative work such as drugs of abuse screening as collection is noninvasive and the concentrations of many drugs and other poisons, and their metabolites, tend to be higher than in blood, thereby facilitating analyte detection. Further aspects related to samples and sampling are discussed in Chapter 2.

1.3.2 Choice of analytical method

In responding to a given analytical problem many factors must be considered (Peng and Chiou, 1990). It may seem self-evident that the method used should be appropriate for the intended analysis. In practice, the choice of method depends on several factors including: the circumstances under which an analysis is requested (i.e. the question being asked), the sample to be analyzed, the nature of the analyte (if known), the expected concentration of any analyte(s), the time available for the analysis, the apparatus available, and the training and experience of the analyst.

The nature of the sample and the expected concentration of any analyte(s) are obvious influences on the choice of method. It may be possible to measure the concentration of a known substance in a relatively pure solution directly using a simple technique such as UV spectrophotometry. However, if the sample is a piece of postmortem tissue such as liver then a wholly different approach will be required. Typically, a representative portion of the tissue will have to be homogenized and the analyte obtained in a relatively pure form by liquid–liquid solvent extraction of the homogenate at an appropriate pH. Further purification or extract concentration steps may be needed prior to instrumental analysis. In the case of organic poisons this will usually be by a chromatographic method such as GC or HPLC because both qualitative and quantitative information can be obtained during the course of the analysis. The choice of instrument may influence the choice of sample preparation procedure, although this is not always the case.

For optically active (chiral) drugs (Table 1.4), the desired clinical activity often resides predominantly in one isomer. The other isomer may be pharmacologically inactive or have different properties from its enantiomer, so administration of a racemate (a 50:50 mixture of enantiomers, Table 1.5) is the same as giving different compounds as far as the body is concerned. The supply of optically active compounds as pure enantiomers (optical isomers) is sometimes indicated by the name used (dexamfetamine, dextropropoxyphene, levorphanol), but this does not always apply (hyoscine, morphine, physostigmine). Moreover, it is thought that up to 25 % of currently used drugs are chiral and are supplied as racemates, usually without any indication of the fact. Atropine is the approved name for (±)-hyoscinamine, for example. Whilst there are few clear indications for providing chiral methodology for routine analytical toxicology at present, it should be noted that chromatographic methods have made a major contribution to the development of pharmacology and therapeutics by providing methods to separate enantiomers on a preparative scale and in biological samples.

1.3.3 Method implementation and validation

Whatever method is used for a given analysis it must be validated, that is it must be shown to be 'fit for purpose'. Understanding method validation is clearly important not only when developing a method, but also when implementing a method for routine use. A number of terms important in understanding method validation are given in Table 1.6. A fundamental starting point in any assay is obtaining certified pure reference material, or at least the best approximation to such material that can be attained. When preparing primary standards, particular attention should be paid to the M_r of salts and their degree of hydration (water of crystallization). Analytical results are normally reported in terms of free acid or base and not of a salt.

Obviously the method must possess adequate sensitivity for the task in hand. The limit of sensitivity is a term often used to describe the limit of accurate measurement, but this is better defined as the lower limit of quantification (LLoQ). The limit of detection (detection limit, 'cut-off value') is a better term for limit of sensitivity.

Quantitative methods must also have good precision (reproducibility) and accuracy (the results must reflect the true concentration of the analyte). Selectivity (freedom from interference, specificity) is important when a single species is to be measured, but broad specificity may be useful when screening for the presence of a particular class of compounds as discussed above.

Table 1.4 Summary of chiral nomenclature

Number of chiral centres	If n = number of optical centres there will be 2^n isomers. Molecules with 2 optical centres can exist as four molecules: two diastereoisomers (diasteromers), each consisting of two enantiomers, (i.e., there are *two pairs of enantiomers*). The exception to this is if two molecules have a *plane of symmetry* (a plane that divides a molecule into two parts, each a mirror image of the other) and therefore cancel out their net optical rotation. In such cases they are known as meso forms
Nomenclature	Enantiomers possess a unique property in that they rotate plane-polarized light in equal and opposite directions. This is the basis of the (+)/(−) or d/l notation, the former being preferred as it avoids confusion with D/L, however it does not unequivocally distinguish between enantiomers because some molecules may change rotation on forming salts. The notation tells nothing about the absolute configuration (i.e., the spatial arrangement) of the atoms
Rotation of plane polarized light	Rotates to the right: dextrorotatory (+ or d); rotates to the left: levorotatory (− or l).
Fischer	The absolute stereochemistry in the Fischer notation gives the absolute spatial arrangement by reference to D-glyceraldehyde. The letters D or L are used (not to be confused with lower case d or l). The Fischer convention is still used for carbohydrates and amino acids. The original choice of D-glyceraldehyde was arbitrary, but was proved correct by X-ray crystallography
Cahn–Ingold–Prelog (CIP)	The CIP system is the definitive method of assigning absolute configuration. The letters *R* and *S* indicate spatial arrangements as follows: 1. Assign values to the substituent groups by highest atomic number[a] 2. Point the lowest value away 3. If the remainder go from high to low clockwise then *R* (rectus) 4. If the remainder go from high to low anticlockwise then *S* (sinister) 5. In case of a tie go to the next atoms along The Fischer convention cannot be simply converted to the CIP system, that is *R* does not always equate to D. All naturally occurring α-amino acids in mammalian proteins are L. Using the CIP system, cysteine and cystine are *S* the others, without sulfur, are *R*

[a]The rules are in fact more detailed: Highest atomic number > highest atomic mass > *cis*- prior to *trans*- > like pairs (RR) or (SS) prior to unlike > lone pairs which are considered an atom of atomic number 0. [N.B. R* indicates a single enantiomer obtained, but with unknown stereochemistry].

Table 1.5 Some terms used in stereochemistry

Absolute stereochemistry	The absolute spatial configuration of the atoms of a molecule
Chiral	Hand-like, that is left- and right-handed mirror images
Enantiomer	One mirror image form of a pair of optically active compounds
Epimers	Optically active molecules with more than two chiral centres differing at only one chiral centre
Epimerization	Partial racemization of one chiral centre in a molecule with two or more chiral centres
Diastereoisomers	Structures with more than one chiral centre such that they are not mirror images (enantiomers)
Inversion	Conversion of one enantiomer to the other
Meso	Optically inactive isomer in which the optical activity of chiral centres are balanced
Racemate	Equimolar mixture of both enantiomers of an optically active compound
Racemization	Conversion of a single enantiomer to a racemate

The recovery of the analyte, that is how much of the compound of interest is recovered from the sample matrix during an extraction, for example, is important if sensitivity is limiting, but need not be an issue if the LLoD, accuracy and precision of the assay are acceptable.

Ideally, whatever the methodology employed, quantitative assay calibration should be by analysis of standard solutions of each analyte (normally 6–8 concentrations across the calibration range) prepared in the same matrix and analyzed as a batch along with the test samples. A graph of response against analyte concentration should be prepared and used to calculate the analyte concentration in the sample (so-called 'external standard' method).

Any quantitative analysis is a measurement and, in common with all measurements, has associated errors, both random and systematic. In chromatographic and other separation methods the 'internal standard' method is often used to reduce the impact of systematic errors such as variations in injection volume or evaporation of extraction solvent during the analysis. Thus, a known amount of a second compound (the internal standard) that behaves similarly to the analyte during the analysis, but elutes at a different place on the chromatogram or is otherwise detected independently of the analyte is added at an appropriate stage in the analysis. Subsequently, the detector response of the analyte relative to the response of the internal standard is plotted against analyte concentration when constructing a calibration graph.

1.3.4 Quality control and quality assurance

Once an analytical method has been validated and implemented it is important to be able to show that the method continues to perform as intended. In qualitative work, known positive and negative specimens should normally be analyzed at the same time as the test sample. A negative control ('blank') helps to ensure that false positives (owing to, e.g. contaminated reagents or glassware)

Table 1.6 Terms used when reporting method validation

Term	Notes
Accuracy	The difference between the measured value and the accepted ('true') value
Calibration range	The range of concentrations between the highest and lowest calibration standards. This should encompass the range of concentrations found in the test samples
Coefficient of variation (CV)	An obsolete term for RSD
Higher limit of quantification (HLoQ)	The highest concentration that can be quantified. Not always quoted, but important in assays with a clear upper 'cut-off', for example immunoassays and fluorescence assays
Internal standard	A second compound, not the analyte, added at an appropriate stage in the assay to correct for systematic errors in the analysis
Limit of detection (LoD)	The smallest amount of analyte that can be detected. Usually defined as some multiple (e.g. 5) of the baseline noise (signal to noise ratio = 5) or multiple of the SD of the blank signal
Linearity	A definable and reproducible relationship between a physicochemical measurement (e.g. UV absorption) and the concentration of the analyte. Not necessarily a straight line
Lower limit of quantification (LLoQ)	The lowest concentration that can be measured within defined limits. Usually a concentration for which the precision and accuracy have been set arbitrarily, for example RSD $< 20\,\%$
Precision	The scatter of measured values about a mean value
	Usually quoted as RSD – within-assay and between-assay precision is commonly given
Relative standard deviation (RSD)	The standard deviation of replicate measurements expressed as a percentage of the mean value:
	$\text{RSD} = \text{SD}/\text{Mean} \times 100\,\%$
	Useful when comparing precision at different concentrations
Selectivity	The ability to distinguish between the analyte and some other compound
Signal-to-noise (S/N) ratio	Strictly, the response to the analyte divided by the amplitude of random electronic noise of the detection system. In practice, the background signal due to interfering compounds is often greater than the electronic noise

are not obtained. Equally, inclusion of a true positive serves to check that the reagents have been prepared properly and have remained stable.

In quantitative work, assay performance is monitored by the systematic analysis of internal quality control (IQC) samples, independently prepared standard solutions of known composition prepared in the same matrix as the samples and not used in assay calibration. Plotting the results for the IQC samples on a chart allows the day-to-day performance of the assay to be monitored and gives warning of any problems as they arise. When new batches of calibration and IQC samples are prepared it is prudent to ensure comparability of the results obtained with those given by an earlier batch, or with the results obtained using external QC material.

Participation in appropriate external quality assessment (EQA) or proficiency testing (PT) schemes is also important (Wilson, 2002). In such schemes, (sometimes lyophilized) plasma, serum, whole blood, or urine specimens are sent to a number of participating laboratories. After reconstitution in deionized water if appropriate, the specimens are analyzed as if they were real samples and the results are reported before the true or target concentrations are made known.

1.4 Applications of analytical toxicology

Cases in which toxicological analyses are requested tend to fall into: (i) emergency and general hospital toxicology, including 'poisons screening' and (ii) more specialized categories such as forensic toxicology, screening for drugs of abuse, therapeutic drug monitoring (TDM) and occupational/environmental toxicology. However there is considerable overlap between all of these areas.

1.4.1 Clinical toxicology

The specialized nature of analytical toxicological investigations dictates that facilities are concentrated in centres that are often remote from the patient. Frequently routine clinical chemical tests will be performed at one site, whilst more complex toxicological analysis will be performed by a different department, possibly at a different location. The toxicology laboratory will usually undertake a range of analyses in addition to emergency toxicology. Despite this, the importance of direct liaison between the physician treating the patient and the analytical toxicologist cannot be overemphasized. Ideally, this liaison should commence before any specimens are collected as some analytes, toxic metals, for example, require special precautions in specimen collection (Section 11.2). At the other extreme, residues of samples held in a clinical chemistry laboratory or by other departments, for example in the emergency department (ED) refrigerator, can be invaluable if the possibility of poisoning is raised in retrospect.

Nowadays, toxicology screening is normally performed using immunoassays and/or TLC and temperature programmed capillary GC-MS. The concept of STA (Section 1.2.1) is valuable, but sometimes overstates the case for absolute reproducibility of retention data. In real life many factors (clinical and circumstantial evidence, availability of a particular poison, past medical history, occupation, number of peaks present on the chromatogram, selective detector responses, etc.) are considered before reporting results.

The range of analyses that can be offered by specialized laboratories, sometimes on an emergency basis, usually encompasses several hundred poisons. 'Poisons screens' must use reasonable amounts of commonly available samples (20–30 mL urine, 2–5 mL plasma). If any tests are to influence immediate patient management, the (preliminary) results should be available within 2–3 h of receiving the specimens (1 h in the case of paracetamol). In some cases the presence of more than one poison, for example, may complicate the analysis and examination of further specimens from the patient may be required.

A quantitative analysis carried out on whole blood or plasma is usually needed to confirm poisoning unequivocally, but this may not be possible if laboratory facilities are limited, or if the compound is particularly difficult to measure. It is important to discuss the scope and limitations of the tests performed with the clinician concerned and to maintain high standards of laboratory practice, especially when performing tests on an emergency basis. It may be better to offer no result rather than misleading data based on unreliable tests. Clinicians often treat poisoned patients on the basis of suspicion and history rather than await the results of a laboratory test, but may change their approach once they have the result. The treatment of paracetamol poisoning is an example.

Circumstantial evidence of the compound(s) involved in a poisoning episode is often ambiguous and thus, on the rare occasions when an analysis for 'poisons' is indicated, it is advisable to perform a 'poisons screen' routinely in all but the simplest cases. Similarly, the analysis should not end after the first positive finding because additional, hitherto unsuspected compounds may be present. One exception is provided by sublethal carbon monoxide poisoning, which can be difficult to diagnose even if carboxyhaemoglobin measurements are available – circumstantial evidence of poisoning may prove invaluable in such cases. Of course, a 'positive' result on a 'poisons screen' does not of itself confirm poisoning because such a result may arise from incidental or occupational exposure to the poison in question or the use of drugs in treatment.

Blood is often the easiest specimen to obtain from an unconscious patient and is needed for many quantitative measurements. Urine is also a valuable specimen as drug or metabolite concentrations tend to be higher than in blood and relatively large volumes are usually available. However, some compounds, such as many benzodiazepines, are extensively metabolized prior to excretion and then blood plasma is the specimen of choice for detecting the parent compound. Quantitative measurements in urine are generally of little use in emergency toxicology. All poisons screens have limitations (Wiley, 1991). Thus, of the drugs commonly used to treat depression, lithium has to be looked for specifically, whilst those monoamine oxidase inhibitors (MAOIs) which act irreversibly, such as phenelzine, have a prolonged action in the body even though plasma concentrations are very low after overdosage. Any drug that is not bound to the enzyme may be excreted rapidly and may be difficult to detect except in a urine specimen obtained soon after the event. Tricyclic antidepressants are very lipophilic and thus urinary concentrations, even after fatal poisoning, may be below the LLoD of the analytical method if death has occurred relatively soon after the ingestion.

1.4.2 Forensic toxicology

Toxicological investigations of deaths (including fatal road traffic accidents) are often undertaken if there is a possibility that drugs or other poisons may have been involved. These include instances where deliberate poisoning, including self-poisoning, is a possibility, especially if death has occurred in children or whilst in police custody, or when decomposition has taken place to such an extent that it is difficult to glean much information as to the cause of death from a conventional postmortem examination. Non-fatal incidents where toxicological investigations may be useful include collapse whilst in custody, alleged offences under Road Traffic Acts involving ethanol or other drugs, allegations of poisoning of relatives or pets, doping in sex offences and other cases of assault. It may also be important to analyze samples from a suspect for the presence of drugs such as ethanol, which may have altered his or her perception or behaviour during the course of a crime.

The specimens available may range from fresh blood to decomposing tissues recovered from a partial skeleton, while the quantity available may range from a kilogram of liver to a dried bloodstain. Breath analysis was introduced in the United Kingdom for the detection of the intoxicated motorist in 1967, initially in the form of indicator tubes based on the reduction of potassium dichromate and more recently in the form of evidential breath ethanol instruments. Difficult areas still include screening for a wide range of compounds which could affect driving performance in, say, 2 mL of whole blood while leaving sufficient sample for a quantitative measurement, and detecting drugs used in sexual or other assaults (drug facilitated sexual assault, DFSA).

The primary role of the Coroner in England and Wales and of the Procurator Fiscal in Scotland is to exclude criminal acts as a possible cause of death. The data derived from such courts may be invaluable in monitoring the incidence of fatal poisoning. The importance of adequately documenting all acute poisoning incidents both in the hospital notes and in the laboratory records becomes clear when it is remembered that even an apparently trivial case may eventually be reviewed in detail in a coroner's court. Required documentation includes correctly recorded patient and sample details, the date and time of collection of samples, details of physical examination, nature and timing of treatment, particularly drug treatment, results of investigations (including units), and conversations with poisons information services and the laboratory. The laboratory should fully document all analyses and keep copies of all the reports issued. Samples should be kept, appropriately stored, for example at −20 °C, until the conclusion of the case.

In assessing the evidence of the analytical toxicologist the courts are concerned especially with the experience of the analyst, the origin and condition of the samples, and the analytical methods used. The ability to prove continuous and proper custody of the specimen is important. Normally, a minimum of two unrelated analytical methods should be employed before a tentative identification is accepted. The results should be presented together with sufficient information to ensure accurate interpretation of the findings by a coroner, magistrate, judge and/or jury. There is always the possibility of an independent examination by a further expert instructed by another party in the case.

1.4.3 Drug abuse screening

The value of blood, breath or urinary measurements in the diagnosis of ethanol abuse and in monitoring abstinence is clear. Screening for drugs of abuse in urine is also valuable in monitoring illicit drug taking in dependent patients and guards against prescribing controlled drugs for patients who are not themselves drug dependent. Some substances disappear from biological samples very rapidly and, depending on the time between administration and sample collection, the parent compound may not be detectable. Sometimes, however, metabolite identification can be used to demonstrate that a particular drug has been taken. Other samples, such as saliva (oral fluid) and sweat, are being used increasingly to test for drugs of abuse. Analysis of hair, long used to assess chronic exposure to toxic metals, can also provide a history of exposure to illicit drugs and other organic poisons (Baumgartner *et al.*, 1979; Kintz, 2004).

Testing for drugs of abuse may also be valuable in the psychiatric assessment of patients presenting with no overt history of drug abuse. In addition, the diagnosis of maternal drug abuse, either during pregnancy or postpartum, can be important in the management of the neonate. The need for drug abuse screening of personnel in sensitive positions (armed forces, security services, pilots, drivers) or those applying for such positions ('employment' and 'pre-employment' screening, respectively), has become accepted in recent years. The detection of illicit or performance-enhancing drug use in sport has also assumed importance. In animal sports the definition of an illicit compound is much easier than in man and can include any substance not normally derived from feedstuffs.

Urine is the specimen of choice in many cases, not only because the concentrations of the compounds of interest tend to be higher than in blood, but also because it is by far the easiest specimen to obtain, especially from patients likely to have damaged veins. Moreover, human urine presents less of a hazard than blood to laboratory staff. The illicit drugs encountered most commonly in the United Kingdom are opiates [mainly heroin (diacetylmorphine, diamorphine)], benzodiazepines, cocaine, amfetamines including metamfetamine and methylenedioxymetamfetamine (MDMA, 'ecstasy') and cannabis. In the United States, abuse of cocaine either as the hydrochloride or as the free base ('crack') is relatively common, and a range of additional compounds may also be encountered, including dextropropoxyphene (propoxyphene), fentanyl and phencyclidine (PCP, 'angel dust').

The purity of 'street' drugs varies widely – heroin may be between 2 and 95 % pure, for example. Overdosage, either with excessively pure 'street' drug or with drug 'cut' with a particularly toxic compound, is a further cause of acute poisoning 'epidemics'. Compounds such as atropine, strychnine, lidocaine, chloroquine, quinine and barbiturates may be used to 'cut' street drugs. Serious acute poisoning may occur if tolerance to heroin or methadone has been reduced through abstinence. Methadone is widely used to treat opioid addiction, although buprenorphine is becoming more commonly employed in this role. Other opioids such as codeine, dihydrocodeine and pethidine (meperidine) also occur.

The availability of a variety of immunoassay kits has proved invaluable, especially in employment and pre-employment screening when large numbers of negative results are to be expected and high sensitivity is required. However, confirmation of positive results with MS linked either to capillary GC or HPLC is essential. In clinical samples TLC can be used to resolve drugs such as morphine from compounds such as codeine and pholcodine that are available in over-the-counter preparations. TLC requires a minimum of apparatus and is generally cost effective. It is also amenable to batch sample processing, but is labour intensive, analyte capacity is low, and interpretation of results can be anything but straightforward. Capillary GC, GC-MS, or HPLC-MS is used to detect and identify amfetamines, and increasingly to confirm TLC and immunoassay results.

Ingestion of laxatives and diuretics in order to produce weight loss is not uncommon and can be difficult to diagnose. Collection of serial urine samples over several days is advisable. Detection of the abuse of osmotic laxatives such as lactulose and bulk-formers such as bran is not possible analytically. The covert ingestion or administration of anticoagulants is well documented but difficult to diagnose.

1.4.4 Therapeutic drug monitoring (TDM)

The measurement of plasma concentrations of drugs given in therapy is useful in assessing adherence and for compounds for which pharmacological effects cannot be assessed easily and for which the margin between adequate dosage and overdosage is small (Hallworth and Watson, 2007). The availability of a variety of immunoassay and other kits means that many TDM assays can be performed more conveniently by such means than by chromatographic methods. However, chromatographic assays are still important in the case of antipsychotic, antiretroviral and immunosuppressant drugs, for amiodarone, where it has proved impossible to produce an antibody which does not cross react significantly with thyroxine and tri-iodothyronine, and in general where active metabolites should be measured as well as the parent compound. Examples include carbamazepine/carbamazepine-10,11-epoxide, clozapine/norclozapine, procainamide/*N*-acetylprocainamide and amitriptyline/nortriptyline.

1.4.5 Occupational and environmental toxicology

The monitoring of occupational or environmental exposure to toxic substances is an important area. Metal ions such as lead and also some organochlorine pesticides such as chlordane and dieldrin have long half-lives in the body and thus accumulation can occur with prolonged exposure to relatively low concentrations. The manufacture of drugs can also present a hazard to those involved via dermal or inhalational absorption. The abuse of alcohol and of controlled drugs is of much current concern in occupational medicine, especially as regards screening for drug or substance abuse amongst potential employees and amongst, for example, operators of heavy machinery and pilots, as discussed above.

Control of occupational exposure to toxic metals, volatile solvents and some other poisons is an integral part of industrial hygiene and has been achieved, in part, by monitoring ambient air concentrations of the compound(s) under investigation. However, an individual's work pattern and attention to safety procedures may greatly influence exposure and 'biological effect' monitoring, where clinical chemical parameters such as blood zinc protoporphyrin are measured as an indicator of lead exposure, is required practice in certain occupations. Not all poisons are amenable to effect monitoring and so 'biological' monitoring is performed widely. This involves measuring blood, urinary, or breath concentrations of a compound and possibly of its metabolites. The American Conference of Governmental Industrial Hygienists (ACGIH) has set guidance values for 38 substances in the United States (ACGIH, 2005), and the Deutsche Forschungsgemeinschaft (DFG) in Germany has introduced values for 63 chemicals (DFG, 2005).

The investigation of the accidental release of chemicals into the workplace or into the environment (so-called chemical incidents) is a topic of current interest. Examples include the Bhopal disaster in India when methyl isocyanate was released into the atmosphere and the Camelford incident in the United Kingdom, in which aluminium sulfate was accidentally added to the local drinking water supply. Toxicological analyses can be valuable, not only in providing evidence of the nature and magnitude of an exposure, but also in demonstrating that no significant exposure has occurred, thereby allaying public apprehension. Clearly, the early collection of appropriate biological samples is essential. In the absence of information to the contrary it is wise to collect 10 mL whole blood (2×5 mL EDTA) and at least 25–50 mL urine (no preservative) from exposed or possibly exposed individuals. The time and date of sampling and the patients' full names should be recorded on the samples and also on a separate record sheet. The samples should be stored at either 4 °C or −20 °C until the appropriate analyses can be arranged. If the incident is investigated in retrospect then samples may exist in a local hospital laboratory, for example.

One area which has been neglected somewhat is that of food-derived poisons. Botulinum toxin and other toxins of microbiological origin are usually considered together with food poisoning. Poisoning from other naturally occurring poisons, which include atropine from *Atropa belladonna*, solanine from potatoes and cyanide from *Cassava* and from apple pips, also occurs (de Wolff, 2004). Here analysis of the foodstuff rather than biological samples can be more helpful in establishing the diagnosis in individual patients. Acute pesticide poisoning sometimes occurs after ingestion of contaminated produce and again analysis of the foodstuff can be helpful (Stinson *et al.*, 1993). The increasing use of herbal or other 'natural' remedies is an area of especial concern.

1.5 Summary

It is impossible to divorce the study of the analytical methods used in performing toxicological analysis on biological and related samples from the study of toxicology itself, especially clinical and forensic toxicology. By the same token, the laboratory can do nothing to help in the diagnostic process unless someone, be it a clinician, pathologist, or some other person, first suspects poisoning and ensures that specimens are collected and sent for analysis. However, appropriate sample collection and handling is not always straightforward and indeed is a subject in its own right.

References

ACGIH (American Conference of Governmental Industrial Hygienists) (2005) *TLVs and BEIs Based on the Documentation of the Threshold Limit Values for Chemical and Physical Agents and Biological Exposure Indices 2005*. ACGIH, Cincinnati, OH.

Baumgartner, A.M., Jones, P.F., Baumgartner, W.A. and Black, C.T. (1979) Radioimmunoassay of hair for determining opiate-abuse histories. *J Nucl Med*, **20**, 748–52.

Boone, C.M., Franke, J.P., de Zeeuw, R.A. and Ensing, K. (1999) Evaluation of capillary electrophoretic techniques towards systematic toxicological analysis. *J Chromatogr A*, **838**, 259–72.

Braithwaite, R.A. and Brown, S.S. (1988) Clinical and sub-clinical lead poisoning: A laboratory perspective. *Human Toxicol*, **7**, 503–13.

Coley, N. G. (1998) Forensic chemistry in 19th century Britain. *Endeavour* **22**, 143–7.

Conway, E.J. (1947) *Microdiffusion Analysis and Volumetric Error*, 2nd edn, Crosby, Lockwood, London.

Curry, A.S., Walker, G.W. and Simpson, G.S. (1966) Determination of ethanol in blood by gas chromatography. *Analyst* **91**, 742–3.

DFG (Deutsche Forschungsgemeinschaft) (2005) *List of MAK and BAT values 2005*. Commission for the investigation of health hazards of chemical compounds in the work area. Report 41. Wiley VCH, Weinheim.

Delves, H.T. and Campbell, M.J. (1988) Measurement of total lead concentrations and lead isotope ratios in whole blood by use of inductively coupled plasma source mass spectrometry. *J Anal At Spectrom*, **3**, 343–8.

Flanagan, R.J., Ruprah, M., Meredith, T.J. and Ramsey, J.D. (1990) An introduction to the clinical toxicology of volatile substances. *Drug Saf*, **5**, 359–83.

Hallworth, M. and Watson, I. (2007) *Therapeutic Drug Monitoring and Laboratory Medicine*, ACB Venture Publications, London.

Jones, A.W. (1998) Historical developments and present status of forensic toxicology in Sweden. *Nordic Rettsmedisin*, **4**, 35–44.

Kintz, P. (2004) Hair analysis, in: *Clarke's Analysis of Drugs and Poisons*, 3rd edn (eds A.C. Moffat, D. Osselton and B. Widdop), Pharmaceutical Press, London, pp. 124–133.

Machata, G. (1975) The advantages of automated blood alcohol determination by head space analysis. *Z Rechtsmed*, **75**, 229–34.

Marsh, J. (1836) Account of a method of separating small quantities of arsenic from substances with which it may be mixed. *Edin N Philosoph J*, **21**, 229–36.

Mitchell, C.A. (1938) *Forensic Chemistry in the Criminal Courts*, Institute of Chemistry of Great Britain and Ireland, London.

Moffat, A.C., Smalldon, K.W. and Brown, C. (1974) Optimum use of paper, thin-layer and gas-liquid chromatography for the identification of basic drugs. I. Determination of effectiveness for a series of chromatographic systems. *J Chromatogr*, **90**, 1–7.

Niyogi, S.S. (1981) Historical overview of forensic toxicology, in *Introduction to Forensic Toxicology* (eds R.H. Cravey and R.C. Baselt), Biomedical Publications, Davis, pp. 7–24.

Orfila, M.P. (1821) *A General System of Toxicology, or, A Treatise on Poisons, drawn from the Mineral, Vegetable, and Animal Kingdoms, considered as to their relations with Physiology, Pathology, and Medical Jurisprudence*, Vol **1** & **2**, 2nd edn. Trans. Waller JA, E Cox, London.

Owen, P., Pendlebury, A. and Moffat, A.C. (1978) Choice of thin-layer chromatographic systems for the routine screening for neutral drugs during toxicological analyses. *J Chromatogr*, **161**, 187–93.

Peng, G.W. and Chiou, W.L. (1990) Analysis of drugs and other toxic substances in biological samples for pharmacokinetic studies. *J Chromatogr*, **531**, 3–50.

Plenck, J.J. (1781) *Elementa Medicinae et Chirurgiae Forensis*, Graeffer, Wien, p. 36.

Reinsch, H. (1841) On the action of metallic copper on solutions of certain metals, particularly with reference to the detection of arsenic. *Philosoph Mag*, **19**, 480–83 (first published J Praktische Chem 1841; 19).

Stinson, J.C., O'Gharabhain, F., Adebayo, G., Chambers, P.L. and Feely, J. (1993) Pesticide-contaminated cucumber. *Lancet*, **341**, 64.

Watson, K. (2004) *Poisoned Lives: English Poisoners and Their Victims*, Hambledon and London, London.

Wiley, J.F. (1991) Difficult diagnoses in toxicology: Poisons not detected by the comprehensive drug screen. *Pediatr Clin North Am*, **38**, 725–37.

Wilson, J. (2002) External quality assessment schemes for toxicology. *Forensic Sci Int*, **128**, 98–103.

de Wolff, F.A. (2004) Natural toxins, in *Clarke's Analysis of Drugs and Poisons* (eds A.C. Moffat, M.D. Osselton and B. Widdop), 3rd edn, Pharmaceutical Press, London, pp. 189–201.

2 Sample Collection, Transport, and Storage

2.1 Introduction

In analytical toxicology no matter how complex the equipment and careful the analysis, the results may be rendered worthless if sample collection, transport, and storage have not been performed with the analysis in mind. Thus, it is important to be familiar with the nature and stability of the analyte(s), the nature of the sample matrix, and the circumstances under which the analysis is to be performed. Proper documentation of the history of the sample (origin, mode of collection, transport, storage, and the like) is essential.

The analyte concentration in the specimen is generally assumed to be representative of the concentration in the particular fluid or tissue sampled. Whole blood, plasma (the fluid obtained on centrifugation of anticoagulated whole blood), or serum (the fluid remaining when blood has clotted) are widely used in clinical work. This is because not only is blood relatively easy to collect, but also a quantitative analysis can often give useful information as to the magnitude of exposure and hence the severity of poisoning. Excretions (exhaled air, urine) or secretions (saliva, bile) are often less useful as regards interpretation of quantitative data, but can be very useful in qualitative work.

Variations in bioanalytical measurements may be subject-dependent and reflect normal physiological changes, whilst others may reflect sample collection and handling procedures (Table 2.1). Postmortem specimens are a special problem because, generally, information on the analyte concentration in blood *at the time of death* is required. Postmortem blood concentrations may not accurately reflect perimortem blood concentrations for several reasons. Haemolysis is common, for example, whilst haemostasis may lead to changes in the cellular composition of the 'blood' being sampled. There is also the possibility of contamination during collection, for example, with stomach contents, and of leakage of analyte from adjacent tissues. The leakage of intracellular potassium into plasma, which begins soon after death, is such an example.

2.2 Clinical samples and sampling

2.2.1 Health and safety

Biological samples may contain infective agents and must be handled with care, especially if originating from drug abusers, and must always be treated as if they are infective. The major common risks are associated with tuberculosis, hepatitis B, and human immunodeficiency virus (HIV). Urine is least likely to be infective. It is thought very likely that following solvent extraction or other robust sample preparation procedures, infective agents will be inactivated, except for variant Creutzfeldt–Jakob Disease (CJD), but in homogenous assays, such as immunoassays,

Table 2.1 Types of variables affecting clinical samples

Variable	Example(s)
(i) Physiological	
Age	Markers of bone turnover such as collagen cross-linkages are increased in childhood
Sex	Sex hormones
Body weight	Urinary creatinine increases with muscle mass
Recent food intake	Plasma glucose, triglycerides, and so on, increase after a normal meal. May delay and/or reduce absorption of some drugs, but increase absorption of others
Diet	Malnutrition or fasting will reduce serum albumin, urea, and phosphate, amongst other parameters
Circadian variation	Some analytes (e.g. cortisol, iron) show diurnal changes in plasma concentration
Menstrual cycle	Plasma concentrations of leutinizing hormone (LH), follicle stimulating hormone (FSH), oestradiol and progesterone vary with the cycle
Pregnancy	Plasma concentrations of human chorionic gonadotropin (HCG), oestradiol, and other biochemicals vary throughout pregnancy
Psychological changes	Venepuncture or hospitalization may increase plasma concentrations of stress-related compounds such as catecholamines
Physiological changes	Posture may affect measurements such as plasma aldosterone and albumin. Exercise can change the blood concentrations of compounds such as lactate
Drugs	Drug treatment may alter concentrations of some plasma constituents even in apparently healthy subjects (e.g. trimethoprim increases serum creatinine)
(ii) Sample/analyte variations	
Incorrect specimen	Value differences between plasma and serum, venous and arterial blood, random and 24 h urine samples (e.g. potassium higher in serum than in plasma)
Incorrect collection	The absence of an appropriate enzyme inhibitor may allow continued enzyme action such as catabolism of glucose or neuropeptides. Failure to acidify urine will decrease urinary catecholamines. Collection into EDTA will decrease plasma calcium. Contamination with ethanol, 2-propanol, or other alcohol used as a disinfectant prior to venepuncture may invalidate ethanol assay
Haemolysis	Red cell lysis may lead to changes in plasma constituents, particularly potassium, phosphate, and some enzymes and may interfere with the analytical method
Cellular contamination	The presence of platelets following incorrect centrifugation will apparently raise plasma serotonin
Incorrect/excessive storage	Some compounds may oxidize even when frozen [e.g. formation of adrenochrome from 5-hydroxyindoleacetic acid (5-HIAA)], or be subject to bacterial degradation (e.g. amino acids in acidified urine)
Collection during an infusion	Collection near to an infusion site will give misleading concentrations of the compound being infused or dilute other blood constituents
Drug treatment	Drugs or metabolites may interfere in the assay

samples may continue to be infective after incubation, even though diluted. Indeed, incubation may increase the titre of the infective agent.

Staff in regular contact with potentially infective materials must be properly trained in the safe handling and disposal of biological samples. Such staff should be vaccinated against hepatitis B, polio, tuberculosis, and tetanus and possibly other diseases in specific countries. Sample handling should be performed with due attention to preventing droplets splashing into the eyes and minimizing aerosol formation (wear eye protection, perform mixing and other procedures in a fume cupboard or microbiological safety cabinet, always use either sealable centrifuge tubes or a centrifuge with sealable rotors). Screw-capped sample tubes are preferable to those with push-in stoppers as there is less risk of aerosol formation when opening the tube.

2.2.2 *Clinical sample types*

Clinical samples can be divided into: (i) blood and related fluids, (ii) body fluids other than blood, (iii) excretory fluids/residues, and (iv) other clinical specimens (Table 2.2). A range of additional specimens may be collected for toxicological purposes. Special precautions will be needed with unstable analytes. Most compounds measured in urine can be considered stable for at least a few hours at room temperature as the urine may already have been held at body temperature for some time before it was voided.

2.2.2.1 Arterial blood

Arterial blood is normally collected by an experienced medical practitioner – it is a relatively dangerous procedure – for the measurement of blood gases and is not usually submitted for toxicological analysis. Capillary blood, which closely approximates to arterial blood, can be obtained by pricking the heel, finger or ear lobe; this procedure is most often performed on small children.

2.2.2.2 Venous blood

Venous blood is obtained by venepuncture of (usually) the median cubital vein of an arm remote from any infusion site. Either a hypodermic needle and syringe (1–50 mL) or a commercial vacuum-sampling system such as a Vacutainer may be employed. A tourniquet can be used to distend the vein prior to venepuncture, but should be released immediately prior to sampling. For repeated sampling, a small cannula may be inserted into a vein in the arm or hand, which allows venous access via a rubber septum. However, maintaining patency may be a problem, and there may be risks of: (i) inducing haemolysis and (ii) of specimen contamination due to use of anticoagulant solutions with such devices.

Following venepuncture, blood should be transferred into an appropriate container as soon as practicable. Some basic analytes and quaternary ammonium compounds, for example tricyclic antidepressants and paraquat, and aluminium bind to glass. Plastic tubes are thus preferred and are also less likely to shatter than glass, especially if frozen. On the other hand, if volatile solvents or anaesthetic gases, for example, are to be analyzed then glass is preferred if available (Section 2.3).

If blood has been collected into a syringe, it is essential that the syringe needle is removed and the blood allowed to flow gently into the collection tube in order to prevent haemolysis. This should be followed by gentle mixing to ensure contact with the anticoagulant if one is being used.

Table 2.2 Some clinical sample types

Blood and related fluids

Blood ('whole blood') is the fluid that circulates through the arteries, capillaries and veins. The adult human body contains some 5–6 litres of blood. It is composed of plasma and blood cells. Normally venous blood is obtained (Section 2.2.2.1). If whole blood is to be analyzed, then the sample should be collected into an appropriate anticoagulant, mixed, and then frozen in order to lyse the cells before the analysis [N.B. Occult blood is altered blood found only in trace amounts particularly in faeces. It is not used as an analytical sample]

Blood cells include red cells (erythrocytes) and white cells (lymphocytes, leucocytes, platelets, etc.). All may be harvested from freshly collected blood with appropriate procedures (Section 2.2.2.4)

Cerebrospinal fluid (CSF) is a plasma ultrafiltrate (i.e. its composition is that of plasma except that high M_r proteins are absent) that surrounds the elements of the central nervous system (CNS). It is obtained by lumbar puncture (needle aspiration from the spinal cord) and is usually collected into sterile tubes.

Cord blood is blood obtained from the umbilical cord at parturition. Normally venous cord blood is obtained in order to reflect neonatal, as opposed to placental, blood. It may be possible to obtain plasma or serum depending on the volume available

Plasma is the liquid portion of blood (Section 2.2.2.3)

Serum is the pale yellow fluid remaining when whole blood has clotted. Its composition is generally the same as plasma except that fibrinogen and factors associated with the clotting process are absent (Section 2.2.2.2)

Body fluids other than blood

Amniotic fluid is the fluid that surrounds the foetus in the amniotic sac

Aqueous humour is the watery fluid occupying the space between the cornea and the iris of the eye

Breast milk is the protein and fat-rich fluid produced by nursing mothers. The first expression of breast milk (colostrum) is especially rich in protein

Gastric aspirate is an acidic aqueous fluid containing digestive enzymes, food, and so on, obtained by aspiration from the stomach

Lymph is a yellowish fluid derived from the lymph glands

Peritoneal fluid is the fluid that accumulates in the peritoneum

Saliva is the viscous, clear secretion of the mucous glands in the mouth. It is related in composition to plasma, but also contains some digestive enzymes

Semen is produced by the testes and the prostate gland, and consists of seminal fluid, which may be obtained from semen by centrifugation, and spermatozoa

Synovial fluid is the clear, viscous, lubricating liquid that fills the synovium (the membrane that surrounds a joint and creates a protective sac)

Tears are the clear watery secretion of the tear ducts of the eye

Vaginal fluid is the viscous secretion of the vagina

Vitreous humour is the transparent, viscous fluid contained behind the lens in the eye

Table 2.2 *(Continued)*

Excretory fluids/residues

Bile is the thick yellow-green fluid secreted by the liver via the gall bladder into the intestine

Exhaled (expired) air generally contains less oxygen and more carbon dioxide and water vapour than ambient air, but may contain other volatile metabolic products

Faeces are the brown, semisolid residues of the digestive process (Section 2.2.2.7)

Sweat is the aqueous fluid excreted by the pores of the skin

Urine is the yellow/yellow-green fluid produced by the kidney. It consists mainly of water, salts, urea, creatinine, and other metabolic products (Section 2.2.2.5)

Other samples

Bezoars are stones or concretions found in the alimentary tract of animals

Bronchoalveolar lavage (BAL) is obtained by washing the bronchi/alveoli with an appropriate solution and aspirating the resulting fluid

Calculi ('stones') are hard crystalline deposits formed in various body cavities such as the kidney

Dialysis fluid (extracorporeal or peritoneal) is the fluid remaining or recovered after dialysis has been performed

Gastric lavage is a specimen obtained by washing the stomach with an appropriate solution and aspirating the resulting fluid

Hair (head, axilliary, or pubic) is sometimes used to assess recent exposure to poisons such as drugs or heavy metals

Nails or nail clippings (finger or toe) are sometimes used to assess exposure to drugs or heavy metals

Nasal swabs are fluid collected onto cotton swabs from inside the nose

Oral fluid is a mixture of saliva, gingival crevicular fluid (fluid from the tooth/gum margin), cellular debris, blood, mucus, food particles, and other material collected from the mouth

Stomach contents may be (i) gastric aspirate, (ii) gastric lavage, (iii) vomit, or (iv) the residue in the stomach at autopsy (Section 2.2.2.6)

Stomach wash-out (SWO): see Gastric lavage

Tissue specimens are either obtained surgically or at postmortem. Tissue obtained from an aborted foetus and/or placenta may sometimes be presented for analysis. A biopsy sample is a small sample of a tissue obtained by a specialist sampling technique

Vomit reflects the composition of gastric aspirate

Even mild haemolysis will invalidate a serum iron or potassium assay, and may invalidate plasma or serum assays for other analytes concentrated in red cells such as chlortalidone. Leaving plasma or serum in contact with red cells can cause changes due to enzymatic activity or redistribution of an analyte between cells and plasma. In general, plasma or serum should be separated from the blood cells as soon as possible. If necessary, whole blood can be stored at -20 °C or below, but freezing will lyse most cell types.

Table 2.3 Anticoagulants for *in vitro* use

Anticoagulant	Concentration (mL^{-1} blood)	Comment
Lithium heparin	10–20 units	General biochemistry
Sodium heparin	10–20 units	General biochemistry
Sodium fluoride (with either EDTA or oxalate)	1–2 mg 6–10 mg	Glucose (inhibits glycolysis) General anticoagulant
Sodium citrate	3 mg	Clotting studies – not recommended for other purposes as the aqueous solution dilutes the specimen
EDTA	2 mg	Haematology (stabilizes readily oxidized compounds)

It is important to use serum or the anticoagulant recommended for a particular measurement (Table 2.3) and not to substitute an alternative without careful consideration. Sodium citrate tubes contain 0.5 or 1 mL of the anticoagulant in aqueous solution and so are unsuitable for quantitative work. Furthermore, dilution of the sample may reduce the degree of plasma protein binding and consequently the plasma:red cell distribution of the analyte. It should be ensured that lithium heparin anticoagulant is not used if plasma lithium is to be measured. Heparin too has been known to interfere in drug analysis.

2.2.2.3 Serum

When whole blood is allowed to stand (15 min, room temperature) in a plain tube (no anticoagulant) a clot forms that will retract sufficiently to allow serum to be collected. For many analyses serum is preferred to plasma because it produces less precipitate (of fibrin) on freezing and thawing.

2.2.2.4 Plasma

Separation of plasma from anticoagulated whole blood normally requires centrifugation, as do many of the phase separation procedures discussed in Chapter 3. The inter-relationship of centrifuge rotor diameter, speed of centrifugation and relative centrifugal force (g-force) is set out in Box 2.1. Swing-out rotors are preferred for separating liquid phases.

On centrifugation of anticoagulated whole blood (2000 g, 10 min, 2–8 °C if necessary), it will separate into three layers: the bottom layer (normally 45 % or thereabouts by volume) consists of red cells; a thin intermediate layer of white cells and platelets called the ‘buffy coat’ is the next layer; and the upper, aqueous, straw-coloured layer is the plasma (about 50 % v/v). Provided the analyte is stable, anticoagulated whole blood can be kept at room temperature or refrigerated (2–8 °C) for two days or so before harvesting plasma.

More plasma than serum can be separated from whole blood. Some commercial tubes contain agents such as plastic beads or a gel that sits at the interface between the cells and the plasma to aid plasma collection. Gel separators have caused problems with some drug analyses

Box 2.1 Calculating relative centrifugal force

- The relative centrifugal force (RCF, g) depends upon the speed of the centrifuge in revolutions per minute (RPM) and the effective radius of rotation, r
- The radius of rotation varies along the length of the centrifuge tube
- RCF may be quoted as maximum, minimum or average
- Conversion tables and nomograms for each rotor are normally supplied by the manufacturer of the centrifuge
- Modern centrifuges have the facility to set the RCF directly
- The RCF will be maximal at the bottom of the tube
- RPM for a required RCF can be calculated from:

$$\text{RPM} = \sqrt{\frac{\text{RCF}}{(1.118 \times 10^{-6})r}}$$

where r is in mm
- RCF from RPM is given by:

$$\text{RCF} = (1.118 \times 10^{-6})r(\text{RPM})^2$$

(Karppi *et al.*, 2000; Berk *et al.*, 2006), although reformulated gels have been claimed to have little effect on therapeutic drug measurements (Bush *et al.*, 2001). However, this work has not been extended to other analytical toxicology tests and tubes containing gel separators are therefore best avoided. The use of such tubes will invalidate many trace element analyses (Chapter 11) and may impair analyses for solvents and other volatiles.

2.2.2.5 *Blood cells*

To collect erythrocytes, heparinized blood should be centrifuged (2000 g, 10 min), the plasma, buffy coat and top 10 % of erythrocytes (mainly reticulocytes) removed, and the remaining erythrocytes carefully washed with isotonic, buffered saline to remove trapped plasma. The cells may be used directly or frozen, either to cause haemolysis, or for storage. Platelets are usually isolated by the slow centrifugation (e.g. 300 g, 15 min) of anticoagulated whole blood to yield platelet-rich plasma, which is recentrifuged (2000 g, 10 min) to harvest the platelets. Other white blood cells are most commonly obtained by centrifugation through media of appropriate density (according to the manufacturer's instructions) or isolated by solid-phase antibody techniques.

Erythrocyte:plasma distribution

If measurement of red cell:plasma distribution is to be performed, it is easier to add the analyte to a portion of 'blank' heparinized whole blood and after allowing time for equilibration and controlling the pH (the pH of blood tends to fall *in vitro* as oxygen is lost), to obtain plasma from one portion of the blood and to freeze and thaw a second portion (to give haemolyzed whole blood) and compare the results. Admittedly this gives the plasma:whole blood ratio, but it is technically far simpler than preparing washed erythrocytes. The analyte erythrocyte concentration can be calculated if the haematocrit (the proportion of erythrocytes in blood) is known:

$$C_E = \frac{C_B - C_P(1 - H)}{H} \tag{2.1}$$

where C_E = erythrocyte concentration, C_B = whole blood concentration, C_P = plasma concentration, and H = haematocrit.

2.2.2.6 *Urine*

Different urine specimens, for example random, early morning, end-of-shift, 24 hour, may be collected in the course of metabolic or other studies. In metabolic studies, it is important to note the time of the beginning and the end of the collection period so that the rate of urine production can be calculated. A random urine sample is a midstream specimen – any preservative, such as 2 mol L^{-1} hydrochloric acid is added afterwards. Fresh urine is yellow/yellow-green in colour, but on storage in acidic solution the colour changes to yellow/brown and even to dark brown due to oxidation of urobilinogen to urobilin. Crystals, particularly of uric acid and calcium oxalate, may form causing turbidity.

When random, early morning, or end-of-shift specimens are collected, it is common practice to relate certain analytical results to a 'fixed' urinary constituent such as creatinine, which is considered to be excreted at a relatively constant rate in normal subjects. However, as creatinine is derived from creatine, there are situations, such as muscle wasting or in bodybuilders dosing with creatine, when this is not strictly true.

The concentrations of many drugs and metabolites, and of some endogenous constituents, will remain the same in acidified urine for over a week at room temperature, and for up to a month at 2–8 °C. Unacidified urine undergoes microbiological attack and many changes occur, including the complete loss of amino acids. For long-term storage acidified urine can be frozen (−20 °C), but it may be necessary to centrifuge the sample to remove any precipitate formed during storage prior to any analysis.

2.2.2.7 *Stomach contents*

This specimen encompasses vomit, gastric aspirate and gastric lavage fluid as well as the contents of the stomach at postmortem. The nature of this sample can be very variable and additional procedures such as homogenization followed by filtration and/or centrifugation may be required to produce a liquid amenable to analysis.

2.2.2.8 *Faeces*

The analysis of faeces is rarely performed in clinical chemistry, but sometimes drug and possibly metabolite analysis may be required in pharmacokinetic and metabolism studies. Analyses may also be requested if, for example, the question of drug leakage from ingested packets of drug antemortem is raised. Unlike plasma, urine, and other fluid samples, faeces are not homogeneous, and thus it is often necessary to analyze the whole sample or homogenize the whole sample and prove that the fraction taken for analysis is representative of the whole. It may take more than a day before an orally administered drug or a drug metabolite appears in faeces.

2.2.2.9 *Tissues*

Histology specimens are usually collected into a preservative such as formalin (aqueous formaldehyde solution). Such pretreatment must be borne in mind if toxicological analyses are requested subsequently. Samples of tissue obtained postmortem are normally kept at 4 °C prior to analysis.

2.3 Guidelines for sample collection for analytical toxicology

Many analytical toxicology procedures require collection of blood, urine, stomach contents, and 'scene residues', that is, material such tablet bottles found at the scene of an incident (Table 2.4). Samples of other appropriate fluids and tissues should also be collected as detailed below, especially when investigating deaths, but may not be required for analysis unless special investigations are required or decomposition is advanced (Forrest, 1993; Skopp, 2004; Flanagan *et al.*, 2005). However, such samples should be retained (4 or −20 °C) in case they are needed.

Table 2.4 Sample requirements for general analytical toxicology

Sample	Notes[a]
Whole blood *or*	10 mL (lithium heparin or EDTA tube – use fluoride/oxalate if ethanol suspected; plastic tube if paraquat suspected; glass or plastic tube with minimal headspace if carbon monoxide or other volatiles suspected)[b]
Plasma/serum	5 mL (send whole blood if volatiles, metals and some other compounds suspected – see above and Table 11.1)
Urine[c]	20–50 mL (plain bottle, no preservative[d])
Gastric contents[e]	25–50 mL (plain bottle, no preservative)
Scene residues[f]	As appropriate
Other samples	Vitreous humour (maximum available, collect separately from both eyes), bile (2 mL) or liver (about 5 g) can substitute for urine in postmortem work. Other tissues (brain, liver, kidney, lung, subcutaneous fat – 5 g) may also be valuable, especially if organic solvents or other volatile poisons are suspected

[a] Smaller volumes may often be acceptable, for example in the case of young children
[b] In postmortem work collect from femoral or other peripheral vein ensuring no contamination from central or cavity blood – collect one portion into 2 % (w/v) sodium fluoride and another into a plain tube
[c] Normally the only sample that is required for drugs of abuse screening
[d] Sodium fluoride (2 %, w/v) should be added if ethanol is suspected and blood is not available
[e] Includes vomit, gastric lavage (stomach washout, first sample), etc.
[f] Tablet bottles, drinks containers, aerosol canisters, and so on – pack entirely separately from biological samples, especially if poisoning with volatiles is a possibility

There are special considerations in sample collection and storage for metal/trace element analysis (Chapter 11).

There is little information on drug distribution within solid tissues in man; collection of approximately 5 g specimens from several sites in organs such as the brain is recommended if the whole organ is available. For liver use the right lobe (Section 2.3.7). The advantages/disadvantages of various specimens are detailed in Table 2.5. An example of a request form designed to accompany specimens submitted for toxicological investigation has been provided (Flanagan *et al.*, 2005).

If poisoning is suspected, a 10 mL blood sample (lithium heparin or EDTA tube) should be taken from an adult (proportionally less from a young child) as soon as possible, for example after

Table 2.5 Advantages and disadvantages of different sample types in analytical toxicology

Specimen	Advantage	Disadvantage	Comment
Blood (plasma/serum or whole blood)	Detect parent compound. Interpretation of quantitative data	Limited volume. Low concentrations of basic drugs and some other poisons	Interpretation of quantitative results from postmortem blood may be difficult
Urine	Often large volume. High concentrations of many poisons	Not always available. Quantitative data not often useful	Standard sample for drugs of abuse screening
Gastric aspirate (stomach contents, stomach wash-out, vomit, etc.)	May contain large amounts of poison, particularly if ingested	If available, variable sample. No use if inhalation or injection	Ensure no cross contamination of other specimens during transport/storage
Saliva/oral fluids	Non-invasive. Qualitative information on exposure to many drugs	Variable sample hence little use for quantitative work. Low concentrations of many analytes	Different pattern of metabolites to blood or urine for many analytes
Hair/nails or nail clippings	Usually available even if decomposition advanced	High sensitivity needed. Only gives exposure data for the days/weeks/months before death	Easy to store (room temperature)
Exhaled air	Non-invasive. Large volume available	Need live patient. Analyte must be volatile	Mainly used to assess ethanol ingestion and in carbon monoxide poisoning
Scene residues (tablet bottles, aerosol cans, etc. near patient)	May contain large amounts of poison	May not have been the poison taken	Ensure no cross contamination of other specimens during transport/storage

Table 2.5 (*Continued*)

Specimen	Advantage	Disadvantage	Comment
Vitreous humour	May be used instead of urine if latter not available	Limited volume but normally two specimens	Analysis may be valuable to help interpret postmortem blood data
Additional tissues (liver, brain, lung, kidney, etc.)	May contain large amounts of poison. If available then large quantity	Interference in analysis. Quantitative data not always easy to interpret	Analysis may be valuable to help interpret postmortem blood data

admission to hospital. In addition, 2 mL of blood should be collected in a fluoride/oxalate tube if ethanol is suspected. Note that tubes of this type for clinical use contain only about 0.1 % (w/v) fluoride (Table 2.3), whereas about 2 % (w/v) fluoride (40 mg sodium fluoride per 2 mL blood) is needed to inhibit fully microbial action in such specimens. Addition of fluoride also helps to protect other labile drugs such as clonazepam, cocaine, and nitrazepam from degradation. If possible, the retention of an unpreserved blood sample is advisable. The use of disinfectant swabs containing alcohols should be avoided, as should heparin anticoagulant solutions that contain phenolic preservatives.

Information recorded on the sample container at the time the sample is collected should include the names (first and family or last name), patient/subject/animal number, the date and time of collection, collection site, and the sample type (including a note of any preservative), and any other appropriate information. The date and time of receipt of all specimens by the laboratory should be recorded and a unique identifying number assigned in each case. The documentation that should be provided with the samples is summarized in Box 2.2.

Box 2.2 Information that should accompany a request for general toxicological analysis

- Name, address and telephone number of clinician/pathologist and/or Coroner's officer, and address to which the report and invoice are to be sent. A postmortem (reference) number may also be appropriate
- Circumstances of incident (including copy of sudden death report if available)
- Past medical history, including current or recent prescription medication, and details of whether the patient suffered from any serious potentially infectious disease such as hepatitis, tuberculosis, or HIV
- Information on the likely cause, and estimated time, of ingestion/death, and the nature and quantity of any substance(s) implicated
- If the patient has been treated in hospital, a summary of the relevant hospital notes should be supplied to include details of emergency treatment and drugs given, including drugs given incidentally during investigative procedures
- Note of occupation/hobbies
- A copy of any preliminary pathology report, if available

2.3.1 *Sample collection and preservation*

In general, biological specimens should be stored at 4 °C before transport to the laboratory. Exceptions to this include hair and nail, which are stable at room temperature, and filter-paper adsorbed dried blood, which is a convenient way of storing and transporting blood samples for specified analyses if refrigerated transport and storage is not feasible (Croes *et al.,* 1994). Dried blood stains and other dried forensic specimens may, of course, be handled similarly (Hammond, 1981).

Each specimen bottle should be securely sealed to prevent leakage, and individually packaged in separate plastic bags. Particular attention should be paid to the packaging of samples to be transported by post or courier in order to comply with current health and safety regulations. Sample volumes or amounts smaller than those indicated in Table 2.4 are often sufficient to complete the analyses required. Submission of very small samples may, however, result in reduced sensitivity and scope of the analyses undertaken, but nevertheless such samples should always be forwarded to the laboratory. Any residual specimen should be kept at −20 °C or below until investigation of the incident has been concluded.

In postmortem work, the use of disposable hard plastic (polystyrene) Sterilin tubes is recommended. If these are not available, then containers with secure closures appropriate to the specimen volumes should be used. Some laboratories provide specimen containers for collecting postmortem blood and urine specimens. It may be important to note if urine was obtained by use of a catheter. Suitable packaging for sending specimens by post may also be supplied. When death has occurred in hospital and poisoning is suspected, any residual antemortem specimens should be obtained as a matter of urgency from the hospital pathology laboratory (not only chemical pathology and haematology, but also immunology, transfusion medicine, and virology departments may be a source of such specimens) and submitted for toxicological analysis in addition to postmortem specimens. Note that the availability of ante- or peri-mortem specimens *does not* negate the need to collect postmortem specimens.

All organ and tissue samples, and any tablet bottles or scene residues, should be placed in separate containers to avoid any chance of cross contamination. Sampling through tissues containing high concentrations of analyte may lead to contamination of the sample.

Sample integrity is of prime concern if there are medicolegal implications as evidence may have to be produced in court. Precautions to ensure sample integrity include: (i) proper sample labelling, (ii) use of tamper-proof containers, (iii) collection of samples such as hair, nail, and femoral blood before opening the body, and (iv) proper accompanying documentation (chain-of-custody documents, Section 2.4). Samples collected for clinical purposes (or even for the coroner) are often not of 'evidential' quality, but such samples may be all that is available. DNA testing may be used to establish the origin of samples where there has been concern over sample integrity.

2.3.2 *Blood (for quantitative work)*

In analytical toxicology, plasma or serum is normally used for quantitative assays. However, some poisons such as carbon monoxide, cyanide and many other volatile organic compounds, lead and other heavy metals, and some drugs, such as chlortalidone, are found primarily in or associated with erythrocytes and thus haemolyzed whole blood should be used for such measurements.

The space above the blood in the tube ('headspace') should be minimized if carbon monoxide, solvents, or other volatiles are suspected.

Provided that the samples have been collected and stored correctly, there are usually no significant differences in the concentrations of poisons between plasma and serum. However, if a compound is not present to any extent within erythrocytes then using lysed whole blood will result in approximately a two-fold dilution of the specimen. A heparinized or EDTA whole blood sample will give either whole blood, or plasma as appropriate. The immunosuppressives ciclosporin, sirolimus, and tacrolimus are special cases because redistribution between plasma and erythrocytes begins once the sample has been collected and so the use of haemolyzed whole blood is indicated for the measurement of these compounds.

In order to maximize the reliability of measurements performed on postmortem blood, it is recommended that: (i) the interval between death and the postmortem examination is minimized, (ii) the body/samples are stored at 4 °C before the examination/after collection, (iii) blood is collected from two distinct peripheral sites, preferably the femoral veins, after tying off the vein proximally to the site of sampling, and (iv) a preservative [about 2 % (w/v) fluoride] is added to a portion of the blood sample/the sample from one vein, and to urine. The exact site of blood sampling should be recorded, as should the time of sampling and (approximate) time of death if known.

If sufficient sample is obtained, this should be divided between unpreserved and preserved (fluoride) tubes, otherwise the entire sample should be preserved unless there is a possibility of poisoning with fluoride or compounds giving rise to fluoride *in vivo*, such as fluoroacetate. If only heart or cavity blood is available this should be clearly stated. The value of giving as full a clinical, occupational, or circumstantial history as possible, together with a copy of the postmortem report, if available, when submitting samples for analysis cannot be overemphasized. Not only might this help target the analysis to likely poisons, but also the interpretation of any analytical results may be greatly simplified.

2.3.3 Blood (for qualitative analysis)

Postmortem blood (about 20 mL) for qualitative analysis only should be taken from the heart (preferably right atrium), inferior vena cava, or another convenient large vessel. The precise sampling site must be recorded on the sample tube. The blood should be free-flowing.

2.3.4 Urine

Urine is useful for 'poisons screening' as it is often available in large volumes and may contain higher concentrations of drugs or other poisons, or metabolites, than blood. The presence of metabolites may sometimes assist identification of a poison if chromatographic techniques are used. A 50 mL specimen from an adult, collected in a sealed, sterile container, is sufficient for most purposes. No preservative should be added. The sample should be obtained as soon as poisoning is suspected, ideally before any drug therapy has been initiated. However, some drugs, such as the tricyclic antidepressants (amitriptyline, imipramine, etc.), cause urinary retention, and a very early specimen may contain insignificant amounts of drug. Conversely, little poison may remain in specimens taken many hours or days after exposure even though the patient may be very ill, for example as in acute paracetamol poisoning.

Table 2.6 Some possible causes of coloured urine

Colour	Possible Cause
Yellow/brown	Bilirubin, haemoglobin, myoglobin, porphyrins, urobilin
	Anthrone derivatives (e.g. from aloin, aloe, cascara, senna, rhubarb, etc.)[a], bromsulfthalein[a], carotenes, chloroquine, congo red[a], cresol, flavins (yellow/green fluorescence), fluorescein, mepacrine, methocarbamol (on standing), methyldopa (on standing), nitrobenzene, nitrofurantoin, pamaquine, phenolphthalein[a], primaquine, quinine, santonin[a]
Red/brown	Bilirubin, haemoglobin, myoglobin, porphyrins, urobilin
	Aminophenazone, anisindione[a], anthrone derivatives[a], bromsulfthalein[a], cinchophen, congo red[a], cresol, deferoxamine[b], ethoxazene, furazolidone, furazolium, levodopa (black on standing), methocarbamol, methyldopa, niridazole, nitrobenzene, nitrofurantoin, phenacetin, phenazopyridine, phenindione[a], phenolphthalein[a], phenothiazines, phensuximide, phenytoin, pyrogallol, rifampicin, salazosulfapyridine, santonin[a], sulfamethoxazole, warfarin
Blue/green	Bile, biliverdin, indican (on standing)
	Acriflavine (green fluorescence), amitriptyline, azuresin, copper salts, ingido carmine, indomethacin, methylene blue[b], nitrofural, phenylsalicylate, resorcinol, toluidine blue[b], triamterene (blue fluorescence)
Black[c]	Blood (on standing), homogentisic acid, indican (on standing), porphobilin
	Cascara (on standing), levodopa (on standing), phenols including propofol, pyrogallol, resorcinol, thymol

[a] pH dependent
[b] Sometimes used to treat poisoning
[c] Some urinary bacteria possess an enzyme able to convert a tryptophan metabolite into a substance that interacts with plastic of urine collection bags to produce indirubin (red) and indigo (blue) giving an intense purple/black colour. Although dramatic, purple urine bag syndrome is harmless and disappears after treatment of the infection (Beunk *et al.*, 2006)

High concentrations of some drugs or metabolites can impart characteristic colours to urine (Table 2.6). Strong smelling poisons such as camphor, ethchlorvynol, and methylsalicylate can sometimes be recognized in urine because they are excreted, in part, unchanged. Acetone may arise from metabolism of 2-propanol. Chronic therapy with sulfa-drugs such as a sulfonamide may give rise to yellow or green/brown crystals in neutral or alkaline urine. Phenytoin, primidone, and sultiame may give rise to crystals in urine following overdose. Characteristic colourless crystals of calcium oxalate may form at neutral pH after ingestion of ethylene glycol, oxalic acid, or water-soluble oxalates. Urine fluorescence may be due to fluorescein added to car antifreeze (often contains ethylene glycol and/or methanol) and possibly to other products to aid leak detection.

For postmortem work, if possible, 2×25 mL urine samples should be collected in sterile plastic containers, one with preservative (2 %, w/v fluoride). If only a small amount of urine is available, all should be preserved with fluoride (but see note on fluoride poisoning above) in a plain 5 mL plastic or glass tube. Boric acid or thiomersal [thimerosal; sodium 2-(ethylmercuriothio)benzoate] containers should *not* be used because of sample contamination with borates and mercury, respectively. Urine specimens collected postmortem are valuable in screening for drugs or poisons, particularly illicit drugs, and are often used for quantitative ethanol analysis to corroborate the results of a blood analysis (Section 17.6.4.8).

2.3.5 Stomach contents

Stomach wash-out (gastric lavage) is rarely performed nowadays in treating acute poisoning. However, if a sample of stomach contents is obtained soon after a poisoning episode, large amounts of poison may be present while metabolites are usually absent. When investigating possible poisoning, it is important to obtain the *first* sample of any lavage fluid because later samples may be very dilute. A representative portion (about 50 mL) without preservative should be taken for analysis. However, all stomach contents should be retained and the volume noted. If the blood concentration is difficult to interpret, most notably in postmortem work, it can be helpful to measure the amount of poison present in the stomach.

Stomach contents are especially useful if poison(s) which are not easy to measure reliably in blood, such as cyanide, have been taken orally. However, great care is needed if cyanide salts or phosphides, for example aluminium phosphide, are thought to have been ingested, particularly on an empty stomach, because highly toxic hydrogen cyanide or phosphine gas may be released due to reaction with stomach acid. Additionally, the presence of these and other volatile materials can lead to cross contamination of other biological specimens unless due precautions are taken.

With stomach contents (and also scene residues – Section 2.3.14), characteristic colours or smells (Table 2.7) may indicate a variety of substances. Many other compounds (e.g.

Table 2.7 Smells associated with particular poisons[a]

Smell	Possible Cause
Almonds	Cyanide
Cloves	Oil of cloves
Fruity	Alcohols (including ethanol), esters
Garlic	Arsenic, phosphine
Mothballs	Camphor or naphthalene
Nail-polish remover	Acetone, butanone
Pears	Chloral
Petrol	Petroleum distillates (may be vehicle in pesticide formulation)
Phenolic (carbolic soap)	Disinfectants, cresols, phenols
Shoe polish	Nitrobenzene
Stale tobacco	Nicotine
Sweet	Chloroform and other halogenated hydrocarbons

[a] *Care* – specimens containing cyanides may give off hydrogen cyanide gas (prussic acid), especially if acidified – stomach contents are often acidic. Not everyone can detect hydrogen cyanide by smell. Similarly sulfides evolve hydrogen sulfide and phosphides evolve phosphine – the ability to smell hydrogen sulfide (rotten egg smell) is lost at higher concentrations

ethchlorvynol, methyl salicylate, paraldehyde, phenelzine) also have distinctive smells. Very low or high pH values may indicate ingestion of acids or alkali, while a green/blue colour suggests the presence of iron or copper salts.

Examination using a polarizing microscope may reveal the presence of tablet or capsule debris. Starch granules used as 'filler' in some tablets and capsules may be identified by microscopy using crossed polarizing filters where they appear as bright grains marked with a dark Maltese cross. If distinct tablets or capsules are observed, these should be placed in individual containers (e.g. Sterilin tubes). Such items and any plant remains or specimens of plants thought to have been ingested, should be examined separately. The local poisons information service or pharmacy will normally have access to publications or other aids to the identification of legitimate and sometimes illicit tablets/capsules by weight, markings, colour, shape, and possibly other physical characteristics. Identification of such material by reference to a computerized product database (Ramsey, 2004) may be possible.

2.3.6 Saliva/oral fluids

Whilst not normally considered in emergency clinical or postmortem work, there is much interest in the collection of saliva or oral fluid from live individuals because collection is non-invasive and reflects current drug or alcohol usage (Schramm *et al.*, 1992a; Aps and Martens, 2005). However, reliable saliva collection requires a co-operative individual and even then is not without problems. Some drugs, medical conditions, or anxiety, for example, can inhibit saliva secretion and so the specimen may not be available from all individuals at all times. Because saliva is a viscous fluid it is less easily poured or pipetted than plasma or urine – routine dilution with aqueous collection buffer is advocated by some authors to minimize this problem (Cozart collector). If dilution with buffer is performed the dilution must be factored into any quantitative report. For qualitative work, such as testing for drugs of abuse, oral fluid (Table 2.2) is normally collected. Oral fluid collection for forensic purposes has the advantage that sampling can be carried out while the donor is under observation, hence it is more difficult to adulterate or substitute the specimen (Spiehler, 2004).

Unstimulated normal human salivary glands do not secrete saliva. However, many stimuli will cause salivation and even during sleep there is usually sufficient stimulation to elicit a very small flow of saliva (typically 0.05 mL min^{-1}). Spitting is usually sufficient to elicit a flow of saliva of about 0.5 mL min^{-1}. Although saliva may be collected from, for example, the parotid gland by cannulation of the glandular duct, the collection of mixed whole saliva is normally the only practical alternative.

In healthy subjects, gingival crevicular fluid may constitute up to 0.5 % of the volume of mixed saliva: this proportion may be markedly increased in patients with gingivitis. Plasma exudate from minor mouth abrasions may also contribute. Therefore, subjects should not brush their teeth or practice other methods of oral hygiene for several hours before saliva is collected.

Chewing paraffin wax, Parafilm, rubber bands, pieces of PTFE or chewing gum will usually elicit a salivary flow of 1–3 mL min^{-1}. Use of acid lemon drops or a few drops of 0.5 mol L^{-1} citric acid are amongst chemical stratagems adopted to stimulate salivary flow. The saliva should be allowed to accumulate in the mouth until the desire to swallow occurs before being expelled into the collection vessel. Obviously stimulating salivary flow can facilitate relatively large volume collection in a short time. Moreover, the pH of physically stimulated saliva is about 7.4, whereas the pH of unstimulated saliva shows a larger variability that may affect the secretion of weak acids and bases. However, any physical or chemical stimulus used during the collection must not absorb or modify the compounds to be measured, nor must it introduce interfering factors into the assay procedure. Paraffin wax and Parafilm, for example, may absorb highly lipophilic molecules, and use of acidic stimulants such as citric acid can result in changes in salivary pH that can alter the secretion rate of ionizable compounds.

2.3.6.1 Collection devices for saliva/oral fluids

In the Salivette (Sarstedt) a dental-cotton (polyester) roll is chewed for 30–45 s with or without further stimulation. The device is available with or without citrate. The saliva-soaked roll is placed in a container and closed with a plastic stopper. The container is centrifuged (3 min, 1000 *g*) inside a polystyrene tube. During centrifugation the saliva passes from the dental-cotton roll into the lower part of the tube from whence it is collected. Cellular and other debris are retained at the bottom of the tube in a small sink (Figure 2.1).

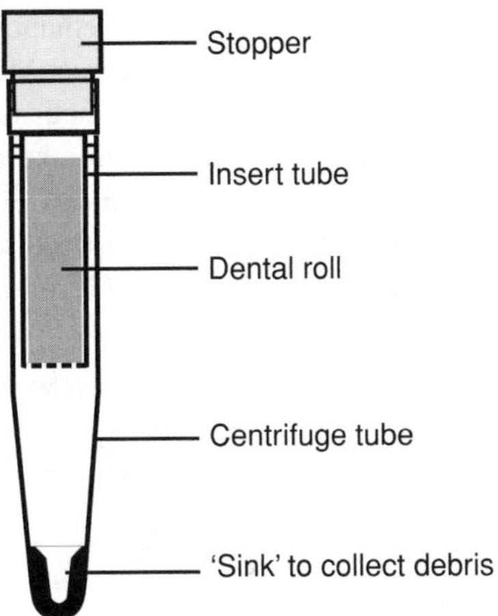

Figure 2.1 Example of a sample collection device for saliva (Salivette, Sarstedt).

The advantage of the Salivette over many other devices is that it reliably absorbs a relatively large volume of saliva (1.5 mL), although a disadvantage is that the dental cotton interferes with some assays, such as that for testosterone (Dabbs, 1991). Another collection device (OraSure) absorbs only 1.0 mL and, moreover, collects oral fluid rather than saliva as the collection pad is placed between cheek and gums (Thieme *et al.*, 1993).

Oral-diffusion-sink devices can be used to collect saliva ultrafiltrate (Wade, 1992). The device contains an analyte 'trap' such as an insoluble β-cyclodextrin and hence a concentration gradient is maintained that promotes diffusion of the analyte from saliva into the device. Clearly this device yields exposure data over a longer period than that achieved with conventional oral fluid collection, but it has been used mainly for studies of hormone rather than drug secretion. Schramm *et al.* (1990) have also developed an *in situ* device to collect saliva ultrafiltrate based on the principle of an osmotic pump. Use of this device to measure salivary phenytoin and carbamazepine (Schramm *et al.*, 1991) and cotinine (Schramm *et al.*, 1992b) has been described, but the method has not gained widespread acceptance.

A range of devices is used for oral fluid collection for testing for drugs of abuse and for alcohol (Spiehler, 2004).

2.3.7 *Sweat*

Collection of sweat has been suggested as a means of testing for abused drugs. Sweat may be collected as liquid perspiration or forehead wipes can be used (Kintz *et al.*, 2000). Alternatively, patches attached to the skin can be employed. Sweat collection is non-invasive and commercially available sweat patches may be worn for an extended period of time (10–14 days or so). Sweat patches can detect drug use that occurred shortly before the patch was applied and whilst the device remains in contact with the skin.

2.3.8 *Exhaled air*

Measurement of concentrations of volatile substances in exhaled (expired) air by infrared or other devices is of course essential in roadside testing for ethanol and valuable in assessing exposure to poisons such as carbon monoxide (Chapter 13). Direct MS of exhaled air can also detect many compounds including volatile and possibly i.v. anaesthetics several days postexposure. However, the use of these techniques is limited by the need to take breath directly from live subjects or patients (Harrison *et al.*, 2003). Similarly, collection of exhaled air into either an impervious plastic (Tedlar or PTFE) bag, or via a special device (Figure 2.2) can facilitate the analysis of a number of volatiles and metabolites via subsequent GC or GC-MS analysis.

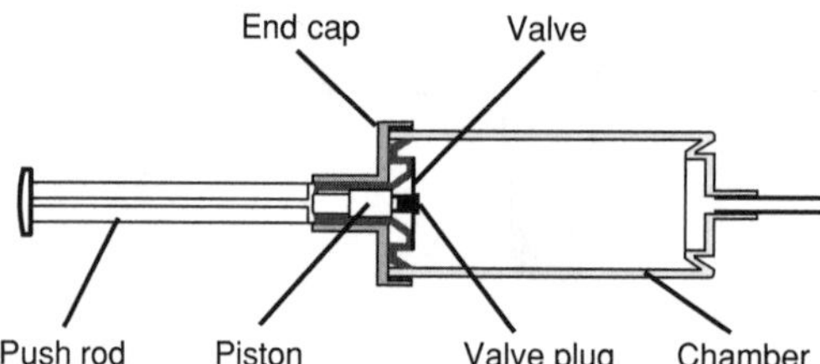

Figure 2.2 Example of a sample collection device for capturing breath samples for volatile solvent analysis [reprinted from Dyne, D., Cocker, J. and Wilson, H.K. (1997) A novel device for capturing breath samples for solvent analysis. Sci Total Environ, **199,** 83–9, with permission from Elsevier].

2.3.9 *Cerebrospinal fluid*

Cerebrospinal fluid (CSF) collected via needle aspiration is sometimes used to assess exposure to centrally acting drugs (Shen *et al.,* 2004), and may be submitted for analysis if a possible drug administration error is under investigation. As with vitreous humour and synovial fluid, CSF is normally within a relatively protected environment, and thus may also provide a valuable sample for corroborative ethanol measurement, for example, in the event that other samples are not available.

2.3.10 *Vitreous humour*

Vitreous humour can sometimes be obtained even if a corpse has been extensively burnt or damaged, if putrefaction is beginning to occur, or if samples such as urine are not available. This specimen may be especially useful when investigating diabetes- or insulin-related deaths, and for the analysis of alcohols, digoxin, lithium, and some other compounds. Vitreous humour is essentially a salt solution with very little protein, and thus any poisons or metabolites present can often be extracted as though they were in solution in buffer. Samples should be collected from each eye separately, and sodium fluoride preservative (2 %, w/v) added. Care must be taken during sampling because use of excessive suction can cause a significant change in the concentration of several analytes (Forrest, 1993), and there is always the possibility of postmortem change, as with other postmortem samples. The presence of concurrent vitreous disease must also be considered (Parsons *et al.*, 2003). Vitreous samples may be replaced with isotonic sodium chloride for cosmetic reasons.

2.3.11 Synovial fluid

Synovial fluid collected via needle aspiration has been used, for example, to assess the uptake of nonsteroidal inflammatory drugs into their likely site of action (Day *et al.*, 1999). As with CSF and vitreous humour, collection of synovial fluid may also be helpful in the event of traumatic death or extensive decomposition as it is a relatively protected environment.

2.3.12 Liver

Liver is easily collected postmortem and readily homogenized. It may contain large amounts of drugs and metabolites, and may be the primary specimen submitted for analysis if blood is not available. A portion (10–20 g) of unfixed (unpreserved) wet tissue should be collected. The sample should be taken from the right lobe if possible to reduce the risk of contamination with bile and because diffusion of poison from the stomach is less likely than in the left lobe (Pounder *et al.*, 1996a, 1996b). An analysis may, in some cases, help to establish whether acute or chronic exposure has occurred, but sometimes the analysis can do little more than establish exposure in the absence of reliable information to aid in the interpretation of quantitative results (Section 17.6).

2.3.13 Other tissues

Other tissue samples may be useful when investigating deaths where volatile substances such as solvents or gases are implicated. Brain, subcutaneous fat, lung (apex), spleen, and kidney are the most useful; 10–20 g wet weight of unfixed tissue should be collected into separate containers. The specimen should be placed in a specimen jar or nylon bag (volatile substance abuse- or anaesthetic-related deaths) and deep-frozen prior to transport to the laboratory, taking care not to overfill sample containers (overfilled tubes may break when frozen).

Measurement of brain concentrations of certain poisons may be useful in specific instances, for example such measurements are said to be helpful when investigating possible cocaine-related deaths. Spleen is rich in erythrocytes and hence may provide a valuable alternative specimen in which to measure carboxyhaemoglobin saturation if blood is not available (Vreman *et al.*, 2006).

2.3.14 Insect larvae

Analysis of blowfly (*Calliphora vicina*) and other insect larvae that feed on rotting flesh may facilitate detection of many drugs originally present in human or animal tissues, although quantitative extrapolations are unreliable (Pounder, 1991; Tracqui *et al.*, 2004). Pupae may be preserved for years, but drug and metabolite concentrations in postfeeding and pupating larvae are much lower than in feeding larvae, suggesting that the larvae metabolize and eliminate drugs during development (Sadler *et al.*, 1995; Campobasso *et al.*, 2004). Metabolism of nordazepam to oxazepam by blowfly larvae has been observed (Pien *et al.*, 2004).

2.3.15 Keratinaceous tissues (hair and nail)

Many metal ions, drugs and their metabolites are sequestered in hair and nail as they are formed, and are not metabolized further. These samples may be useful if chronic exposure is suspected, for example in deaths related to drug abuse (particularly opiates and methadone) where establishing recent drug use is very important, and for poisons that may have been eliminated from other commonly sampled fluids and tissues before sample collection. Lidocaine, a common adulterant in illicit heroin and cocaine, has been monitored in hair (Sporkert and Pragst, 2000). Hair may also survive longer after burial than other tissues (Kintz, 2004). If a single exposure to a poison is suspected, as in drug-facilitated sexual assault (DFSA, date rape), but the suspected agent is not detected in blood or urine, waiting for 1–2 months for head hair to grow and then performing segmental analysis may reveal the presence of the drug (Kintz *et al.*, 2003).

Head hair (a bunch of hairs the thickness of a pencil, about 100–200 hairs) should be either plucked or tied at the root end with cotton thread and then cut approximately 2 mm from the scalp at the vertex posterior (crown) of the head (Box 2.3; Figure 2.3). Make sure the

Box 2.3 Protocol for collection of head hair for testing for illicit drugs

- The ideal sample is collected from the vertex (the crown) of the head by cutting approximately 2 mm from the scalp
- Take a sample of hair about the thickness of a pencil (100–200 hairs)
- Pinch the hair tightly with the fingers and tie with cotton thread at the root end before cutting
- Cut the sample as close as possible to the scalp, making sure the scissors are level with the scalp
- Still holding the sample tightly, align the cut root ends of the sample and carefully place flat on a piece of aluminium foil with the cut root ends projecting about 15 mm *beyond* the end of the foil
- Mark the root end of the foil and fold the foil around the hair and pinch tightly to keep in place
- Fold the foil again in half lengthwise
- Place the sample in a tamper-proof envelope. Complete and sign the request form, making sure that the donor also signs if necessary. If there are special instructions that do not appear on the form, but are felt relevant, make a note on a separate sheet and enclose with the sample

scissors are level with the scalp and not at an angle. Tying with thread helps preserve the alignment necessary for segmental analysis. The sample should be laid aligned in aluminium foil, with the

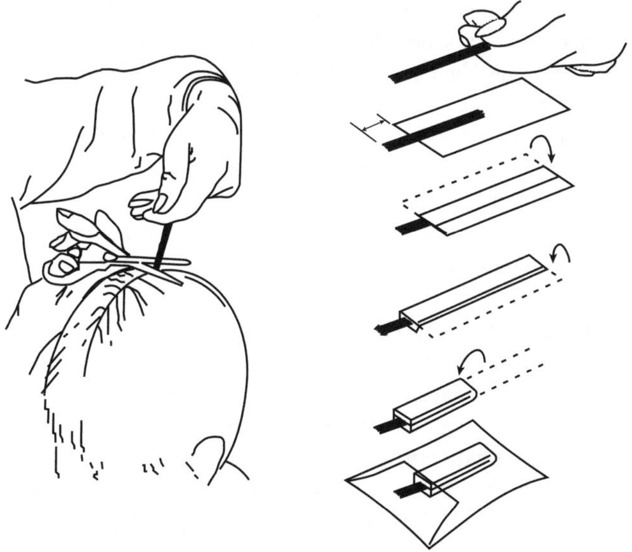

Figure 2.3 Schematic of head hair collection.

proximal end clearly identified. Pubic or axilliary hair may be substituted if no head hair remains, or if the head hair has been excessively bleached or permed.

Nail clippings may be used to monitor uptake of antifungal drugs such as itraconazole (Badcock and Davies, 1990). In postmortem work, whole nails should be lifted from the fingers or toes. This provides an even longer potential window for detecting exposure than hair. However, relatively little is known about the mechanisms of uptake and retention or drugs and metabolites in nail. In addition, the slower growth rate of nail, especially toe nail, as compared to hair makes segmental analysis, and hence interpretation, more difficult (Drummer and Gerostamoulos, 2002).

2.3.16 Bone and bone marrow

Bone marrow may be useful in poison identification in exhumations where all soft tissue has degenerated (Watterson, 2006). For nortriptyline, a bone marrow:blood ratio of 30 has been demonstrated experimentally after five days of nortriptyline treatment (Winek *et al.*, 1993). Bone itself may be useful if chronic poisoning by arsenic or lead, for example, is suspected.

2.3.17 Injection sites

Possible injection sites should be excised, packed individually and labelled with the site of origin. Appropriate 'control' material (i.e. from a site thought to not be an injection site) of similar composition should be supplied separately.

2.3.18 'Scene residues'

Such material, which may include tablets, powders, syringes, infusion fluids, and so forth, may give valuable information as to the poison(s) involved in an incident, and should be packaged separately from any biological samples. This is especially important if volatile compounds are involved. If police attend a scene these materials may find their way to a forensic laboratory rather than accompanying the patient. All items should be labelled and packed with care. Scene residues may be particularly valuable in deaths involving medical, dental, veterinary, or nursing personnel who may have access to agents that are difficult to detect once they have entered the body. Investigation of deaths occurring during or shortly after anaesthesia should include the analysis of the anaesthetic(s) used, including inhalational anaesthetics, in order to exclude an administration error. Needles must be packaged within a suitable shield to minimize the risk of injury to laboratory and other staff.

2.4 Sample transport and storage

It is usually advisable to contact the laboratory by telephone in advance to discuss urgent or complicated cases. Most specimens, particularly blood and urine, may be sent by post if securely packaged in compliance with current regulations. However, if legal action is likely to be taken on the basis of the results, it is important to be able to guarantee the identity and integrity of the specimen from when it was collected through to the reporting of the results. Thus, such samples should be protected during transport by the use of tamper-evident seals and should, ideally, be submitted in person to the laboratory by the coroner's officer or other investigating personnel. *Chain of custody* is a term used to refer to the process used to maintain and document the history of the specimen (Box 2.4).

Box 2.4 Chain of custody documents

- Name of the individual collecting the specimen
- Name of each person or entity subsequently having custody of it, and details of how it has been stored
- Dates/times the specimen was collected or transferred
- Specimen or postmortem number
- Name of the subject or deceased
- Brief description of the specimen
- Record of the condition of tamper-evident seals

Fully validated assays must include data on the stability of the analyte under specified storage conditions. In the absence of other information, biological specimens should be stored at 2–8 °C prior to analysis, if possible, and ideally any specimen remaining after the analysis should be kept at 2–8 °C for 3–4 weeks in case further analyses are required. In view of the medicolegal implications of some poison cases (e.g. if it is not clear how the poison was administered or if the patient dies) then any specimen remaining should be kept (preferably at −20 °C) until investigation of the incident is concluded.

With regard to drugs, some compounds such as clonazepam, cocaine, nifedipine, nitrazepam, thiol drugs, and many phenothiazines and their metabolites are unstable in biological samples at room temperature. Exposure to sunlight can cause up to 99 % loss of clonazepam in serum after 1 h at room temperature. Covering the outside of the sample tube in aluminium foil is a simple precaution in such cases. *N*-Glucuronides such as nomifensine *N*-glucuronide are unstable and may be present in plasma at high concentration; on decomposition the parent compound is reformed. Some compounds unstable in whole blood or plasma are listed in Table 2.8.

Table 2.8 Some drugs, metabolites and other poisons unstable in whole blood or plasma

Volatile Compound(s)	Non-volatile Compound(s)
Aerosol propellants, anaesthetic gases, carbon monoxide, ethanol, ethchlorvynol, hydrogen cyanide, mercury, methanol, nicotine, OP nerve agents, organic solvents, paraldehyde, volatile nitrites ('amyl nitrite', etc.)	Acyl glucuronide metabolites, amiodarone, bupropion, carbamate esters (physostigmine, pyridostigmine), ciclosporin[a], cyanide ion, 1,4-dihydropyridines (nifedipine), esters (aspirin, benzocaine, cocaine, diltiazem, heroin, methylphenidate, 6-monoacetylmorphine, pethidine, procaine, suxamethonium), glyceryl trinitrate (GTN) and other organic nitrites and nitrates, *N*-glucuronide metabolites (nomifensine), insulin, proinsulin C-peptide, lysergic acid diethylamide (LSD), nitrobenzodiazepines (clonazepam, loprazolam, nitrazepam), nitrophenylpyridines (nifedipine, nisoldipine), olanzapine, *N*-oxide metabolites (nomifensine), *S*-oxide metabolites, paracetamol (acetaminophen), peroxides and other strong oxidizing agents, phenelzine, phenothiazines[b], quinol metabolites (4-hydroxypropranolol), rifampicin, sirolimus[a], *N*-sulfate metabolites (minoxidil), tacrolimus[a], thalidomide, thiol- (sulfydryl)-containing drugs (captopril), thiopental, zopiclone

[a]Redistributes between plasma and red cells on standing – use whole blood. [b]Particularly those without an electron-withdrawing substituent at the 2-position.

Solid sodium fluoride (2 %, w/v) may be added to inhibit microbial and some other degradative enzymes as discussed above (Section 2.3). Esters (including carbamates and organophosphates) may be rapidly hydrolyzed by plasma esterases, including cholinesterase. If physostigmine or cocaine is to be measured accurately, the blood should be drawn into tubes containing an excess of neostigmine. Storage at −20 °C or below is recommended if the analysis cannot be performed immediately and if the stability of the analyte is unknown. However, even this may not be ideal because *N*- and *S*-oxides may be reduced to the parent compounds. Quinols such as 4-hydroxypropranolol and some sulfur-containing compounds such as olanzapine, on the other hand, are readily oxidized and stabilization by addition of a reducing agent such as ascorbate or sodium metabisulfite is necessary, but in the case of olanzapine, for example, there is the possibility of reducing the *N*-oxide by such an addition.

Storage at −5 to −70 °C should be accompanied by basic precautions to preserve sample integrity (Box 2.5). The requirements of the local ethics committee and other guidance on the retention and storage of clinical samples must be complied with.

Box 2.5 Guidance on freezer storage of samples

- Do not freeze whole blood if plasma or serum is to be analyzed
- Ensure that labelling is waterproof
- Ensure tubes are tightly sealed and well filled, but do not overfill tubes, especially glass tubes
- Do not keep too long to minimize freeze-drying effects
- Keep a record of freezer contents
- Keep a continuous record of freezer temperature (e.g. by use of the Tiny Talk data logger: http://www.omniinstruments.co.uk/aquisition/Gemini/manbot.htm#TinyTalk, accessed 15 July 2007)
- Fit an alarm in case of freezer failure

2.5 Common interferences

Plasticizers, particularly phthalates, may originate from plastic bags used to store transfusion blood, infusion tubing, and from soft plastic closures for blood tubes. Such compounds are often retained on reversed phase HPLC systems (Section 8.5.2) and show good absorption at 254 nm and below. Polyvinylchloride (PVC), for example, can contain up to 40 % (w/w) di-2-ethylhexylphthalate and concentrations of this latter compound of up to about 0.5 g L^{-1} have been reported after storage of plasma in PVC bags for 14 days (Dine *et al.*, 1991). A further consideration is that postmortem specimens may contain putrefactive bases such as phenylethylamines and indole, which may interfere in the analysis of amfetamines and other stimulants. Hexanal may arise from breakdown of fatty acids. GC retention data and mass spectra of a number of plasticizers, pollutants, and other substances that may be encountered in toxicological analyses have been reported (Pfleger *et al.*, 2007). Modern blood-collection tubes may contain a range of additives including surfactants, which may interfere in immunoassays, for example (Stankovic and Parmar, 2006).

Drugs may arise from unexpected sources including food and over-the-counter (OTC) medicines. Quinine may originate from ingestion of tonic water, for example, caffeine from caffeinated beverages (tea, coffee, cola) and some proprietary stimulants, chloroquine and related compounds from malaria prophylaxis, and pholcodine and other opiate analogues from cough and cold cures. Morphine is a constituent of some antidiarrhoeal preparations. A caffeine metabolite, paraxanthine, may be a problem in HPLC theophylline assays (Rowe *et al.*, 1988). Lidocaine-containing gel is commonly used as a lubricant during procedures such as bladder catheterization or bronchoscopy, and measurable plasma concentrations of lidocaine and some metabolites may be attained. The alkaloids emetine and cephaeline, and their metabolites, have been detected in stomach contents, plasma, and in urine after syrup of ipecacuanha (ipecac) was given to induce vomiting, especially in children, although this form of treatment is no longer recommended practice.

Sedatives such as pethidine (meperidine) may be given prior to computerized tomographic (CT) scans, lumbar puncture, or other investigations. Neuromuscular blocking agents such as atracurium, which is metabolized to laudanosine, and vecuronium, may be encountered in samples from patients undergoing mechanical ventilation. 1,3-Propanediol (propylene glycol) is used as a vehicle in i.v. infusions. Benzoic acid, which is metabolized to hippuric acid, is used as a preservative in some drugs and foods. The antibiotic metronidazole is often encountered in samples from hospitalized patients. Iodinated hippuric acids are used as X-ray contrast media. Alcohols may originate from skin cleansing swabs. Such compounds and also drugs given in emergencies, anticonvulsants, for example, may not be recorded on record sheets. Some compounds or their metabolites may have very long plasma half-lives. Chlorpromazine metabolites, for example, have been reported in urine many months after stopping therapy.

Contamination with trace elements is a particularly difficult area (Chapter 11). Contamination with volatiles, such as solvents used in the laboratory, must be guarded against if one of the solvents in question is to be tested for in a biological or related sample. Glassware and other items must be kept clean and tested regularly for contamination via internal quality control (IQC) procedures.

2.6 Summary

Although not in the immediate control of the laboratory, every effort must be made to ensure appropriate priority is given to sample collection and handling because if this is not done properly all subsequent effort is wasted. Care in sample collection is especially important in postmortem and overt medicolegal work, but even in clinical work effort in providing advance information to clinicians and pathologists on sample requirements (site of collection, volume, addition of sodium fluoride, etc.) and feedback on the problems that will arise when mistakes are made can only prove beneficial.

References

Aps, J.K. and Martens, L.C. (2005) Review: The physiology of saliva and transfer of drugs into saliva. *Forensic Sci Int*, **150**, 119–31.

Badcock, N.R. and Davies, A. (1990) Assay of itraconazole in nail clippings by reversed phase, high performance liquid chromatography. *Ann Clin Biochem*, **27**, 506–8.

Berk, S.I., Litwin, A.H., Du, Y., Cruikshank, G., Gourevitch, M.N. and Arnsten, J.H. (2006) False reduction in serum methadone concentrations by BD Vacutainer serum separator tubes (SSTTM). *Clin Chem*, **52**, 1972–4.

Beunk, J., Lambert, M. and Mets, T. (2006) The purple urine bag syndrome. *Age Ageing*, **35**, 542.

Bush, V., Blennerhasset, J., Wells, A. and Dasgupta, A. (2001) Stability of therapeutic drugs in serum collected in vacutainer separator tubes containing a new gel (SST II). *Ther Drug Monit*, **23**, 259–62.

Campobasso, C.P., Gherardi, M., Caligara, M., Sironi, L. and Introna, F. (2004) Drug analysis in blowfly larvae and in human tissues: a comparative study. *Int J Legal Med*, **118**, 210–4.

Croes, K., McCarthy, P.T. and Flanagan, R.J. (1994) Simple and rapid HPLC of quinine, hydroxychloroquine, chloroquine, and desethylchloroquine in serum, whole blood, and filter paper-adsorbed dry blood. *J Anal Toxicol*, **18**, 255–60.

Dabbs, J.M. (1991) Salivary testosterone measurements: collecting, storing, and mailing saliva samples. *Physiol Behav*, **49**, 815–7.

Day, R.O., McLachlan, A.J., Graham, G.G. and Williams, K.M. (1999) Pharmacokinetics of nonsteroidal anti-inflammatory drugs in synovial fluid. *Clin Pharmacokinet*, **36**, 191–210.

Dine, T., Luyckx, M., Cazin, M., Brunet, C., Cazin, J.C. and Goudaliez, F. (1991) Rapid determination by high performance liquid chromatography of di-2-ethylhexyl phthalate in plasma stored in plastic bags. *Biomed Chromatogr*, **5**, 94–7.

Drummer, O.H. and Gerostamoulos, J. (2002) Postmortem drug analysis: analytical and toxicological aspects. *Ther Drug Monit*, **24**, 199–209.

Flanagan, R.J., Connally, G. and Evans, J.M. (2005) Analytical toxicology: guidelines for sample collection postmortem. *Toxicol Rev*, **24**, 63–71.

Forrest, A.R.W. (1993) Obtaining samples at postmortem examination for toxicological and biochemical analysis. *J Clin Pathol*, **46**, 292–6.

Hammond, M.D. (1981) Detection of drugs in blood stains. *Anal Proc*, **18**, 299–303.

Harrison, G.R., Critchley, A.D., Mayhew, C.A. and Thompson, J.M. (2003) Real-time breath monitoring of propofol and its volatile metabolites during surgery using a novel mass spectrometric technique: a feasibility study. *Br J Anaesth*, **91**, 797–9.

Karppi, J., Akerman, K.K. and Parviainen, M. (2000) Suitability of collection tubes with separator gels for collecting and storing blood samples for therapeutic drug monitoring (TDM). *Clin Chem Lab Med*, **38**, 313–20.

Kintz, P. (2004) Value of hair analysis in postmortem toxicology. *Forensic Sci Int*, **142**, 127–34.

Kintz, P., Cirimele, V., Jamey, C. and Ludes, B. (2003) Testing for GHB in hair by GC/MS/MS after a single exposure. Application to document sexual assault. *J Forensic Sci*, **48**, 195–200.

Kintz, P., Cirimele, V. and Ludes, B. (2000) Detection of cannabis in oral fluid (saliva) and forehead wipes (sweat) from impaired drivers. *J Anal Toxicol*, **24**, 557–61.

Parsons, M.A., Start, R.D. and Forrest, A.R. (2003) Concurrent vitreous disease may produce abnormal vitreous humour biochemistry and toxicology. *J Clin Pathol*, **56**, 720.

Pfleger, K., Maurer, H.H. and Weber, A. (eds) (2007) *Mass Spectral and GC Data of Drugs, Pollutants, Pesticides and their Metabolites*, 3rd edn, Wiley-VCH, Weinheim.

Pien, K., Laloup, M., Pipeleers-Marichal, M., Grootaert, P., De Boeck, G., Samyn, N., Boonen, T., Vits, K. and Wood, M. (2004) Toxicological data and growth characteristics of single postfeeding larvae and puparia of *Calliphora vicina* (Diptera: Calliphoridae) obtained from a controlled nordiazepam study. *Int J Legal Med*, **118**, 190–3.

Pounder, D.J. (1991) Forensic entomo-toxicology. *J Forensic Sci Soc*, **31**, 469–72.

Pounder, D.J., Adams, E., Fuke, C. and Langford, A.M. (1996b) Site to site variability of postmortem drug concentrations in liver and lung. *J Forensic Sci*, **41**, 927–32.

Pounder, D.J., Fuke, C., Cox, D.E., Smith, D. and Kuroda, N. (1996a) Postmortem diffusion of drugs from gastric residue: an experimental study. *Am J Forensic Med Pathol*, **17**, 1–7.

Ramsey, J.D. (2004) Solid dosage form identification, in *Clarke's Analysis of Drugs and Poisons* (eds A.C. Moffat, D. Osselton and B. Widdop), 3rd edn, Pharmaceutical Press, London, pp. 249–258.

Rowe, D.J., Watson, I.D., Williams, J. and Berry, D.J. (1988) The clinical use and measurement of theophylline. *Ann Clin Biochem*, **25**, 4–26.

Sadler, D.W., Fuke, C., Court, F. and Pounder, D.J. (1995) Drug accumulation and elimination in *Calliphora vicina* larvae. *Forensic Sci Int*, **71**, 191–7.

Schramm, W., Annesley, T.M., Siegel, G.J., Sackellares, J.C. and Smith, R.H. (1991) Measurement of phenytoin and carbamazepine in an ultrafiltrate of saliva. *Ther Drug Monit*, **13**, 452–60.

Schramm, W., Pomerleau, O.F., Pomerleau, C.S. and Grates, H.E. (1992b) Cotinine in an ultrafiltrate of saliva. *Prev Med*, **21**, 63–73.

Schramm, W., Smith, R.H., Craig, P.A. and Kidwell, D.A. (1992a) Drugs of abuse in saliva: a review. *J Anal Toxicol*, **16**, 1–9.

Schramm, W., Smith, R.H., Craig, P.A., Paek, S.H. and Kuo, H.H. (1990) Determination of free progesterone in an ultrafiltrate of saliva collected *in situ*. *Clin Chem*, **36**, 1488–93.

Shen, D.D., Artru, A.A. and Adkison, K.K. (2004) Principles and applicability of CSF sampling for the assessment of CNS drug delivery and pharmacodynamics. *Adv Drug Deliv Rev*, **56**, 1825–57.

Skopp, G. (2004) Preanalytic aspects in postmortem toxicology. *Forensic Sci Int*, **142**, 75–100.

Spiehler, V. (2004) Drugs in saliva, in *Clarke's Analysis of Drugs and Poisons*, 3rd edn (eds A.C. Moffat, D. Osselton and B. Widdop), Pharmaceutical Press, London, pp. 109–123.

Sporkert, F. and Pragst, F. (2000) Determination of lidocaine in hair of drug fatalities by headspace solid-phase microextraction. *J Anal Toxicol*, **24**, 316–22.

Stankovic, A.K. and Parmar, G. (2006) Assay interferences from blood collection tubes: a cautionary note. *Clin Chem*, **52**, 1627–8.

Thieme, T., Fitchen, J., Bartos, F., Salinsky, M., Green, T. and Holden, W. (1993) Therapeutic drug monitoring using oral samples collected with the OraSure device. *Ann N Y Acad Sci*, **694**, 337–9.

Tracqui, A., Keyser-Tracqui, C., Kintz, P. and Ludes, B. (2004) Entomotoxicology for the forensic toxicologist: much ado about nothing? *Int J Legal Med*, **118**, 194–6.

Vreman, H.J., Wong, R.J., Stevenson, D.K., Smialek, J.E., Fowler, D.R., Li, L., Vigorito, R.D. and Zielke, H.R. (2006) Concentration of carbon monoxide (CO) in postmortem human tissues: effect of environmental CO exposure. *J Forensic Sci*, **51**, 1182–90.

Wade, S.E. (1992) Less-invasive measurement of tissue availability of hormones and drugs: diffusion-sink sampling. *Clin Chem*, **38**, 1639–44.

Watterson, J. (2006) Challenges in forensic toxicology of skeletonised human remains. *Analyst*, **131**, 961–5.

Winek, C.L., Morris, E.M. and Wahba, W.W. (1993) The use of bone marrow in the study of postmortem redistribution of nortriptyline. *J Anal Toxicol*, **17**, 93–8.

3 Sample Preparation

3.1 Introduction

Although in analytical toxicology simple colour tests (Section 4.2) and some immunoassays and enzyme-based assays, for example, may often be performed directly upon the specimen (homogenous assay), some form of treatment of the sample is normally necessary prior to the analysis, even if this only consists of adding an internal standard. The complexity of the sample preparation procedure used will depend to a large extent on the nature of the sample, the nature of the drug or poison to be analyzed (including whether it is unstable or extensively metabolized), whether a chromatographic method is to be used and, if so, the chosen method of detection. Clearly all these are factors are interdependent (Figure 3.1).

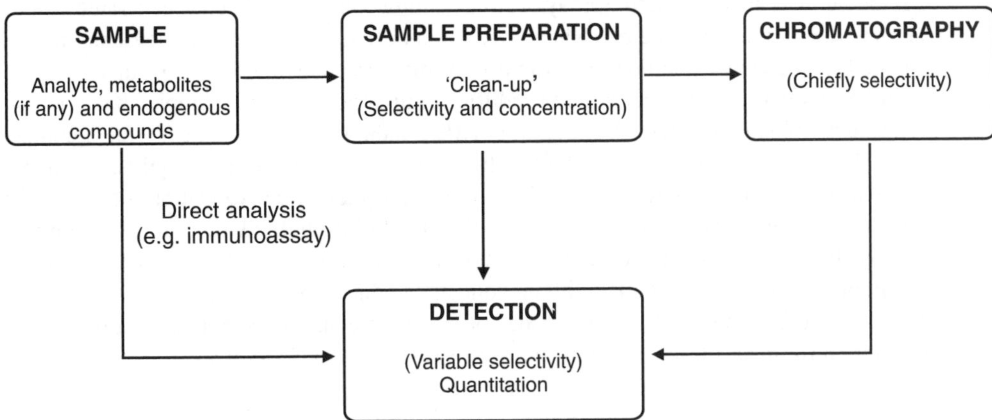

Figure 3.1 Schematic diagram of the steps that may be involved in sample preparation.

Additional aims of sample preparation may be removal of insoluble residues and interfering compounds, and sometimes concentration or even dilution of the analyte to adjust sensitivity (Box 3.1). Judicious choice of solvent extraction conditions, including pH-controlled back-extraction of weak electrolytes into aqueous solution, sometimes followed by re-extraction into solvent, can improve both selectivity and sensitivity (Smith, 2003). Traditionally sample preparation has been performed beforehand ('off-line'), but there is now much interest in performing the sample preparation and analysis steps automatically ('on-line') to minimize errors and reduce (labour) costs, especially in HPLC.

The method chosen for sample preparation depends on the overall analysis strategy. If the analyte is thermally labile then GC is usually inappropriate, as is evaporation of an extraction solvent at an elevated temperature. If the analyte concentration is high, or a particular assay is very sensitive, then sample preparation may be minimal. On the other hand, trace analysis may require a

Box 3.1 Aims of sample preparation

Achieve compatibility of analyte with the analytical system via:

- Solubilization/homogenization of solids/emulsions, etc.
- Disruption of protein binding
- Addition of internal standard(s)
- Removal of insoluble residues/interfering compounds
- Concentration/dilution of analyte
- Stabilization, decomposition and/or derivatization of analyte to improve extraction, chromatographic, and/or detection properties
- Hydrolysis of conjugates

complex assay procedure with multiple concentration and clean-up steps. Urine and bile may contain higher concentrations of compounds of interest and fewer insoluble residues than whole blood, plasma or serum, and as a result sample preparation may sometimes be simplified. This being said, providing the assay is 'fit for purpose' (Section 1.3.3), any method should be as technically simple as possible, not only to minimize costs, but also to maximize reliability and reproducibility.

If the samples are for a routine assay or from a multicentre study that has generated hundreds of samples, then throughput and reproducibility are important as well as accuracy, and time spent in method optimization is justified. If a drug is to be measured only occasionally, then consideration should be given to adapting a method used for a similar analyte, for example a GC method with a change in column oven temperature, or an HPLC method with a change in eluent composition or UV detector wavelength. If a method includes TLC or GC then liquid–liquid extraction (LLE) should be considered. In addition, LLE is usually appropriate if derivatization is to be performed, as many such reactions require the absence of water. Solid-phase extraction (SPE), on the other hand, is often suitable for HPLC methods as the drug is usually eluted from the SPE cartridge with a hydrophilic solvent compatible with an aqueous methanol or acetonitrile (i.e. reversed-phase) HPLC eluent (Section 8.5.2). Some modes of sample preparation are listed in Box 3.2. Sample preparation for metals/trace elements analysis is discussed in Chapter 11.

Box 3.2 Modes of sample preparation

(a) Solids/tissues
- Physical disruption (homogenization, sonication, heat/microwave)
- Chemical disruption (enzymes, acid/base)

(b) Liquids
- Direct analysis after off-line treatment (filtration, internal standard addition, protein precipitation)
- Direct 'on-line' analysis
- Headspace analysis (also purge/trap, microdiffusion)
- Liquid-liquid extraction (LLE) (direct or via derivative formation)
- Solid phase extraction (SPE, also known as sorbent extraction, SE)
- Solid phase microextraction (SPME) of liquid or headspace
- Liquid phase microextraction (LPME)
- Supercritical fluid extraction (SFE) of solid matrix

In plasma many drugs bind to albumin and many basic drugs also bind to α_1-acid glycoprotein, a cell-breakdown product with no known role in plasma. Protein-bound analytes must be released during the course of the analysis if 'total' (free + protein-bound) drug is to be measured. Analysis of standard solutions of the analyte prepared in analyte-free plasma is a simple way of checking the effect of protein binding on analyte recovery. Equilibrium dialysis or ultrafiltration may be used to separate 'free' from bound drug thus allowing the 'free fraction' to be measured (Section 3.3).

It is important to prevent or, failing this, to understand any reactions occurring during sample preparation. The cholinesterase inhibitor physostigmine, for example, is rapidly hydrolyzed if extracted at pH 9.5 or above, whilst *N*-desmethylpropoxyphene (norpropoxyphene), the major plasma metabolite of dextropropoxyphene, rearranges to an amide at pH 11 or above. Bupropion decomposes under strongly basic conditions. Many *N*-oxides are reduced to the parent compounds at pH 12 and above. HPLC has an advantage over GC in that the possibility of thermal decomposition during the analysis is minimized, and refrigerated autosampler trays and analytical columns may be used if decomposition at room temperature is a consideration.

The potential for loss of analyte during sample preparation such as adsorption onto the extraction vessel must also be remembered. Other factors being equal, miniaturization and the use of the minimum number of extract transfer steps will, in general, give the best recovery.

Plasma, serum and other fluids are normally sampled by pipetting, a positive displacement pipette being used for particularly viscous fluids. Whole blood or postmortem blood is usually analyzed after freezing and thawing to disrupt erythrocyte membranes, sedimentation of cell debris before sampling being an option. As with other viscous fluids, use of a positive displacement pipette is recommended, or alternatively specimens can be apportioned by weight. In some jurisdictions blood ethanol in relation to driving a motor vehicle is defined by weight rather than by volume, for example. If severe dehydration of a postmortem specimen is suspected, haemoglobin can be measured and the analytical result corrected to a 'normal' haematocrit. Tissue specimens are normally analyzed wet after removing any surface blood, by blotting gently with filter paper, for example.

3.2 Modes of sample preparation

3.2.1 Direct analysis/on-line sample preparation

Direct injection of blood after dilution with aqueous internal standard solution has been used for many years in the GC analysis of ethanol and other low molecular weight volatile compounds using columns packed with a porous polymer such as Chromosorb 101 or Porapak Q, or a modified carbon black (Section 7.6.2.2). A large (5- or 10-fold) dilution with internal standard solution is used to facilitate use of aqueous calibration standards and QA solutions; sensitivity is not a problem. However, in general, GC and many HPLC assays require protein removal and/or LLE or SPE prior to the analysis. This being said, a number of other approaches to the direct HPLC analysis of plasma or serum have been investigated.

Internal surface reverse-phase (ISRP) chromatography is a form of restricted access media (RAM) HPLC that uses column/eluent combinations which permit protein to elute without precipitation whilst retaining and separating hydrophobic compounds (Hagestam and Pinkerton, 1985). However, the eluent pH has to be about 7 and the concentration of the organic component has to be low to minimize the risk of protein precipitation (Section 8.5.2). Similar considerations apply to the direct injection of plasma or serum onto unmodified silica HPLC columns. This latter approach has been used in the analysis of some antimicrobial drugs and in other applications, but has not found wide acceptance.

Alternatively, one or more precolumns, containing appropriate stationary phases, may be placed in-line with the analytical column. Computer controlled column switching of eluent flow between the columns is a method of on-line sample preparation; usually, internal standard addition is performed off-line beforehand. Such procedures have been investigated as part of automated analytical sequences for basic drugs (Binder *et al.*, 1989). More recent ideas include the use of a restricted access media (RAM) extraction column to remove protein and other unwanted high M_r compounds [an initial methanol concentration of 5 % (v/v) is sufficient to release protein bound analytes without causing protein precipitation] (Rbeida *et al.*, 2004). In some cases a second column packed with a mixed-mode stationary phase (MMP) is used to further purify the analyte(s) and any internal standard prior to elution onto the analytical HPLC column (Georgi and Boos, 2004). Another approach is to use a RAM extraction column in conjunction with so-called 'turbulent flow' chromatography (Thermo Fisher Scientific) to allow direct injection of plasma (Vintiloiu *et al.*, 2005)

A further approach (Automated Sequential Trace Enrichment of Dialysates, ASTED) was the use of in-line microdialysis together with a trace-enrichment cartridge (TEC) prior to automated analysis by conventional HPLC (Cooper *et al.*, 1988). In the commercial development (Gilson), typically drug protein binding was disrupted by appropriate pH and buffer selection prior to the recipient stream passing to the TEC and then to the HPLC. This approach was particularly attractive as the TEC did not become contaminated readily as the analyte has been 'donated' to the dialysate stream. However, the capital cost was considerable, the recovery sometimes depended on the sample matrix, and the instrument is no longer available commercially.

3.2.2 Protein preciptation

Despite attempts to introduce on-line systems for plasma or serum HPLC assays, generally off-line sample preparation is used. Plasma protein precipitation with analysis of the supernatant resulting after centrifugation is perhaps the simplest approach, the procedures used being derived from sample preparation methods employed prior to UV/visible spectrophotometry as outlined in Chapter 4. In some cases the supernatant is analyzed directly by HPLC or by LC-MS, whilst in others further manipulation(s) such as LLE or SPE may be needed. The possibility of loss of analyte(s) with the precipitate must be considered.

The efficiency of various plasma protein removal procedures has been compared (Table 3.1). Other such mixtures include aqueous zinc sulfate (5 % w/v):methanol (100 + 43), 5-sulfosalicylic acid (3.2 % w/v) in water:methanol (1 + 1) and methanol:acetonitrile (1 + 5). If strongly acidic reagents are used, the analyte and any internal standard must be stable at low pH values. Brief cooling to −20 °C before centrifugation may enhance protein precipitation. When evaluating such a procedure it is always advisable to add a second portion of precipitation reagent to the supernatant obtained initially to ensure that no protein remains in solution. Supernatant filtration using a suitable system is a precaution sometimes adopted to minimize the risk of injecting particles, but the possibilities of loss of analyte on the filter and of introducing contaminants must be assessed for each assay.

Methanol containing 0.2 % (v/v) concentrated hydrochloric acid (2 volumes) when added to plasma or serum (1 volume) followed by vortex-mixing and high-speed centrifugation (10 000 *g* or so, 30 s) gives efficient protein precipitation (Flanagan and Ruprah, 1989). However, the sensitivity is less than if it had been possible to inject the sample directly because of the three-fold

Table 3.1 The effectiveness of some common protein removal procedures (Blanchard, 1981)

(i) *Human plasma*

Precipitant	Precipitant (v/v) required to remove 99 % of protein
10 % (w/v) Trichloroacetic acid	0.2
6 % (w/w) Perchloric acid	0.7
Acetonitrile	1.3
Acetone	1.4
Ethanol	3.0
Methanol	4.0
Saturated ammonium sulfate	about 3.0
Ultrafiltration	not applicable[a]
Heat (80 °C)	not applicable[b]
Dialysis	not applicable[c]

(ii) *Red cell extract*[d]

Precipitant	Precipitant (v/v) required to remove 97 % of protein
5 % (w/v) Trichloroacetic acid	4.5
10 % (w/v) Trichloroacetic acid	3.8
15 % (w/v) Trichloroacetic acid	3.3

[a] Provided there is no leakage ultrafiltration will remove 98 % of protein
[b] Only poor removal of protein (about 95 %, 10 min)
[c] Recipient stream is protein free, depends on membrane porosity
[d] Human red cells (initial protein content 311 g L^{-1})

dilution, and selectivity is dependent on the chromatographic separation and the selectivity of detection because all solutes are injected. The high proportion of methanol in the supernatant also means that it is not normally possible to inject more than 10–20 μL onto the HPLC column unless an eluent with a similarly high organic solvent content is being used, as unacceptable band spreading will result (Section 5.4.2.3).

If an organic solvent is used as the protein precipitant, an increase in the amount of the 'extract' that may be injected, with consequent enhancement of sensitivity, may be achieved by transferring the supernatant to a clean tube and evaporating to dryness under a stream of compressed air or nitrogen. This obviously takes time, but solvent:water mixtures are easier to evaporate than water alone (azeotropic effect). Compounds such as potassium carbonate may be added to plasma:organic solvent mixtures to 'force' an organic layer to form, thus simplifying extract concentration after removal. Evaporation to dryness may be necessary prior to some derivatization reactions (Section 3.7).

Box 3.3 Microdiffusion

- Simple means of purifying volatile analytes prior to colorimetric analysis
- Enclosed system: sample/standard and 'releasing' reagent added to opposite sides of outer well, 'trapping' reagent added to inner well
- Contents of outer well mixed – released volatile product allowed to diffuse into 'trapping' solution
- Principal disadvantages:
  - Diffusion cells may be difficult to obtain
  - Time taken for diffusion to complete
  - Limited to volatile analytes (room temperature)

3.2.3 Microdiffusion

Microdiffusion (Box 3.3) is a form of sample purification that relies on the liberation of a volatile compound, for example hydrogen cyanide in the case of cyanide salts, from the test solution held in one compartment of an enclosed system. The volatile compound is subsequently 'trapped' using an appropriate reagent (sodium hydroxide solution in the case of hydrogen cyanide) held in a separate compartment of the specially constructed (Conway) apparatus (Figure 3.2).

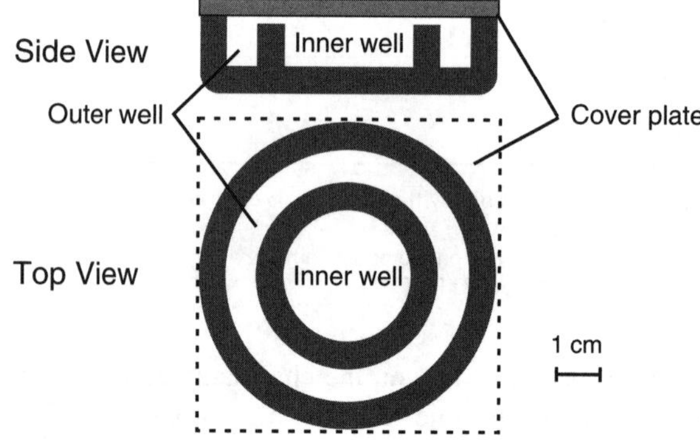

Figure 3.2 Apparatus for volatile analytes: Conway microdiffusion.

The cells are normally allowed to stand for 2–5 h (room temperature) for the diffusion process to be completed, but sometimes shorter incubation times can be used. The analyte concentration is subsequently measured in a portion of the 'trapping' solution either spectrophotometrically, or by visual comparison with standard solutions analyzed concurrently in separate cells. The Conway apparatus may be made from glass, but polycarbonate must be used with fluorides as hydrogen fluoride etches glass. Lightly smearing the cover with petroleum or silicone grease ensures an air-tight seal. In order to carry out a quantitative assay at least eight cells are needed: 'blank' sample, a minimum of three calibrators, test sample (in duplicate) and positive control sample (in duplicate). It is important to clean the diffusion apparatus carefully after use.

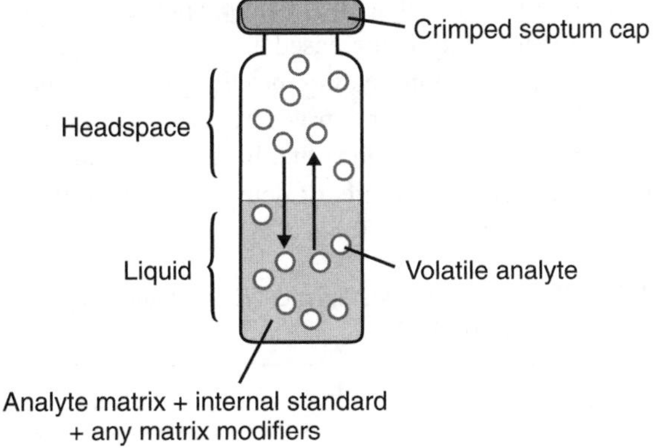

Figure 3.3 Apparatus for volatile analytes: GC headspace vial.

3.2.4 Headspace and 'purge-and-trap' analysis

The principle underlying GC headspace analysis is that in a sealed vial at constant temperature equilibrium is established between volatile components of a liquid or solid sample in the vial and the gas phase above it – the 'headspace' (Figure 3.3). After allowing due time for equilibration (normally 15 min or so) a portion of the headspace may be withdrawn via a rubber septum using a gas-tight syringe and injected onto the GC column. Advantages of this method are that the risk of contamination of the column with non-volatile residues is virtually eliminated and that automation is relatively simple. An internal standard or other additive such as a matrix modifier (e.g. a concentrated salt solution used to enhance partitioning of less volatile sample components into the gas phase) may be added prior to the incubation, and quantitative analyses may be performed after constructing a calibration graph. This technique is widely used in the analysis of ethanol and other volatile substances in biological samples and in the pharmaceutical industry for measuring solvent residues in tablets, amongst other applications.

Generally, headspace vials are available in sizes of 6, 10 and 20 mL – a vial large enough to ensure an adequate headspace or phase ratio without excessive dilution of the components of interest in the headspace should be used. The phase ratio should normally be at least 50 % of the sample volume. Headspace vials may be round- or flat-bottomed. Round-bottomed vials are stronger and may work more reliably with an autosampler. Round-bottomed vials also tend to withstand higher pressures, and hence are more suited to work at elevated temperatures and if derivatization is to be performed. Headspace vials must be correctly crimped – the cap should not turn after application to the vial and septa suitable for the incubation temperature and the analyte(s) must be used. Poor quality septa may 'bleed' (release volatile components into the headspace), and, on the other hand, may adsorb analyte(s) from the headspace.

Sample preparation aims to maximize the concentration of volatile components in the headspace and minimize unwanted contamination from other compounds in the sample. The equilibrium distribution of an analyte between the sample phase and the gas phase in the vial is denoted by the partition coefficient (K):

$$K = \frac{C_s}{C_g} \tag{3.1}$$

where C_s is the concentration of analyte in the sample phase and C_g is the concentration of analyte in the gas phase. Thus, sensitivity is increased as K decreases, that is compounds that have low K values will tend to partition more readily into the gas phase and have relatively low limits of detection. An example of this is hexane in water: at 40 °C, hexane has a K value of 0.14 in an air–water system. Conversely, compounds that have high K values partition less readily into the gas phase and have relatively high limits of detection. For example, at 40 °C, ethanol has a K value of 1355 in an air–water system (Table 3.2).

Table 3.2 Partition coefficients (K) of some common solvents in an air-water system (40 °C)

Solvent	K	Solvent	K
Cyclohexane	0.077	Butyl acetate	31.4
Hexane	0.14	Ethyl acetate	62.4
Tetrachloroethylene	1.48	Butanone	139.5
1,1,1-Trichloroethane	1.65	Butanol	647
o-Xylene	2.44	2-Propanol	825
Toluene	2.82	Ethanol	1355
Benzene	2.90	Dioxane	1618
Dichloromethane	5.65		

K can be lowered by changing the incubation temperature or by changing the composition of the sample matrix. In the case of ethanol in water, K can be lowered from 1355 to 328 by raising the incubation temperature from 40 to 80 °C, although when measuring blood ethanol by headspace GC, an incubation temperature of 60 °C or so is used to minimize injection of water vapour.

K may also be changed by adding salts to the analyte matrix or by altering the phase ratio. High salt concentrations in aqueous samples decrease the solubility of polar organic volatiles and promote transfer into the headspace, that is give lower K values. However, the magnitude of the salting-out effect is not the same for all compounds. Compounds with K values that are already relatively low, for example, show very little change in K after adding a salt to an aqueous sample. Generally, volatile polar compounds in polar matrices (aqueous samples) will experience the largest shifts in K and have higher responses after the addition of a salt. Common substances used to decrease matrix effects include ammonium chloride, ammonium sulfate, sodium chloride, sodium citrate, sodium sulfate and potassium carbonate. A practical consideration is that it is not easy to add reproducible amounts of solid substances to the vials.

K is also dependent on the phase ratio, β, the relative volume of the headspace compared to the volume of the sample in the sample vial:

$$\beta = \frac{V_g}{V_s} \quad (3.2)$$

where V_s is the volume of sample phase and V_g is the volume of gas phase. Sensitivity is increased as β is minimized, that is lower values of β (larger sample sizes) will yield higher responses for volatile compounds. However, changes in β will not always yield the increase in response needed to improve sensitivity. When β is decreased by increasing sample size, compounds with high K values partition less into the headspace compared to compounds with low K values and yield correspondingly smaller changes in C_g. Samples that contain compounds with high K values need to be optimized to provide the lowest K value before changes are made in β. Thus, partition coefficients and phase ratios together determine the final concentration of volatile compounds in the headspace.

The concentration of volatile compounds in the gas phase can be expressed as:

$$C_g = \frac{C_0}{K + \beta} \tag{3.3}$$

where C_g is the concentration of volatile analytes in the gas phase and C_0 is the original concentration of volatile analytes in the sample. Striving for the lowest values for both K and β will result in higher concentrations of volatile analytes in the gas phase and therefore better sensitivity.

In addition to working with the partition coefficient (K), the phase ratio (β) and derivatization reactions, sensitivity in headspace GC can also be improved by simply increasing the size of the headspace sample that is withdrawn from the sample vial for analysis. Increasing the sample size also means that the amount of time it takes to transfer the sample to the column will increase in proportion to the column volumetric flow rate. Sample size can be increased only to the point that increases in peak width, caused by longer sample transfer times, do not affect chromatographic separations.

Larger sample sizes and longer transfer times can be offset to an extent by using cryogenic cooling and sample refocusing at the head of the GC column. In the 'purge-and-trap' method volatile compounds are liberated from liquid samples or suspensions by bubbling with an inert carrier gas such as nitrogen or helium and subsequently either condensed in a receiver cooled usually with solid carbon dioxide, or adsorbed on a cartridge filled with a material such as Tenax-GC (Section 7.3.1). Thermal desorption and other static headspace preconcentration techniques have been reviewed (Kolb and Ettre, 1997).

3.2.5 *Liquid–liquid extraction*

Traditionally, purification of lipophilic analytes such as many centrally acting drugs has been performed by extracting the biological sample with an inert, water-immiscible organic solvent at an appropriate pH. Generally it is best to use the least polar solvent that will effectively extract the analyte – selecting a solvent with too much 'extracting power' may reduce the selectivity of the assay by promoting the extraction of interfering compounds. On the other hand, use of solvents such as butyl acetate or methyl *tert*-butyl ether (MTBE) in poisons screening or in generic methods prior to HPLC, for example, has the advantage that drugs of differing polarity often can be analyzed using similar methodology. Properties of the ideal extraction solvent are summarized in Box 3.4.

Some form of mechanical mixing of the aqueous and organic phases is normally necessary. Of the methods available, vortex mixing is the quickest and the most efficient method for relatively

Box 3.4 Properties of the ideal extraction solvent

- Good 'extracting power'
- Low solubility in water
- Less dense than water
- Moderate volatility to facilitate removal by evaporation, but not so volatile as to evaporate during sample preparation
- Stable/inert (contains no added stabilizers)
- Low flammability
- Low toxicity (not only by inhalation, but also via dermal absorption)
- Inexpensive and readily available in required purity
- No UV absorption or electrochemical activity
- No response/deleterious effects on NPD/ECD/MS

small volumes. Rotary mixers capable of accepting tubes up to 30 mL volume are valuable for performing relatively large volume extracts of plasma/serum, urine, or stomach contents, and serve to minimize the risk of emulsion formation – vigorous shaking is unnecessary. Use of strongly acidic or basic extraction conditions may also be a factor in promoting emulsion formation hence more moderate conditions should be used whenever possible.

Centrifugation in a bench-top centrifuge capable of accepting test tubes of up to 30 mL volume and attaining speeds of 2000–3000 rpm is normally effective in promoting phase separation such that most of an organic extract can be removed. The centrifuge should have sealed buckets and be 'flash proof' to minimize the risk of explosion from ignition of solvent vapour. Use of sealed tubes and rotor units also serves to minimize the risks associated with centrifugation of potentially infective specimens. Regular centrifuge cleaning and maintenance programmes are important.

The relative centrifugal force (RCF, g; Box 2.1) is not usually critical provided that sufficient force is applied to achieve phase separation. Swing-out rotors are preferable to angled ones as this ensures that the interface between the organic and aqueous phases is at right angles to the tube, facilitating removal of the solvent. It is important that the tubes do not break or distort during centrifugation. Thus, the exact RCF and time of centrifugation will depend on local factors, such as the type of tube and the centrifuge being used. Plastic tubes and organic solvents may introduce plasticizers and other contaminants. Use of high grade solvents is generally advisable.

After centrifugation, freezing the aqueous layer (-20 to -60 °C) in order to simplify phase separation by pouring is advocated by some authors. Alternatively, the solvent extract may be removed using a disposable Pasteur pipette.

Once the extract has been removed, the extraction solvent is often vaporized under a stream of compressed air, or nitrogen if analyte oxidation is a concern, prior to reconstitution of the dried extract in a suitable volume of an appropriate solvent. Back-extraction of acids or bases into aqueous solution of appropriate pH can be used to remove neutral interferences. Concentration of the analyte to improve sensitivity can be achieved at the solvent evaporation stage and extraction solvents such as ethyl acetate, with strong chromophores at lower UV wavelengths, or dichloromethane, which would otherwise interfere with the HPLC or GC detector, can be used and then removed. On the other hand, solvent evaporation is time consuming, interferences may be concentrated, and relatively volatile analytes may be lost unless precautions are taken (e.g. acidifying the extract to promote salt formation in the case of volatile bases such as amfetamine).

It may also be necessary to add a basic compound to the reconstituting solvent in order to remove adsorbed basic analytes from glass tubes.

Some commonly used extraction solvents are listed in Table 3.3. Mixtures of solvents may be used for specific purposes. Dichloromethane:2-propanol (9 + 1) (RD > 1), for example, has long been used to extract morphine and other opiates while mixtures such as dichloromethane:heptane (1 + 1) (RD < 1) are useful if a chlorinated solvent is needed, but an upper layer is required to simplify extract removal. Alternatively, inorganic salts can be added in sufficient quantity to the aqueous phase to increase the density until the organic phase forms the upper layer. Care should be taken when adding salts (e.g. to 'salt out' analytes or to adjust the pH) that the phases are not inverted accidentally. A further complication is that some hydrochloride salts of basic drugs are soluble in organic solvents, especially chlorinated solvents such as chloroform, dichloromethane and chlorobutane. Sulfate and phosphate salts are much less soluble in organic solvents, hence sulfuric acid is preferred if performing back-extraction into water as part of a sample preparation procedure.

Table 3.3 Some widely used extraction solvents

Solvent	RD	BPt (°C)	UV Cut-off (nm)	Flammable	Dielectric constant	Polarity[a]	Solubility in water (g L^{-1})
Butyl acetate[b]	0.88	125	255	Yes	5.01	—	7.0
Chloroform[c]	1.49	61	245	No	4.81	4.4	8.0
Cyclohexane	0.78	81	210	Yes	2.03	0	0.0
1,2-Dichloroethane	1.25	83	230	Yes	10.65	3.7	8.7
Dichloromethane	1.32	40	235	Yes	8.93	3.4	13
Ethyl acetate[b]	0.90	77	255	Yes	6.02	4.3	83
Heptane	0.68	98	210	Yes	1.92	0	0.5
MTBE[b]	0.74	55	220	Yes	4.5	—	48
Petroleum ether[d]	0.65	40–60	210	Yes	about 2	—	0.0
Toluene	0.87	111	285	Yes	2.38	2.3	0.53
2,2,4-Trimethylpentane	0.69	99	210	Yes	1.9	0.4	0.0

[a]According to Snyder (1974) [b]Hydrogen acceptor [c]Hydrogen donor [d]Boiling range 40–60 °C – mixture of pentanes, hexanes, etc. – other boiling ranges available

The inhalational toxicity and other hazards associated with use of some solvents should not be ignored. Benzene, for example, is a proven human carcinogen, whilst occupational exposure to hexane or to 2-hexanone (butyl methyl ketone) is associated with the development of peripheral neuropathy (Table 3.4). 'Isohexane' (Fisher Scientific), a mixture of hexane isomers containing less than 5 % (v/v) *n*-hexane, is a safer alternative to hexane itself. Ammonium hydroxide too is unpleasant, if not overtly hazardous, and loss of ammonia from stored reagent may in time render the reagent unreliable – use of buffers such as tris(hydroxymethyl) aminomethane [tris(hydroxymethyl)methylamine, 'Tris'] or sodium borate can give pH values in the 9–11 region

Table 3.4 Especial hazards associated with the use of some solvents

Solvent	Hazard
Benzene	Human carcinogen
Carbon disulfide	Neurotoxin
Carbon tetrachloride, chloroform (also 1,2-dichloropropane, 1,1,2,2-tetrachloroethane)	Hepatorenal toxins; known carcinogens
Dichloromethane	Carboxyhaemoglobinaemia
Diethyl ether	Highly flammable; may form explosive peroxides
Di-isopropyl ether	May form explosive peroxides
Hexane, 2-hexanone	Peripheral neurotoxins
Trichloroethylene	Cardiotoxin

and has the advantage over phosphate buffers in that bacterial growth is inhibited, thus facilitating longer-term storage at room temperature.

In general more 'polar' solvents will extract a larger number of compounds. Unfortunately, 'polarity' is not a simple property, but rather a composite of different physical characteristics, including the ability to form hydrogen bonds, dipole moment, and dielectric constant. Alkanes do not form hydrogen bonds, have little or no dipole moment and low dielectric constants, and are nonpolar. Alcohols, which can form hydrogen bonds and have high dielectric constants, are polar (Table 3.3). The nature of the analyte may contribute to the degree of extraction. Analytes that can form hydrogen bonds, either as acceptor or donor, will be better extracted into solvents that support such bonding. Alcohols such as 2-propanol or 3-methylbutanol (isoamyl alcohol) may be added to aprotic solvents to encourage hydrogen bonding and thus to enhance the extraction of relatively water-soluble analytes, and to reduce adsorption of basic drugs onto glassware.

Solvents are sometimes ranked in order of their ability to elute analytes in adsorption chromatography – an elutropic series. In TDM and in pharmacokinetic studies the least polar solvent that will extract the compounds of interest should be used as this will generally give the cleanest extracts. In drug screening, higher polarity extraction solvents should be chosen to increase the likelihood of extracting a larger range of potential analytes.

Simple LLE with direct analysis of the extract has been used for many years prior to TLC or GC (Figure 3.4). LLE has been replaced to a certain extent by SPE for HPLC sample preparation, but a well-defined LLE system is robust and cost effective. HPLC eluents that use a high proportion of an organic component also allow solvent extracts to be analyzed directly (Figure 3.5; Section 8.8.2). Repeating dispensers fitted with gas-tight Luer-fitting glass syringes and stainless steel needles may be used for solvent and reagent additions. Use of clear glass test tubes (60 × 5 mm i.d., Dreyer tubes) as extraction vessels simplifies extract removal via a fine-tipped plastic Pasteur pipette and minimizes the risk of contamination with the aqueous phase. Use of a high-speed centrifuge gives rapid phase separation and minimizes the problem that may be posed by emulsion formation.

This microextraction approach is simple, inexpensive and suitable for emergency work, and has been used for hundreds of thousands of analyses since it was introduced in the early 1970s.

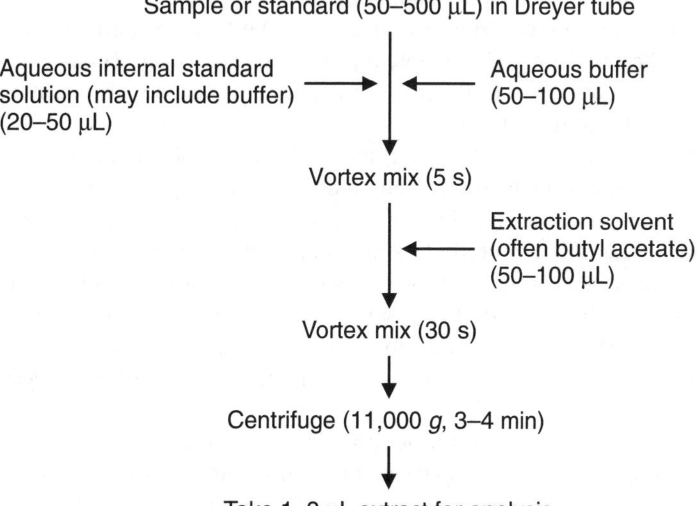
Sample or standard (50–500 μL) in Dreyer tube
Aqueous internal standard solution (may include buffer) (20–50 μL)
Aqueous buffer (50–100 μL)
Vortex mix (5 s)
Extraction solvent (often butyl acetate) (50–100 μL)
Vortex mix (30 s)
Centrifuge (11,000 g, 3–4 min)
Take 1–2 μL extract for analysis

Figure 3.4 Flow diagrams exemplifying microextraction procedures prior to GC.

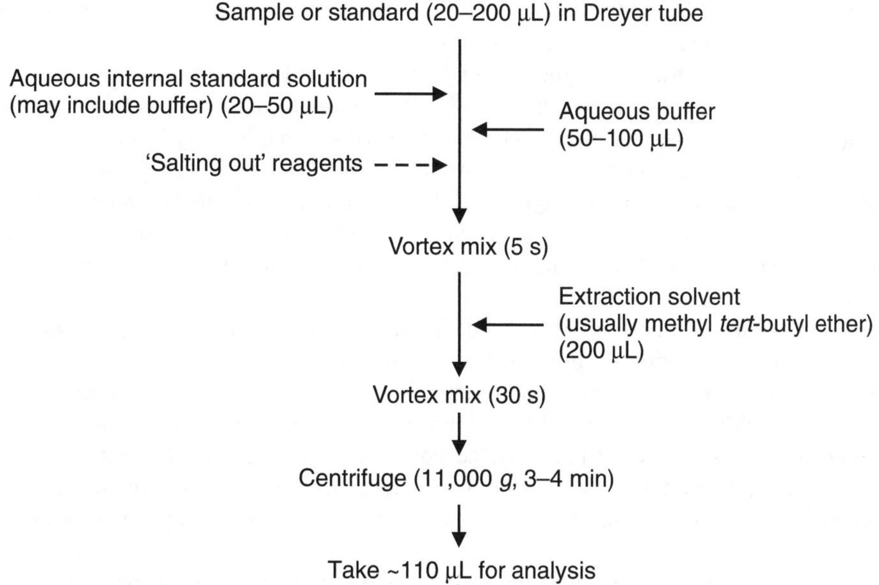
Sample or standard (20–200 μL) in Dreyer tube
Aqueous internal standard solution (may include buffer) (20–50 μL)
Aqueous buffer (50–100 μL)
'Salting out' reagents
Vortex mix (5 s)
Extraction solvent (usually methyl tert-butyl ether) (200 μL)
Vortex mix (30 s)
Centrifuge (11,000 g, 3–4 min)
Take ~110 μL for analysis

Figure 3.5 Flow diagrams exemplifying microextraction procedures prior to HPLC.

However, it is not amenable to automation and hence is labour intensive if large numbers of samples are to be assayed. Moreover, care must be taken to minimize the risk of glass tubes breaking in the centrifuge. The sample injection step, however, can be automated. For example, extracts may be transferred to 0.5 mL (30 × 8 mm i.d.) capped disposable polypropylene tubes in an appropriate autosampler tray.

As with headspace GC, 'salting out' relatively water-soluble analytes is a further possibility. This relies on increasing the ionic strength of the aqueous phase thus encouraging partition of a relatively water-soluble analyte into an immiscible organic solvent. Common substances used in this way include ammonium chloride, ammonium sulfate, sodium chloride, sodium citrate, sodium sulfate and potassium carbonate. However, with LLE caution is needed because an emulsion may ensue if excess salt is added. The assay of atenolol in breast milk is an example when prior extraction with MTBE can be used to remove lipophilic contaminants, before the addition of sodium chloride to facilitate atenolol extraction into a second volume of MTBE (Flanagan *et al.*, 1988).

If derivatization is to be performed then performing the extraction using a solvent that is compatible with the derivatization reagent may eliminate the need for a solvent evaporation step. A further consideration when choosing an extraction solvent is whether it is likely to interfere in the analysis. It is difficult to remove all traces of solvent and even a small residual amount can seriously affect the detection limit. Chlorinated solvents are best avoided, not only on safety and environmental grounds, but also because they poison electron-capture detectors in GC. In addition, some solvents may not be suitable because they would react with the analyte, for example a ketone such as butanone (methyl ethyl ketone, MEK) will react with primary amines. Some solvents may contain traces of decomposition products, for example phosgene in chlorinated hydrocarbons, and aldehydes and ketones in alcohols.

Antioxidants such as hydroquinone or pyrogallol are added to some ethers to limit the formation of explosive peroxides. Antioxidants are highly electroactive molecules that may react with the analyte and will adversely affect ED methods in HPLC, particularly if the antioxidant has been concentrated by evaporation of extraction solvent. Diethyl ether is highly volatile and flammable, and is best avoided. Ethanol is frequently added to stabilize chlorinated solvents such as chloroform and dichloromethane, and may react with derivatizing agents. Even ethanol stabilization may not prevent decomposition (Cone *et al.*, 1982). HPLC-grade dichloromethane is available with pentene as stabilizer.

Butyl acetate (Rutherford, 1977) and MTBE (Flanagan *et al.*, 1988; Flanagan *et al.*, 2001) give efficient extraction of many drugs and metabolites from plasma at an appropriate pH and form the upper layer thus simplifying extract removal for analysis. These solvents do not interfere in NPD, in ECD, or in MS and the extracts are generally suitable for direct analysis on compatible systems, except that butyl acetate cannot be used directly with HPLC-UV (MTBE has a relatively low UV cut-off, 215 nm). Unlike other ethers, such as diethyl and di-isopropyl, MTBE does not form peroxides at ambient temperature and thus additives such as quinones are unnecessary.

A simple procedure developed to measure the antipsychotic olanzapine in plasma at the concentrations attained during chronic treatment with this drug has been described (Flanagan *et al.*, 2001). Ascorbate can be added with the aim of stabilizing olanzapine during the extraction if this is thought necessary. This relatively simple scheme can be applied to the extraction of many other basic drugs prior to GC or HPLC assay, and has the advantage of concentrating the analyte prior to injection, thus maximizing sensitivity.

3.2.5.1 Theory of pH-controlled liquid–liquid extraction

The extraction of a weak acid or base into an organic solvent is a function of the pH of the aqueous solution, the pK_a of the analyte, and the partition coefficient describing the distribution of the analyte between the solvent and water. Under ideal conditions, the fraction extracted, F, is given by:

$$F = [1 + V_{aq}/(V_{org} \cdot APC)]^{-1} \tag{3.4}$$

where V_{aq} and V_{org} are the volumes of the aqueous and organic phases, respectively, and *APC* is the apparent partition coefficient of the analyte. The *APC* is a function of the true partition coefficient, *TPC* (i.e. the partition coefficient of the non-ionized, extracted form). For a base this becomes:

$$TPC = [1 + 10^{(\mathrm{p}K_a - \mathrm{pH})}] \cdot APC \tag{3.5}$$

and for an acid:

$$TPC = [1 + 10^{(\mathrm{pH} - \mathrm{p}K_a)}] \cdot APC \tag{3.6}$$

Thus, the fraction extracted can be calculated from knowledge of pH, pK_a and *TPC*.

The data of Figure 3.6 have been derived for five imaginary bases (pK_a = 8.0) with differing lipophilicities (*TPC* = 0.1–1000) by combining Equations (3.4) and (3.5). The fraction ionized was calculated from the Henderson–Hasselbalch equation:

$$\mathrm{pH} = pK_a + \log[\mathrm{acid}]/[\mathrm{base}] \tag{3.7}$$

(broken lines, Figure 3.6). The predicted extraction is influenced by the total partition coefficient (*TPC*), as well the volume ratio of organic to aqueous phase. For the base with *TPC* = 1, the maximum that can be extracted is 50 %, even when the base is completely non-ionized. The degree

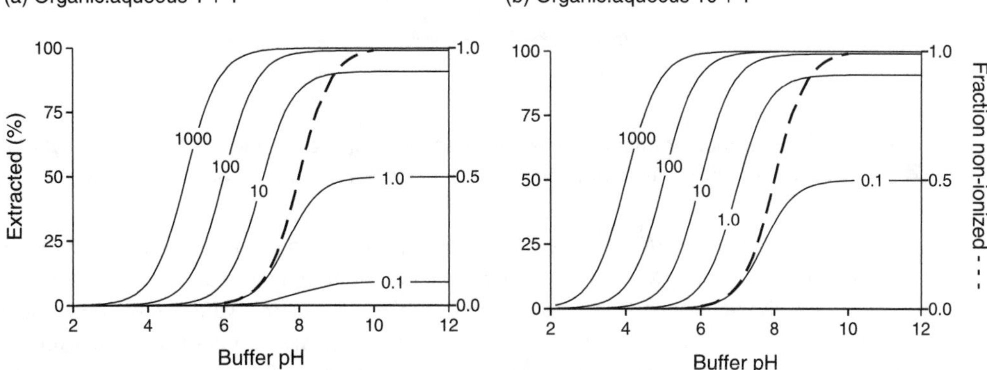

Figure 3.6 Simulated extraction curves for bases (pK_a = 8.0) with solvent:water partition coefficients of 0.1, 1, 10, 100 and 1000: (a) with equal volumes of aqueous and organic phases, (b) with 10-fold increase in volume of organic phase.

of extraction of this base could be increased by using a higher volume of organic phase relative to the aqueous one, for example 10:1 [Figure 3.6(b)]. However, there are practical limits to the extent to which this can be done; in the case of the base with $TPC = 0.1$ it is only 50 % extracted with a 10-fold increase in the volume of organic solvent and under these circumstances, a solvent with more 'extracting power' should be evaluated. The extraction of the base with $TPC = 10$ can be increased by increasing the volume of the organic phase. The most lipophilic base ($TPC = 1000$) is almost completely extracted, even when it is 99 % ionized, at pH 6. Furthermore, such a compound will require low pH buffers to ensure complete back-extraction, particularly into a relatively small volume of aqueous phase.

Typically, the extraction pH should be approximately 2 units more (or less, in the case of acids) than the pK_a of the analyte when extracting relatively polar compounds. Under these conditions the analyte will be <1 % ionized, and, as can be seen from Figure 3.6, there is little to be gained from adjusting the pH to more than 2 units above the pK_a for any of the compounds. Use of more extreme pH values increases the risk of decomposing the analyte. The simulated curves of Figure 3.6 predict that very lipophilic compounds can be extracted from aqueous solutions even when they are largely ionized in the aqueous phase and this is borne out in practice. Imipramine, for example, can be extracted from pH 7 buffer even though it is >99.6 % ionized at this pH. The more polar imipramine metabolites require higher pH values (i.e. less ionization of the analyte) to promote efficient extraction (Figure 3.7). This is due partly to the higher pK_a and partly to the reduced partition coefficient. In the case of the phenolic metabolite of imipramine, the extraction efficiency falls at pH values above 10 as ionization of the phenol is promoted. Clearly, where metabolites or congeners are a consideration, the extraction pH may need to be a compromise; in this example, pH 10 would be optimum for all the compounds.

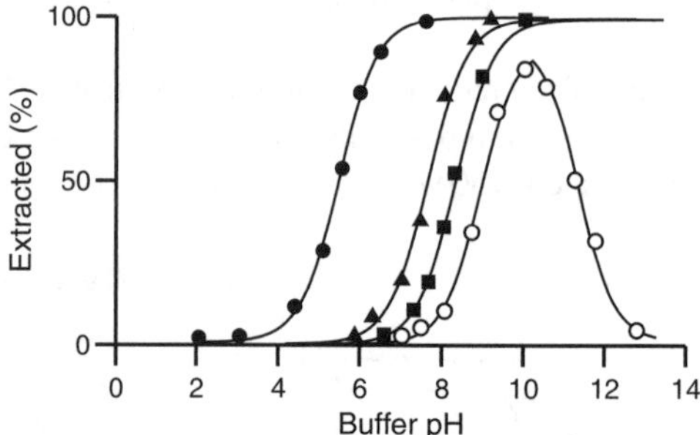

Figure 3.7 Extraction of imipramine (•, pK_a 9.5), desipramine (▲, pK_a 10.2), didesmethylimpramine (■, pK_a 9.8) and 2-hydroxyimipramine (o, pK_a 9.5 and 10.7) from aqueous buffer into heptane with pH.

Strong acids have low pK_a values, whilst the converse is true for strong bases (Table 3.5). It is not possible from the pK_a alone to know whether a compound is an acid or a base. Indeed, for some complex compounds with several ionizable groups, pK_a values have been measured, but it has not proved possible to assign them unequivocally. Assignment is normally on the basis of the chemical groups present and the salts that the compound forms, for example thiopental (p$K_a = 7.6$) must be an acid because it is available as the sodium salt. Comparison with related compounds may help in understanding how a molecule ionizes, for example nordazepam has two ionizable groups, but diazepam has only one.

Table 3.5 Some pK_a values of acidic, basic and amphoteric compounds

Compound	Acid		Base		Compound	Acid		Base
Amoxicillin	2.4	9.6		7.4	Lidocaine			7.9
Amfetamine				9.8	Pethidine (meperidine)			8.7
Aspirin	3.5				Morphine	9.9		8.0
Chloramphenicol	11.0				Naproxen	4.2		
Clozapine			3.7	7.6	Nordazepam	12.0		3.5
Cocaine				8.6	Paracetamol	9.5		
2,4-D	2.7				Phenytoin	8.3		
Debrisoquine				11.9	Salicylic acid	3.0	13.4	
Diazepam				3.3	Sulfadimidine	7.7		
Fluphenazine			3.9	8.1	Warfarin	5.0		

Although knowledge of the pK_a may be helpful in choosing a pH at which to perform an extraction, some literature values may be misleading. The quoted pK_a of chloramphenicol, 11.03 (http://www.daylight.com/meetings/emug00/Sayle/pkapredict.html, accessed 30 December 2005), is very different from the value of 5.5 given elsewhere (Reynolds, 1993). Chloramphenicol, an amide, is expected to be a very weak acid and a pK_a of 11 is consistent with this. An explanation for this huge discrepancy (>300,000-fold difference in ionization constant) is that the lower value is for the ionization of the carboxylic acid group in the chloramphenicol ester, chloramphenicol sodium succinate. Similarly, pK_a values for medazepam of 4.4 (25 °C) (Barrett *et al.*, 1973) and 6.2 (37 °C) (le Petit, 1976) have been described. The difference is too large to be ascribed to a difference in temperature and because the partitioning characteristics of medazepam were inconsistent with a value of 4.4, the pK_a was re-evaluated and found to be 6.3 at 25 °C (Whelpton, 1989).

Other discrepancies and errors in pK_a may arise from the way in which these have been measured. The pK_a values of weak electrolytes that show marked spectral (UV) shifts as a function of pH and are moderately soluble in aqueous buffers, are relatively easy to measure, whereas for lipophilic compounds it may be necessary to use limiting aqueous solubility methods (Green, 1967). It may be possible to titrate water-soluble compounds, but values from nonaqueous titrations, even when the results have been extrapolated to zero organic solvent concentration, should be used with caution. Furthermore, when measuring pK_a values in alkaline buffers, special care is needed to exclude carbon dioxide and to use pH electrodes which have been designed to give accurate readings at high pH values.

When the pK_a is uncertain then this can be derived from iterative curve fitting of the experimental data (*APC* versus pH) to Equations (3.5) or (3.6) to solve for *TPC* and pK_a (Whelpton, 2004). The *APC* values of [^{3}H]-morphine as a function of pH were measured and the data [Figure 3.8(a)] fitted to the equations described above using Graph Pad Prism v3, to give $TPC = 0.62$, p$K_a1 = 7.77$, p$K_a2 = 9.69$ (solid line). The experimental data, as percent extracted, and the fitted line, are shown in Figure 3.8(b). The advantage of this approach is that once the parameters are known,

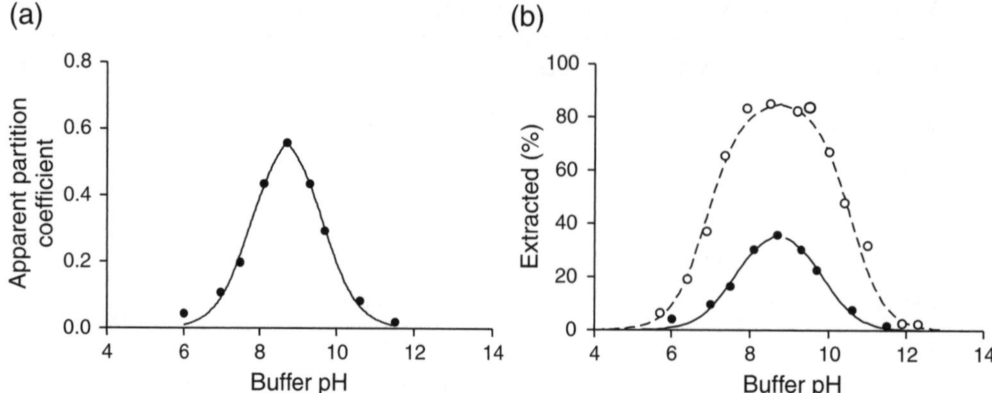

Figure 3.8 Partitioning of [^{3}H]-morphine between buffer and an equal volume of toluene:butanol (9 + 1): (a) APC versus pH (•) and the fitted data (solid line), (b) proportion extracted (%) versus pH (•). Comparison of predicted (broken line) and actual extraction assayed by HPLC (○) for a 10:1 volume ratio of organic solvent:aqueous phase.

the extraction recoveries can be predicted for any organic:aqueous volume ratios at any pH value. The agreement between the predicted curve for an organic:aqueous ratio of 10 + 1 (broken line) with experimental (HPLC) data for extraction of morphine using such a ratio of phases was good [Figure 3.8(b)]. A day of experimentation allows optimization of extraction pH, solvent volume, and pH required for back-extraction into a predefined volume of the aqueous phase. If a method is to be modified, the effect of a change in experimental conditions can be calculated in a few minutes.

3.2.5.2 *Ion-pair extraction*

Ionized lipophilic analytes can be extracted into organic solvents providing an ion of the opposite charge (counter ion) is extracted to maintain electrostatic neutrality. For a cation, C^+_{aq}, being extracted as an ion-pair with a suitable anion, A^-_{aq},

$$C^+_{aq} + A^-_{aq} = CA_{org} \tag{3.8}$$

the distribution ratio, D_C, is given by:

$$D_C = [CA_{org}]/[C^+_{aq}] = E_{CA}[A^-_{aq}] \tag{3.9}$$

where E_{CA} is the equilibrium constant. Thus, the distribution is a function not only of the nature of the analyte and the extraction solvent, but also of the nature and concentration of the counter ion. Partitioning can be further affected by introducing a substance with which the analyte is able to complex. This makes ion-pair extraction versatile, but also complicated (Schill, 1976). Sodium octylsulfate is one of the ion-pairing agents advocated, but precipitation of plasma protein with acetonitrile and evaporation to dryness of the acetonitrile extract was required before addition of the ion-pairing agent in one such method (Hoogewijs and Massart, 1984).

Ion-pair extraction may be used with SPE (Section 3.2.6), employment of alkyl-modified or strong cation (SCX) or anion (SAX) exchange-modified silica stationary phases resulting in a simplified approach to that achieved with LLE *per se*. As an example, Lukaszewski (1985) used ion-pairing with potassium iodide to extract the quaternary ammonium compound paraquat from 5 mL urine into chloroform:propanol (4 + 1) prior to pyrolysis GC, but prewashing with ammoniacal dichloromethane:propanol (4 + 1) was required, and the extracted paraquat had to be taken up in water, washed several times with diethyl ether, and finally the water evaporated before reconstitution in ethanol. In contrast, Gill *et al.* (1983) simply added alkalinized urine (1 mL) to pretreated octadecylsilyl-modified silica SPE cartridges, and after washing with water and then with methanol, eluted paraquat with acidified methanol (5 mL), evaporated the extract to dryness, and analyzed a portion of the residue using HPLC.

It is worth noting that aberrant partitioning in LLE can sometimes be explained by unintentional ion-pair extraction, the counter ions originating from the buffer or biological matrix. Therefore, unless ion-pair extraction is required, solvents that promote ion-pair extraction are best avoided. Chloroform, dichloromethane, ethyl acetate and 4-methyl-2-pentanone (methyl isobutyl ketone, MIBK) have been used as solvents for ion-pair extraction. Hydrocarbons are less likely to extract ion-pairs than solvents that are hydrogen donors, such as chloroform and alcohols. If ion-pairing is suspected then addition of a different counter ion (e.g. bromide or iodide) will change the degree of extraction.

An indication of the lipophilicity of a compound can be obtained from log P values, where P is the octanol:water partition coefficient. However, P is sometimes difficult to measure accurately, particularly when log $P > 3$. For very lipophilic acids or bases it may be necessary to measure the partitioning between a buffer and octanol, the pH and the volume of the buffer being chosen so that approximately 50 % of the compound is extracted (partitioned). The partition coefficient of the unionized species (P) is calculated from knowledge of the buffer pH and analyte pK_a. Many log P values are not experimental values, but have been calculated from the chemical formula. The starting point for such calculations is a structurally similar molecule whose log P value is known, and then increments ('pi') are added or subtracted for various substituents. This approach requires that the initial choice is suitable, and there are sometimes discrepancies between calculated and measured values. Note that P values are for partitioning between *water* and octanol. Some quoted 'log P' values have been measured at pH 7.4, and are nowadays referred to as log D values.

3.2.5.3 Liquid–liquid extraction columns

LLE can be facilitated and/or automated by the use of polypropylene columns containing diatomaceous earth as a solid support for the aqueous phase. These should not be confused with SPE columns (Section 3.2.6) as the aqueous sample is retained in the column matrix and the extracting solvent (RD > 1) allowed to run through the column. A wide variety of LLE extraction columns is available, including 96 well-plates (ISOLUTE HM-N, IST, Argonaut Technologies. http://www.biotage.com/DynPage.aspx?id=2390, accessed 1 January 2006). Columns such as Tox-Elut have been widely used in preparing urine extracts for drug screening. An advantage of this approach is that a wide range of basic compounds, including morphine, and also weak acids such as barbiturates, can be extracted in a single step (Box 3.5).

3.2.6 Solid-phase extraction

Extraction of compounds by adsorption onto solid materials such as activated charcoal, Florisil, or ion-exchange resins followed by washing with water and elution of compounds of interest

Box 3.5 Tox-Elut columns for TLC sample preparation

- LLE using prebuffered diatomaceous earth columns has advantages of simplicity and can extract acids, bases, neutral and amphoteric compounds (including morphine) in one step
- Use 20 mL Tox-Elut, pH 9 (Varian – www.varianinc.com)
- Add 20 mL urine followed by 30 mL dichloromethane:2-propanol (9 + 1). Evaporate extract to dryness (compressed air or nitrogen), and reconstitute in 0.1 mL methanol
- Disadvantages:
  - Cost of extraction tubes (but save time)
  - Lose some of selectivity of acid/base extracts, but use multiple spotting/developing solvents/detection reagents

using methanol, for example, is not a new idea. However, use of siliceous or other materials with relatively close particle size distribution (15 to 100 μm) in disposable plastic syringe barrels permits sequential extraction, clean-up, and finally reproducible elution of drugs and other analytes at relatively low pressures. A range of barrel geometries, reservoir volumes, bed sizes and materials (glass columns, stainless steel frits) are available. Samples may be loaded using positive pressure (e.g. from a syringe), by drawing through with a vacuum, or by centrifugation using a swing-out rotor system.

Advantages of SPE are that batch processing can be simplified and that the extracts may contain fewer interfering compounds than with LLE. A further feature when screening for unknowns is that a range of analytes can be extracted either sequentially, or simultaneously. However, SPE columns are relatively expensive and it may not be possible to retain very water-soluble analytes such as zwitterions and quaternary ammonium compounds, on bonded-phase columns unless there are residual silanol groups available for ionic interaction. Problems with SPE of physostigmine (Box 3.6) were traced to a batch of cyanopropyl columns that had a higher carbon loading than usual, resulting in masking of the silanol groups. In many cases simple LLE with an appropriate solvent can be used to purify a lipophilic compound more quickly than with SPE. On the other hand, analyte concentration may often be achieved more easily with SPE than with LLE, while

Box 3.6 Solid-phase extraction of physostigmine in plasma for HPLC (Hurst and Whelpton, 1989)

Method

- Affix Bond-Elut CN column (100 mg, 1 mL; Varian) to vacuum manifold and wash with (i) methanol (2 × 1 mL), (ii) purified water (2 × 1 mL) and (iii) 0.1 mol L^{-1} dipotassium hydrogen orthophosphate, pH 9.2 (1 mL)
- Centrifuge plasma sample (1000 *g*, 5 min) and add 0.5 mL to column
- Wash with purified water (3 × 1 mL) and centrifuge (500 *g*, 1 min) to remove excess water
- Add methanol (50 μL) and centrifuge (500 *g*, 1 min)
- Fit Luer needle to column and elute with minimum volume (typically 250 μL) methanol into appropriate centrifuge tubes during centrifugation (500 *g*, 2 min)
- Cap tubes, vortex-mix (5 s), centrifuge (12,000 *g*, 3 min), and inject 100 μL extract onto the HPLC column

use of SPE columns to concentrate an analyte from a solvent extract may provide a quicker and possibly safer alternative to solvent evaporation. In this context it is helpful if the volume used in the final elution step is as small as possible.

The commonest SPE mode used is reversed-phase with C_8 or C_{18} materials. Recoveries exceeding 95 % with correspondingly low RSDs are possible, and subnanogram quantities of analyte may be assayed. Frequently, the compounds of interest can be eluted in a small volume of fluid. Even if the resulting eluate contains interfering substances, the reduced sample volume allows easier sample handling during any subsequent purification steps.

SPE is particularly useful for drugs and other poisons that are difficult to extract by LLE, for example quaternary ammonium compounds (pyridostigmine, (+)-tubocurarine, paraquat), and other hydrophilic compounds such as glucuronides. Because ionized molecules can be isolated by SPE, it is possible to develop methods that do not require such extreme pH values as required by LLE, which makes SPE suitable for isolating drugs that are not stable under extremes of pH. Furthermore, it should be possible to perform SPE in an inert atmosphere by using syringes to load and elute samples, or nitrogen to provide positive pressure to SPE columns to facilitate elution. Once the analyte has been retained, a range of clean-up procedures may be used. For reversed-phase materials, water may remove residual hydrophilic materials, including proteins and buffer. Water–organic solvent mixtures may be used to remove interferences. If the minimum volume of solvent is used to elute the analyte then further sample concentration prior to analysis may be unnecessary (Figure 3.9).

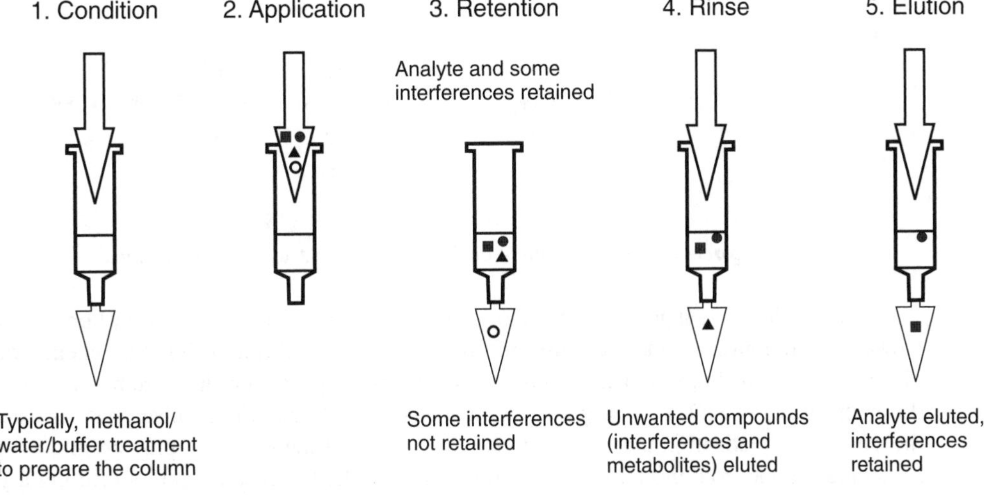

Figure 3.9 Schematic diagram of a solid-phase extraction procedure.

In addition to unmodified silica, a range of bonded-phase materials analogous to those used as HPLC packings is available (Table 3.6). Columns packed with alkyl-modified materials are used most frequently. The stationary phase is attached by reaction of the appropriate chlorosilane with a surface silanol moiety on the base silica under anhydrous conditions in the same way as with HPLC packings and bonded capillary GC columns (Figure 3.10).

Table 3.6 Some alkylsilyl-modified silica column packings for use in SPE and HPLC

Bonded Phase	Name	Bonded Phase	Name
$—(CH_2)_{17}CH_3$	*n*-Octadecyl (ODS, C_{18})	$—(CH_2)_3NH_2$	Aminopropyl
$—(CH_2)_7CH_3$	*n*-Octyl (C_8)	$—(CH_2)_3CN$	Cyanopropyl, nitrile (CN)
$—(CH_2)_5CH_3$	*n*-Hexyl (C_6)	$—CH_2COOH$	Carboxymethyl (WCX)
$—CH_2CH_3$	Ethyl (C_2)	$—(CH_2)_3\text{-}\overset{\oplus}{N}(CH_3)_3$	Trimethylaminopropyl (SAX)
$—CH_3$	Methyl (C_1)	$—(CH_2)_3\text{-}N(CH_2CH_3)_2$	Diethylaminopropyl (WAX)
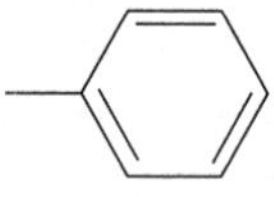 $—C_6H_5$ (phenyl ring)	Phenyl	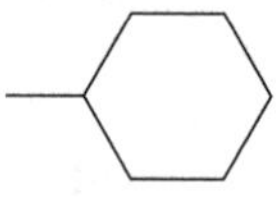 $—C_6H_{11}$ (cyclohexyl ring)	Cyclohexyl
$—(CH_2)_3OCH_2CH(OH)CH_2OH$	Diol	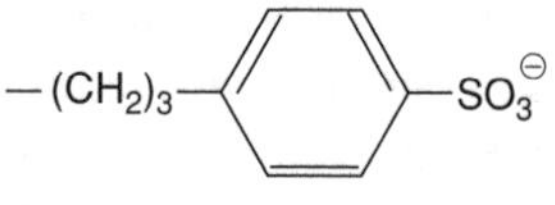 $—(CH_2)_3—C_6H_4—SO_3^{\ominus}$	4-Sulfophenylpropyl (SCX)

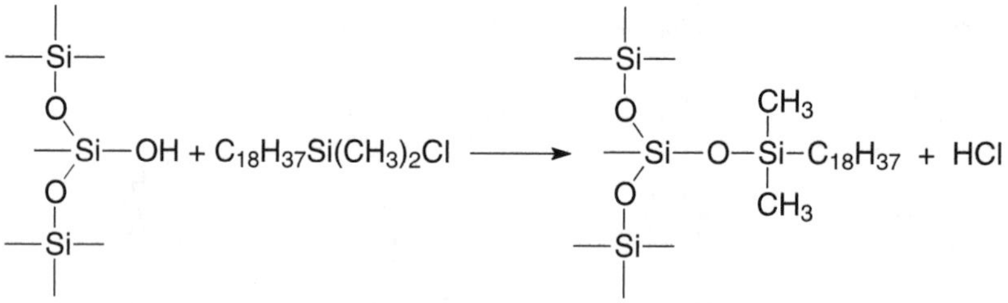

$$\equiv Si\text{–}OH + C_{18}H_{37}Si(CH_3)_2Cl \longrightarrow \equiv Si\text{–}O\text{–}Si(CH_3)_2\text{–}C_{18}H_{37} + HCl$$

Figure 3.10 Reaction of chlorodimethyloctadecylsilane with silica.

Generally, the bonded phase is linked to a surface silanol moiety via a silyloxy bond (Si–O–Si). However, different manufacturers use different base silicas, different bonding chemistries and different phase loadings, and it is thus not surprising that their products sometimes give very different results when used in a particular analysis, even though the bonded phase is ostensibly the same. The realization that interactions with residual silanol groups play an important, and sometimes subtle, role in many SPE methods, has resulted in manufacturers offering a greater choice of phases, with popular columns being available in 'end capped' (i.e. trimethylchlorosilane-treated to minimize residual silanols) and 'non-end capped' versions.

Phenylboronic acid (PBA)-based SPE columns (Figure 3.11) are unusual in that the analytes are retained by covalent bonding. The boronate moiety has a high specificity for 1,2- or 1,3-aliphatic, 1,2-aromatic or *cis*-1,2-alicyclic diols such as catechols, nucleic acids, some low M_r proteins, and carbohydrates. Retention is by formation of 5- or 6-membered ring phenylboronates under neutral or basic conditions. PBA columns have proved to be effective in the isolation of catecholamines from biological fluids and for the separation of RNA from DNA. Elution is achieved under acidic conditions, for example using methanolic hydrochloric acid. PBA columns have been used to

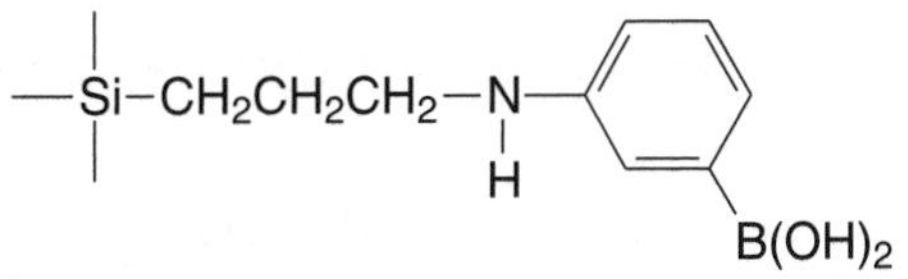

$$-\overset{|}{\underset{|}{Si}}-CH_2CH_2CH_2-\underset{H}{\underset{|}{N}}-C_6H_4-B(OH)_2$$

Figure 3.11 Bonding of phenylboronic acid in SPE columns.

extract the antiviral drug ribavirin from serum prior to HPLC assay (Svensson *et al.*, 2000), and have also been evaluated for glucuronide extraction from urine (Tugnait *et al.*, 1991). The columns are expensive, but can be reused.

So-called 'mixed-mode' columns exploit two or more retention mechanisms to bring about the desired result. The use of nonpolar and ion-exchange modifiers on one column, for example, allows efficient clean-up to remove nonpolar impurities without the risk of prematurely eluting the compounds of interest. Using a strong cation exchange (SCX)-modified phase to retain a basic drug allows washing with methanol to remove interferences, prior to analyte elution with alkaline ethyl acetate. Adoption of 96-well plate technology and robotics increases off-line throughput. The compatibility of SPE materials with HPLC systems means that on-line extraction prior to liquid chromatography is an effective option for high throughput analysis.

Despite much effort, the development of SPE extraction protocols remains largely empirical (Van Horne, 1985). Indeed, different protocols may be required for urine, plasma and whole blood. Haemolyzed whole blood can be assayed after centrifugation to remove cell debris, but postmortem blood may require dilution in buffer or analyte-free plasma before application to the column. Little work has been done on SPE for tissue digests except in the case of hair digests.

The steps in developing protocols for physostigmine assay (Box 3.6) have been described (Hurst and Whelpton, 1989). The method can be used with minor modification for the analysis of whole blood or urine. A number of protocols for other analyses (tocainide and mexiletine, amiodarone and desethylamiodarone, diazepam and metabolites, ciclosporin, acid/neutral/basic drugs, Δ^9-carboxytetrahydrocannabinol) have been detailed (Harkey, 1989). A procedure for extracting acidic and neutral, and basic drugs from urine designed for use with GC is given in Box 3.7.

SPE is not without its practical difficulties (Whelpton and Hurst, 1988). Particles in the sample may cause blockage – it is advisable to filter or centrifuge samples (*after* any pH adjustment) before they are applied to the SPE column. A white gelatinous precipitate is often formed when urine is made alkaline, so methods in which buffer and sample are both added to the column reservoirs are best avoided. Note that because of the trace enrichment capabilities of SPE, dilution of the sample before addition to the column is a practical option if this minimizes the risk of blockage. Impurities in deionized water and in buffer solutions may be retained, so it is important to use high purity materials. Sometimes it is helpful to purify reagents and other elution solvents by filtration through suitable SPE columns before use.

Recoveries in SPE may differ, not only between sample types, but also within a sample type. With some drugs, recoveries from urine may be affected by the nature of the specimen, concentrated samples having a higher content of salts that can act as counter ions in SPE. Reduced recoveries with plasma or serum may be indicative of plasma protein binding competing with binding to the solid phase. For example, the retention of buprenorphine from plasma on C_{18} SPE columns was 33, 90 and 95 %, at flow rates of 10, 1 and 0.1 mL min^{-1}, respectively. When an aqueous solution of buprenorphine hydrochloride was applied to the column, the recovery was

Box 3.7 SPE of drugs from urine prior to GC (Harkey, 1989)

Method

- Add 5 mL urine to 2 mL phosphate buffer (0.1 mol L^{-1}, pH 6.0) in a glass test tube and adjust to pH 5.5–6.5 using 0.1 mol L^{-1} aqueous sodium hydroxide or 1 mol L^{-1} aqueous acetic acid
- Insert SPE column (Clean-Screen, Worldwide Monitoring Corporation, Morrisville, USA) into vacuum manifold and wash with 1 mL methanol and 1 mL phosphate buffer (0.1 mol L^{-1}, pH 6.0)
- Attach an 8 mL fritted reservoir to the top of the extraction column and add urine. Gently dry column under vacuum
- Wash with 1 mL phosphate buffer (0.1 mol L^{-1}, pH 6.0) followed by 0.5 mL aqueous acetic acid (1 mol L^{-1})
- Dry column under vacuum (5 min) and wash with 1 mL heptane. Elute acidic and neutral drugs with 4 × 1 mL dichloromethane
- Evaporate eluate to dryness under a stream of nitrogen (30–40 °C). Reconstitute extract in 0.1 mL ethyl acetate and inject 1–2 μL into the GC column
- Wash columns with methanol (1 mL) and elute basic drugs with 2 mL methanolic ammonium hydroxide (2 % v/v)
- Add 3 mL deionized water and 0.2–0.3 mL dichloromethane to eluate. Vortex-mix (15 s) and inject 1–2 μL of the dichloromethane layer onto the GC column

>97 %, irrespective of the flow rate. Increasing the capacity of the column, reducing the flow rate to allow equilibration with the solid phase, disrupting protein binding, for example by adding a small amount of ethanol or diluting the sample, say, 10-fold with water, may be necessary. An alternative approach to avoiding variable recoveries due to protein binding is to combine LLE and SPE (Whelpton *et al.*, 1990).

The AASP (Advanced Automated Sample Processor) was a dedicated automated SPE extraction system (Varian), but is no longer available. However, an automated system, which can be used with most SPE columns (Automatic Sample Preparation with Extraction Columns, ASPEC; Chen *et al.*, 1993) is still manufactured (Gilson).

Molecularly imprinted polymers (MIPs) – an approach referred to as 'predetermined selectivity' – have been developed. Once polymerized, if high assay sensitivity is to be achieved the template molecule must be removed from the polymer by extensive washing, otherwise there is the danger of residual template molecule interfering in the assay. It has been suggested that a homologue of the drug to be analyzed should be used as the template molecule and, providing drug and homologue can be separated in a subsequent chromatographic procedure, then leaching of traces of template from the column will not be a problem. However, if such an approach works, then the specificity of the MIP must be questioned. A further problem is that retention by MIPs is affected by the nature of the incubation medium. It is possible to have MIPs custom synthesized and, although this requires an initial outlay, the added specificity when coupled with large numbers of samples to be analyzed, may make this approach cost effective (Andersson, 2000). Other specialized approaches include the use of antibodies bound to SPE columns (immunoaffinity SPE) (Stevenson, 2000).

3.2.7 *Solid-phase microextraction*

Solid-phase microextraction (SPME), initially introduced for the analysis of environmental samples, has been used increasingly for the extraction of drugs and other lipophilic analytes from biological matrices either directly, or after derivatization (Furton *et al.*, 2000; Lord and Pawliszyn, 2000a; Mills and Walker, 2000). In its simplest form, a spring-loaded solid probe [originally a fused silica fibre; Figure 3.12(a)] coated with a polymer film is inserted through a septum into a vial containing the sample (usually aqueous liquid or headspace), and after an appropriate period (normally 20–30 min) the fibre is retracted [Figure 3.12(b)]. Analyte or derivatized analyte partitions between the sample and the liquid phase supported on the fibre, usually polydimethylsiloxane

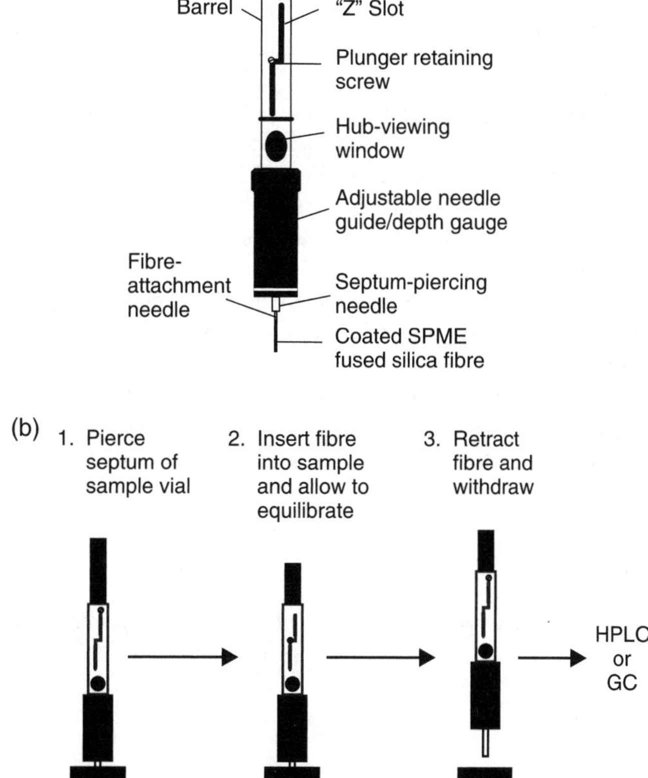

Figure 3.12 Solid-phase microextraction: (a) spring-loaded solid probe coated with a polymer film, (b) mode of use of the probe.

Table 3.7 Some SPME liquid phases that may be useful in analytical toxicology

Fibre coating	Application
7 μm PDMS	Higher M_r compounds
30 μm PDMS	Intermediate M_r compounds
100 μm PDMS	Volatiles and other low M_r compounds
85 μm Polyacrylate	Polar semivolatile compounds
65 μm PDMS/PDVB	Volatiles, amines and nitroaromatics
65 or 70 μm Carbowax/PDVB	Alcohols and other polar compounds
50/30 μm PDVB/Carboxen	Range of volatilities
75 or 85 μm Carboxen/PDMS	Gases and other low M_r compounds
50 μm/30 μm PDVB/Carboxen/PDMS	Semi-volatile compounds

(PDMS), PDMS-polydivinylbenzene (PDVB), polyacrylate or other adsorbent material (Table 3.7, Figure 3.13). Some form of agitation is needed with liquid samples. As an alternative a magnetic stir-bar sealed in glass and coated with PDMS can be used.

Unlike LLE, the aim is not to extract all the analyte from the sample, as SPME is an equilibrium extraction process; indeed the analyte concentration of the sample may be little affected if only a very small proportion of analyte is removed. Subsequently, extracted analytes are either thermally desorbed in the case of GC/GC-MS, or injected via a modified sample loop in the case of HPLC. Loaded probes may be sealed for transport to the laboratory. An added advantage is that there is no injection solvent.

Obvious attractions of SPME are that no extraction solvents are required and all the material that is extracted by the probe may be analyzed directly. The probe may be reconditioned and reused 50–100 times. Metal SPME fibres are now available (Supelco). These are more robust and

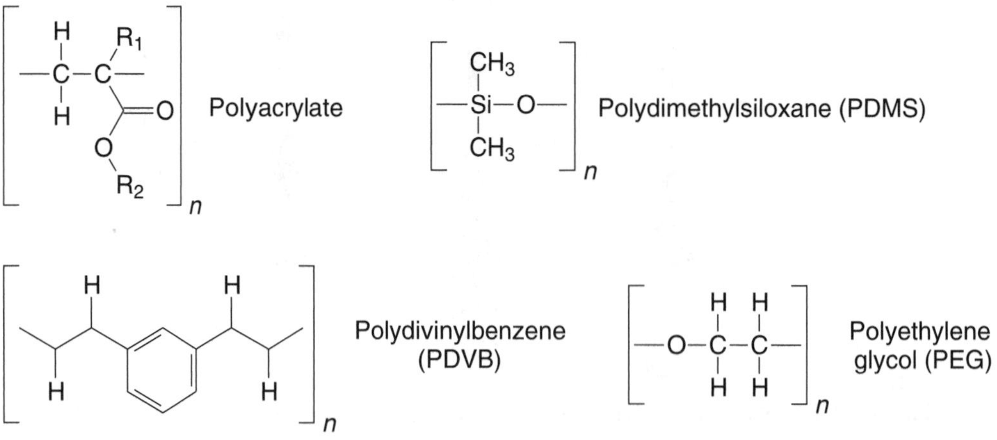

Polyacrylate

Polydimethylsiloxane (PDMS)

Polydivinylbenzene (PDVB)

Polyethylene glycol (PEG)

Figure 3.13 Structures of some SPME absorbents. Carboxen (Supelco) is a carbonaceous resin.

are said to give longer life than the fused silica fibres used originally. The problem of variation in extraction conditions has been addressed by preloading of the extraction phase with an internal standard, and then measuring both analyte uptake and internal standard loss from the extraction phase (Wang *et al.*, 2005).

In headspace SPME the distribution of the analyte between the three phases is given by:

$$C_0 V_s = C_h^\infty V_h + C_s^\infty V_s + C_f^\infty V_f \tag{3.10}$$

where C_0 is the initial concentration of the analyte, C_h^∞, C_s^∞ and C_f^∞ are the equilibrium concentrations of the analyte in the headspace, solution and fibre coating, respectively, and V_h, V_s and V_f are the volumes of the headspace, solution and fibre coating, respectively. If the fibre is placed in the solution then the headspace term is omitted.

When acids or bases are extracted it is normally necessary to buffer the pH to increase the proportion of non-ionized drug to optimize the extraction. Salting out may be employed to increase the proportion of drug extracted. Differences in the ionic strength of urine samples may lead to variable recovery unless the samples are brought to similar ionic strength by adding salt. Increasing the temperature usually reduces the time to reach equilibrium, but may also reduce the amount extracted. Potential disadvantages include competition between drug and endogenous compounds for the fibre, particularly when the mechanism is adsorption rather than partitioning.

A well-chosen internal standard (preferably isotopically labelled if using MS detection, Chapter 10) may be required to ensure that the method is sufficiently robust. It is important to define the time required for equilibrium, and protein binding is likely to have more of an effect on recovery than when practically all the analyte is extracted into an organic phase as in LLE. Plasma protein binding not only reduces the amount of analyte extracted, but also increases the time for equilibration (Lord and Pawliszyn, 2000b; Figure 3.14).

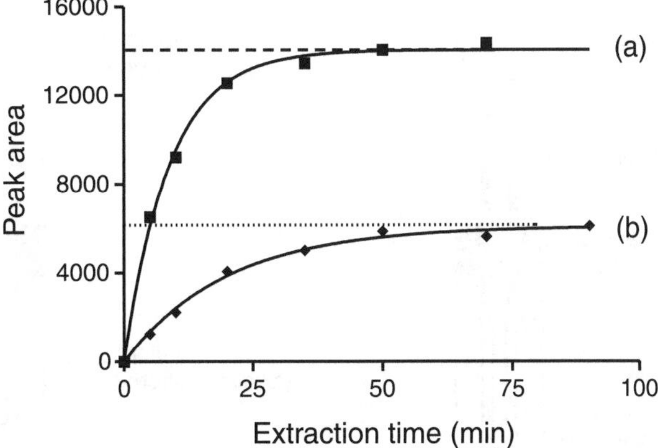

Figure 3.14 Equilibration time profile for SPME of lidocaine from plasma after buffering 1:1 to pH 9.5 (a) with and (b) without deproteinization. The solid lines represent the least-squares fit to: Peak area = Maximum Peak area $[1 - \exp(-kt)]$ (redrawn from Koster EH, Wemes C, Morsink JB, de Jong GJ. Determination of lidocaine in plasma by direct solid-phase microextraction combined with gas chromatography. J Chromatogr B, 2000; **739**: 175–82, with permission from Elsevier).

When the data of Figure 3.14 are fitted to exponential curves:

$$X_t = X_m[1 - \exp(-kt)] \quad (3.11)$$

where: X_t is the amount extracted at time t, X_m is the maximum amount that can be extracted at equilibrium and k is the first-order rate constant, then the half-lives (derived from $t_{0.5} = 0.693/k$) with, and without deproteination, are 6.1 and 13.8 min, respectively. Because >95 % X_m is attained in five half-lives, then with deproteination equilibration was reached in approximately 30 min, whereas to reach the same degree of equilibration in the non-deproteinated sample would take 70 min. If the time for desorption and analysis is similar to, or longer than, the time required for extraction, then one sample can be extracted while the previous one is being analyzed.

Single-drop microextraction (SDME) using a nonvolatile or immiscible solvent and thin-film microextraction using PDMS sheet are further variants on the microextraction theme. Each has unique features (Dietz *et al.*, 2006). Notably, the use of thin PDMS membranes has advantages of faster and more sensitive extraction when compared with conventional thick-film formats (Bruheim *et al.*, 2003).

3.2.8 Liquid-phase microextraction

In liquid-phase microextraction (LPME), analytes are extracted from biological samples such as whole blood or plasma (0.1–4 mL) into an acceptor solution contained in the lumen of a disposable porous hollow fibre (polypropylene tubing crimped at one end) supported in a sealed glass vial (Pedersen-Bjergaard and Rasmussen, 2005; Figure 3.15). Internal standardization can be used as with LLE and direct GC, HPLC, or CE analysis of the acceptor solution eliminates further extract handling. For GC, the tubing is prewashed with acetone and extraction is into a water-immiscible solvent such as dodecyl acetate, dihexyl ether, or octanol (10–50 μL) contained in the lumen of the tube. Conventional pH manipulation of the sample can be used if necessary and extraction times

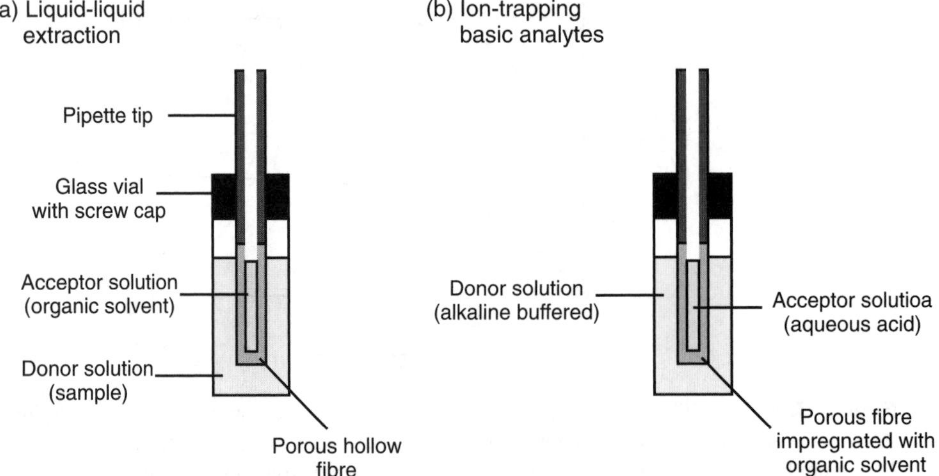

Figure 3.15 Schematic of two modes of liquid-phase microextraction (LPME).

are 15–45 min, not a problem for batch analysis as only mild shaking is needed. Because of: (i) the high analyte capacity of the acceptor liquid and (ii) the substantial phase ratio difference that can be achieved, analyte enrichment of 10–200-fold is feasible (Rasmussen *et al.*, 2000; Basheer *et al.*, 2004).

For HPLC or CE, prewashing is not used, and the extraction conditions can be further manipulated by impregnating the tubing with immiscible solvent and then placing an appropriate aqueous acceptor solution in the lumen – analyte trapping can then be achieved [Figure 3.15(b)] by an analogous mechanism to that by which basic drugs appear in stomach contents after i.v. injection (Section 15.3.1.2). Even ion-pairing of polar analytes with trapping of ionized species in buffer (Ho *et al.*, 2005) and electrokinetic migration (Pedersen-Bjergaard and Rasmussen, 2006) have been demonstrated.

3.2.9 *Supercritical fluid extraction*

Above a certain temperature (T_c) and pressure (P_c), a vapour does not exist as a gas, a solid or a liquid, but as a supercritical fluid (Figure 3.16). These fluids have densities, diffusivities and solvent strengths similar to liquids, but viscosities comparable to gases, which means mass transfer is faster and extraction times are reduced compared with liquids. Use of supercritical fluids in sample preparation has advantages of mild conditions, no thermal degradation of analyte and that the solvent readily evaporates after extraction. It is ideal for powdered samples such as soils, plant material and hair, but is generally unsuitable for liquid or wet matrices unless the sample is adsorbed on, for example, silica gel prior to the extraction. Moreover, the cost of the equipment is high (Smith, 1999).

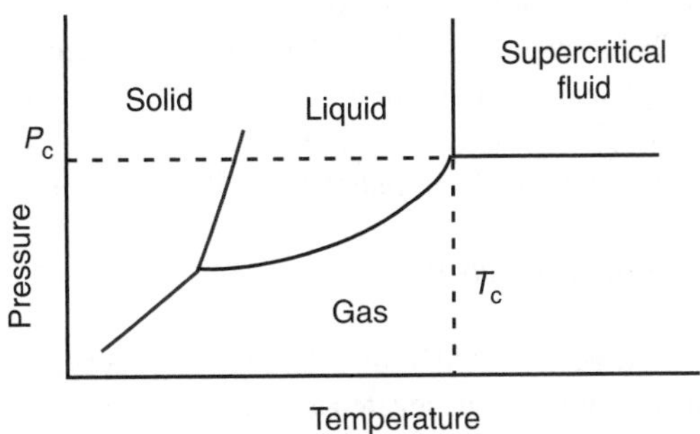

Figure 3.16 Phase diagram for a fluid.

Most supercritical fluid extraction (SFE) processes are quite simple. A sample is placed in the extraction thimble and supercritical fluid is pumped through the thimble (Figure 3.17). The extracted analytes are trapped as the fluid flows through a restricting nozzle. The fluid is vented from the collection trap, allowing the solvent to either escape or be recompressed for future use. In reality the process is a leaching operation, but it is commonly referred to as extraction. Carbon dioxide is normally the supercritical fluid of choice as it is inexpensive, nontoxic (although an asphyxiant gas), available with high purity and has a low critical temperature (31.3 °C). Typical operating conditions with supercritical carbon dioxide are 28,000 kPa (4000 psi) and 50 °C.

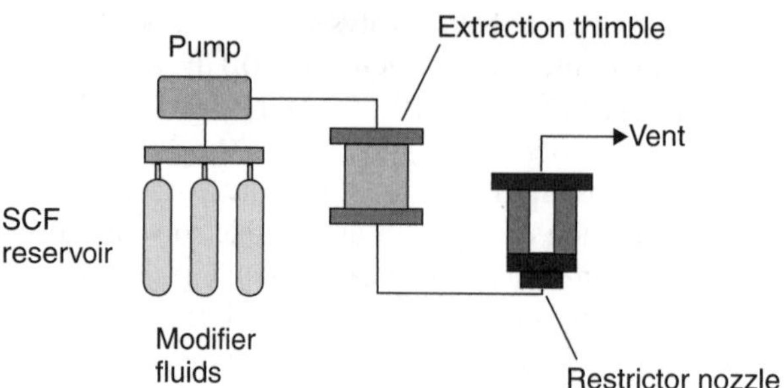

Figure 3.17 Supercritical fluid extraction.

The dominant factors that govern the SFE of an analyte are the solubility of the analyte in the fluid, the mass transfer kinetics of the analyte from the matrix to the fluid, and interactions between the fluid and the matrix. Altering the properties of the solvent system can change the extraction efficiency. Temperature and pressure are important influences. For example, raising the pressure increases the density of the fluid and often increases solubility. Obviously changing the solvent may provide different solvent properties. Supercritical water gives different extraction efficiencies as compared to ethylene or carbon dioxide. Anhydrous ammonia is different again. Adding a small amount of a second material (cosolvent) such as a small amount of methanol to carbon dioxide enhances the solubility of hydrophilic compounds. Addition of hydrogen chloride or ammonia renders the fluid acidic or basic, respectively.

Interfacing SFE to HPLC is not easy, but interfacing with GC is simple with use of a focusing trap, and with TLC the fluid can be sprayed onto the plate. SFE has been used to extract opiates (Edder *et al.*, 1994), and cocaine and benzoylecgonine (Morrison *et al.*, 1998) from hair prior to GC-MS. SFE followed by GC-MS has also been used to extract methylenedioxyamfetamine (MDA), methylenedioxymetamfetamine (MDMA) and methylenedioxyethylamfetamine (MDEA) from hair. Mephentermine was the internal standard (Allen and Oliver, 2000). SFE has been used to extract barbiturates from plasma (Spell *et al.*, 1998).

3.2.10 Accelerated solvent extraction

Accelerated solvent extraction (ASE), also known as subcritical fluid extraction, uses organic solvents under high pressures and at elevated temperatures (normally 100–140 °C) in an automated system similar to that used in SFE (Figure 3.18). Microwave-assisted extraction is similar except

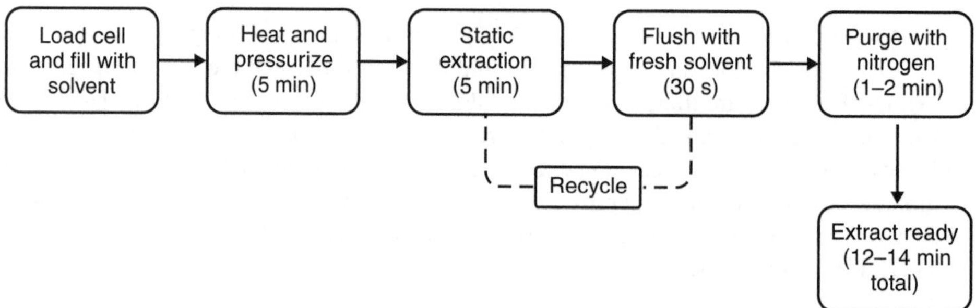

Figure 3.18 Outline of accelerated solvent extraction (ASE).

that microwave heating replaces a conventional oven. Solid samples are placed in the extraction vessel and brought to operating pressure by pumping solvent or a mixture of solvents into the vessel. The vessel is heated to the desired temperature and when the pressure in the vessel reaches a preset value, a pneumatic valve opens and allows the extract to be transferred to the liquid trap (Figure 3.19). Fresh solvent can be pumped into the extraction vessel and the process repeated if necessary.

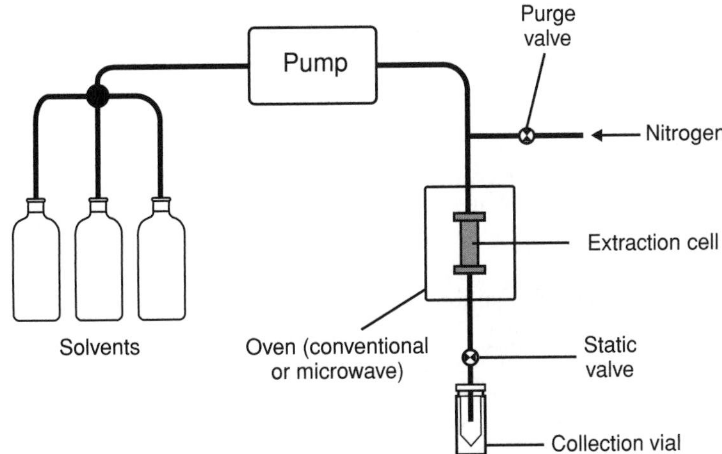

Figure 3.19 Schematic diagram of apparatus for accelerated solvent extraction.

ASE can be useful because supercritical carbon dioxide does not possess the solvent strength needed to extract efficiently some polar analytes from complex matrices even after adding modifiers such as methanol. In addition, ASE uses less solvent than conventional Soxhlet extraction techniques and is faster. The higher temperatures used also make it easier for the solvent to overcome intermolecular interactions of the analyte and matrix effects, but of course this could be a disadvantage with a labile analyte. ASE is not a selective extraction technique, however, and is essentially confined to solid matrices. There may also be a problem with loss of more volatile analytes when the extract is being transferred to the collection trap. Another disadvantage is there are no instruments currently available that can be interfaced directly to a chromatographic system.

3.3 Measurement of non-bound plasma concentrations

It is sometimes necessary to measure the concentration of non-bound ('free') drug in plasma or serum, usually as part of TDM of drugs such as phenytoin. The plasma 'free fraction' is often in equilibrium with saliva, but equilibrium with the cerebrospinal fluid (CSF) concentration requires that the drug is able to cross the blood–brain barrier. For strongly (>90 %) plasma protein-bound analytes, 'free' plasma, saliva and CSF concentrations are often very low and thus 'free' plasma and CSF drug measurements are challenging and require high-sensitivity analytical methods, especially as sample volume is often limited.

Spectroscopic methods have been used to investigate the interactions between ligands and dilute solutions of purified proteins, but for plasma, bound and non-bound drug are separated normally by ultrafiltration or equilibrium dialysis and the concentration of the analyte in each fraction measured independently. There are a number of issues to be borne in mind when planning and conducting plasma protein binding studies (Box 3.8). Most methods perturb binding to some extent. SPME of plasma or serum can be used to measure non-bound drug directly providing the proportion extracted is small and does not perturb the binding (Lord and Pawliszyn, 2000b). The problem of proteins leaking though the membrane applies to both ultrafiltration and equilibrium dialysis and should be monitored by protein assay.

Box 3.8 Measurement of plasma protein binding

(a) Ultrafiltration
- Relatively quick
- Change in protein concentration
- Adsorption of the analyte onto the apparatus
- Protein leakage

(b) Equilibrium dialysis
- Often slow to reach equilibrium
- Decomposition of the analyte and microbial growth
- Adsorption of the analyte onto the dialysis membrane
- Protein leakage
- Dilution of the sample

For highly protein-bound drugs, measuring the non-bound drug concentration presents an analytical challenge. For some drugs, immunoassays may be sensitive enough for routine analysis of the non-bound fraction and kits are commercially available that contain suitable calibrators for quantifying the total and non-bound concentrations. As phenytoin is approximately 90 % bound in plasma the concentration of the calibrators for the free drug need only be an order of magnitude less than the calibrators for measuring the total concentration. There is some doubt as to whether all the antibodies that are offered by some suppliers are specific enough for the purpose (Roberts *et al.*, 2001). However, for drugs that are >99 % bound it may not be possible to obtain accurate concentrations using standard laboratory methods given that analyte and protein concentrations must be those attained under physiological conditions. Under these conditions it may be necessary to use radiolabelled drug. A very small quantity of high specific-activity labelled drug is added to plasma, incubated to ensure equilibration with non-labelled analyte, and the free and bound fractions are then separated for radioactive counting, usually by liquid scintillation spectrometry (Section 12.2.3.1). Clearly, this approach is more suited to a research laboratory.

3.3.1 Ultrafiltration

Several ultrafiltration devices are available commercially. The filtration membranes used are made from a variety of materials and have M_r cut-offs in the range $M_r = 10{,}000$–$30{,}000$. Most devices are designed to be centrifuged to provide the filtration pressure. Ultrafiltration should not be confused with ultracentrifugation in which protein-bound and 'free' analyte can be separated as layers, often with the aid of a density gradient. This latter technique is particularly useful for investigating binding to lipoproteins which can be separated as layers floating on the surface of plasma, the density of which has been adjusted with potassium bromide (Michael-Titus *et al.*, 1995).

Ultrafiltration is more convenient and quicker than equilibrium dialysis, but the protein concentration in the retentate increases during filtration, potentially increasing the proportion of analyte bound. To minimize this problem as small a volume of ultrafiltrate should be collected as practicable, and the volume of retentate made up with an appropriate buffer solution periodically during centrifugation. Binding of the analyte to the filtration membrane (common with many lipophilic drugs) will reduce the concentration in the filtrate. Control experiments to ascertain the magnitude of this problem should be conducted. It is good practice to collect serial samples of filtrate for analysis to ensure that the sample is representative of the non-bound concentration. Some filtration membranes need to be soaked in water or buffer before use and as a result the first ultrafiltrate collection(s) may be markedly dilute.

3.3.2 Equilibrium dialysis

Equilibrium dialysis is the 'gold standard' for protein binding studies. However, it is not without problems. The time for equilibration can be several hours, during which there may be microbial growth, particularly with plasma at 37 °C, possibly leading to changes in protein and analyte concentrations and analyte binding. If an antibiotic is added it cannot be assumed that it does not interfere with the binding. An advantage of dialysis is that the problem of adsorption of analyte to the membrane and apparatus is largely overcome by measuring the concentrations on either side of the membrane. Adsorption will reduce the concentrations in the donor and recipient (dialysate) solutions, but at equilibrium the non-bound concentration in solution will be the same on either side of the membrane [Figure 3.20(a)]. Thus, it is possible to calculate the fraction bound or free and to relate this to the initial plasma concentration of the analyte.

Equilibrium dialysis cells [Figure 3.20(b)] are often home-made using Perspex or similar material, or simply consist of cellulose dialysis tubing (Visking tubing) tied at both ends, floating in buffer solution [Figure 3.20(c)]. The cells have the advantage that both sides of the membrane can be sampled periodically to test for equilibration. Visking tubing (1 cm wide when flat) should be soaked in buffer until it is soft, cut into suitable lengths and a double knot tied at one end. Plasma (0.5–1 mL) is pipetted into the tubing, which is sealed with a second double knot, ensuring that there is an air bubble to aid mixing. The simple expedient of using double knots reduces the likelihood of protein leakage. The dialysis tubes should be briefly rinsed with buffer to remove any excess sample, dried with a tissue and placed in tubes (such as screw-capped glass tubes) containing buffer. The capped tubes are gently shaken until the non-bound drug has equilibrated across the membrane, as ascertained by serial sampling of the buffer solution. The dialysis sack is removed and cut open for sampling, a process facilitated by supporting the sack in a glass test tube of appropriate dimensions [Figure 3.20(c)].

On the protein side of the membrane there is bound and non-bound analyte so the concentration is the total concentration, C_t, and that on the buffer side is the non-bound or free concentration, C_f, so the fraction free is:

$$\text{Fraction free} = \frac{\text{concentration in dialysate buffer } (C_f)}{\text{concentration in dialysis sac } (C_t)} \tag{3.12}$$

The protein-bound concentration is $(C_t - C_f)$ and so the fraction bound, β is:

$$\beta = \frac{C_t - C_f}{C_t} \tag{3.13}$$

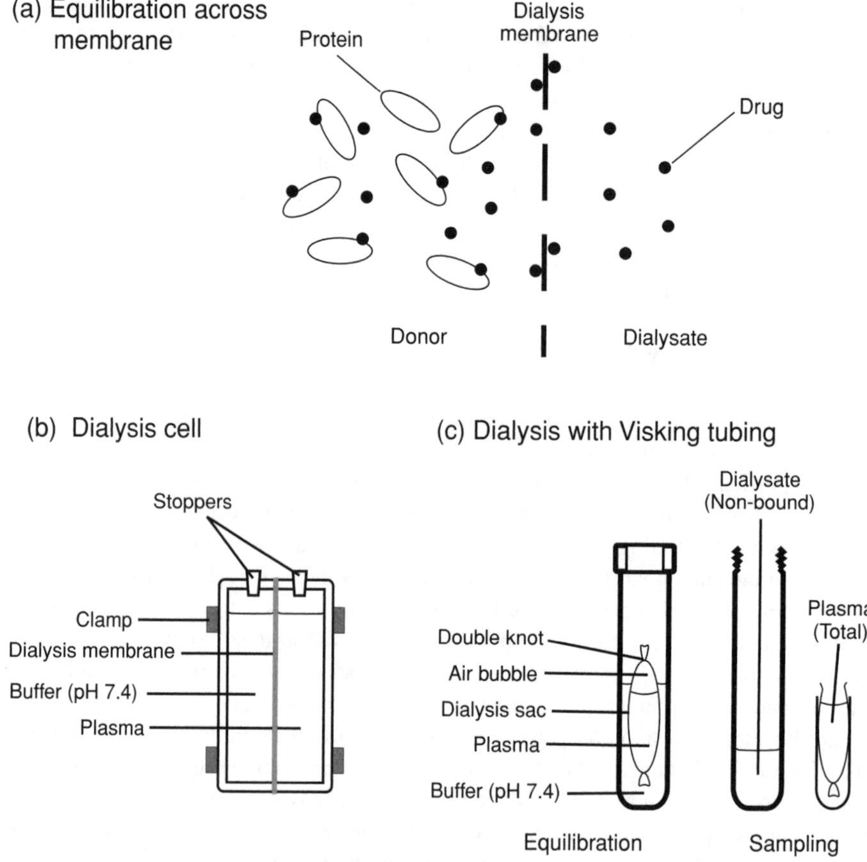

Figure 3.20 Diagrammatic representation of equilibrium dialysis: (a) equilibration of a drug either side of a dialysis membrane, (b) dialysis cell and (c) using Visking tubing. Analyte, but not protein or protein-bound analyte, diffuses through the dialysis membrane until the nonbound concentrations on either side of the dialysis membrane are equal.

Because of analyte being transferred to the dialysate and possibly lost by adsorption to the apparatus, C_t is not the same as the initial plasma concentration. However, provided that the fraction bound does not markedly change with changes in total concentration, then the non-bound concentration in the original plasma sample can be derived from Equation (3.13).

3.4 Hydrolysis of conjugated metabolites

Cleavage of conjugates is an important step in toxicological analysis, especially of urine. Many drugs and metabolites (e.g. benzodiazepines, laxatives, opiates and steroids) are excreted in urine and in bile predominantly as conjugates with D-glucuronic acid or with sulfate, or sometimes with both. Whilst sulfates are esters, glucuronides may be ethers (acetals), esters (acylals), or *N*- or *S*-glucuronides.

In order to maximize sensitivity in drug/poison screening, and, if appropriate, in order to measure such conjugates indirectly, that is in conjunction with independent measurement of unconjugated analyte, either selective or non-selective hydrolysis of the sample can be undertaken prior to further sample processing. In quantitative work, calibration or QC samples containing the conjugate of interest should be carried through the chosen procedure in order to monitor the efficiency of hydrolysis.

Incubation with strong mineral acid, such as an equal volume of 5 mol L^{-1} hydrochloric acid (15–30 min, 100 °C at atmospheric pressure, or in a microwave or pressure cooker), is cheap and gives rapid, but non-selective, hydrolysis of conjugates. Use of a domestic microwave is potentially hazardous, but commercial instruments offering temperature control are available (Milestone Scientific). Quartz hydrolysis vessels used as inserts are said to abolish the risk of memory effects from adsorption of analytes onto the PTFE vessels supplied with the instrument. All strong acid hydrolysis procedures are likely to introduce additional, hitherto unseen compounds into sample extracts even if due care is taken to ensure effective neutralization/buffering during sample preparation. Destruction of labile analytes as well as conjugates may be advantageous as in the colour test for urine paracetamol (Section 4.2), but may lead to impaired selectivity as in the hydrolysis of certain benzodiazepines to 5-chloro-2-methylaminobenzophenone (Section 6.5).

In contrast, incubation with β-glucuronidase (EC 3.2.1.31) and/or arylsulfatase (EC 3.1.6.1) (15 h, 35 °C) can give selective hydrolysis of conjugates under relatively mild conditions. Generally cleaner extracts result, but the incubation time is longer (usually overnight), there may be matrix effects from the enzyme solution (Meatherall, 1994), and the enzyme preparations are relatively expensive. There is a range of enzymes available. Those from *Escherichia coli* or bovine liver lack arylsulfatase and therefore preparations of the digestive juice of *Helix pomatia*, which contain both enzymes, are widely used. β-Glucuronidase and arylsulfatase from *Patella vulgata* are also marketed. The enzymes exhibit maximum activities at different pH values. The optimal pH for the *H. pomatia* enzyme is pH 4.5 whilst that for *E. coli* is about 6.8. Furthermore, the *H. pomatia* enzyme can be incubated at temperatures up to 45 °C, which can reduce the incubation time markedly. Some temperature-tolerant preparations allow incubation up to 60 °C (2 h incubation).

Control experiments should be performed to ensure that the incubation conditions are optimum with regard to the degree of hydrolysis and stability of the analytes. It should also be borne in mind that prolonged incubation of potentially infective samples may increase the titre of the infective agent. Finally, note that β-glucuronidase is specific for β-D-glucuronides and some glucuronides rearrange during incubation, and therefore cannot be hydrolyzed enzymatically.

There have been many attempts to prepare immobilized enzyme preparations, including preparations for on-line cleavage of conjugates in pre- or post-column reactors in HPLC. The use of columns packed with purified glucuronidase and arylsulfatase co-immobilized on an agarose gel matrix in the hydrolysis of conjugates in urine has been reported (Figure 3.21). Even the relatively stable conjugates of morphine could be completely hydrolyzed within 60 min in authentic urine samples. Matrix effects and effects of high concentrations of paracetamol conjugates, for example, were minimal as a large excess of each enzyme was used. The columns could be reused at least 70 times if rinsed with methanol:acetate buffer (2 + 8), and were stable on storage for at least 12 weeks.

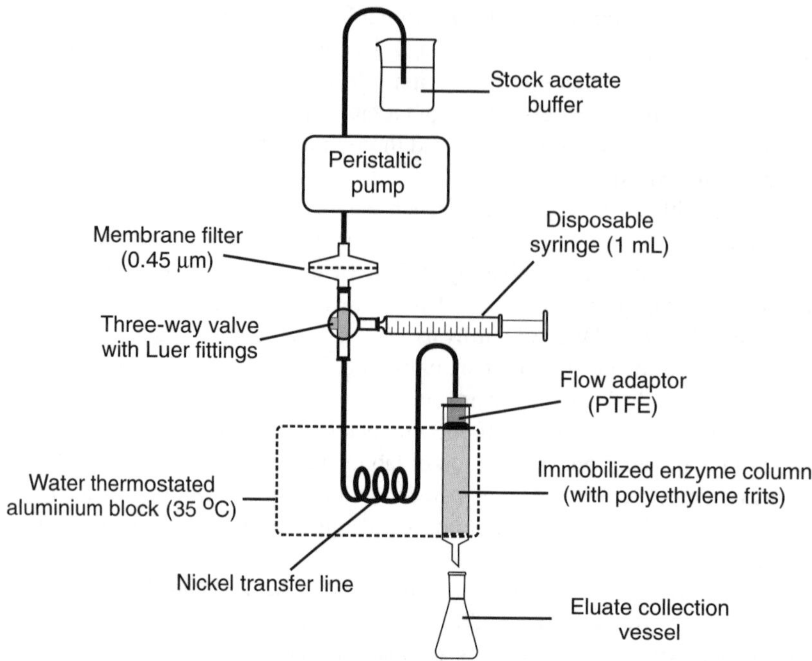

Figure 3.21 Apparatus for the cleavage of conjugates with immobilized glucuronidase and sulfatase columns. Modified after Toennes and Maurer (1999), by permission of the American Association for Clinical Chemistry.

3.5 Extraction of drugs from tissues

Conventionally measurements in portions of organs such as liver and brain have been performed via mechanical homogenization using a polytetrafluorethylene (PTFE)-in-glass homogenizer, and/or acid digestion on, say, 5 g wet weight of tissue prior to solvent extraction at an appropriate pH. Other tissues such as lung or muscle generally require use of a cutting or mincing action. However, digestion (12–16 h, room temperature) of small amounts (say 100 mg) of tissue with collagenase, protease, or lipase often gives much improved recovery when compared to conventional procedures and has the advantage that, once the digest has been prepared, analogous methodology and calibration standards to those used with plasma can be employed (Osselton *et al*., 1977; de Groot and Wubs, 1987). It is obviously important to ensure that the enzymatic hydrolysis does not destroy the analyte(s), hydrolyze metabolites to reform the analyte(s), or introduce other interferences. An internal standard can be added to the homogenization buffer or at a later stage.

Various enzyme-based digestion procedures have been reviewed (Shankar *et al*., 1987). Although papain gave the highest recoveries of added drug from liver homogenate, Subtilisin A gave similar recoveries for most compounds. Moreover, only a short incubation time (1 h) was required (Table 3.8). A procedure developed to measure lidocaine in tissue specimens after treatment with Subtilisin A using a longer digestion period prior to HPLC analysis of a solvent extract of the digest is outlined in Box 3.9. There has been tentative interest in using ASE (Section 3.2.9)

Table 3.8 Recovery of added drug from liver homogenate by various digestion procedures (Shankar *et al.*, 1987)

Drug	Recovery (%)					
	Papain	Subtilisin A	Neutrase	Trypsin	Acid hydrolysis	Stas-Otto
Chlordiazepoxide	96	60	58	40	56	47
Chlorpromazine	76	67	62	50	47	38
Diazepam	90	80	87	71	66	41
Diphenhydramine	67	57	61	54	39	—
Imipramine	78	82	81	55	54	34
Nitrazepam	80	74	38	57	52	49
Oxazepam	92	85	66	63	73	64
Promethazine	88	61	54	50	44	32

Box 3.9 Subtilisin A digestion for HPLC of lidocaine in tissue (Monkman *et al.*, 1989)

Reagent

- Lyophilized subtilisin A (Novo Nordisk, Windsor, UK) (2 g L^{-1}) in sodium dihydrogen orthophosphate/disodium hydrogen orthophosphate buffer (7 mmol L^{-1}, pH 7.4).

Method

- Dissect as out 100 mg wet weight portions of tissue, remove excess fluid on filter paper, add tissue to preweighed 10 mL tapered glass tubes and record the exact weights
- Add 2 g L^{-1} subtilisin A (1.0 mL), seal the tubes with ground-glass stoppers, and incubate in a water bath (50 °C, 12–16 h)
- Cool the tubes, vortex-mix (5 s), transfer 0.2 mL portions to 60 × 5 mm i.d. glass tube (Dreyer tube)
- Add 20 μL Tris buffer (2 mol L^{-1}, pH 11.0) and 20 μL internal standard solution (3 mg L^{-1} aqueous bupivacaine) and vortex-mix (5 s)
- Add methyl *tert*-butyl ether (MTBE) (200 μL) and vortex-mix (30 s)
- Centrifuge (11,000 *g*, 3 min)
- Inject 100 μL of the MTBE extract onto the HPLC column

in the analysis of drugs and other poisons in tissues and related samples (Coopman *et al.*, 1998; Abend *et al.*, 2003).

3.5.1 *Hair analysis for drugs and organic poisons*

Human hair consists of protein (65–95 %, mainly keratin), water (15–35 %) and lipid (1–9 %) together with some trace components including trace elements and heavy metals (Chapter 11). Drugs and metabolites are incorporated in the structure of hair as it is synthesized in the hair follicle, but may also arise due to diffusion from skin and sebum, an oily secretion of the sebaceous glands that helps to preserve the flexibility of hair. It is important to be able to differentiate compounds incorporated into hair that arise from within the body from substances that may contaminate the hair once it has grown away from the skin. Thus, after measurement of the length of the hair sample and segmentation if required (normally segments no less than 10 mm in length are used), most hair analysis protocols incorporate a decontamination step prior to the actual analysis (Box 3.10).

Box 3.10 Steps in the analysis of drugs and metabolites in hair

Pre-analytical

- Decontamination with detergents such as shampoos, surgical scrubbing solutions, surfactants [0.1 % sodium dodecylsulfate (SDS) for example]; phosphate buffer; organic solvents such as acetone, diethyl ether, methanol, ethanol, dichloromethane, hexane, or pentane (various volumes, various contact times)
- Analysis of washings to monitor removal of any surface contamination
- Measurement of length from scalp/body end and segmentation if required
- Accurate weighing of dried hair segments and fragmentation (fine cutting, ball mill)

Sample preparation

- For immunoassay
  - Incubation in aqueous buffer
- For GC-MS, HPLC-MS, HPLC-MS-MS
  - Incubation with internal standard solution, and
  - Incubation in acidic or basic aqueous solution, and LLE, SPE, or SPME, with derivatization if necessary, or
  - Incubation in organic solvent (often dichloromethane, methanol or acidified methanol) and LLE, SPE, or SPME, with derivatization if necessary, or
  - Enzyme digestion and LLE, SPE, or SPME, with derivatization if necessary

Assay calibration and reporting results

- Incubate 'blank' hair with known amounts of analyte across the anticipated concentration range
- Calculation of results (e.g. $\mu g\ g^{-1}$hair)

Generally, a single washing step is used, although a second identical wash is sometimes performed. If external contamination is found by analyzing the initial washings, repeated washing/analysis cycles may demonstrate that any surface contamination has been removed prior to the actual analysis. It has been suggested that the analyte concentration in the hair after washing should show a 10-fold increase over the concentration in the last wash for a positive result to be accepted (Baumgartner and Hill, 1992).

After the washing stages, the dried hair or hair segments may be chopped finely with a razor blade or pulverized in a ball mill prior to solubilization of drug and any metabolite(s) remaining in the hair. The choice of sample preparation procedure and the precise conditions used (such as the pH and the molarity of aqueous incubation solutions, and the duration and temperature of incubation) depends on: (i) the analyte(s) and (ii) the analysis system to be used (RIA, GC-MS, etc.) (Box 3.10). If the sample is incubated in sodium hydroxide solution solubilization of the hair is complete, but compounds such as cocaine and 6-monoacetylmorphine (6-MAM) will be hydrolyzed. Alternatively, incubation in hydrochloric acid (0.1 mol L^{-1} for 16 h at room temperature, 45 °C, or 56 °C; 0.6 mol L^{-1} for 30 min at 120 °C) will minimize loss of such analytes (Cirimele *et al.*, 1996.

Use of enzymatic digestion with, for example, proteinase K (Offidani *et al.*, 1993), Biopurase (Fujii *et al.*, 1996) or β-glucuronidase/arylsulfatase (Moeller *et al.*, 1993), is a further option. Extraction with organic solvent has involved incubation in an ultrasound bath (45 °C, 6 h or so) and has the advantage that GC-MS or derivatizaton followed by GC-MS can be carried out directly following solvent evaporation. Hydrolysis of labile analytes is minimal. Use of HPLC-MS-MS can simplify sample preparation even further.

Detection of drug metabolite(s) in hair that cannot be explained by surface contamination or hydrolysis of surface contaminants provides evidence that positive findings have arisen as a result of systemic exposure (Cone *et al.*, 1991). Cocaethylene and norcocaine are examples, but there is still debate as to the problems posed by external contamination, particularly for 'crack' (cocaine free base), cannabis and heroin when smoked. It has been suggested that it is not possible to remove contamination from such sources by conventional washing procedures (Blank and Kidwell, 1995). Similarly, seepage from a corpse may contaminate hair and of course may also contain metabolites.

The use of SFE in hair analysis was discussed above (Section 3.2.8). Even if SFE does permit rapid and efficient extraction of drugs and metabolites, all the considerations as to prewashing, segmentation, internal standard addition, derivatization for GC-MS, and assay calibration ('spiked' hair), still remain (Box 3.11).

Box 3.11 SFE of amfetamines in hair prior to GC-MS (Allen and Oliver, 2000)

- Wash hair sequentially with 2 × (i) sodium dodecyl sulfate solution, (ii) dichloromethane, (iii) methanol and (iv) deionized water
- Sonicate (15 min, room temperature) and dry (room temperature)
- Cut finely and weigh 50 mg portions into analysis vials
- Add methanol (3 mL) and internal standard solution (mephentermine, 1 mg L^{-1} in methanol) to the vial and seal
- Prepare calibration standards by adding a methanolic solution of the appropriate analyte (10 μg g^{-1} hair) to analyte-free hair before sealing the vial
- Sonicate (15 min) and leave vials open for the methanol to evaporate (12–16 h, room temperature)
- SFE extraction solvent dichloromethane:2-propanol (90 + 10) pumped at 10 % in 90 % carbon dioxide, flow-rate 2 mL min^{-1}
- SFE dynamic extraction mode (260 bar, 70 °C, 30 min)
- Add 50 μL pentafluoropropionic anhydride:ethyl acetate (1 + 1) to the dried SFE extract

3.6 Derivatization

Generally derivatization should be avoided as it adds an extra step or steps to the analytical procedure and thus may increase the possibility of error. This being said, in GC derivative formation is performed to achieve satisfactory chromatography, to improve the detection characteristics of an analyte and sometimes to provide additional evidence of compound identity. Derivatization may be used to shorten or lengthen the retention time as required, and also to permit the separation

Box 3.12 Summary of reasons for derivatization in GC

- To improve resolution and reduce tailing of polar compounds (–OH, –COOH, $–NH_2$, –SH and other functional groups) and thereby increase sensitivity and selectivity
- To adjust retention and thereby improve resolution
- To facilitate enantiomer resolution
- To analyze relatively nonvolatile compounds
- To increase detectability by introducing detector-specific moieties (e.g. $–COCF_3$)
- To improve stability of compounds (e.g. in mass fragmentation)
- To confirm analyte identity

of enantiomers (Box 3.12). In HPLC, derivative formation, although sometimes used to stabilize an analyte, is seldom needed to achieve satisfactory chromatography except when performed to permit the separation of enantiomers. Derivatization is thus used mainly to enhance the selectivity and sensitivity of detection (Talwar *et al.*, 1999).

The choice of reagent and hence the reaction conditions will be based on the functional group to be derivatized, the presence of other functional or labile groups in the molecule, and the reason for performing the derivatization. Generally, three types of reactions are performed, *viz*. silylation, acylation and alkylation. In GC, silylation generally reduces retention, whereas acylation usually increases retention. The order of elution for derivatives of a homologous series will be the same as the underivatised parent compounds.

Aspects of derivatization have been reviewed (Blau and Halket, 1993). An obvious factor is that the analyte must be amenable to derivatization. It is also important that any internal standard undergoes the same derivatization reaction as the analyte. Use of a stable isotope-labelled internal standard in MS work is ideal (Section 10.7). The derivatization reaction may be carried out during the extraction or other sample preparation procedure, on a dried residue, in an autosampler vial, on a SPME fibre in a headspace vial, or after injection onto the GC column ('on-column'), and indeed post-column in HPLC. If derivatization is to be performed prior to the injection, then the reaction needs to be rapid and quantitative, as otherwise problems may arise in assessing the optimum reaction time.

There may be a number of confounding issues: (i) the need to remove excess derivatizing agent; (ii) the risk of decomposition of the analyte, the derivatizing agent, or other reagents, and of the derivative itself; and (iii) the risk of interference in the assay by contaminants, breakdown products, or reaction by-products. 'On-column' derivatization in GC may not suffer from these drawbacks and has the advantage that the analyte and derivatizing reagent can be injected directly. However, the analyte may not be amenable to this technique. Finally, care must be taken to ensure that artefacts such as acetylation of morphine to 6-monoacetylmorphine or diamorphine, or decomposition of diamorphine or 6-monoacetylmorphine to morphine, do not occur during, or as a result of, the derivatization process used (Kushnir *et al.*, 1999).

Derivatization for GC, for HPLC, and for MS is discussed further in Chapters 7, 8 and 10.

3.7 Summary

Guidelines for sample preparation in analytical toxicology are summarized in Table 3.9. The myriad of different approaches to analytical toxicology problems might seem at first sight

Table 3.9 Guidelines for sample preparation

Aim to	Avoid
• Keep the analysis as simple as possible	• Unnecessary additives (e.g. preservatives)
• Use high quality reagents, solvents, etc.	• Use of internal standards if these reduce precision
• Understand the clinical pharmacology, metabolism and toxicology of the analyte	• Hazardous solvents and, if possible, toxic/hazardous reagents
• Use the smallest sample possible compatible with the required sensitivity/selectivity	• Using more sample than is needed
• Increase sample size/injection volume to increase sensitivity if needed	• Using an extraction solvent that is too 'powerful' for the application
• Use external standards in colour tests, UV/Visible spectrophotometry and TLC, and internal standards in GC, HPLC and MS	• Using unnecessary extremes of heat or pH
	• SPE if LLE is satisfactory
	• Developing oversensitive methods

confusing, but with experience and in the knowledge of the instrumentation and expertise available in a particular centre, then certain standard approaches to similar problems will soon be developed. Those given here reflect personal experience.

Other approaches, which bring together a number of the themes developed above include those described for STA in urine and plasma, respectively. The hydrolysis of only a portion of a urine specimen removes the risk of destroying acid-labile analytes whilst enhancing the likelihood of detecting compounds principally excreted as conjugates or other acid-labile metabolites (Box 3.13). The solvent mixture has very high 'extracting power' and with this even relatively water-soluble analytes such as morphine will be extracted – similarly the extraction pH optimizes the extraction of any phenols by ensuring the highest proportion of the unionized, extractable form. The formation of acetylated derivatives to ensure reasonable GC behaviour by use of acetic anhydride requires catalysis to ensure a good yield of the derivatives of interest. Pyridine, although toxic, has the advantage over non-toxic catalysts such as *N*-methylimidazole in that it is volatile and can be removed by evaporation together with excess acetic anhydride prior to GC-MS analysis.

Box 3.13 Sample preparation for systematic qualitative toxicological analysis in urine by GC-MS (Theobald *et al.*, 2005)

- Reflux portion of urine (2.5 mL) for 15 min with 12 mol L^{-1} hydrochloric acid (1 mL)
- Add 2.3 mol L^{-1} aqueous ammonium sulfate (2 mL) + 10 mol L^{-1} aqueous sodium hydroxide (1.5 mL) to give a pH of 8–9
- Add to unhydrolyzed portion of urine (2.5 mL) and extract with 5 mL dichloromethane:2-propanol:ethyl acetate (1 + 1 + 3)
- Evaporate to dryness (reduced pressure, 70 °C) and treat with 0.1 mL acetic anhydride: pyridine (3 + 2) (5 min, microwave 440 W)
- Evaporate to dryness (reduced pressure, 70 °C) and reconstitute in 100 μL methanol

The approach advocated for plasma is inherently more complicated than that for urine because: (i) plasma protein is present, (ii) analyte concentrations are often much lower and (iii) quantitation is required (if possible) and hence more analytical options are needed (Box 3.14). Sodium sulfate is used to increase the ionic strength of the aqueous phase in the simple extraction thus enhancing extractability of the analytes, whilst the sequential neutral and basic extractions are used to enhance the range of compounds extracted. The use of diethyl ether is not generally recommended now on health and safety grounds.

Box 3.14 Sample preparation for systematic toxicological analysis in plasma by GC-MS and/or HPLC-ACPI

(a) General procedure for GC-MS (underivatized) and HPLC-MS-APCI (Maurer *et al.*, 2002)
- To plasma (1 mL) add 0.1 mL methanolic [^{2}H]-trimipramine (10 mg L^{-1}) and 5 mL saturated aqueous sodium sulfate
- Extract with 5 mL diethyl ether:ethyl acetate (1 + 1) and centrifuge
- Transfer extract to pear-shaped flask and add 0.5 mL aqueous sodium hydroxide (1 mol L^{-1}) to aqueous phase
- Re-extract with 5 mL diethyl ether:ethyl acetate (1 + 1) and centrifuge
- Evaporate combined extracts to dryness and reconstitute in methanol (0.1 mL)
- For LC-MS evaporate portion (50 μL) of methanolic extract to dryness and take up in acetonitrile (50 μL)

(b) High-sensitivity procedure for HPLC-MS-APCI (Maurer *et al.*, 2004)
- To plasma (0.5 mL) add 2 mL purified water, 0.05 mL methanolic ^{2}H-trimipramine (10 mg L^{-1}) and mix (rotary mixer, 15 s)
- Pre-condition SPE columns (Isolute Confirm HCX 130 mg, 3 mL) with methanol (1 mL) and purified water (1 mL)
- Load sample mixture onto SPE column and wash with purified water (1 mL), 10 mmol L^{-1} aqueous hydrochloric acid (1 mL) and methanol (2 mL)
- Dry under vacuum and elute into autosampler vials with 1.5 mL methanol:aqueous ammonium hydroxide (RD 0.88) (98 + 2)
- Evaporate to dryness under nitrogen (56 °C) and reconstitute in methanol (50 μL)

The Confirm HCX mixed-mode (reversed-phase C_8 and SCX) SPE column was used because this gave very clean extracts, even from haemolyzed blood samples. Assay development time is likely to have been greater than with the LLE procedure, and although the established procedure is elegant, the number of steps involved and the cost of the columns must be borne in mind when deciding to use SPE.

References

Abend, A.M., Chung, L., McCollum, D.G. and Wuelfing, W.P. (2003) Development and validation of an automated extraction method (accelerated solvent extraction) and a reverse-phase HPLC analysis method for assay of ivermectin in a meat-based chewable formulation. *J Pharm Biomed Anal*, **31**, 1177–83.

Allen, D.L. and Oliver, J.S. (2000) The use of supercritical fluid extraction for the determination of amphetamines in hair. *Forensic Sci Int*, **107**, 191–9.

Andersson, L.I. (2000) Molecular imprinting for drug bioanalysis. A review on the application of imprinted polymers to solid-phase extraction and binding assay. *J Chromatogr B*, **739**, 163–73.

Barrett, J., Smyth, W.F. and Davidson, I.E. (1973) An examination of the acid base equilibra of 1,4-benzodiazepines by spectroscopy. *J Pharm Pharmacol*, **25**, 387–93.

Basheer, C., Lee, H.K. and Obbard, J.P. (2004) Application of liquid-phase microextraction and gas chromatography-mass spectrometry for the determination of polychlorinated biphenyls in blood plasma. *J Chromatogr A*, **1022**, 161–9.

Baumgartner, W.A. and Hill, V.A. (1992) Hair analysis for drugs of abuse: decontamination issues, in *Recent Developments in Therapeutic Drug Monitoring and Clinical Toxicology* (ed. I. Sunshine), Dekker, New York, pp. 577–597.

Binder, S.R., Regalia, M., Biaggi-McEachern, M. and Mazhar, M. (1989) Automated liquid chromatographic analysis of drugs in urine by on-line sample cleanup and isocratic multi-column separation. *J Chromatogr*, **473**, 325–41.

Blanchard, J. (1981) Evaluation of the relative efficacy of various techniques for deproteinizing plasma samples prior to high-performance liquid chromatographic analysis. *J Chromatogr*, **226**, 455–60.

Blank, D.L. and Kidwell, D.A. (1995) Decontamination procedures for drugs of abuse in hair: are they sufficient? *Forensic Sci Int*, **70**, 13–38.

Blau, K. and Halket, J. (1993) *Handbook of derivatives for chromatography*, 2nd edn, Wiley, New York.

Bruheim, I., Liu, X. and Pawliszyn, J. (2003) Thin-film microextraction. *Anal Chem*, **75**, 1002–10.

Chen, X.H., Franke, J.P., Ensing, K., Wijsbeek, J. and de Zeeuw, R.A. (1993) Pitfalls and solutions in the development of a fully automated solid-phase extraction method for drug screening purposes in plasma and whole blood. *J Anal Toxicol*, **17**, 421–6.

Cirimele, V., Kintz, P. and Mangin, P. (1996) Comparison of different extraction procedures for drugs in hair of drug addicts. *Biomed Chromatogr*, **10**, 179–82.

Cone, E.J., Buchwald, W.F. and Darwin, W.D. (1982) Analytical controls in drug metabolic studies. II. Artifact formation during chloroform extraction of drugs and metabolites with amine substituents. *Drug Metab Dispos*, **10**, 561–7.

Cone, E.J., Yousefnejad, D., Darwin, W.D. and Maguire, T. (1991) Testing human hair for drugs of abuse. II. Identification of unique cocaine metabolites in hair of drug abusers and evaluation of decontamination procedures. *J Anal Toxicol*, **15**, 250–5.

Cooper, J.D.H., Turnell, D.C., Green, B. and Varillon, F. (1988) Automated sequential trace enrichment of dialysates and robotics. A technique for the preparation of biological samples prior to high-performance liquid chromatography. *J Chromatogr*, **456**, 53–69.

Coopman, V.A., Cordonnier, J.A. and Vandecasteele, P.G. Accelerated solvent extraction: a new tool for toxicological screening of postmortem samples? in Proceedings of the SOFT-TIAFT Joint Meeting 5–9 October 1998 Albuquerque, NM (http://www.tiaft.org/tiaft98/fri/p/f_p_36.html, accessed 31 December 2005).

Dietz, C., Sanz, J. and Camara, C. (2006) Recent developments in solid-phase microextraction coatings and related techniques. *J Chromatogr A*, **1103**, 183–92.

Edder, P., Staub, C., Veuthey, J.L., Pierroz, I. and Haerdi, W. (1994) Subcritical fluid extraction of opiates in hair of drug addicts. *J Chromatogr B*, **658**, 75–86.

Flanagan, R.J., Bhamra, R., Walker, S. Monkman, S.C. and Holt D.W. (1988) Measurement of cardioactive drugs in biological samples by HPLC. *J Liquid Chromatogr*, **11**, 1015–40.

Flanagan, R.J., Harvey, E.J. and Spencer, E.P. (2001) HPLC of basic drugs on microparticulate strong cation-exchange materials—a review. *Forensic Sci Int*, **121**, 97–102.

Flanagan, R.J. and Ruprah, M. (1989) HPLC measurement of chlorophenoxy herbicides, bromoxynil, and ioxynil, in biological specimens to aid diagnosis of acute poisoning. *Clin Chem*, **35**, 1342–7.

Fujii, J., Higashi, A. and Nakano, M. (1996) Examination of stability of anticonvulsants in a protease solution and assay of anticonvulsants in hairs. *Biol Pharm Bull*, **19**, 1614–7.

Furton, K.G., Wang, J., Hsu, Y.L., Walton, J. and Almirall, J.R. (2000) The use of solid-phase microextraction–gas chromatography in forensic analysis. *J Chromatogr Sci*, **38**, 297–306.

Georgi, K., Boos, K.-S. (2004) Control of matrix effects in bioanalytical MS/MS using on-line multidimensional solid-phase extraction. *LC-GC Europe*, **17**, 21–4.

Gill, R., Qua, S.C. and Moffat, A.C. (1983) High-performance liquid chromatography of paraquat and diquat in urine with rapid sample preparation involving ion-pair extraction on disposable cartridges of octadecyl-silica. *J Chromatogr*, **255**, 483–90.

Green, A.L. (1967) Ionization constants and water solubilities of some aminoalkylphenothiazine tranquillizers and related compounds. *J Pharm Pharmacol*, **19**, 10–6.

de Groot, G. and Wubs, K.L. (1987) A simple enzymic digestion procedure of intact tissue samples in pharmacokinetic drug analysis. *J Analyt Toxicol*, **11**, 175–8.

Hagestam, I.H. and Pinkerton, T.C. (1985) Internal surface reversed phase silica supports for liquid chromatography. *Anal Chem*, **57**, 1757–63.

Harkey, M.R. (1989) Bonded phase extraction in analytical toxicology, in *Analytical Aspects of Drug Testing* (ed. D.G. Deutsch), Wiley, New York, pp. 59–85.

Ho, T.S., Reubsaet, J.L., Anthonsen, H.S., Pedersen-Bjergaard, S. and Rasmussen, K.E. (2005) Liquid-phase microextraction based on carrier mediated transport combined with liquid chromatography-mass spectrometry. New concept for the determination of polar drugs in a single drop of human plasma. *J Chromatogr A*, **1072**, 29–36.

Hoogewijs, G. and Massart, D.L. (1984) Development of a standardized analysis strategy for basic drugs, using ion-pair extraction and high-performance liquid chromatography. VI. Drug level determination in saliva. *J Chromatogr*, **309**, 329–37.

Hurst, P.R. and Whelpton, R. (1989) Solid phase extraction for an improved assay of physostigmine in biological fluids. *Biomed Chromatogr*, **3**, 226–32.

Kolb, B. and Ettre, L.S. (1997) *Static Headspace-Gas Chromatography: Theory and Practice*, Wiley, London.

Kushnir, M.M., Crockett, D.K., Nelson, G. and Urry, F.M. (1999) Comparison of four derivatizing reagents for 6-acetylmorphine GC-MS analysis. *J Anal Toxicol*, **23**, 262–9.

Lord, H. and Pawliszyn, J. (2000a) Evolution of solid-phase microextraction technology. *J Chromatogr A*, **885**, 153–93.

Lord, H. and Pawliszyn, J. (2000b) Microextraction of drugs. *J Chromatogr A*, **902**, 17–68.

Lukaszewski, T. (1985) The extraction and analysis of quaternary ammonium compounds in biological material by GC and GC/MS. *J Anal Toxicol*, **9**, 101–8.

Maurer, H.H., Kratzsch, C., Kraemer, T., Peters, F.T. and Weber, A.A. (2002) Screening, library-assisted identification and validated quantification of oral antidiabetics of the sulfonylurea-type in plasma by atmospheric pressure chemical ionization liquid chromatography-mass spectrometry (APCI-LC-MS). *J Chromatogr B*, **773**, 63–73.

Maurer, H.H., Tenberken, O., Kratzsch, C., Weber, A.A. and Peters, F.T. (2004) Screening for, library-assisted identification and fully validated quantification of twenty-two beta-blockers in blood plasma by liquid chromatography-mass spectrometry with atmospheric pressure chemical ionization. *J Chromatogr A*, **1058**, 169–81.

Meatherall, R. (1994) Optimal enzymatic hydrolysis of urinary benzodiazepine conjugates. *J Anal Toxicol*, **18**, 382–4.

Michael-Titus, A.T., Whelpton, R. and Yaqub, Z. (1995) Binding of temoporfin to the lipoprotein fractions of human serum. *Br J Clin Pharmacol*, **40**, 594–7.

Mills, G.A. and Walker, V. (2000) Headspace solid-phase microextraction procedures for gas chromatographic analysis of biological fluids and materials. *J Chromatogr A*, **902**, 267–87.

Moeller, M.R., Fey, P. and Wennig, R. (1993) Simultaneous determination of drugs of abuse (opiates, cocaine and amphetamine) in human hair by GC/MS and its application to a methadone treatment program. *Forensic Sci Int*, **63**, 185–206.

Monkman, S.C., Armstrong, R., Flanagan, R.J., Holt D.W., and Rosevear S. (1989) High performance liquid chromatographic measurement of lignocaine in tissue samples following transabdominal placental biopsy. *Biomed Chromatogr*, **3**, 88–91.

Morrison, J.F., Chesler, S.N., Yoo, W.J. and Selavka, C.M. (1998) Matrix and modifier effects in the supercritical fluid extraction of cocaine and benzoylecgonine from human hair. *Anal Chem*, **70**, 163–72.

Offidani, C., Strano Rossi, S. and Chiarotti, M. (1993) Improved enzymatic hydrolysis of hair. *Forensic Sci Int*, **63**, 171–4.

Osselton, M.D., Hammond, M.D. and Twitchett, P.J. (1977) The extraction and analysis of benzodiazepines in tissues by enzymic digestion and high-performance liquid chromatography. *J Pharm Pharmacol*, **29**, 460–2.

Pedersen-Bjergaard, S. and Rasmussen, K.E. (2005) Bioanalysis of drugs by liquid-phase microextraction coupled to separation techniques. *J Chromatogr B*, **817**, 3–12.

Pedersen-Bjergaard, S. and Rasmussen, K.E. (2006) Electrokinetic migration across artificial liquid membranes New concept for rapid sample preparation of biological fluids. *J Chromatogr A*, **1109**, 183–90.

le Petit, G.F. (1976) Medazepam pKa determined by spectrophotometric and solubility methods. *J Pharm Sci*, **65**, 1094–5.

Rasmussen, K.E., Pedersen-Bjergaard, S., Krogh, M., Ugland, H.G. and Gronhaug, T. (2000) Development of a simple in-vial liquid-phase microextraction device for drug analysis compatible with capillary gas chromatography, capillary electrophoresis and high-performance liquid chromatography. *J Chromatogr A*, **873**, 3–11.

Rbeida, O., Christiaens, B., Hubert, P., Lubda, D., Boos, K.S., Crommen, J. and Chiap, P. (2004) Evaluation of a novel anion-exchange restricted-access sorbent for on-line sample clean-up prior to the determination of acidic compounds in plasma by liquid chromatography. *J Chromatogr A*, **1030**, 95–102.

Reynolds, J.E.F. (ed.) (1993) *Martindale: The Extra Pharmacopoeia*, 30th edn. Pharmaceutical Press, London.

Roberts, W.L., Annesley, T.M., De, B.K., Moulton, L., Juenke, J.M. and Moyer, T.P. (2001) Performance characteristics of four free phenytoin immunoassays. *Ther Drug Monit*, **23**, 148–54.

Rutherford, D.M. (1977) Rapid micro-method for the measurement of diazepam and desmethyldiazepam in blood plasma by gas-liquid chromatography. *J Chromatogr*, **137**, 439–48.

Schill, G. (1976) Ion-pair solvent extraction and chromatography, in *Assay of Drugs and other Trace Compounds in Biological Fluids*, (ed. E. Reid), North-Holland, Amsterdam, pp. 87–102.

Shankar, V., Damodaran, C. and Sekharan, P.C. (1987) Comparative evaluation of some enzymic digestion procedures in the release of basic drugs from tissue. *J Anal Toxicol*, **11**, 164–7.

Smith, R.M. (2003) Before the injection–modern methods of sample preparation for separation techniques. *J Chromatogr A*, **1000**, 3–27.

Smith, R.M. (1999) Supercritical fluids in separation science—the dreams, the reality and the future. *J Chromatogr A*, **856**, 83–115.

Snyder, L.R. (1974) Classification of solvent properties of common liquids. *J Chromatogr*, **92**, 223–230.

Spell, J.C., Srinivasan, K., Stewart, J.T. and Bartlett, M.G. (1998) Supercritical fluid extraction and negative ion electrospray liquid chromatography tandem mass spectrometry analysis of phenobarbital, butalbital, pentobarbital and thiopental in human serum. *Rapid Commun Mass Spectrom*, **12**, 890–4.

Stevenson, D. (2000) Immuno-affinity solid-phase extraction. *J Chromatogr B*, **745**, 39–48.

Svensson, J.O., Bruchfeld, A., Schvarcz, R. and Stahle, L. (2000) Determination of ribavirin in serum using highly selective solid-phase extraction and high-performance liquid chromatography. *Ther Drug Monit*, **22**, 215–8.

Talwar, D., Watson, I.D. and Stewart, M.J. (1999) Routine analysis of amphetamine class drugs as their naphthaquinone derivatives in human urine by high-performance liquid chromatography. *J Chromatogr B*, **735**, 229–41.

Theobald, D.S., Fehn, S. and Maurer, H.H. (2005) New designer drug 2,5-dimethoxy-4-propylthiophenethylamine (2C-T-7): studies on its metabolism and toxicological detection in rat urine using gas chromatography/mass spectrometry. *J Mass Spectrom*, **40**, 105–16.

Toennes, S.W. and Maurer, H.H. (1999) Efficient cleavage of conjugates of drugs or poisons by immobilized beta-glucuronidase and arylsulfatase in columns. *Clin Chem*, **45**, 2173–82.

Tugnait, M., Ghauri, F.Y., Wilson, I.D. and Nicholson, J.K. (1991) NMR-monitored solid-phase extraction of phenolphthalein glucuronide on phenylboronic acid and C18 bonded phases. *J Pharm Biomed Anal*, **9**, 895–9.

Van Horne, K.C. (ed.) (1985) *Sorbent Extraction Technology*, Analytichem International, Harbor City, California.

Vintiloiu, V., Mullett, W.M., Papp, R., Lubda, D. and Kwong, E. (2005) Combining restricted access material (RAM) and turbulent flow for the rapid on-line extraction of the cyclooxygenase-2 inhibitor rofecoxib in plasma samples. *J Chromatogr A*, **1082**, 150–7.

Wang, Y., O'Reilly, J., Chen, Y. and Pawliszyn, J. (2005) Equilibrium in-fibre standardisation technique for solid-phase microextraction. *J Chromatogr A*, **1072**, 13–7.

Whelpton, R. (2004) Iterative least squares fitting of pH-extraction curves using an object-orientated language. *Chromatographia*, **59**, S193-S196.

Whelpton, R. (1989) Iterative least squares fitting of pH-partition data. *Trends Pharmacol Sci*, **10**, 182–3.

Whelpton, R., Dudson, P., Cannell, H. and Webster, K. (1990) Determination of prilocaine in human plasma samples using high-performance liquid chromatography with dual-electrode electrochemical detection. *J Chromatogr B*, **526**, 215–22.

Whelpton, R. and Hurst, P.R. (1988) Solid-phase extraction: personal experiences, in *Bioanalysis of Drugs and Metabolites especially Anti-inflammatory and Cardiovascular* (eds E. Reid, J.D. Robinson and I.D. Wilson), Plenum, New York, pp. 289–294.

4 Colour Tests, and Spectrophotometric and Luminescence Techniques

4.1 Introduction

The use of colour tests in analytical toxicology has its origins in inorganic qualitative analysis. Some simple colour tests for poisons still find a place under certain circumstances, but chromogenic reactions are most valuable when used in the visualization of developed TLC plates. UV/visible absorption and spectrophotofluorimetry were also amongst the early techniques applied in qualitative and quantitative toxicological analyses, and whilst still used in their own right, these techniques now also find great application as detection systems in HPLC. UV spectrophotometry is also important for monitoring reaction rates in immunoassays or in monitoring NAD/NADH- or NADP/NADPH-linked enzyme reactions. Atomic absorption and emission spectrophotometry and other spectroscopic techniques remain important in trace element and toxic metal analysis (Chapter 11).

4.1.1 Historical development

Conjugated organic compounds absorb UV (190–400 nm) or visible (400–700 nm) light – the greater the degree of conjugation the greater the degree of absorption at longer wavelengths. Thus, highly conjugated compounds such as β-carotene and haemoglobin absorb in the visible part of the spectrum and are coloured. Less conjugated aromatic compounds absorb in the UV region of the spectrum. Observation of colour, either native or after chemical treatment, may give useful qualitative information. However, major advances followed fundamental discoveries in spectroscopy, the study of the absorption of electromagnetic radiation by chemical compounds, and spectrometry, the quantitative study of this phenomenon.

The absorbance of light by materials was first explored by the German mathematician Johann Heinrich Lambert (1728–1777) who discovered that for monochromatic radiation (in practice radiation of a narrow band of wavelengths) the amount of light absorbed was directly proportional to the path length of the light through the material and was independent of the intensity of the incident light. The German astronomer Wilhelm Beer (1797–1850) expanded this work and found that, for dilute solutions, there was a linear relationship between analyte concentration and absorbance.

The clinical chemist Henry Bence Jones (1813–1873) used an emission spectroscope to detect lithium in urine and in other tissues including lenses removed from the eyes of cataract patients (Jones, 1865a, 1865b) and, working with August Dupré (1835–1907), used fluorescence analysis to detect quinine in blood and other tissues (Jones and Dupré, 1866). Samples were extracted with dilute sulfuric acid, which was made alkaline with sodium hydroxide and extracted with diethyl ether. The ether was left to evaporate and the residue dissolved in sulfuric acid prior to fluorescence measurement using an induction spark as the light source. As this extraction failed to remove background fluorescence, control samples were taken through the procedure for comparison with samples from treated guinea pigs.

The introduction of the first commercially-available UV spectrophotometer, the Beckman DU, by Arnold O. Beckman (1900–2004), the inventor of the pH meter, and Howard H. Cary (1908–1991) extended the useable wavelength range from the visible into the UV region allowing many more compounds to be detected and measured (Cary and Beckman, 1941). The impetus for the development of this instrument was military interest in vitamin A, a colourless breakdown product of β-carotene. However, groups such as that led by Bernard B. Brodie (1907–1989) went on to use the instrument to develop quantitative assays for many drugs. Brodie established some of the basic rules for the successful measurement of drugs and other poisons in biological specimens (Brodie *et al.*, 1947), many of which are still applicable today.

As with the development of practical UV spectrophotometers, military interest, this time in antimalarial drugs, also encouraged progress in spectrophotofluorimetry. Brodie and Sidney Udenfriend (1918–2001) used a simple Coleman filter fluorimeter to quantify quinacrine in plasma, but it was clear that if the excitation wavelength could be varied many more molecules could be analyzed using this technique. Robert L. Bowman (1916–1995) developed the first double monochromator spectrophotofluorimeter, which, with the collaboration of the America Instrument Company, was exhibited at the 1956 Pittsburgh Analytical Instrument Conference as the AMINCO-Bowman SPF (Udenfriend, 1995).

4.2 Colour tests

Many drugs and other poisons give characteristic colours with appropriate reagents if present in sufficient amounts, and in the absence of interfering compounds. A few of these tests are, for practical purposes, specific, but usually compounds containing similar functional groups will also react, and thus interference from other poisons, metabolites or contaminants is to be expected. Further complications are that colour description is very subjective, even in people with normal colour vision, while the colours produced usually vary in intensity or hue with concentration, and may be unstable. Recording the results of the tests is impossible without photography (Box 4.1).

Box 4.1 Colour tests

- Very quick/cheap: add reagent(s) and observe colour
- Usually homogenous assay – apart from some immunoassays most other toxicology tests require some form of sample pretreatment
- Mainly useful for urine/gastric contents/scene residues
- Usually carried out in clear glass test tubes, but a white, glazed tile is better (uniform background; uses less reagent)
- Must always analyze a reagent blank and positive control with samples
- Subjective – people vary in how they perceive/describe colours; also colour may vary in intensity
- Enormous range of tests available, but most have poor selectivity

The reagents for these homogenous assays typically contain strong acids or alkalis or use potentially hazardous organic chemicals. Appropriate safety precautions must be observed. Many tests can be performed satisfactorily in clear glass test tubes. However, the use of a 'spotting' tile (a white glazed porcelain tile with a number of shallow depressions or wells in its surface) gives a uniform background against which to assess any colours produced and also minimizes the volumes of reagents and sample used. Colour tests feature prominently in the IPCS Basic Analytical Toxicology manual (Flanagan *et al.*, 1995) where common problems and sources of interference in particular tests are emphasized. The colours given by some commonly used tests are also illustrated. Badcock (2000) and Jeffery (2004) have listed some further tests.

When performing colour tests it is always important to analyze concurrently with the test sample:

1. A 'reagent blank', that is an appropriate sample, that is known *not* to contain the compound(s) of interest. If the test is to be performed on urine then 'blank' (analyte-free) urine should be used, otherwise deionized water is generally adequate.
2. A known positive sample at an appropriate concentration. If the test is to be performed on urine, then ideally urine from a patient or volunteer known to have taken the compound in question should be used. However, this is not always practicable and then 'spiked' urine ('blank' urine to which has been added a known amount of the compound under analysis) should be used.

Colour tests are useful in that a minimum of equipment and expertise is required. However, the reagent has to be available and stable. Sensitivity is limited and the tests are usually applicable only to urine and/or other samples containing relatively large amounts of poison, such as stomach contents. It is possible to extract the poison and apply the colour test to the residue, though this is rarely done for biological fluids. False negatives are a risk even when a test is used for its intended purpose in appropriate samples. Positive results with poisons such as paracetamol or paraquat serve to indicate the need for a quantitative measurement in plasma. Other tests such as the Forrest test for imipramine-type compounds and the FPN test for phenothiazines (Table 4.1) are less useful and only indicate the need for confirmatory chromatographic analysis. These same reagents can be utilized as location agents following TLC and some have been developed for use in quantitative measurements.

Sometimes the analytical toxicology laboratory will be asked to identify pharmaceutical dosage forms (tablets, etc.) or plant material either intact or in stomach contents. The World Health Organization (WHO) (1986, 1991, 1998) has produced a series of manuals giving details of simple colour tests to confirm the identity of pharmaceutical preparations and more recently herbal medicines. These tests are aimed at confirming the identity of formulations, many of which have low intrinsic toxicity, prior to clinical use and a range of reagents is needed in order to perform the tests.

4.3 UV/visible spectrophotometry

When a compound is irradiated with electromagnetic radiation of an appropriate wavelength it will absorb energy. This absorbed energy may be emitted as radiation at a less energetic (longer) wavelength (fluorescence or phosphorescence), dissipated as heat, or give rise to a photochemical reaction. Gamma- and X-rays occupy the short wavelength (high energy) region of the spectrum. Ultraviolet (UV), visible, infrared and microwave radiations come next and finally, radio waves.

Table 4.1 Some commonly used colour tests [see Flanagan *et al.* (1995) for further details]

Test	Analyte(s)	Fluid	Limit of sensitivity (mg L^{-1})	Additional compounds detected
o-Cresol/ ammonia	Paracetamol	Urine	1[a]	Aniline,[b] benorylate, ethylenediamine,[c] nitrobenzene,[b] phenacetin
Diphenylamine	Oxidizing agents	Gastric contents[d]	10	Bromates, chlorates, iodates, nitrates, nitrites, peroxides
Dithionite	Paraquat (blue)	Urine[e]	1	—
	Diquat (yellow-green)		5	
Diphenylamine	Ethchlorvynol	Urine	1	—
Forrest	Imipramine	Urine	25	Clomipramine, desipramine, trimipramine
FPN[f]	Phenothiazines	Urine	25[g]	—
Fujiwara	Trichloro-compounds	Urine	1[h]	Chloral hydrate, chloroform, dichloralphenazone, trichloroethylene, triclofos
Trinder's[i]	Salicylates	Plasma Urine Gastric contents[d]	10	Aloxiprin,[b] aminosalicylate, aspirin,[b] benorylate,[b] methylsalicylate,[b] salicylamide[b]

[a] As *p*-aminophenol; [b] After metabolism or hydrolysis; [c] From aminophylline, for example; [d] Includes vomit, stomach wash-out, scene residues and similar samples; [e] Can be used on plasma in severe cases; [f] Ferric chloride/perchloric acid/nitric acid; [g] As chlorpromazine; [h] As trichloroacetate; [i] Ketones interfere

The absorption of the different types of radiation produces different effects. UV and visible light excite electrons from their ground state to higher energy (excited) states. Wavelengths (λ) of maximum absorbance are denoted λ_{max}. Infrared radiation (IR) induces molecular (bond) vibrations, whilst microwaves are used to induce electron spin transformations in electron spin resonance spectroscopy (ESR). Nuclear magnetic resonance (NMR) spectroscopy uses wavelengths between those of radio and television waves to detect nuclear electron spin flips.

4.3.1 *The Beer–Lambert law*

In UV/visible spectrometry, the analyte absorbs some of the incident light energy as a result of electrons in molecules being excited to higher energy levels.

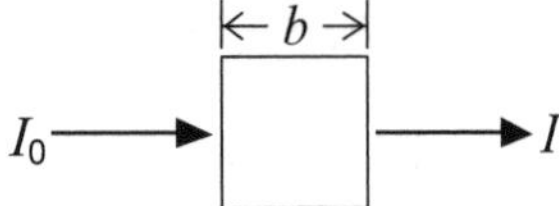

If the incident light intensity is I_0 and the transmitted light energy is I, then transmission T is:

$$T = \frac{I}{I_0} \tag{4.1}$$

The law that governs the relationship between the intensity of light entering and leaving a cell is the Beer–Lambert Law. This states that, for a solution of an absorbing solute in a transparent solvent, the fraction of the incident light absorbed is proportional to the number of solute molecules in the light path (Beer's Law) and the path length (Lambert's Law):

$$\text{Absorbance} = \log\left(\frac{I_0}{I}\right) = \varepsilon cb \tag{4.2}$$

where I_0 is the incident light intensity, I is the transmitted light intensity, c the solute concentration (mol L^{-1}), ε is the molar absorptivity, previously known as the extinction coefficient (L cm^{-1} mol^{-1}), and b is the path length (cm).

The molar absorptivity is a fundamental property of the analyte, but it also depends on temperature, wavelength and solvent. Absorbance (A) is linearly related to both solute concentration and path length for dilute solutions only. In older textbooks it was known as optical density (OD) or extinction (E), but these terms are now obsolete. The specific absorbance ($A_{1\,\%,1\,\text{cm}}$) is the absorbance of a 1 % (w/v) solution of the solute in a 1 cm path length cell, and is usually written in the shortened form A_1^1. There are a number of reasons why deviations from the Beer–Lambert Law may become apparent (Box 4.2).

Box 4.2 Beer–Lambert Law: Factors causing non-linearity

- Analyte concentration
  - At high concentrations (>0.01 mol L^{-1}) electrostatic interactions between species reduces absorbance
  - Refractive index changes at high analyte concentration
- Interactions/decomposition
  - Analyte dissociates/associates or reacts with solvent
  - Particles in light path
  - Phosphorescence or fluorescence in the sample
  - Oxygen absorption becomes limiting at wavelengths below about 205 nm
- Instrumental
  - Polychromatic (not monochromatic) incident light. The incident slit is of finite width, and thus the incident light is of more than one wavelength. Because ε varies with wavelength, this leads to deviations in Beer's Law. For a given slit width the variation in ε will be greater on the slope of an absorption peak than at a maximum, hence the effect is reduced by performing the measurement at λ_{max}
  - The effect of stray light from reflected radiation in the monochromator reaching the exit slit becomes more apparent at higher absorbance

4.3.2 Instrumentation

Spectrophotometers can be divided into two basic types: single beam or double beam. The simplest of the single beam designs (Figure 4.1) are often referred to as colorimeters when their operation is limited to the visible region of the electromagnetic spectrum. The light source in such instruments is a tungsten filament lamp and the analytical wavelength is selected using a suitably coloured filter. The light beam passes through the sample cuvette held in a light-tight sample chamber and the transmitted light intensity is measured with a photocell. After amplification the signal is monitored with a meter. Early colorimeters were calibrated in linear transmission, absorbance being superimposed as a logarithmic scale on the meter.

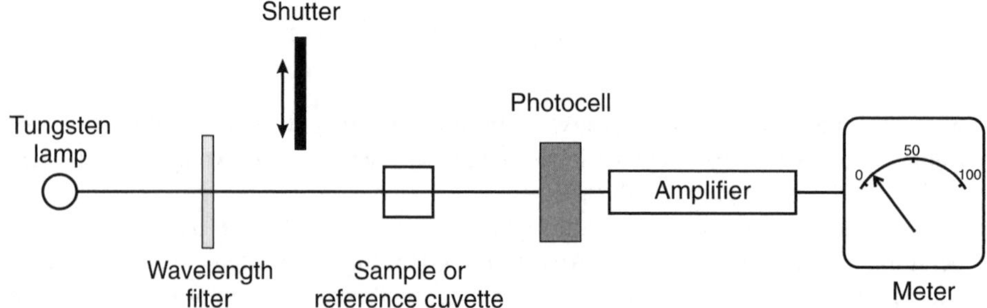

Figure 4.1 The basic components of a single beam colorimeter or UV spectrophotometer.

To measure the absorbance of a sample solution:

- Select the wavelength
- Close the shutter and adjust the meter to zero transmission
- Place the reference cuvette in the sample chamber and adjust the meter to 100 % transmission (zero absorbance)
- Replace the reference cuvette with the sample cuvette and record the absorbance.

The introduction of hydrogen (later deuterium) lamps and monochromators (quartz prisms or diffraction gratings) extended the useable wavelength range into the UV region. Tungsten–halogen filament lamps, in which the presence of a small amount of iodine assists in redepositing vaporized tungsten onto the filament thus prolonging lamp life, are also useable into the far UV.

Single-beam spectrophotometers have the advantage of high energy and good signal-to-noise ratios, and are suitable for quantitative measurements at a single wavelength. Spectra can be obtained manually, but at each wavelength the zero and 100 % transmission must be set as detailed above, making the process tedious in the extreme – the relatively complex absorption spectrum of benzene would take a skilled operator an entire day's concentrated effort to produce. Double-beam spectrophotometers overcome this problem by automatically compensating for differences in incident light intensities during the scan.

In optical-null instruments the light beam is split into sample and reference beams by a rotating semicircular mirror (often referred to as the beam chopper or simply 'chopper'). The beams, which consist of pulses of light that are out of phase, are directed through the sample or reference compartments and then recombined before being directed to a single photomultiplier tube (Figure 4.2).

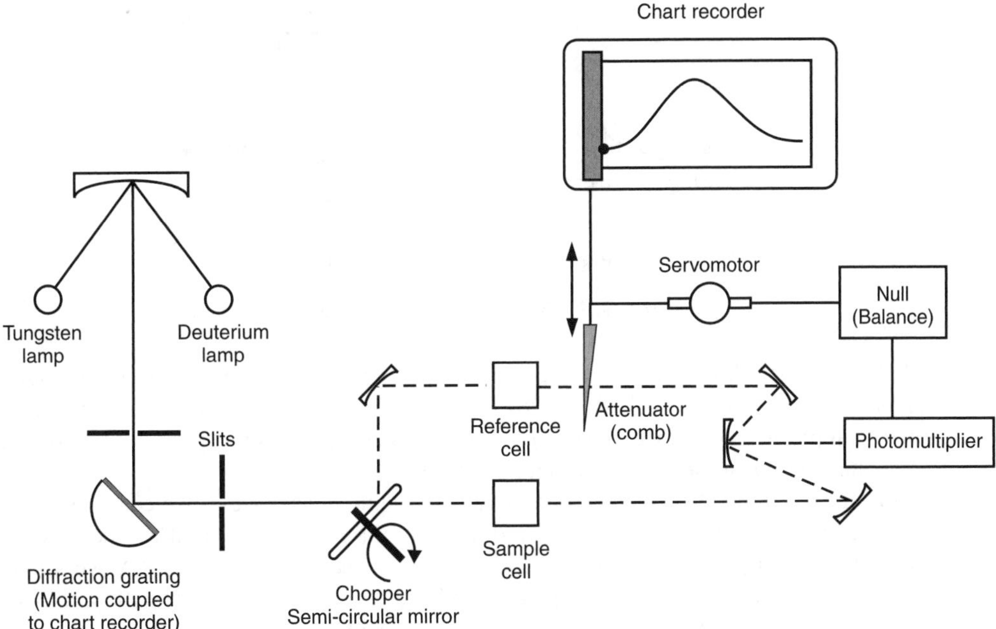

Figure 4.2 Diagram of a typical optical-null double-beam spectrophotometer.

When the absorbances in the sample and reference beams are identical, then the photomultiplier tube 'sees' a continuous beam of light. However, if the sample has a higher absorbance than the reference sample, the beam is no longer continuous, but stepped. The optical-null circuitry drives a servomotor that moves an attenuator (comb) into the reference beam until the signals are balanced and the movement of the attenuator is recorded as the spectrum. Wavelength accuracy is maintained by linking the recorder movement with the monochromator movement. At high absorbance ($A = 2$), transmission is 1 % and little, if any, light reaches the detector, the recorder pen becomes 'dead', and no readings can be made. The UV cut-off values of some common solvents are shown in Table 3.5.

Modern double-beam instruments use two photomultiplier tubes and ratio the signals. However, background electrical noise can be limiting at high absorbance values, so like the optical-null instruments, no measurements can be made under these circumstances. Modern single-beam instruments use microprocessors, so that background scans can be stored and subtracted from the sample data.

In a diode array detector (DAD, Figure 4.3) the detector is a linear array, typically 3–7 cm long, of over 1000 diodes (in modern instruments the wavelength resolution is of the order of 1 nm per diode, wavelength accuracy ± 1 nm and sensitivity better than 10^{-4} au). Light from the source passes through the sample to a diffraction grating, which disperses the beam onto the array; each diode records the absorbance at a defined wavelength. Thus, a spectrum is recorded almost instantaneously, which makes the diode array detector particularly useful for HPLC, as several spectra can be scanned across an eluting peak.

With all types of spectrophotometer it is important to ensure that the monochromator is aligned correctly. This can be checked by observing the absorbance maxima (λ_{max}) of a known reference

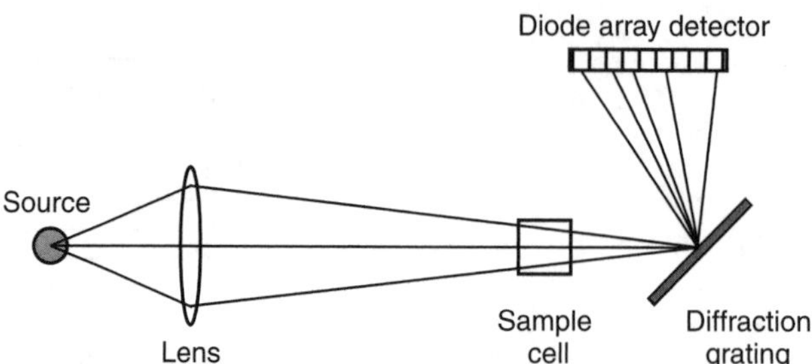

Figure 4.3 Diagram of a linear diode array detector.

solution or material. A holmium oxide glass filter, for example, has major peaks at a number of important wavelengths (241.5, 279.4, 287.5, 287.5, 333.7, 360.9, 418.4, 453.2, 536.2 and 637.5 nm). A simple method of checking photometric accuracy is to measure the absorbance of an acidic potassium dichromate solution (Flanagan *et al.*, 1995).

It is important that the cuvettes used in the spectrophotometer are of the correct specification and that they are scrupulously clean. Traditionally, glass cells were used for measurements in the visible region (>400 nm), and fused silica or quartz cells were used for UV work (190–400 nm). However, disposable plastic cells are available for working from 220–900 nm. Normally, cells of 1 cm path length are employed, but 2 or 4 cm path length cells can sometimes be used to enhance sensitivity. Cells for spectrophotometry usually have two optical faces and two ground or ribbed faces, and should be handled by the ground or ribbed faces only. The optical faces may be wiped with a soft cloth or photographic lens tissue if necessary. Samples should be introduced with the aid of a suitable pipette, taking care not to scratch the inside surface of the cell. Quartz cells should be marked, for example on a ground face, to ensure that they are always placed in the spectrophotometer in the same orientation. After use such cells should be cleaned carefully and allowed to dry.

Double-beam spectrophotometers have the advantage that 'background' absorbance from reagents, solvents and so on, can be allowed for by including a 'blank' (analyte-free) extract in the reference position. Normally an extract of 'blank' plasma or serum will be used to provide the solution used in the reference cell, but purified water may be used in certain assays. A matched pair of cells is necessary for use in double-beam instruments. When filled with identical solutions, the absorbance readings should be within ± 1 %. Pairs of matched cells may be purchased and should be stored together in a dust-free environment.

4.3.2.1 Derivative spectrophotometry

Rather than plotting absorbance against wavelength, a plot of the first (or higher derivative) versus wavelength can reveal spectral detail that would otherwise be lost:

$$\frac{\mathrm{d}^n A}{\mathrm{d}\lambda^n} = \frac{\mathrm{d}^n \varepsilon}{\mathrm{d}\lambda^n}.c.b \qquad (4.3)$$

where n is the order of the derivative [the other symbols are as defined for Equation (4.2)]. Because sensitivity is dependent on the rate of change of ε ($\mathrm{d}\varepsilon/\mathrm{d}\lambda$) rather than ε itself, the technique is suited to analytes with sharp absorbance peaks. Consequently when the analyte signal is difficult to quantify because of interferences, as is often the case with extracts of biological samples, derivative spectrophotometry may be advantageous, particularly when the width of the overlapping peak is greater than twice the width of the analyte peak. Generally, the peak-to-peak heights are proportional to the concentration of the analyte. Note, however, that spectral information tends to degrade with increasing order of derivative, particularly when the spectrum is poorly defined (noisy) to start with, and there is also an obvious increase in the cost of more complex instruments.

Derivative spectra can be obtained using electronic circuits to convert the analogue absorbance (zero-order) signal to the derivative of the appropriate order, or digital signals can be transformed mathematically. A third method for first-order spectra is to use wavelength modulation. Scanning the spectrum with alternate beams from two monochromators, with wavelengths set 2–3 nm apart, will produce something very close to the derivative spectrum (ΔA is approximately the slope at the particular wavelength).

The use of derivative spectrophotometry has been shown to improve the specificity and sensitivity of the dithionite method for paraquat (Figure 4.4), but, in the main, derivative spectrophotometry has not been adopted widely, possibly because of the rapid growth of HPLC.

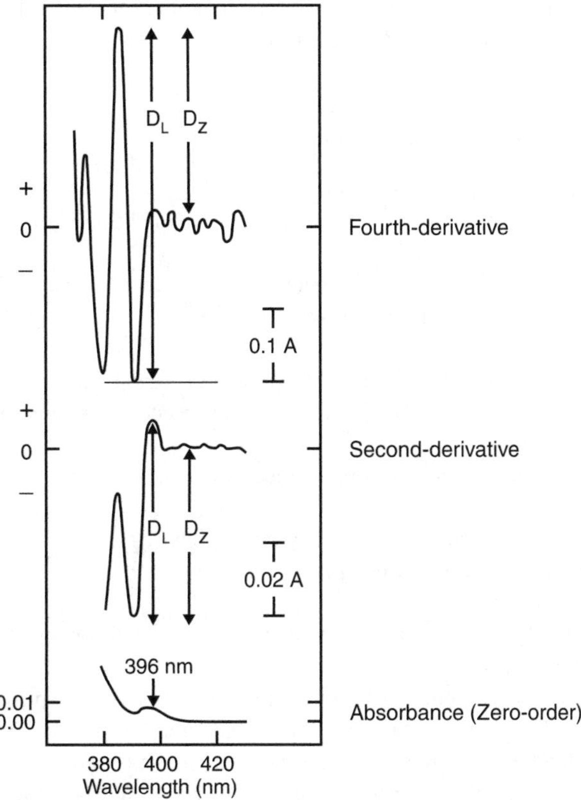

Figure 4.4 Zero-order, second- and fourth-derivative spectra of paraquat after dithionite reduction. Redrawn from Fell, A.F., Jarvie, D.R. and Stewart, M.J. (1981) Analysis for paraquat by second- and fourth-derivative spectroscopy. Clin Chem, **27**, 286–92, by permission of the American Association for Clinical Chemistry.

4.3.3 Spectrophotometric assays

Spectrophotometry may be used in both qualitative and quantitative work. However, one of the major problems in attempting to apply spectrophotometry *per se* to complex matrices is that it is not an inherently selective technique. Most endogenous compounds absorb in the UV, if not in the visible part of the spectrum, and even those with low molar absorptivities may be present in such excess compared to the concentration of the analyte that selectivity and hence sensitivity is a problem.

While spectrophotometric assays generally lack selectivity, some information as to the purity of a sample extract may often be obtained by examining the UV absorption spectrum using a scanning spectrophotometer; manual scanning can be performed on simpler instruments. UV spectra of extracts of urine, stomach contents or scene residues can give useful qualitative information, and can be used as an adjunct to more specific and sensitive drug screening procedures, although even here if a HPLC-DAD system is in use in a laboratory it may well be easier to use such a system, at least in the first instance. UV absorption spectra of many compounds of interest are given in Moffat *et al.* (2004), but care is needed to ensure that the pH/solvent combination employed is the same as that used to produce the reference spectrum. Wavelength scanning at different pH values can be useful in identifying the constituents of tablets. Some aspects of the use of spectrophotometry in analytical toxicology are summarized in Box 4.3.

Box 4.3 Visible and UV Spectrophotometry

Visible spectrophotometry (colorimetry)

- Visible spectrophotometry the first really reliable quantitative method used for drug/metabolite analysis in blood
- Aim is to perform a chromogenic reaction directly in a sample with minimum treatment
  - Can use protein precipitation or microdiffusion
- Problems:
  - Relatively few analytes undergo convenient chromogenic reactions
  - Relatively poor selectivity/sensitivity (interference from metabolites/other drugs)

UV spectrophotometry

- Double-beam spectrophotometers or linear diode array instruments give the most reliable results
- Can perform UV spectrometry directly on sample extracts
  - Care: use silica or appropriate plastic cuvettes
  - Generally poor selectivity
- Use of double-beam scanning instruments (210–400 nm) or linear diode arrays can aid in analyte identification
  - Many poisons have either a poorly defined UV absorption spectrum or do not absorb in the UV, although compounds that have characteristic spectra can be readily identified
  - Always likelihood of interference if more than one drug/metabolite present
  - Complex spectra may be resolved using derivative spectrophotometry
- UV spectrophotometry finds greatest application as detector in HPLC, monitoring reaction rate in immunoassays or in NAD/NADH- or NADP/NADPH-linked enzyme-based assays

It may be possible with techniques such as protein precipitation, solvent extraction, possibly with back-extraction at an appropriate pH, or microdiffusion, to develop a spectrophotometric assay for a particular analyte in a particular biological fluid. Concentration of the analyte into a smaller volume during solvent extraction, for example, may give adequate sensitivity provided there are no interfering contaminants. This being said, many spectrophotometric methods suffer from interference from metabolites to a greater or lesser extent. Other drugs are further potential sources of interference. These may have been prescribed or obtained over-the-counter, and may be chemically unrelated to the analyte, although chemically related compounds and any metabolites are more likely to interfere. Compounds from the same pharmacological class as the analyte may be co-prescribed and may be chemically similar to the analyte, hence metabolism, extraction, derivatization and absorbance spectra are all likely to be similar.

Historically, UV methods for measuring 5,5′-disubstituted barbiturates in plasma or urine were based on the fact that these compounds show markedly different absorption at different pH values (Figure 4.5). Measurements at pH 10 and pH 14, for example, were advocated in order to provide selectivity over salicylates and sulfonamides (Broughton, 1956), and differentiation in long- and short-acting barbiturates was described by repeating the measurements after alkaline hydrolysis under controlled conditions. However, even then numerous interferences and anomalies became apparent (Broughton, 1963; Tompsett, 1969).

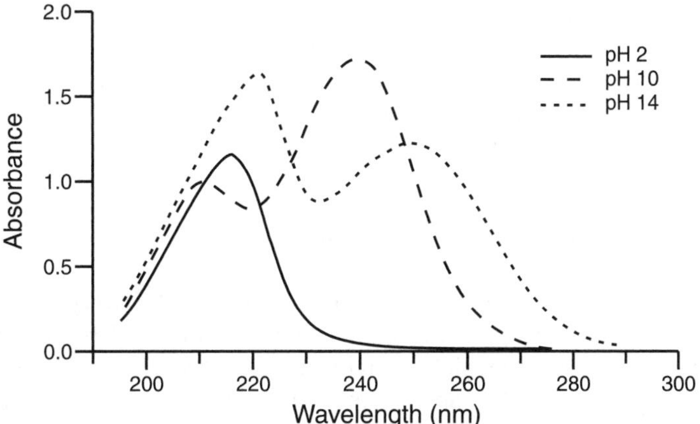

Figure 4.5 UV Spectrum of barbital (50 mg L^{-1}) in aqueous buffers.

Although spectrophotometric methods often suffer from a lack of selectivity when metabolites are present, this is not always the case. Of three spectrophotometric methods for paracetamol that were compared with HPLC, two methods that included a hydrolysis step (paracetamol is mainly metabolized by glucuronide and sulfate conjugation) overestimated the true concentration of parent drug by up to sevenfold. However, the third method, based on the relatively specific nitration of paracetamol, gave values that were in good agreement with HPLC (Stewart *et al.*, 1979). Thus, despite the cautions listed above, quantitative spectrophotometry can be used successfully when an analyte is present at relatively high concentrations, when a characteristic colour reaction can be employed or when the analyte can be isolated in relatively pure form prior to the analysis, as illustrated in the further examples below.

4.3.3.1 Salicylates in plasma or urine

The principle of this test (Box 4.4) is the colour which often develops when phenols are treated with a ferric iron [iron(III)] salt. Different phenols give a range of colours, salicylate producing a purple-red colour. Plasma proteins are precipitated with mercuric chloride [mercury(II) chloride] so that the colour in the supernatant can be measured directly. The mercuric chloride may be omitted if only urine samples are to be assayed. Any interference from colour in the sample is reduced by the large dilution incorporated in the procedure.

Box 4.4 Measurement of salicylates in plasma or urine

Reagents

- Trinder's reagent [mercuric chloride (10 g) + 1 mol L^{-1} hydrochloric acid (12 mL) + iron(III) nitrate nonahydrate [$Fe(NO_3)_3{\cdot}9H_20$] (4 g) in 200 mL deionized water]
- Calibration solutions containing salicylic acid (0, 200, 400 and 800 mg L^{-1}) in deionized water. Store at 4 °C

Method

- Add Trinder's Reagent (5 mL) to plasma/serum or calibration solution (1 mL)
- Vortex-mix (30 s) and centrifuge (3000 rpm, 10 min)
- Measure the absorbance of the supernatant layer (double-beam spectrophotometer, 540 nm) against plasma 'blank'
- Construct graph of absorbance versus salicylate concentration in the calibrators, and calculate the plasma salicylate concentration
- Limit of quantification: 50 mg L^{-1} salicylate

4.3.3.2 Carboxyhaemoglobin (COHb) in whole blood

The half-life of COHb in blood is usually very short, especially if oxygen is given, so COHb assays are rarely indicated clinically. However, spectrophotometric COHb assay is simple if potentially interfering oxygenated haemoglobin is removed (Box 4.5). Haemoglobin is normally present in blood as deoxygenated (HHb) and oxygenated (O_2Hb) forms together with a small amount of

Box 4.5 Measurement of carboxyhaemoglobin

Method

- Add 25 mL 0.1 % (v/v) ammonium hydroxide to 0.2 mL blood
  - Portion A – saturate with pure CO
  - Portion B – saturate with oxygen to dissociate COHb
  - Portion C – test sample
- Add sodium dithionite, mix and read absorbance at 500–600 nm versus reagent blank
- Measure absorbance at 540 (COHb max), and 579 nm (isobestic point) and calculate the ratio A_{540}/A_{579} for A, B and C
- Calculate % COHb saturation

methaemoglobin (MetHb), that is to say oxidized haemoglobin, in which the iron is present as iron(III) (Fe^{3+}). In the assay, oxyhaemoglobin, which has a spectrum similar to that of COHb and MetHb are reduced to HHb with sodium dithionite. COHb is unaffected and hence retains its characteristic spectrum, which can be measured spectrophotometrically. Absorbance readings are made at 579 nm (the isobestic point, i.e. the wavelength at which both HHb and COHb have the same absorbance) and 540 nm, the wavelength where the difference in absorbance, $A_{\mathrm{COHb}} - A_{\mathrm{HHb}}$, is greatest (Figure 4.6). Note how the 100 % and 0 % COHb values are obtained by treating portions of the sample with either carbon monoxide or oxygen, respectively. Commercial systems such as the IL 682 Co-Oximeter, http://www.gmi-inc.com/Products/IL682Coox.htm (accessed 20 December 2005) are available and aim to compensate for the presence of other haemoglobin species such as sulfhaemoglobin in blood obtained postmortem (Mayes, 1993; Widdop, 2002).

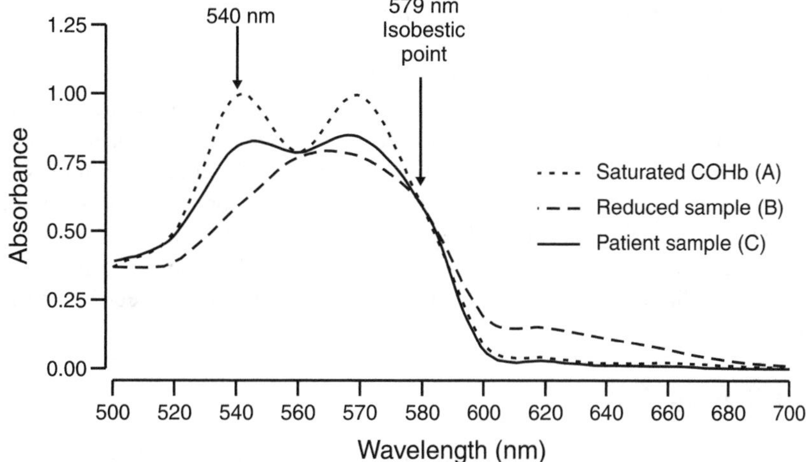

Figure 4.6 Measurement of carboxyhaemoglobin (note the wavelength expansion between 560–580 nm).

4.3.3.3 Cyanide in whole blood by microdiffusion

Cyanide ion is present in blood mainly in erythrocytes, hence whole blood is used in cyanide assay, but care has to be taken to avoid interference from thiocyanate present largely in plasma. Cyanide ion is non-volatile at room temperature, but is converted into volatile hydrogen cyanide under acidic conditions; thiocyanate is unaffected. If acidification of a biological sample such as blood containing cyanide is performed in a Conway microdiffusion cell (Section 3.2.3), the hydrogen cyanide that is released can be trapped in alkali as cyanide ion in the centre well. Reaction with 3-nitrobenzaldehyde/1,2-dinitrobenzene gives a coloured product that can be removed from the cell and the intensity of the colour measured spectrophotometrically, using the mixture from the 'blank' analysis in the reference position (Box 4.6). Assay calibration is by analysis of standard sodium or potassium cyanide solutions analyzed in separate cells (Mayes, 1993).

Box 4.6 Cyanide by microdiffusion

Method

- Place 3.6 mol L^{-1} sulfuric acid (1 mL) in Conway outer well
- Place sample, blank, or standard (0.5 mL) at opposite side of outer well
- Place 0.5 mL of 3-nitrobenzaldehyde solution + 0.5 mL of 1,2-dinitrobenzene solution (both 0.05 mol L^{-1} in 2-methoxyethanol) + 0.1 mL aqueous sodium hydroxide (0.5 mol L^{-1}) in the centre well
- Seal the cell, gently mix the contents of the outer well, and allow to stand (20 min, room temperature)
- Add 1 mL aqueous methanol (1 + 1) to the centre well, transfer contents to spectrophotometer cells and measure absorbance (560 nm)

4.3.3.4 Colorimetric measurement of sulfonamides

This assay is based on the Bratton–Marshall reaction undergone by primary aromatic amines. After protein precipitation with trichloroacetic acid (TCA), sulfadimidine (sulfamethazine) is diazotized using sodium nitrite and coupled with *N*-naphthylethylenediamine (NED). Excess nitrite ion has to be removed by reaction with ammonium sulfamate before NED is added. *N*-acetylsulfadimidine, which does not react with nitrite, is measured as sulfadimidine after acid hydrolysis. This method can be used to assess acetylator status (Section 17.2.1) and is convenient and inexpensive when large numbers of samples are to be assayed; it is not necessary to use *N*-acetylsulfadimidine calibration standards. HPLC offers greater precision, however (Whelpton *et al.*, 1981).

4.4 Luminescence

The term luminescence is applied to the phenomenon whereby light is emitted when 'excited' molecules return to a lower energy, ground state. The three major subdivisions of luminescence are fluorescence, phosphorescence and chemiluminescence. With fluorescence and phosphorescence, molecules are excited by absorption either of incident electromagnetic radiation (light), which is then re-emitted (usually as light of a longer wavelength), or by heat, such as via a flame, as in flame emission photometry (Chapter 11), respectively. Chemiluminescence, as its name implies, results when light is emitted from molecules excited as a result of a chemical reaction at ambient or near ambient temperatures without the introduction or generation of heat.

4.4.1 Fluorescence and phosphorescence

When molecules absorb light, the energy from the light is transferred to the electrons in the molecules so that the molecules are promoted from the lower energy, ground state, to a higher energy, excited state. The molecules usually remain in the excited state for a very short period of time before they return to the ground state. As the molecules return to the lower energy state they must lose energy. Most molecules lose this energy in the form of molecular vibration by dissipation as heat, but some molecules lose some of this energy via the emission of light. The emitted light is usually of a longer (less energetic) wavelength than that of the absorbed light. This is because excited molecules lose some of the absorbed energy by radiationless transitions, such as collisions with other molecules, or the electrons return to higher energy levels in the ground state, and so the energy of the emitted light is correspondingly less.

Fluorescence is the term used when light is emitted almost instantaneously, usually within 10^{-9} or 10^{-12} s of the energy being absorbed. Resonance fluorescence describes the situation when the frequency of the emitted light is the same as the absorbed light.

Phosphorescence occurs when an excited electron in the singlet state (paired electrons having opposite spin) crosses to a triplet state (paired electrons having the same spin). These transitions are known as 'forbidden transitions' because their probability is so low. The electron must return to the excited singlet state before it can return to the singlet ground state and emits light as it does so, and thus phosphorescence is much longer lived than fluorescence, usually of the order of several seconds. Phosphorescence assays are not widely used in analytical toxicology, one problem being the need on occasions to cool the samples with liquid nitrogen (77 K) (Hadley *et al.*, 1974), although here has been recent interest in kinetic phosphorescence assays for uranium in biological samples (Ejnik *et al.*, 2000).

4.4.1.1 Intensity of fluorescence and quantum yield

The fraction of excited molecules that emit light is the quantum yield or quantum efficiency (ϕ) of fluorescence. Highly fluorescent molecules will have quantum yields approaching 1. For dilute solutions, the intensity of the fluorescence, I_f, is a function of ϕ, the absorbance, and the intensity of the incident light, I_0. The absorbance is given by Beer's Law so:

$$I_f = 2.3 I_0(\varepsilon c b)\phi \tag{4.4}$$

From Equation (4.4), it is clear that the size of the signal is proportional to the intensity of the incident light as well as the concentration of the analyte, *c*, and as a consequence, the fluorescence, unlike absorbance, is 'relative' and not absolute. The concentration of an analyte cannot be calculated from the fluorescent intensity in the way that it can be calculated from absorbance and molar absorptivity in spectrophotometry. Analyte sample concentrations must always be obtained by comparison with the analyte concentrations of standard solutions. Fluorescence is reported as 'relative intensity' or 'fluorescence (arbitrary units)'.

Equation (4.4) only holds true for dilute solutions. Quenching, caused by absorption of the incident or emitted light by other compounds, or by the analyte itself (internal absorption), reduces fluorescent efficiency, as does quenching caused by molecular interactions (collisions) leading to non-radiative relaxation of excited molecules to the ground state. A very important consequence of quenching is that as the concentration of analyte is increased, the intensity of the fluorescence will reach a maximum and then decline. Thus, it is possible for two samples, one dilute and one concentrated, to show the same degree of fluorescence. If a high degree of quenching is suspected, the sample should be diluted and re-assayed.

Fluorescence is usually reduced at increased temperatures (increased number of collisions) and is often pH dependent. The ionized and non-ionized forms of a weak electrolyte may both be fluorescent, but it is rare for both species to have the same quantum efficiency. Barbiturates, for example, fluoresce in 0.1 mol L^{-1} sodium hydroxide. At this pH these compounds are ionized and have strong absorptions at approximately 255 nm (Figure 4.5). Phenols such as catechol, however, fluoresce when unionized, that is under acidic conditions (Miles and Schenk, 1973).

4.4.1.2 Instrumentation

Although in fluorescence the emitted light is non-directional, it is measured at right angles to the excitation beam to minimize the amount of incident light reaching the detector (Figure 4.7). Some instruments have both excitation and emission filters, or the instrument may have a combination of a diffraction grating and a filter, or, as in the more expensive instruments, there may be two diffraction gratings.

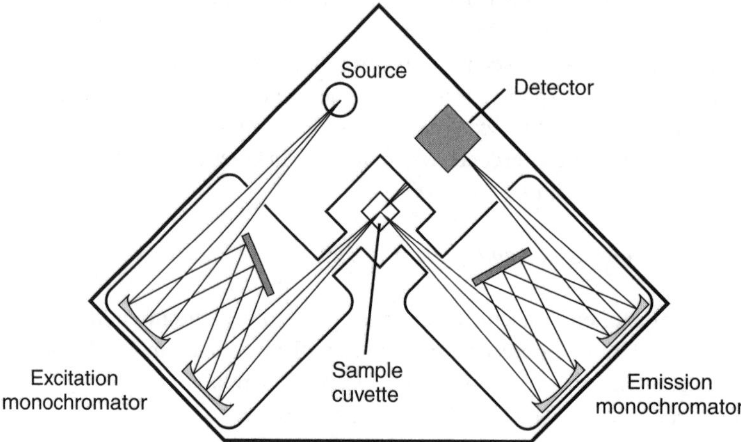

Figure 4.7 Diagram of a double-grating spectrophotofluorimeter (based on the AMINCO-Bowman design).

Xenon lamps are popular as they have high energy output between 350–1000 nm and so are useful general-purpose light sources. Xenon lamps can be used down to 200 nm, but the light intensity is about an order of magnitude lower at this wavelength and for some applications, a deuterium lamp, or even a mercury lamp when the intense line at 254 nm can be used, may be better. Halogen lamps may be used for excitation in the visible region. Because the intensity of the fluorescence is dependent on the intensity of the incident light [Equation (4.4)] it is important in any report to state the type of lamp used. High intensity xenon lamps may be pulsed to reduce the risk of photodecomposition of an analyte.

Sample cuvettes may be square or circular in cross section. The square ones have four optical faces and so these should be handled carefully by the top and not where the light beams pass through them. Although sample holders are often designed to take 1 cm square cuvettes, the area over which the measurements are made is defined by the width of the emission slits, and smaller cuvettes may be used without any loss of fluorescent signal; indeed smaller path length cells have the advantage that there will be less loss of light due to absorption.

Because the incident light intensity is generally much higher than the intensity of the emitted light, scattering of light from particles in the sample solution can seriously affect assay sensitivity. Thus, good analytical technique is important in high sensitivity fluorescence assays. Also, the more similar the excitation and emission wavelengths, the more likely it is that incident light will interfere with the emission signal. The use of dual diffraction gratings and narrow slit widths will reduce this problem.

An indication of the excitation wavelength for a particular analyte can be deduced from the UV spectrum. If the excitation wavelength is set to an absorption peak then the emission wavelength can be scanned until the emission maximum is found. With the emission monochromator set to its λ_{max}, the excitation monochromator can then be scanned to locate the exact excitation λ_{max}. The excitation and emission spectra will now be maximal and equal (Figure 4.8).

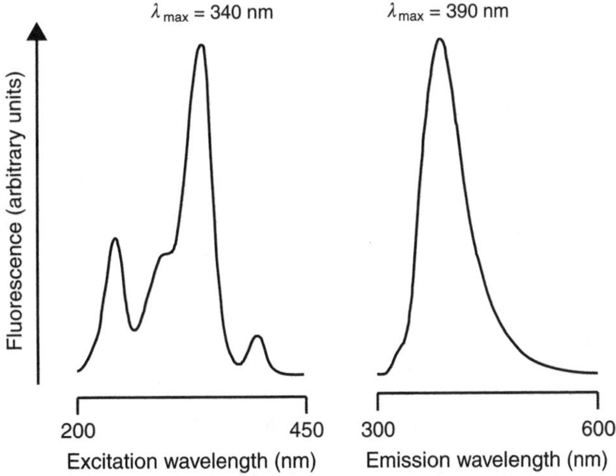

Figure 4.8 Excitation and emission spectra of quinine (1 mg L^{-1}) in deionized water.

4.4.1.3 Fluorescence assays

Fluorimetry gives enhanced sensitivity over spectrophotometric methods for those drugs that are naturally fluorescent. Generally sensitivity is increased by 1 or 2 orders of magnitude because: (i) only a limited number of drugs and other poisons possess natural fluorescence and (ii) both excitation and emission wavelengths can be varied. High quantum efficiencies are associated with compounds that have low energy $\pi^* \rightarrow \pi$ transitions – these include conjugated aromatic compounds (generally the more rigid the structure, the greater the fluorescence). The substitution of an alkyloxy moiety on an aromatic ring usually gives rise to fluorescent properties. Substitution of electron-withdrawing groups (carbonyl, carboxylate, nitro), however, usually reduces the intensity of any fluorescence. Drugs with good fluorophores include several antimalarials and tetracycline antibiotics, propranolol and alkaloids such as ergometrine, LSD and physostigmine (Figure 4.9).

If an analyte does not posses native fluorescence, derivatization to a fluorescent product may be possible if the analyte possesses a reactive (functional) group amenable to chemical reaction. A number of fluorigenic reagents are available, including dansyl chloride and fluorescamine. 1,2-Phthalic dicarboxaldehyde (*o*-phthaldialdhyde, OPA) is particularly useful for primary amines and has been used to enhance detectability in HPLC (Section 8.7.1).

As an example of a fluorescence assay, quinine and its stereoisomer quinidine exhibit strong natural fluorescence under neutral or acidic conditions. Both drugs are relatively lipophilic and thus solvent extraction can be used to separate the parent compounds from fluorescent polar metabolites, notably the 3-hydroxy metabolite, and this forms the basis of sensitive fluorescence

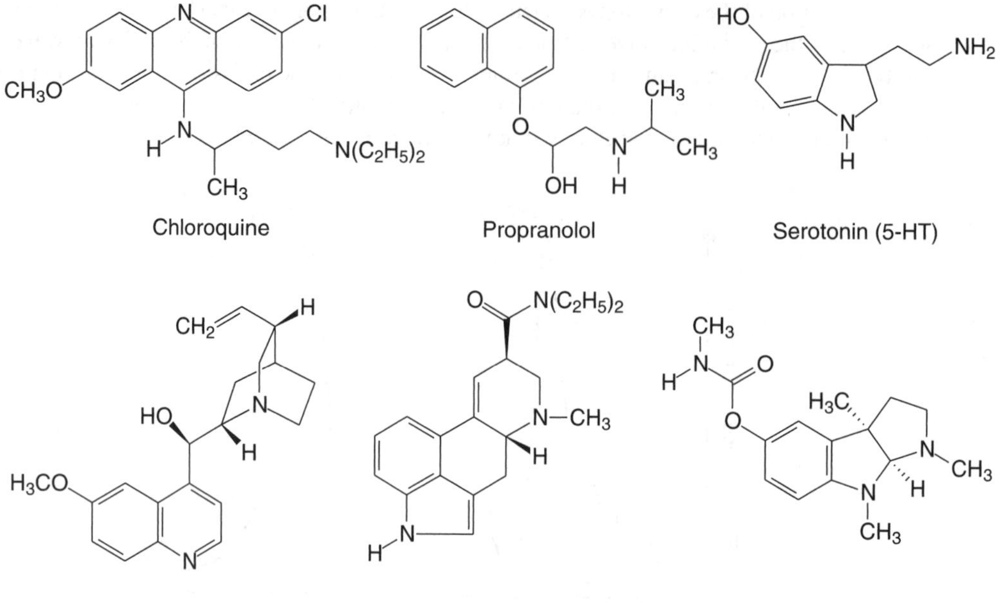

Chloroquine Propranolol Serotonin (5-HT)

Quinine Lysergic acid diethylamide Physostigmine

Figure 4.9 Some compounds that exhibit natural fluorescence.

assays such as that for plasma quinine (Box 4.7). However, commercial quinine and quinidine can contain up to 10 % (w/w) of the respective dihydro derivatives, which have similar extraction and fluorescence properties to the parent compounds, and thus even the combination of solvent extraction and spectrophotofluorimetry is not totally selective in this case. A chromatographic method such as HPLC with fluorescence detection is therefore needed to resolve the dihydro analogues of quinine and quinidine, and hence permit selective measurement.

Box 4.7 Fluorescence assay for plasma quinine (Hall *et al.*, 1973)

Method

- Add analyte-free plasma, sample or standard solution (0.5 mL) to a stoppered glass centrifuge tube, and add 0.1 mol L^{-1} sodium hydroxide solution (1 mL) and toluene (5 mL)
- Rotary mix (5 min) and centrifuge (3000 rpm, 10 min; bench centrifuge)
- Transfer 4 mL supernatant to a second glass tube and add 5 mL sulfuric acid (0.05 mol L^{-1})
- Rotary mix (5 min) and centrifuge (3000 rpm, 10 min; bench centrifuge)
- Discard the toluene layer (supernatant) and measure by spectrophotofluorimetry (excitation 350 nm, emission 450 nm)

4.4.2 *Chemiluminescence*

One of the most widely known examples of chemiluminescence is the light given out by fireflies. Firefly luciferase, an enzyme from the North American firefly, *Photinus pyralis,* which is commonly used in laboratory work, catalyzes the oxidation of D-luciferin with hydrolysis of ATP and emission of light. When the quantity of ATP is limiting, the light emitted is proportional to the amount of ATP present and this forms the basis of chemiluminescent ATP assays. Commonly, it is oxidation of a suitable species that produces an excited state, which decays to the ground state with emission of light. Although chemiluminescence assays are not as widely applicable as spectrophotometric ones, the increased sensitivity (10–100 times) over other luminescence methods and the simple instrumentation required, make them attractive when high sensitivity is needed. Such assays are being used to increase the sensitivities of immunoassays (Chapter 12). Chemiluminescence HPLC detectors are also available (Section 8.3.3).

Luminol (5-amino-2,3-dihydrophthalazine-1,4-dione) is often used in chemiluminescent assays as it can be oxidized by perborate, permanganate, hypochlorite and iodine, but the most useful reaction is that with hydrogen peroxide (Figure 4.10). When excited 3-aminophthalate decays it emits intense blue light (425 nm). Several enzyme-catalyzed reactions produce hydrogen peroxide and so luminol chemiluminescence can be used to monitor such reactions, including enzyme immunoassays (Chapter 12). The reaction can be catalyzed by metals, including iron in haemoglobin, and this forms the basis of using luminol/hydrogen peroxide spray to detect blood stains.

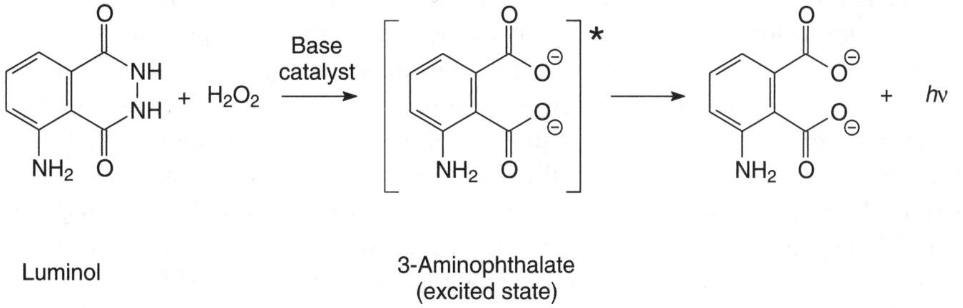

Luminol + H_2O_2 → (Base catalyst) 3-Aminophthalate (excited state) → 3-aminophthalate + $h\nu$

Figure 4.10 Chemiluminescent reaction of luminol with hydrogen peroxide.

The luminol system can be used to quantify those ions that catalyze the reaction (e.g. Fe^{2+} and Cu^{2+}) as the luminescence is proportional to the concentration of the ions. Enhancers such as *p*-iodophenol can increase the light intensity by as much as 2500 times and the light is emitted as a prolonged glow, facilitating measurements which may be made several minutes after the reaction has been initiated (Kricka *et al.*, 1987).

Chemiluminescent reactions can be direct, as illustrated by luminol, or indirect via an energy transfer system as exemplified by peroxyoxalate derivatives. Oxidation of peroxyoxalate produces excited intermediates, such as 1,2-dioxetanedione (Figure 4.11), that react with a fluorophore, which may be the analyte or a fluorescent derivative of the analyte.

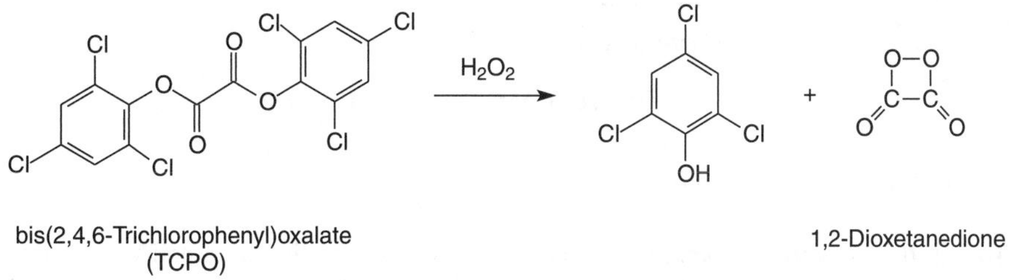

bis(2,4,6-Trichlorophenyl)oxalate (TCPO) → (H_2O_2) 2,4,6-trichlorophenol + 1,2-Dioxetanedione

Figure 4.11 Hydrogen peroxide oxidation of TCPO.

Electrochemiluminescence represents a further refinement that is being used increasingly, particularly with the tris(2,2′-bipyridyl)–ruthenium(III) system (Li *et al.*, 2003). Reduction gives an excited complex, which decays with the emission of light:

$$Ru(bpy)_3^{3+} + \text{reductant} \rightarrow \left[Ru(bpy)_3^{2+}\right]^* + h\nu$$

where bpy = 2,2′-bipyridine. Greenway and Dolman (1999) used electrochemically generated tris(2,2′-bipyridyl)–ruthenium(III) to measure some tricyclic antidepressants and phenothiazines. Tris(2,2′-bipyridyl)–ruthenium(II) and the analyte were electrochemically oxidized simultaneously, the product of the latter acting as the reductant in the chemiluminescence reaction.

Oxidation of some analytes may generate chemiluminescent products. Acidic potassium permanganate has been used to detect several drugs including morphine (Abbott *et al.*, 1986).

4.4.2.1 *Instrumentation*

A basic luminometer is very simple, consisting of a light-tight box in which to place the sample tube and a photomultiplier tube (Figure 4.12). There is no external light source and no need for filters or monochromators, although these may be included in more complex instruments. Refinements might include a mirror behind the sample-tube holder, an injection port so that reagents to start the reaction may be added, and a magnetic stirrer. Replacing the sample tube with a coil of PTFE tubing in front of the photomultiplier tube converts the basic system into one that can be used either for flow injection analysis, or as an HPLC detector. Each chemical event emits one photon of light. During the course of the reaction the photomultiplier tube and electronic circuitry ‘count’ the number of flashes and so the output is a numerical count, although with some instruments the signal can be integrated to give a curve in a manner analogous to radioactivity HPLC detectors. Luminometers that will read 96-well format microplates are available. Because the instrumental requirements are so simple, portable versions are also available for use in environmental applications, for example.

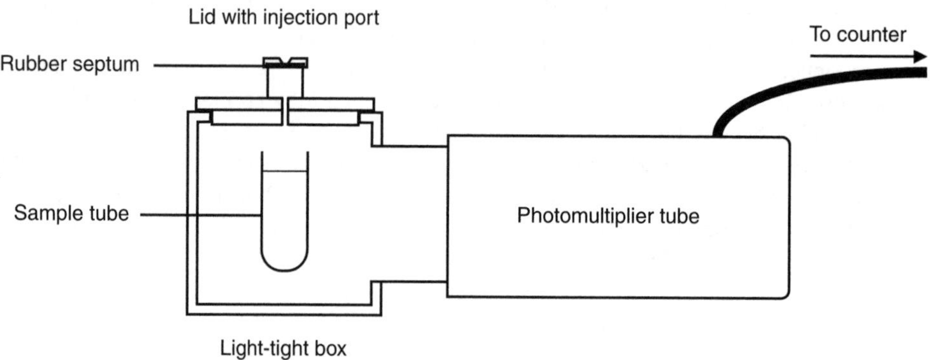

Figure 4.12 A simple luminometer.

4.4.2.2 *Chemiluminescence assays*

To improve selectivity, chemiluminescence assays are usually combined with a separation technique such as chromatography or capillary electrophoresis, or with an immunoassay. The original flow injection analysis (FIA) of morphine (Abbott *et al.*, 1986) was subsequently modified by adding an HPLC step (Abbott *et al.*, 1987). Detection cells consisting of molecular imprinted polymers (MIP) in glass tubes have also been used. Such an approach was used to detect morphine, which was trapped on the MIP and potential interferences were washed off, before permanganate was added to induce chemiluminescence (He *et al.*, 2005).

Enzyme reactions have been used to impart selectivity. The acetylcholine assay in which acetylcholine is hydrolyzed to choline with acetylcholinesterase, and the choline oxidized to betaine and hydrogen peroxide, with the latter being measured via the luminol system described above, is an example (Israël and Lesbats, 1982). Several variations of this approach have appeared including construction of immobilized enzyme reactors for HPLC detection of acetylcholine. Although these can be used with luminol and a chemiluminescence detector, electrochemical detection of the hydrogen peroxide at platinum electrodes is a viable alternative (Flanagan *et al.*, 2005).

Chemiluminescence is important in specific detectors such as the flame photometric detector (Section 7.2.2.6), for the detection of sulfur and phosphorus, and the chemiluminescent nitrogen detector (Section 8.3.4).

4.5 Summary

UV/visible spectrophotometry, spectrophotofluorimetry and chemiluminescence are valuable techniques in analytical toxicology, both in their own right and when used in conjunction with other techniques such as HPLC and immunoassay. Modern microprocessor-controlled spectrophotometers offer many advantages over older single- and double-beam instruments. These include, not only increased stability and sensitivity, but also background correction to facilitate spectral scanning. Nevertheless, with the possible exceptions of HPLC-DAD systems in assessing peak homogeneity and in STA, and FTIR in the GC of solvents and other volatiles, spectral scanning has been largely superseded in STA by GC-MS. With the exception of the urine tests for paracetamol and paraquat, colour tests too have been largely superseded, although such tests may still be valuable if used with due caution when other methods are not available.

References

Abbott, R.W., Townshend, A. and Gill, R. (1986) Determination of morphine by flow injection analysis with chemiluminescence detection. *Analyst*, **111**, 635–40.

Abbott, R.W., Townshend, A. and Gill, R. (1987) Determination of morphine in body fluids by high-performance liquid chromatography with chemiluminescence detection. *Analyst*, **112**, 397–406.

Badcock, N.R. (2000) Detection of poisoning by substances other than drugs: a neglected art. *Ann Clin Biochem*, **37**, 146–57.

Brodie, B.B., Udenfriend, S. and Baer, J.E. (1947) The estimation of basic organic compounds in biological material. 1. General principles. *J Biol Chem*, **168**, 299–309.

Broughton, P.M. (1956) A rapid ultraviolet spectrophotometric method for the detection, estimation and identification of barbiturates in biological material. *Biochem J*, **63**, 207–13.

Broughton, P.M. (1963) Barbiturates giving atypical results on ultraviolet spectrophotometric analysis. *Lancet*, **i**, 1266–7.

Cary, H.H. and Beckman, A.O. (1941) A quartz photoelectric spectrophotometer. *J Opt Soc Am*, **31**, 682–9.

Ejnik, J.W., Hamilton, M.M., Adams, P.R. and Carmichael, A.J. (2000) Optimal sample preparation conditions for the determination of uranium in biological samples by kinetic phosphorescence analysis (KPA). *J Pharm Biomed Anal*, **24**, 227–35.

Flanagan, R.J., Braithwaite, R.A., Brown, S.S., Widdop, B. and de Wolff, F.A. (1995) *Basic Analytical Toxicology*, World Health Organization, Geneva.

Flanagan, R.J., Perrett, D. and Whelpton, R. (2005) *Electrochemical Detection in HPLC. Analysis of Drugs and Poisons*, Royal Society of Chemistry, Cambridge.

Greenway, G.M. and Dolman, S.J. (1999) Analysis of tricyclic antidepressants using electrogenerated chemiluminescence. *Analyst*, **124**, 759–62.

Hall, A.P., Czerwinski, A.W., Madonia, E.C. and Evensen, K.L. (1973) Human plasma and urine quinine levels following tablets, capsules, and intravenous infusion. *Clin Pharmacol Ther*, **14**, 580–5.

He, Y., Lu, J., Liu, M. Du J, Nie F. (2005) Determination of morphine by molecular imprinting-chemiluminescence method. *J Anal Toxicol*, **29**, 528–32.

Hadley, S.G., Muraki, A.S. and Spitzer, K. (1974) The fluorescence and phosphorescence spectra and phosphorescence decay time of harmine, harmaline, harmalol, harmane and norharmane in aqueous solutions and EPA at 77K. *J Forensic Sci*, **9**, 657–69.

Israël, M. and Lesbats, B. (1982) Application to mammalian tissues of the chemiluminescent method for detecting acetylcholine. *J Neurochem*, **39**, 248–50

Jeffery, W. (2004) Colour tests, in *Clarke's Analysis of Drugs and Poisons*, 3rd edn. (eds A.C. Moffat, D. Osselton, and B. Widdop), Pharmaceutical Press, London, pp. 279–300.

Jones, H.B. (1865a) On the rate of passage of cystalloids into and out of the vascular and non-vascular textures of the body. *Proc R Soc Lond*, **14**, 63–64.

Jones, H.B. (1865b) On the rate of passage of cystalloids into and out of the vascular and non-vascular textures of the body. *Proc R Soc Lond*, **14**, 220–223.

Jones, H.B. and Dupré, A. (1866) On a fluorescent substance, resembling quinine, in animals; and on the rate of passage of quinine into the vascular and non-vascular textures of the body. *Proc R Soc Lond Series B*, **15**, 73–93.

Kricka, L.J., Thorpe, G.H.G. and Stott, R.A.W. (1987) Enhanced chemiluminescence enzyme immunoassay. *Pure Appl Chem*, **59**, 651–4.

Li, F., Zhang, C., Guo, X. and Feng, W. (2003) Chemiluminescence detection in HPLC and CE for pharmaceutical and biomedical analysis. *Biomed Chromatogr*, **17**, 96–105.

Mayes, R.W. (1993) ACP Broadsheet No 142: November 1993. Measurement of carbon monoxide and cyanide in blood. *J Clin Pathol*, **46**, 982–8.

Miles, C.I. and Schenk, G.H. (1973) Fluorescence and phosphorescence of phenylethylamines and barbiturates. Analysis of amphetamine and barbiturate preparations. *Anal Chem*, **45**, 130–6.

Moffat, A.C., Osselton, M.D. and Widdop, B. (eds.) (2004) *Clarke's Analysis of Drugs and Poisons*, Vol. 2, 3rd edn, Pharmaceutical Press, London.

Stewart, M.J., Adriaenssens, P.I., Jarvie, D.R. and Prescott, L.F. (1979) Inappropriate methods for the emergency determination of plasma paracetamol. *Ann Clin Biochem*, **16**, 89–95.

Tompsett, S.L. (1969) Interference from the presence of other substances in detecting and determining barbiturates in biological material. *J Clin Pathol*, **22**, 291–5.

Udenfriend, S. (1995) Development of the spectrophotofluorometer and its commercialization. *Protein Sci*, **4**, 542–55.

Whelpton, R., Watkins, G. and Curry, S.H. (1981) Bratton-Marshall and liquid-chromatographic methods compared for determination of sulfamethazine acetylator status. *Clin Chem*, **27**, 1911–4.

WHO (World Health Organization). (1986) *Basic Tests for Pharmaceutical Substances*, WHO, Geneva.

WHO (World Health Organization). (1991) *Basic Tests for Pharmaceutical Dosage Forms*, WHO, Geneva.

WHO (World Health Organization). (1998) *Basic Tests for Drugs: Pharmaceutical Substances, Medicinal Plant Materials and Dosage Forms*, WHO, Geneva.

Widdop, B. (2002) Analysis of carbon monoxide. *Ann Clin Biochem*, **39**, 378–91.

5 Introduction to Chromatography and Capillary Electrophoresis

5.1 General introduction

Chromatography and electrophoresis are separation methods. Normally, a mixture is applied in solution as a narrow initial zone or bolus to an appropriate stationary phase or, in the case of electrophoresis, a column or gel containing an electrolyte. In chromatography, a suitable flowing mobile phase or eluent is introduced to the mixture, or eluent flow is otherwise initiated and the components of the mixture are separated by differential physiochemical interaction with the stationary and mobile phases. The stationary phase may be a porous solid such as silica gel (adsorption chromatography), or an immiscible or nonvolatile liquid held on a suitable solid support (partition chromatography). The technique has many similarities to SPE (Section 3.2.6), which in some ways can be viewed as preparative chromatography. In electrophoresis an electrical potential is applied and separation is by movement of ionized species to the appropriate electrode.

5.1.1 Historical development

The father of chromatography is recognized as the Russian botanist M.S. Tswett (1872–1919). He stated that 'chromatography is a method in which the components of a mixture are separated on an adsorbent column in a flowing system'. In his original experiments (1903), Tswett packed a fine powder such as sucrose into a glass tube to produce a column of the desired height. After extracting green and yellow chloroplast pigments from leaves and transferring them to petroleum ether, he poured a small volume of the extract onto the column. When the pigments had formed a narrow initial band at the top of the adsorbent, fresh solvent (petroleum ether) was added and pressure applied to the top of the column. As the solvent flowed through the column the individual pigments moved at different rates and were eventually separated from each other. The key features of Tswett's technique were the application of the mixture as a narrow initial zone and the development of the chromatogram by application of fresh solvent. Others had employed procedures based on the phenomena of adsorption or partition, but these lacked Tswett's critical development step and therefore did not yield effective resolution of mixtures.

Amongst many subsequent developments, those of A.J.P. Martin (1910–2002) and R.L.M. Synge (1914–1994), who were awarded the 1952 Nobel Prize in Chemistry for the discovery of partition chromatography (Martin and Synge, 1941), and of A.T. James and A.J.P. Martin, who developed gas–liquid partition chromatography (James and Martin, 1952), stand out (Adlard, 2003). Further important work lay in the development of sensitive detectors for GC. The flame-ionization detector (FID) responds to most organic compounds, whilst the electron-capture detector (ECD) shows an enhanced and selective response to compounds containing strong electronegative moieties such as halogen atoms or nitro groups. The nitrogen–phosphorus detector (NPD), also known as the alkali flame-ionization detector (AFID), shows an enhanced and selective response to compounds containing C–N bonds or phosphorus.

In 1948 the Swedish analyst A.W.K. Tiselius (1902–1971) was awarded the Nobel Prize in Chemistry for his pioneering work in developing electrophoresis. Capillary electrophoresis (CE) was pioneered by Hjerten, who had built tube electrophoresis units by 1959 and went on to demonstrate that free-flowing electrophoresis in capillary tubes with UV detection was feasible. Jorgenson and Lukacs (1981) developed the first truly useful CE instrument and showed that it had exceptional resolution. More recently, capillary electrophoresis and related techniques have been used in the analysis of drugs and other poisons.

Chromatography and electrophoresis have developed into a range of modes including paper chromatography (PC), thin-layer chromatography (TLC), ion-exchange chromatography, gel-permeation chromatography (GPC, also known as size-exclusion chromatography, SEC), affinity chromatography, gas chromatography (GC), supercritical-fluid chromatography (SFC), high-performance liquid chromatography (HPLC), capillary electrophoresis (CE, also known as capillary zone electrophoresis, CZE), and capillary electrochromatography (CEC). CE and CEC are hybrid techniques wherein eluent flow is produced by electro-osmosis.

In the 1950s there was great emphasis on the development of TLC methods. At the same time, research in liquid chromatography led to the development of various amino-acid analyzers. These analyzers used columns made of styrene/divinylbenzene polymers that were derivatized to make strong cation-exchange (SCX) materials. After gradient elution (changing the eluent composition with time in a defined way), the column eluate was mixed with ninhydrin, heated and passed through a spectrophotometric cell and the absorbance monitored (570 nm). The analysis of one sample could take 2 days.

In PC and TLC, techniques that are sometimes referred to as planar chromatography or development chromatography, solutions of the analytes are usually applied as small discrete spots or bands. The application solvent is evaporated before the edge of the paper sheet or strip, or thin-layer plate, is placed in the liquid mobile phase, which is drawn along the sheet or plate (the stationary phase) by capillary action. Several analyses may be performed in parallel. All the analytes are detected (normally visualized) at the end of the development process.

In most other forms of chromatography, a sample (or sample extract) is either added to the eluent, which may be a gas, a liquid or a supercritical fluid, such as carbon dioxide, or placed on a support material or otherwise concentrated before the eluent is introduced. The eluent containing the analyte and other components of the sample/sample extract is then allowed or made to flow through or past a stationary phase supported within a column. The mobile and stationary phases are chosen such that different components of the sample have different affinities for each phase. A component that has poor affinity for the stationary phase will pass through the column quite quickly, and *vice versa*. As a result of these differences in mobility, sample components become separated as they travel through the column. This process is called elution chromatography and analytes are detected sequentially as they elute from the column. In both development and elution chromatography sample transport is by continuous addition of mobile phase. Various modifications of these techniques are possible, for example development of a TLC plate in a second dimension using a different mobile phase.

There has been continuous development in many branches of chromatography and in CE since the 1960s, particularly in materials and in the refinement of instrumentation that has resulted in the efficient, reliable and sensitive analytical methods that form the backbone of modern routine laboratory analysis (Ettre, 2002; Issaq, 2002; Grob and Barry, 2004). Manufacturers' catalogues and web sites often contain up-to-date information on newer products, although important experimental details may be lacking. Chromatograms, for example, may have been obtained using pure compounds and/or concentrated solutions may have been injected; in some cases the actual amount injected and the detector sensitivity may not be stated. As well as the formal scientific literature, the trend to publications sponsored by manufacturers or funded from advertising has produced some useful free magazines; *LC GC* is probably the best of these.

5.2 Theoretical aspects of chromatography

Chromatographic theory was studied by Wilson (1940), who discussed the quantitative aspects of chromatography in terms of diffusion, rate of adsorption, and isotherm nonlinearity. The first comprehensive mathematical treatment describing column performance (using the height equivalent to a theoretical plate, HETP), in terms of stationary phase particle size and diffusion, was presented by Martin (1949). However, it was van Deemter *et al.* (1956) who developed the rate theory to describe the separation processes following on from earlier work of Lapidus and Admunson. Giddings and Eyring (1955) first looked at the dynamic theory of chromatography and, from the 1960s onward, examined many aspects of GC and general chromatography theory. A compendium of terms related to chromatography and associated areas, LLE, for example, is available (IUPAC, 1997).

In the simplest forms of GC and of HPLC, the stationary phase is simply a rigid material packed within a column through which the eluent flows, but more usually the stationary phase is coated or bonded directly to a column, or to particles of a rigid support material packed within the column. In the mid 1960s, discussion of the parallels between LC and GC suggested that use of smaller particles in HPLC would lead to better efficiency, hence greater speed of analysis, and thus better sensitivity/selectivity. The problem with columns packed with larger particles was the slow mass transfer of the analyte molecules into and out of the pores of the stationary material. Packings made with smaller particles were thus investigated to improve resolution. A range of suitably small particle size packings based on 10 μm average particle size silica gels (sized with air classification techniques) were soon introduced, and work began on how best to pack these small particles to give efficient columns. Organosilane bonding technologies were also introduced.

5.2.1 *Analyte phase distribution*

The components of a mixture become distributed between the stationary and mobile phases as they are transported through the stationary phase in the fluid mobile phase. Differences between their distribution coefficients (K_D) result in separation:

$$K_D = C_s / C_m \quad (5.1)$$

where C_s is the molar concentration of a component in the stationary phase, and C_m is the molar concentration of the same component in the mobile phase. In development chromatography the distance moved by the analyte in relation to the distance moved by the eluent ('solvent front') is called the retention fraction (R_F). In elution chromatography the time between sample injection and an analyte peak reaching a detector at the end of the column is termed the retention time (t_R).

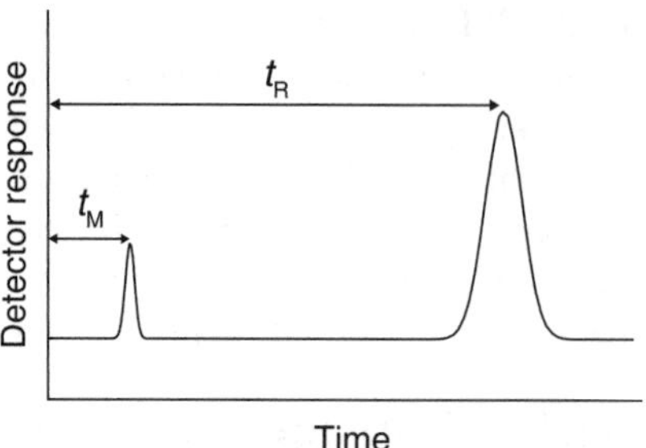

Figure 5.1 Elution chromatography: retention time.

The time taken for the mobile phase itself to pass through the column is called t_M (Figure 5.1). In HPLC, multiplying the retention time by the eluent flow rate (which can usually be considered to be constant) gives the retention volumes, V_M and V_R. V_M is known as the void volume of the column and is the volume of solvent that surrounds the packing material. Because gases are much more compressible than liquids, the average flow rate in a GC column is less than the flow rate measured at the column outlet and the corrected retention volume is given by:

$$V_R^0 = \frac{3}{2}\left[\frac{(P_i/P_o)^2 - 1}{(P_i/P_o)^3 - 1}\right] V_R \qquad (5.2)$$

where P_i and P_o are the inlet and outlet pressures, respectively. A different correction is required if the flow rate is measured at the column inlet rather than the outlet.

The term retention factor (k, sometimes called capacity factor, k') is often used to describe the migration rate of an analyte on a column. The retention factor for analyte A is defined as:

$$k_A = \frac{t_R - t_M}{t_M} \qquad (5.3)$$

Because analyte migration is defined in terms of the number of column volumes, this parameter is independent of column geometry and flow rate. One problem, though, is that it may be very difficult to measure t_M accurately – a small error here can give large errors in k especially with very short columns. In GC, t_M is normally taken as the first deviation of the baseline post-injection (it is assumed that some air gets into the system at the time of injection hence this is known as the 'air peak'), whilst in HPLC with UV detection the first deviation of the baseline following the injection of a small quantity of non-retained solvent such as acetone is used as the t_M marker.

Accurate measurement of t_R, on the other hand, may be difficult with asymmetric peaks. If $k < 1$, accurate measurement of t_R may be difficult, and the analyte peak may be lost in the solvent front that is often seen with biological extracts. On the other hand, values of $k > 20$ are usually associated with long retention times and broad peaks – temperature programming in GC (gradient elution in HPLC) is one way of sharpening late-eluting peaks and thereby extending the range of analytes that can be detected/measured in a given analysis. Ideally, in isocratic (constant eluent composition) HPLC and isothermal (constant column temperature) GC k should be in the range 2–5 or thereabouts.

5.2.2 *Column efficiency*

To obtain optimal sensitivity and selectivity, sharp, symmetrical peaks (so-called because of their appearance on a recording device) are the goal in elution chromatography, whilst small, discrete spots or bands are the aim in development chromatography. Analyte peaks or bands that move most quickly in a chromatographic system tend to be sharper than those that elute later because there is less time for zone broadening (Section 5.2.3). The extent of zone broadening determines the efficiency of a column for a given analyte under the analytical conditions used. For any meaningful comparison to be made between two columns, efficiency must be measured with the same analyte under the same conditions. Efficiency can be expressed either as the number of theoretical plates or plate number (N), or by the height equivalent of a theoretical plate (HETP, or just H). The theoretical plate model of chromatography postulates that the column contains a large number of separate layers or plates, a concept based on bubble-cap fractionation columns, in which the greater the number of plates in a given length, the greater the separation (Figure 5.2). Separate equilibration of the sample between the stationary and mobile phases is said to occur at each plate.

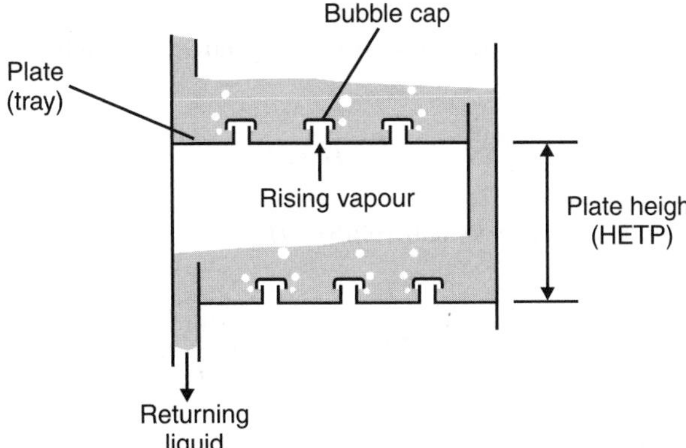

Figure 5.2 Schematic representation of a fractionating column showing the concept of plate height.

The width of a chromatographic peak is influenced by a number of random processes, each of which has its own variance, and so an ideal chromatographic peak will be of Gaussian shape with a variance, σ^2, equal to the sum of the total variances (Figure 5.3). There are two main ways to calculate N. The first is to estimate the 'peak width' ($W_b = 4\sigma$) by constructing tangents to the points of inflexion. The plate number is:

$$N = 16\left(\frac{t_R}{W_b}\right)^2 \tag{5.4}$$

A more practical method is to use the width of the peak at half the peak height ($W_{0.5} = 2.355\sigma$) so:

$$N = 5.54\left(\frac{t_R}{W_{0.5}}\right)^2 \tag{5.5}$$

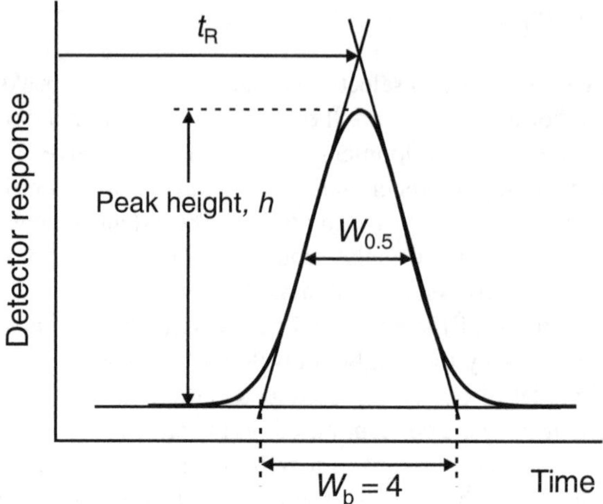

Figure 5.3 Chromatographic efficiency: measurement of *HETP*.

The advantage of HETP or H is that it is independent of column length. If the length of the column is L cm, then:

$$H = L/N \tag{5.6}$$

An efficient column will have a small value of H. Efficiency can also be expressed as plates per metre:

$$\text{Plates m}^{-1} = \frac{100N}{L} \tag{5.7}$$

5.2.3 *Zone broadening*

The plate model assumes that analyte equilibration between the stationary phase and the mobile phase is virtually instantaneous. A more realistic description of the chromatographic process takes account of the time needed for analyte equilibration between the phases. The shape of a chromatographic peak is affected by the rate of elution, but it is also affected by the different paths available to solute molecules as they travel between particles of stationary phase. Considering the various mechanisms which contribute to band broadening gives the *van Deemter equation* (van Deemter *et al.*, 1956):

$$H = A + B/u + Cu \tag{5.8}$$

where u is the average linear velocity of the mobile phase and A, B and C are factors that contribute to band broadening.

5.2.3.1 *Multiple path and eddy diffusion*

Analyte molecules take different paths through the column at random. This will cause broadening of the solute band, because different paths are of different lengths. The spreading is proportional to the particle size (diameter = d_p) and is independent of the mobile phase velocity. Van Deemter introduced a constant (2λ) to account for the inhomogeneity and quality of the packing so his expression for the multipath term contribution was:

$$A = 2\lambda d_p \tag{5.9}$$

For an ideally packed column, $A = d_p$, that is $\lambda = 0.5$.

5.2.3.2 *Longitudinal diffusion*

Considering a band of analyte in a column, the concentration will be lower at the edges of the band than at the centre because the analyte diffuses out from the centre to the edges. The greater the amount of time that the band is in the column, the greater will be the diffusion and hence the band broadening. High eluent velocities will decrease the effect of longitudinal diffusion by reducing the time the analyte is in the column, that is the contribution to H caused by longitudinal diffusion is inversely proportional to u. Because the van Deemter equation was derived for packed column GC, the longitudinal diffusion in the stationary phase was negligible compared to that in the gas phase, so:

$$B = 2\gamma D_m \tag{5.10}$$

D_m is the diffusivity of the analyte in the mobile phase, and γ a correction factor for the geometry of the packing, which must affect the diffusional path.

5.2.3.3 *Resistance to mass transfer*

The analyte takes time to equilibrate between the phases. If the eluent velocity is high and the analyte has a strong affinity for the stationary phase, then the analyte in the eluent will move ahead of the analyte in the stationary phase giving band broadening. The higher the eluent velocity, the larger will be the band broadening. The term used by van Deemter was:

$$C = \frac{8k}{\pi^2(1+k)^2}\frac{d_s^2}{D_s} \tag{5.11}$$

where D_s is the diffusivity of the analyte in the stationary phase and d_s is the thickness of the film of stationary phase. Again, because eddies in the eluent aid mixing, the resistance to mass transfer in the mobile phase was considered negligible and ignored.

Thus, the van Deemter equation can be written:

$$H = 2\lambda d_p + \frac{2\gamma D_m}{u} + \frac{8k}{\pi^2(1+k)^2}\frac{d_s^2}{D_s}u \tag{5.12}$$

Despite the fact that the van Deemter equation was derived for packed column GC, the terms illustrate the importance of:

- particle size (smaller particles give greater efficiency)
- the thickness of the stationary phase (thinner films give greater efficiency)
- eluent velocity.

If the van Deemter equation is to be used to explain chromatographic behaviour in HPLC, then a term for the resistance to mass transfer in the eluent must be included. Like the resistance to mass flow in the stationary phase, this is inversely proportional to u, and so the general form of the equation, although more complex, is the same as Equation (5.8). Van Deemter plots, in which HETP is plotted against eluent velocity, are very useful in ensuring that the mobile phase flow-rate is optimal (Figure 5.4).

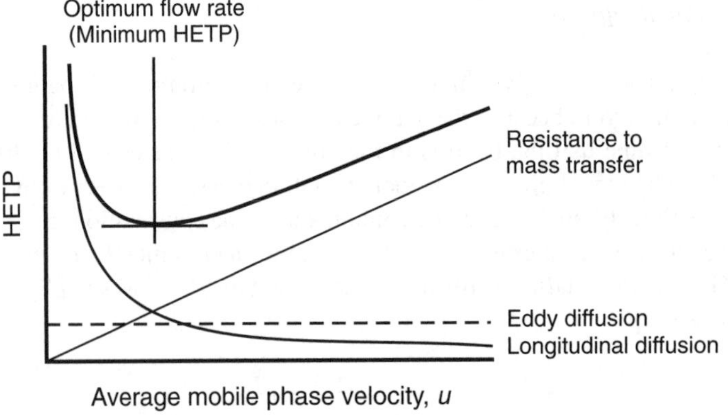

Figure 5.4 Chromatographic efficiency: effect of mobile phase velocity on HETP.

It is important to consider the effects of column diameter on the linear flow rate. If the column diameter is reduced by half then the flow rate (mL min^{-1}) must be reduced to one quarter to maintain the original linear flow.

The Golay equation was formulated to describe dispersion in open tubular or capillary columns (Golay, 1958). Because there is no multipath term, it takes the form:

$$H = B/u + Cu \tag{5.13}$$

For a column of internal radius, r, it is:

$$H = \frac{2D_{\text{m}}}{u} + \frac{(1 + 6k + 11k^2)r^2}{24(1 + k)^2 D_{\text{m}}}u + \frac{K^3 r^2}{6(1 + k)^2 K^2 D_{\text{s}}}u \tag{5.14}$$

where K is the distribution coefficient of the analyte between the stationary and mobile phases. Whereas the size of the packing is a major contributor to band broadening in packed columns, it is clear that for capillary columns the resistance to mass transfer terms are proportional to the square of the radius.

The Giddings equation was derived using a random walk model and takes the general form:

$$H = \frac{A}{1 + (E/u)} + \frac{B}{u} + C \tag{5.15}$$

Unlike the van Deemter equation, this becomes zero when there is no eluent flow, but as the flow increases the term E/u will tend to zero and the equation then takes the same form as the van Deemter equation.

The Knox equation is an empirically derived equation applicable to HPLC (Done *et al.*, 1973) and simpler than that of Giddings. H is replaced by the reduced plate height, h, which equals H/d_p (i.e. the plate height in terms of the particle size):

$$h = A\nu^{1/3} + \frac{B}{\nu} + C\nu \tag{5.16}$$

and ν is the reduced velocity and equals ud_p/D_m. The coefficient A normally ranges from 0.2–1.7 and decreases with increasing homogeneity of the packing. B ranges from 1.6–1.8 and C ranges from 0.05 to 0.03. Usually h has a minimum value of 2–3 at $\nu = 2$–3, so in HPLC, $H = 2$–3 d_p at the optimum. For example, if a 3 μm average particle size stationary phase is used, the optimum H is 6–9 μm, and in the case of a column 25 cm in length, the theoretical plate number that should be obtained is about 40,000.

5.2.4 *Extra-column contributions to zone broadening*

Three factors should be taken into consideration: (i) sample volume, (ii) detector volume and (iii) the connecting tubing between the injection value and the column, and the column and the detector. Band spreading due to the first two is proportional to their respective volumes. Generally, therefore, the volume injected should be kept as small as practicable. If detectors are connected in series the detector with the largest internal cell (swept) volume should be placed last whenever possible. The internal diameter (d) of the connecting tubing can have a major effect, as spreading is proportional to d^4. Narrow-bore PTFE tubing is generally suitable for post-column connections in HPLC.

5.2.5 *Temperature programming and gradient elution*

Another practically important variable in LC and, to an extent in GC, is 'peak capacity', that is the number of peaks that can be represented on a chromatogram. A typical packed HPLC column with $N = 5000$ plates yields a peak capacity between 17 for values of k from 0.2–2 and about 50 for k values from 0.5–20. However, when the sample contains a large number of analytes of differing volatilities or polarities, then temperature programming (GC) or gradient elution (LC) will be required. Isothermal or isocratic analysis of such a mixture would not only require very long run times, but also the late-eluting peaks would be so wide that they would be 'lost' in the baseline noise of the chromatogram. With temperature programming, the analytes are condensed in a tight band at the top of the column and only begin to migrate when the temperature is high enough, that is some analytes are being separated whilst others are stationary. Thus, the time a particular analyte is actually moving in the column is similar to that of any other analyte and so the band spreading due to longitudinal diffusion (normally proportional to $t_R^{0.5}$) is very much reduced for later eluting peaks. The same principle applies to gradient elution LC, but in this case it is the eluent composition that is modified.

5.2.6 Selectivity

The column selectivity factor, α, describes the separation of two species (A and B) on the column. When calculating the selectivity factor, A always elutes before B, that is α is always >1.

$$\alpha = \frac{k_B}{k_A} \tag{5.17}$$

Although α describes the separation of band centres, it does not take into account peak width. Another measure of separation is provided by resolution (R):

$$R = \frac{2[(t_R)_B - (t_R)_A]}{W_A + W_B} \tag{5.18}$$

where W = baseline peak width. Baseline resolution is attained when $R = 1.5$ (Figure 5.5).

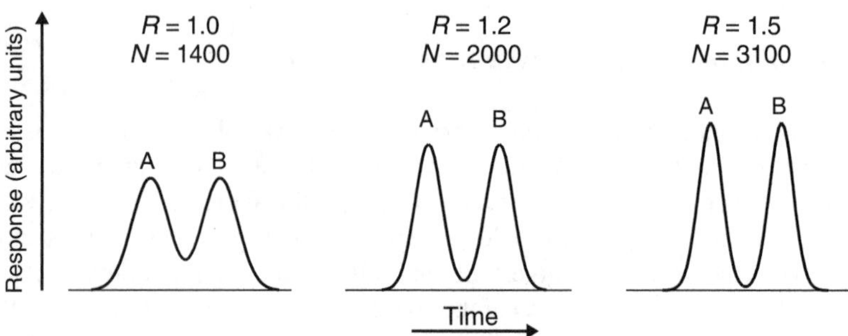

Figure 5.5 Effect of efficiency on resolution of A and B. Values calculated using $k_A = 10$, $\alpha = 1.2$; N calculated for peak B.

It is also useful to relate resolution to the number of theoretical plates (N), the selectivity factor (α) and the retention factors (k) of the two solutes.

$$R = \frac{\sqrt{N}}{4}\left(\frac{\alpha - 1}{\alpha}\right)\left(\frac{k_B}{k_B + 1}\right) \tag{5.19}$$

To obtain good resolution, all three terms of Equation (5.19) must be maximized. Increasing N by lengthening the column will cause some increase in band broadening. Alternatively, N can be increased by decreasing the size of the stationary phase particles, but there are practical limits to this because use of smaller particles increases the pressure needed to maintain eluent flow. In GC especially this is self-limiting as increased gas pressure simply compresses the mobile phase. This is the great theoretical advantage of CE and CEC as eluent flow is not dependent on eluent inlet pressure.

An alternative measure of column efficiency is provided by the separation number (Trennzahl, *TZ*), which is calculated thus (Kaiser, 1976a, 1976b):

$$TZ = \frac{t_{R_B} - t_{R_A}}{W_{0.5_A} + W_{0.5_B}} - 1 \tag{5.20}$$

TZ represents the number of peaks that can be inserted between the peaks of two consecutive hydrocarbon homologues in a chromatogram and is related to resolution:

$$TZ = \frac{R}{1.18} - 1 \tag{5.21}$$

Separation number is very useful in temperature programming in GC and gradient programming in LC where conventional (isothermal or isocratic) measures of column efficiency cannot be used. If $TZ = 0$ an additional peak will not fit between the two peaks of interest, if $TZ = 1$ one peak will fit, and so on.

On a practical level, chromatographic separations can often be improved by changing the column temperature (mainly GC) or the eluent composition (LC). The selectivity factor, α, can also be manipulated to improve separations. Sometimes changing the composition of the stationary phase (mainly GC, but also LC) or using eluent additives such as a species that complexes with one of the analytes in the stationary phase are effective in improving resolution and/or peak shape.

It is important to understand that matrix components or other analytes may affect separations. In TLC co-extracted sample components, for example, may influence the analysis if present at much higher concentrations than other analyte(s). In GLC, HPLC and CE the injection solvent, or the injection volume in the case of headspace GC, or indeed the presence of other analytes in high concentrations, may alter the retention characteristics of other components of a mixture.

5.2.7 *Peak asymmetry*

The ideal chromatographic peak will be a perfect Gaussian shape, reflecting the additive combination of the random motions of the analyte in the system. In practice symmetrical peaks are not obtained, and their deviation from that of a true Gaussian peak – the peak asymmetry – can be measured and compared. To do this, the peak is usually measured as shown in Figure 5.6.

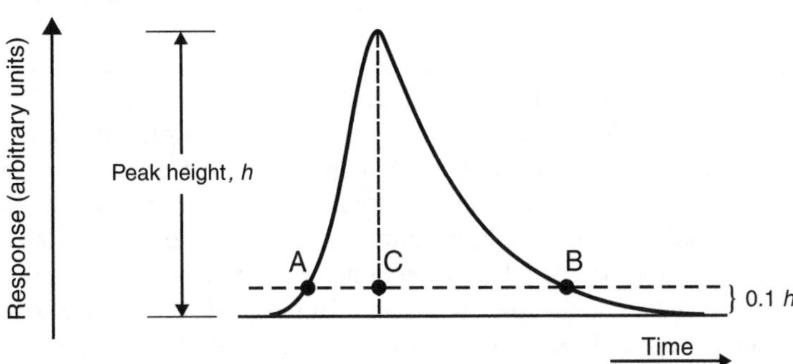

Figure 5.6 Measurement of peak asymmetry.

Here, AC and CB are measured at 10 % of the total peak height (h) above the baseline, and the asymmetry factor, A_s, is calculated thus:

$$A_s = \frac{\text{CB}}{\text{AC}} \tag{5.22}$$

A Gaussian peak will have A_s equal to unity. Tailing peaks, usually caused by adsorption of analyte to the 'reactive sites' on the column or column walls, will produce A_s values greater than 1. Fronting peaks, often a result of overloading the column with analyte or eluent flow through channels within the column, will have $A_s < 1$. Peak shape can be considered poor if A_s exceeds 1.5–2, depending on the type of analysis and the conditions used, due to its impact on the resolution of components and the likely error in the measurement of retention times. Fronting peaks lead to an overestimate of the true retention time, whilst tailing gives an underestimate of the retention time compared with that of a Gaussian peak.

5.3 Measurement of analyte retention

Retention factors (k), absolute retention times (volumes) and retention times relative to the retention of a given compound (internal standard) can be useful ways of recording retention data in GC. However, the Kovats retention index (Kovats, 1961) provides a method of recording retention data that is independent of eluent flow rate, column length, phase loading and operating temperature. Moreover, accurate measurement of t_0 is not required. Straight-chain (normal) hydrocarbons are assigned an index of 100 × the number of carbon atoms in the molecule (e.g. decane = 1000). The retention index of a given analyte at a given column temperature is then calculated by difference from the retention indices of the normal alkanes eluting before and after the analyte (Supelco, 1997). Retention indices can also be calculated from data generated on a temperature program by applying the following formula during individual ramps of the program (van den Dool and Kratz, 1963; Lee and Taylor, 1982):

$$RI_x = 100z + 100\frac{t_{R_x} - t_{R_z}}{t_{R_{(z+1)}} - t_{R_z}} \tag{5.23}$$

where: RI_x = retention index of x, z = n-alkane with z carbon atoms eluting before x, t_{Rx} = retention time of x, t_{Rz} = retention time of z and $t_{R(z+1)}$ = retention time of n-alkane with $z + 1$ carbon atoms eluting after x.

A practical problem in the use of retention index data in GC is that normal hydrocarbons show no response on selective detectors such as ECD or NPD (Section 7.2.2) and do not give particularly good responses on MS, hence attempts have been made to construct retention index schemes based on, for example, trialkylamines (Watts and Simonick, 1987) or nitroalkanes (Aderjan and Bogusz, 1988). However, in practice, alkylamines are not pleasant compounds to work with and such schemes found little favour, an effluent splitter system to an FID being used to give hydrocarbon responses if needed. A detailed discussion of the use of relative retention time and GC retention index data in analytical toxicology is given by Huizer (1991). The analysis of a selected group of compounds structurally similar to the analytes, but with known retention index values, has been proposed as a way of correcting for systematic errors in retention index measurements of unknowns (Franke *et al.*, 1993).

Many attempts have been made to develop a suitable retention index system for HPLC using, for example, homologous series of alcohols, ketones or nitroalkanes (Bogusz *et al.*, 1988; Smith, 1995), but in practice such methods offer no advantage over retention factors and other simpler ways of expressing retention in HPLC.

5.4 Summary

Chromatographic techniques, notably GC and HPLC, and to an extent TLC, are of unrivalled importance in analytical toxicology as discussed in the following chapters, but must be used with due care and attention to detail if reliable results are to be obtained. An appreciation of the theoretical aspects of chromatography as presented here is important in making best use of the resolving power of these systems. However, SFC, once widely advocated for a variety of applications, has been largely discarded. Capillary electrophoretic techniques too, whilst of value in pharmaceutical QC and in separating enantiomers on an analytical scale, as discussed in Chapter 9, at present have neither the sensitivity nor the mechanical strength to provide a robust system for trace analysis.

References

Aderjan, R. and Bogusz, M. (1988) Nitroalkanes as a multidetector retention index scale for drug identification in gas chromatography. *J Chromatogr*, **454**, 345–51.

Adlard, E.R. (2003) Fifty years of gas chromatography. *Chromatographia*, **57**(Suppl): S13–8.

Bogusz, M., Neidl-Fischer, G. and Aderjan, R. (1988) Use of corrected retention indices based on 1-nitroalkane and alkyl arylketone scales for HPLC identification of basic drugs. *J Anal Toxicol*, **12**, 325–9.

van Deemter, J.J., Zuiderweg, F.J. and Klinkenberg, A. (1956) Longitudinal diffusion and resistance to mass transfer as causes of nonideality in chromatography. *Chem Eng Sci*, **5**, 271–89.

Done, J.C., Kennedy, G.J. and Knox, J.H. (1973) The role of particle size and configuration in high speed liquid chromatography, in *Gas Chromatography* (eds. S.G. Perry and E.R. Adlard), Applied Science Publishing, Barking, pp. 145–155.

Van Den Dool, H. and Kratz, P.D. (1963) A generalization of the retention index system including linear temperature programmed gas-liquid partition chromatography. *J Chromatogr*, **11**, 463–71.

Ettre, L.S. (2002) *Milestones in the Evolution of Chromatography*, ChromSource, Franklin, TN.

Franke, J.P., de Zeeuw, R.A. and Bogusz, M. (1993) An overview of the standardization of chromatographic methods for screening analysis in toxicology by means of retention indices and secondary standards. *Fresenius J Anal Chem*, **347**, 67–72.

Giddings, J.C. and Eyring, H. (1955) A molecular dynamic theory of chromatography. *J Phys Chem*, **59**, 416–21.

Golay, M.J.E. (1958) Theory of chromatography in open and coated tubular columns with round and rectangular cross sections, in *Gas Chromatography 1958* (ed. D.H. Desty) Butterworths, London, pp. 36–55.

Grob, R.L. and Barry, E.F. (eds.) (2004) *Modern Practice of Gas Chromatography*, 4th edn. Wiley, New York.

Huizer, H. (1991) The use of gas chromatography for the detection and quantitation of abused drugs, in *The Analysis of Drugs of Abuse* (ed. T.A. Gough) Wiley, Chichester, pp. 23–92.

Issaq, H.J. (ed.) (2002) *A Century of Separation Science*, Dekker, New York.

IUPAC (1997) *Compendium of Analytical Nomenclature. Definitive Rules*, 3rd edn. (http://www.iupac.org/publications/analytical_compendium/, Section 9: Separations, accessed May 24, 2006).

James, A.T. and Martin, A.J.P. (1952) Gas-liquid partition chromatography: the separation and microestimation of volatile fatty acids from formic acid to dodecanoic acid. *Biochem J*, **50**, 679–90.

Jorgenson, J.W. and Lukacs, K.D. (1981) Free-zone electrophoresis in glass capillaries. *Clin Chem*, **27**, 1551–3.

Kaiser, R.E. (1976a) Correct measurement and evaluation of efficiency factors in chromatography—real separator stage factor, separator factor, apportioning efficiency factor. 1. Gas-Chromatography. *Chromatographia*, **9**, 337–49.

Kaiser, R.E. (1976b) True quality of separation in chromatography (GC, HPLC, HPTLC). *Chromatographia*, **9**, 463–7.

Kovats, E. (1961) Zusammenhänge zwischen strucktur und gaschromatographischen daten organischer verbindungen. *Fresenius Z Anal Chem*, **181**, 351–66.

Lee, J. and Taylor, D.R. (1982) Relationships between temperature programmed and isothermal Kovats retention indexes in gas-liquid-chromatography. *Chromatographia*, **16**, 286–9.

Martin, A.J.P. and Synge, R.L.M. (1941) A new form of chromatogram involving two liquid phases. *Biochem J* : **35**, 1358–68.

Martin, A.J.P. (1949) Some theoretical aspects of partition chromatography. *Biochem Soc Symp* : **3**, 4–20.

Smith, R.M. (ed.) (1995) *Retention and Selectivity in Liquid Chromatography*. Journal of Chromatography Library, 57. Elsevier, Amsterdam.

Supelco (1997) *The Retention Index System in Gas Chromatography: McReynolds Constants*. Supelco, Bellefonte, PA (http://www.sigmaaldrich.com/Graphics/Supelco/objects/7800/7741.pdf, accessed 24 March 2006).

Watts, V.W. and Simonick, T.F. (1987) A retention index library for commonly encountered drugs and metabolites using tri-n-alkylamines as reference compounds, nitrogen–phosphorus detectors, and dual capillary chromatography. *J Anal Toxicol*, **11**, 210–4.

Wilson, J.N. (1940) A theory of chromatography. *J Am Chem Soc*, **62**, 1583–91.

6 Thin-Layer Chromatography

6.1 Introduction

Thin-layer chromatography (TLC) involves the movement by capillary action of a liquid phase (generally an organic solvent) through a thin, uniform layer of stationary phase (usually hydrated silica gel, SiO_2) held on a rigid or semirigid support, normally a glass, aluminium or plastic sheet or 'plate' (Poole, 2004). TLC is relatively inexpensive and simple to perform, although a fume cupboard or hood is normally needed, particularly if plates are sprayed rather than being dipped to visualize the analytes. TLC can be a powerful qualitative technique when used together with some form of sample pretreatment such as LLE (Section 3.2.5). However, some separations can be difficult to reproduce, even when they are performed in the same laboratory. The interpretation of results can also be very difficult, especially if several drugs and/or metabolites are present (Box 6.1).

Box 6.1 Application of TLC in analytical toxicology

- Robust qualitative technique
- Batch analysis – many extracts can be analyzed together
- Primarily used for urine, although extracts of other samples can be analyzed
- Non-destructive: can usually recover analyte if plate not treated with visualization reagent
- Does not require complex equipment (but needs fume cupboard or hood)
- Relatively low sample capacity – chromatographic system easily overloaded
- Quantitative measurements are possible using densitometric scanning

TLC of solvent extracts of urine, stomach contents or scene residues is widely used in poison 'screening' procedures and is also recommended for the detection and identification of a number of specific compounds and groups of compounds. In some respects, TLC is an extension of the colour tests discussed in Section 4.1 as the colours formed with various reagents form the basis of compound identification. However, incorporation of: (i) a solvent extraction and concentration step and (ii) a chromatographic step enhances both sensitivity and selectivity. If required, unreacted zones (i.e. zones not treated with a visualization reagent), can be removed and prepared for GC, HPLC or MS analysis.

Generally, it is unwise to rely solely on TLC without corroboration of the results by GC, for example. This is because the resolving power of TLC is limited and the interpretation of the chromatograms obtained is, to an extent, subjective (Ojanperä, 1992). This being said, high-performance TLC (HPTLC) has superior resolving power as compared to conventional TLC (Section 6.8), and the combination of efficient chromatography with colour visualization confers good selectivity. Quantitative work is feasible with a densitometric scanner.

It is important to note that many spray reagents used in analyte visualization are extremely toxic and spraying should *always* be carried out in a suitable fume cupboard or hood. Furthermore, drift from spaying can leave residues on the walls of the cupboard, so lining the cupboard is to be recommended to facilitate periodic cleaning. Disposal of sprayed TLC plates must also be undertaken with due care after recording the results. Some practical points are summarized in Box 6.2.

> **Box 6.2** Practical aspects of TLC in analytical toxicology
>
> - Difficult to obtain reliable TLC plates – pretest a plate from every batch purchased
> - Difficult to automate
> - If developing solvent uses ammonium hydroxide, concentrated ammonia (RD 0.88, 28 % w/v) *must* be used; loss of ammonia from reused solvent is a frequent cause of poor results
> - Use 'saturated' TLC tanks to obtain reproducible results
> - Must score columns on large plates for good results
> - Many spray reagents are very toxic – *care*!!

6.2 Preparation of thin-layer plates

On a standard TLC plate, the stationary phase is normally a uniform film (0.25 mm thick) of silica gel (20 μm average particle size). The usual size of the plate is 20 × 20 cm, although smaller sizes may be used. Plates incorporating a fluorescent indicator are available so that UV absorbing compounds can be located as dark spots on a fluorescent background prior to spraying with visualization reagents, if required. Prior soaking of the plate in methanolic potassium hydroxide and drying may improve the chromatography of some basic compounds when using certain solvent systems, but generally addition of concentrated ammonium hydroxide (RD 0.88) to the mobile phase has the same effect.

TLC plates can be prepared in the laboratory from silica gel containing an appropriate binding agent (silica gel G contains gypsum – hydrated calcium sulfate – as binding agent) by using a simple slurry-spreading apparatus. It is important to ensure that the glass plates (usually 20 × 20 × 0.5 cm) are clean and free from grease before use. The silica gel is mixed with twice its own weight of water to form a slurry, which is then quickly applied to the glass plate using a commercially available spreader to form a film about 0.25 mm thick. Additives such as a fluorescent indicator may be included if required. The plates should be dried in air and should be kept free of moisture prior to use. The quality of such 'home-made' TLC plates should be carefully monitored; activation (i.e. heating at 100 °C for 30 min before use) may be helpful in maintaining performance. Preparing TLC plates by dipping glass plates into a slurry of silica with subsequent drying gives very variable results and is not to be recommended.

Home-made plates generally give silica layers that are much more fragile than those of commercially available plates. In addition, chromatographic performance tends to be much less reproducible with home-made plates. Experience suggests that it is best to standardize on a particular brand of commercially available plates, such as Silica gel 60 F_{254} (E. Merck, Darmstadt, Germany). However, even with commercial plates batch-to-batch variations in retention, and also in sensitivity with certain analyte–spray reagent combinations, may be encountered.

6.3 Sample application

Some commercially available plates are supplied with a special adsorbent layer to simplify application of the sample. Normally, however, the sample is loaded directly onto the silica gel layer. The plate should be prepared by marking the origin by drawing a light *pencil* line at least 2 cm from the bottom of the plate – care should be taken not to disturb the silica surface in any way. A line should then be drawn on the plate 10 cm above the origin to indicate the optimum position of the solvent front; other distances may be used if required. It is advisable when using 20×20 cm plates to score columns approximately 2 cm wide vertically up the plate leaving scored margins on either side of the plate to minimize 'edge effects' as discussed below.

The samples and any standards should be applied at the origin in the appropriate columns. Sample loading should be performed carefully using a micropipette or syringe so as to form 'spots' no more than 5 mm in diameter. If larger spots are produced then resolution will be impaired when the chromatogram is developed. The volume of solvent applied should be kept to a minimum; typically 5–10 μL of solution containing about 10 μg of analyte is suitable. Sample extracts reconstituted as appropriate should be applied first, followed by the required standard(s) or mixtures of standards; applying the sample extracts *before* the standards minimizes the risk of cross contamination. The use of a nonpolar solvent to apply the spots will result in small concentrated spots, as the analytes tend to be adsorbed onto the silica. However, this is not always practicable and methanol will usually prove satisfactory. The spots should be dried with a stream of air or nitrogen prior to chromatography. Warm air blowers should be used with care, having regard for the stability and volatility of analytes such as amfetamines. The plate must be allowed to cool before development is commenced.

6.4 Developing the chromatogram

Glass TLC development tanks, in a range of sizes to suit different TLC plates, are available from many suppliers. Such tanks normally have a ground glass rim, which forms an airtight seal with a glass cover plate. A small amount of silicone lubricant jelly may be used to secure the seal. Some tanks have a well at the bottom to reduce the amount of solvent required. Tanks should be lined with filter paper or blotting paper on three sides and the solvent added at least 30 min before the chromatogram is to be developed to saturate the atmosphere with solvent vapour as this aids reproducible chromatography. Faster separations are possible with saturated tanks, and the results are less dependent on the chromatographic conditions, such as the dimensions and shape of the tank and the ambient temperature.

Some TLC mobile phases consist of a single solvent, but most are mixtures. One of the most widely used mobile phases in analytical toxicology is ethyl acetate:methanol:concentrated ammonium hydroxide ($85 + 10 + 5$), 'EMA'. It is important to prepare mobile phases daily, because the composition may change with time due to evaporation or chemical reaction. In particular, loss of ammonia not only from the mobile phase, but also from reagent bottles once opened causes many problems.

The chromatogram is developed by placing the loaded plate in the equilibrated tank, ensuring that the level of the solvent is above the bottom edge of the silica layer on the plate, but below the level of the 'spots' applied to the plate, and quickly replacing the lid (Figure 6.1). The plate should be observed to ensure that the solvent front is being drawn up uniformly. The mobile phase

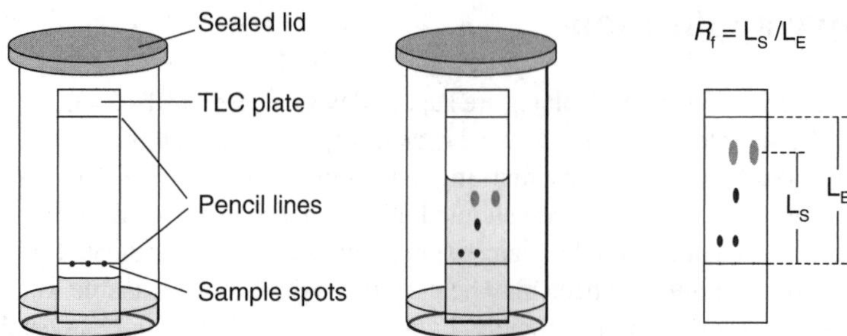

Figure 6.1 Thin-layer chromatography: schematic and calculation of R_f values.

movement is primarily due to capillary forces and, as the stationary phase is dry, the profile of the mobile phase is concave (the shape of an advancing meniscus). Consequently, the solvent front will show a degree of curvature, particularly at the edges of the plate; more serious curvature or bowing may be observed if the atmosphere in the tank is not sufficiently saturated with solvent vapour. This 'edge' effect can be minimized by dividing the plate into about 2 cm columns, as discussed above. The velocity of the mobile phase decreases because of its viscosity and the increasing weight of the mobile phase as it progresses up the plate. Forced-flow systems (Section 6.8.1) overcome this by maintaining a constant flow rate (Figure 6.2).

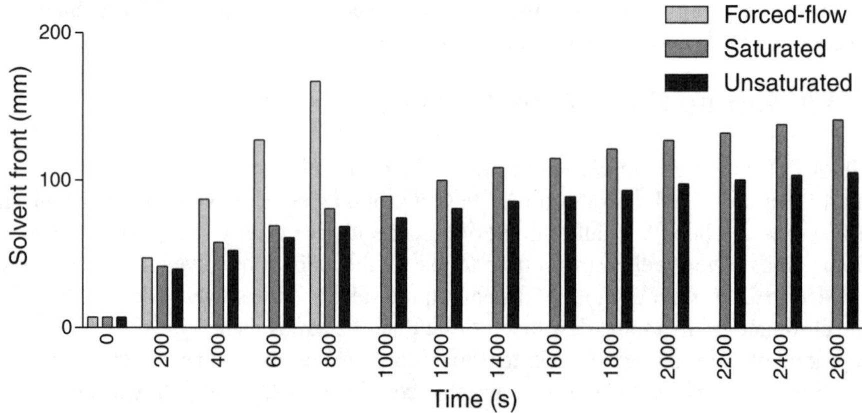

Figure 6.2 Comparison of the migration of the solvent front in saturated and unsaturated chambers. With forced-flow chromatography the solvent front moves at constant rate. Eluent: chloroform. Reprinted from Kalász, H. and Báthori, M. (2001) Pharmaceutical applications of TLC. LC GC Europe, **14**, 311–21, with permission from Advanstar Communications.

The chromatogram should be allowed to develop for the intended distance, usually 10 cm from the origin, after which the plate should be taken from the tank, placed in a fume cupboard or under a fume hood and allowed to dry. Drying may be hastened by blowing warm air (from a hair dryer) over the plate until all traces of solvent have been removed. This can be especially important with ammoniacal mobile phases because the presence of residual ammonia affects the reaction observed with certain spray reagents.

6.5 Visualizing the chromatogram

When the chromatogram has been developed and the plate dried, the chromatogram should be examined under UV light (254 and 366 nm) in a suitable enclosed box and the positions of any fluorescent compounds ('spots') noted. If a fluorescent marker has been added to the silica, many substances present appear as dark areas against a fluorescent background. Spraying or dipping the plates usually quenches the fluorescence of the indicator.

In clinical toxicology the use of chromogenic chemical detection reagents generally gives more useful information. Plates can be 'dipped' in reagent, but unless special precautions are taken, the structure of the silica tends to be lost and the chromatogram destroyed. Thus, the reagent is normally lightly applied as an aerosol using a commercially available spray bottle attached to a compressed air or nitrogen line. Varying the inlet pressure varies the density of the aerosol and thus the amount of reagent reaching the chromatogram in a given time.

When spraying the plate it can be difficult to obtain even coverage, particularly where the plate is standing on a surface and excess spray reagent may be drawn up the plate by capillary action destroying the lower part of the chromatogram. The simple expedient of inverting the plate before spraying overcomes this problem. Glass plates can be used to 'mask' portions of the plate if selected columns are to be sprayed with different reagents (Box 6.3, Figure 6.3). Alternatively, if plastic- or aluminium-backed plates are used then the appropriate regions can be cut up and sprayed separately. The appearance of certain compounds may change with time and so it is important to record results as quickly and carefully as possible, noting any time-dependent changes. A standardized data recording system is valuable for reference purposes – ideally a digital camera should be used and the files archived.

Box 6.3 Visualizing reagents for acidic and basic extracts (see Figure 6.3)

- Mercury(I) nitrate reagent
  - Mercury(I) nitrate (1 g) + water (100 mL)
  - Add concentrated nitric acid until the solution is clear
- Acidified iodoplatinate reagent
  - Platinum(II) chloride (0.25 g) + potassium iodide (5 g) + concentrated hydrochloric acid (5 mL)
  - Make up to 100 mL with water
- Mandelin's reagent
  - Ammonium vanadate (finely powdered, 0.5 g) + concentrated sulfuric acid (100 mL)
  - Shake well before use
- Aqueous sulfuric acid
  - Water (50 mL) + concentrated sulfuric acid (50 mL)
  - Add acid slowly with care and cooling

Various combinations of spray reagents may be used to aid identification of unknown substances and some agents may be useful for identifying functional groups. For example, Dragendorff's reagent gives orange to orange-red spots with amines, in particular tertiary amines and quaternary ammonium compounds, and so traditionally has been used to detect alkaloids. Ninhydrin

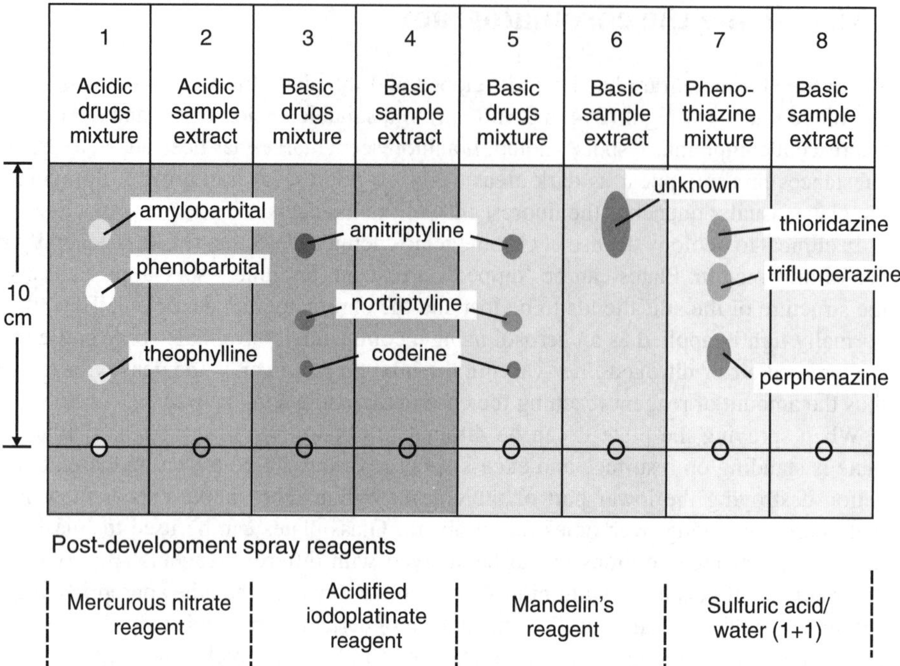

Figure 6.3 Example of differential visualization of a TLC plate.

reacts with primary amines, including amino acids. Secondary amines can be visualized using nitroprusside:acetaldehyde (Table 6.1).

In the example shown in Figure 6.3, areas of the plate were masked with thin glass plates and each region sprayed separately. However sequential spraying with different reagents is a valuable identification technique. Dragendoff's reagent can be sprayed after the plate has been sprayed with ninhydrin and FPN reagent (Table 6.1), for example. Furthermore, quite complex reactions may be performed by spraying sequentially.

Benzodiazepines in urine have been identified after acid hydrolysis to aminobenzophenones. Those with a primary aromatic amine may be localized using the Bratton–Marshall reaction (Box 6.4). Analytes are diazotized with nitrous acid and coupled with *N*-1-naphthylethylene diamine to give highly coloured azo dyes, the excess nitrous acid having been removed by reaction with ammonium sulfamate (Berry and Grove, 1973). Benzophenones from benzodiazepines with N_1-alkyl substituents can be dealkylated, for example by exposure to UV radiation (Schütz, 1989), which appears preferable to heating (Volf, 1998), before reaction with nitrous acid to ensure increased sensitivity (Figure 6.4).

Clearly, the method cannot distinguish between benzodiazepines that are hydrolyzed to a common benzophenone (e.g. diazepam, ketazolam, temazepam). However, the migration of 5-chloro-2-methylaminobenzophenone is different from that of the 2-amino-5-chlorobenzophenone formed from nordazepam and oxazepam. Generally, the spots are purple in colour, but the 7-aminobenzodiazepine metabolites of nitrazepam and clonazepam, for instance, give di-aminobenzophenones, which appear blue with the Bratton–Marshall reagent. Interference from other primary amines such as sulfonamides is minimized by using relatively nonpolar eluents.

Table 6.1 Examples of chromogenic reagents

Reagent	Constituents	Comments
Diazotized 4-nitroaniline	(i) 4-Nitroaniline (0.7 g) in hydrochloric acid (1 mol L^{-1}, 100 mL) (ii) Sodium nitrite (1 g) in water (100 mL) Add 4 mL (i) dropwise to 5 mL ice-cold (ii) and make up to 100 mL with water	Phenols
Dragendorff	(i) Bismuth subnitrate [basic bismuth nitrate, $4BiNO_3(OH)_2BiO(OH)$] (0.85 g) dissolved in glacial acetic acid (10 mL) and water (40 mL) (ii) Potassium iodide (8 g) in water (20 mL) Mix 5 mL (i) and 5 mL (ii) and glacial acetic acid (20 mL). Make up to 100 mL with water	Amines and quaternary ammonium compounds, particularly alkaloids
Folin–Ciocalteau	Available as a stock solution (2 mol L^{-1}) Dilute with water or ethanol (1+1) before use	Phenols
FPN (Ferric Perchloric Nitric)	Iron(III) chloride (5 % w/v, 5 mL), perchloric acid (20 % w/w, 45 mL), nitric acid (50 % v/v, 50 mL)	Chiefly phenothiazines
Ninhydrin	Ninhydrin (0.3 g), pyridine (5 mL), methanol (95 mL)	Primary amines, amino acids
Nitroprusside–acetaldehyde	(i) Sodium nitroprusside (5 g) + acetaldehyde (10 mL) + water to 100 mL (ii) Sodium carbonate (5 g) in water (100 mL) Mix equal volumes of (i) and (ii) before use	Secondary amines

The chromatography of benzophenones and related hydrolysis products of some of the more recently introduced benzodiazepines, including clonazepam, lorazepam, flunitrazepam, prazepam and quazepam, has been described (Schütz, 1989).

6.6 Retention factor (R_f)

Retention factor (R_f, also know as retardation factor) is normally used to record TLC retention data. R_f is defined as:

$$R_f = \frac{\text{Distance the analyte has travelled from the origin}}{\text{Distance the solvent front has travelled from the origin}}$$

Box 6.4 TLC identification of benzodiazepines as benzophenones

- Hydrolysis
  - Mix sample (urine, stomach contents) or standard (10 mL) with concentrated hydrochloric acid (3 mL)
  - Heat 100 °C (30 min)
- Extraction
  - Cool and extract with petroleum ether (40–60 °C fraction)[a]
  - Centrifuge to separate the phases
  - Transfer the organic phase to a clean tube and evaporate to dryness
  - Reconstitute in petroleum ether (40–60 °C fraction) (0.1 mL)
- Chromatography
  - Silica gel plates (prepared as described above)
  - Develop in toluene:glacial acetic acid (97 + 3)[b]
- Visualization
  - Spray the plates sequentially with the following drying between each application:
    (i) aqueous sulfuric acid (Box 6.3)
    (ii) freshly prepared aqueous sodium nitrite solution (10 g L^{-1})
    (iii) aqueous ammonium sulfamate solution (50 g L^{-1})
    (iv) *N*-1-naphthylethylene diamine (10 g L^{-1} in acetone:water, 80 + 20)

[a]Aqueous phase may be made alkaline prior to extraction (Schütz, 1989)
[b]Benzene:glacial acetic acid (97 + 3) used originally

Thus, R_f is a fraction (0.30, 0.75, etc.) and is (reasonably) independent of the length of the chromatogram (distance the solvent front has travelled from the origin). hR_f values ($R_f \times 100$) are even more convenient, especially if the length of the chromatogram is standardized at 10 cm (100 mm) as then hR_f is simply the distance (mm) the analyte has travelled from the origin.

There are many factors that influence the reproducibility of hR_f values including: (i) the TLC plate, (ii) the amount of analyte applied to the plate, (iii) the development distance, (iv) the degree of tank saturation, (v) the ambient temperature and (vi) the effect of coextracted compounds (matrix effects). The influence of some of these factors can be minimized if reference compounds (standards) are analyzed with each sample. When calculating hR_f values for unknown substances, it is usual to calculate a 'corrected hR_f value' from experimentally observed values of sample and reference compounds (Box 6.5). Alternatively, a graph of accepted literature values of hR_f and

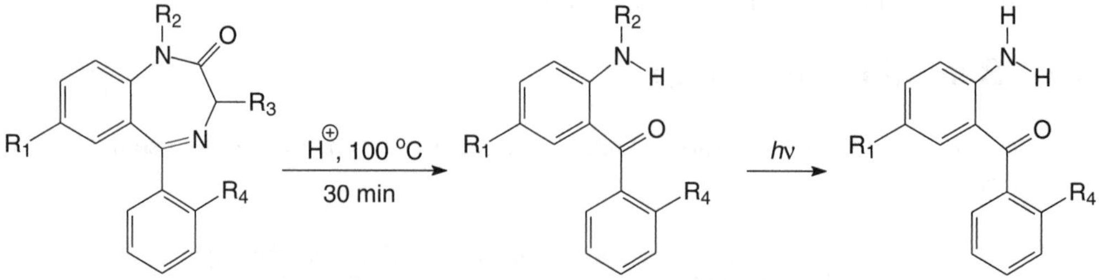

Figure 6.4 Hydrolysis of benzodiazepines and dealkylation of *N*-alkylbenzophenones.

Box 6.5 Calculation of corrected hR_f values

Calculation of corrected hR_f for an unknown analyte, U with an observed hR_f of U_{obs}. If U_{obs} is between two standards A and B, with uncorrected hR_f values of A_{obs} and B_{obs}, respectively, then the corrected value for the unknown is:

$$\text{corrected } hR_f \text{ of unknown } (U) = A_{lit} + \left(\frac{B_{lit} - A_{lit}}{B_{obs} - A_{obs}}\right)(U_{obs} - A_{obs})$$

where A_{lit} and B_{lit} are the accepted literature (database) values for the hR_f of A and B.

observed values for the reference standards (ideally four, equally spaced on the chromatogram) can be plotted against the observed values. The corrected value for the unknown can thus be obtained by interpolation.

A major complication is that the chromatography of compounds that originate from biological extracts is often different from that of the pure substances because of the presence of additional sample components in solvent extracts, for example. It is not practical to prepare standard solutions in every type of matrix that may be encountered, but assuming that there is enough sample, addition of a small quantity of a suspected component to an extract followed by reanalysis may help elucidate if that particular component is present in the extract or not. If the added material and the unknown do not co-chromatograph this is a powerful way of eliminating the suspected component.

The correct interpretation of TLC data requires practice and experience, but there are some basic guidelines (Box 6.6).

Box 6.6 Analytical toxicology – interpretation and storage of TLC data

- Analyze known reference compounds on the same plate as unknowns
  - Spot the extracts first!
- R_f data are not very reproducible
  - Compare results with authentic compound for uncommon drugs/poisons
  - Analyze on same plate as the unknown
- Interpretation depends on the complexity of the case and the experience of the operator
  - Many basic drugs give very similar colour reactions with certain sprays
  - Always correlate with GC, HPLC and immunoassay data
- Difficult to store developed chromatograms – use (digital) photography

Identification of candidate molecules for further investigation can be made by the judicious use of a combination of development systems and visualization reagents. Using five eluent systems, Zingales (1967, 1968) was able to identify 45 psychotropic drugs. However, a problem with using colour as a means of analyte identification is describing a given colour unambiguously. The colour produced by fluphenazine with FPN reagent, for example, has been described as cameo, salmon and orange (and probably many more hues). The perception of a given colour is also likely to change with the amount of analyte present. One attempt to overcome this with the colours produced by phenothiazines with sulfuric acid has been to measure the wavelength maximum (Whelpton, 1997). It is worth noting that metabolism, particularly hydroxylation, changes the colours, for example mesoridazine (a thioridazine metabolite) gives a colour that is very similar to that given by chlorpromazine.

In interpreting TLC results, as in any analytical toxicology results, it is important to understand the nature of the sample and chemistry of the analyte(s). Usually, only small quantities of very lipophilic compounds are excreted in urine, although their metabolites may be present at high concentrations. Thioridazine, for example, is prescribed as a racemate and because sulfoxidation and *N*-oxidation introduce additional chiral centres, this gives rise to diastereoisomeric metabolites, which often chromatograph as characteristic pairs of spots (Flanagan *et al.*, 1995; plate 11).

6.6.1 TIAFT-DFG R_f data compilation

The second edition of the TIAFT-DFG R_f data compilation (TIAFT-DFG, 1992) provides R_f values for some 1800 toxicologically relevant substances on 10 TLC systems that can be used for general screening purposes. There are also data on special systems for the analysis of pesticides and benzodiazepine hydrolysis products (benzophenones), in urine. Much of this information is available from other sources (Moffat *et al.*, 2004).

6.7 Toxi-Lab

Toxi-Lab (Varian, www.varianinc.com/cgi-bin/nav?products/consum/toxilab/, accessed 12 July 2007) is a standardized TLC system that is available in kit form together with a compendium of colour plates showing R_f values, colour reactions and additional information to facilitate interpretation (AB System). Cellulose impregnated with silica takes the place of the conventional TLC plate – a practical advantage is that dipping in visualization reagents can be used, obviating the need for a fume hood.

Over 700 analytes and their metabolites have been documented for ease of identification. The system is kept up to date: citalopram, mirtazapine, olanzapine and quetiapine are amongst the compounds documented in recent years. However, as with TLC in general, problems can arise when attempting to differentiate compounds with similar mobility and colour reactions, especially if more than one compound is present. Moreover, the kit is aimed primarily at the United States market and thus some common United Kingdom drugs are not included in the compendium. The dangers inherent in the inappropriate use of this kit in overdose work have been stressed (Dawling and Widdop, 1988). It should be used with caution for testing for drugs of abuse in urine (Wilson *et al.*, 2001).

6.8 High-performance thin-layer chromatography

HPTLC plates have a smaller average particle size (5–10 μm) and give greater efficiency than conventional plates. 'Reversed-phase' plates, where a hydrophobic moiety (usually C_2, C_8 or C_{18}) is bonded to the silica matrix, are also available. However, HPTLC and reversed-phase plates are more expensive and have a lower sample capacity than conventional plates, an important consideration in overdose toxicology when the amounts of analyte present can easily overwhelm even ordinary TLC plates.

HPTLC plates are typically 10 × 10 cm. HPTLC spots should be as discrete as possible (1–2 mm diameter). Automated spotting systems are advisable and are mandatory for quantitative work.

6.8.1 Forced-flow planar chromatography

Rather than rely on the capillary action to draw eluent up the plate, several ways of 'driving' the flow have been described (Nyiredy, 2003), including rotational planar chromatography (RPC) in which centrifugal force is used, electroplanar chromatography (EPC) which is the TLC equivalent of CEC (Section 9.9.1) and overpressured-layer chromatography (OPLC). This last technique, introduced in the 1970s, has been given a new lease of life since 2000 with the introduction of improved apparatus and a change of name; the acronym now stands for Optimum Performance Laminar Chromatography. With the flow being driven by high-pressure pumps and the possibly of on-line detection, OPLC can be thought of as flat-bed HPLC; indeed the manufacturers refer to the sorbent layer as a column rather than a plate (Figure 6.5).

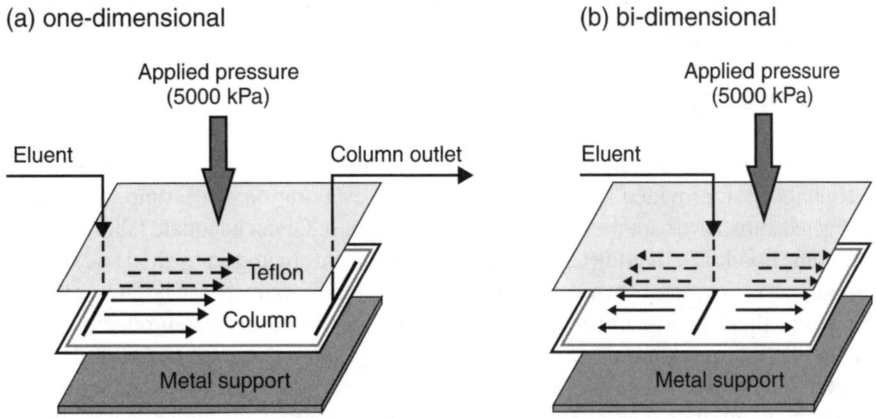

Figure 6.5 Two modes of forced-flow planar chromatography. Reproduced from Bryston, N. and Papillard, D. (2004) An introduction to OPLC. Operation and applications. LC GC Europe, **17**, 41–7.

A basic OPLC unit can be connected in the same way as a cylindrical HPLC column and be used on-line with isocratic or gradient elution. Or it can be used off-line, in much the same way as a conventional TLC plate. Thus, OPLC retains the advantages of TLC, including multiple channels, two-dimensional development plus many innovations, such as being able to perform two-dimensional chromatography on four samples on single plate. A method for screening for basic drugs in 2 mL urine samples by dual-plate OPLC has been described (Pelander *et al.*, 2003).

6.9 Quantitative thin-layer chromatography

Provided adequate care is taken, quantitative TLC can give highly reproducible results, with RSDs of 2–3 %. HPTLC with scanning densitometery, which may be used in several modes including transmittance, fluorescent quenching and fluorescence, is probably the most reliable approach. Automatic spotting ensures precise application of the sample. However, for quantitative work variability in application of the chromogenic reagent can be problematic. Dipping manually is extremely technique dependant and an automated dipper is the only acceptable approach. On-line OPLC (Section 6.8.1) allows continuous quantification with standard HPLC detectors as discussed above. If a densitometer is not available then spots may be scraped from the plates and the analytes eluted for quantification by whichever technique is deemed suitable.

The Iatroscan was introduced as a quantitative TLC system based on quartz rods (150 × 0.9 mm o.d.) coated with silica or alumina (Chromarod). Separations are performed on the rods (supported in a stainless steel frame) and after drying the analytes are quantified by FID. The MK6 version has dual FID/FPD, allowing high-sensitivity detection of sulfur- and phosphorus-containing compounds. Despite the obvious advantage of being able to chromatograph high boiling point compounds, the technique has not been widely adopted.

In metabolic studies, radiolabelled compounds are used frequently, and analytes can be located on TLC plates (e.g. by autoradiography) and removed for scintillation counting, or located and quantified by radioactive scanning. TLC is particularly useful for metabolic and toxicological studies during product development as all the metabolites in an extract will be somewhere on the plate, even if they are in the solvent front or at the origin, assuming that they are not volatile when they may be lost by evaporation. With HPLC there is always the possibly that metabolites have been retained on the column.

6.10 Summary

TLC still finds a place in STA, but rarely in quantitative work. However, TLC does have the advantage that, provided analytes are not lost by evaporation or decomposition on the plate, all extracted substances are present on the chromatogram. Given adequate laboratory facilities, such as a fume hood, TLC has the advantages over elution chromatography that: (i) expensive analytical equipment is in general not needed and (ii) several sample extracts can be analyzed simultaneously. However, there is no substitute for experience, both in ensuring reproducible chromatography and in the interpretation of results. The chromatographic system may be easily overloaded, and whilst some metabolites may give similar colour reactions to the parent compound, for example, others may not, thus making interpretation more difficult. Thus, unless a scanning densitometer is available, recourse to another technique such as GC or HPLC is often necessary, as discussed in the ensuing chapters.

References

Berry, D.J. and Grove, J. (1973) Emergency toxicological screening for drugs commonly taken in overdose. *J Chromatogr*, **80**, 205–20.

Dawling, S. and Widdop, B. (1988) Use and abuse of the Toxi-Lab TLC system. *Ann Clin Biochem*, **25**, 708–9.

Flanagan, R.J., Braithwaite, R.A., Brown, S.S., Widdop, B. and de Wolff, F.A. (1995) *Basic Analytical Toxicology*, World Health Organization, Geneva.

Moffat, A.C., Osselton, D., Widdop, B. (eds) (2004) *Clarke's Analysis of Drugs and Poisons*, 3rd edn, Pharmaceutical Press, London.

Nyiredy, S. (2003) Progress in forced-flow planar chromatography. *J Chromatogr A*, **1000**, 985–99.

Ojanperä, I. (1992) Toxicological drug screening by thin-layer chromatography. *Trends Anal Chem*, **11**, 222–30.

Pelander, A., Ojanperä, I., Sistonen, J., Rasanen, I. and Vuori, E. (2003) Screening for basic drugs in 2-mL urine samples by dual-plate overpressured layer chromatography and comparison with gas chromatography-mass spectrometry. *J Anal Toxicol*, **27**, 226–32.

Poole, C.F. (2004) Thin-layer chromatography, in *Clarke's Analysis of Drugs and Poisons*, 3rd edn (eds A.C. Moffat, D. Osselton and B. Widdop), Pharmaceutical Press, London, pp. 392–424.

Schütz, H. (1989) *Benzodiazepines II. A Handbook. Basic Data, Analytical Methods, Pharmacokinetics and Comprehensive Literature*, Springer-Verlag, Berlin.

The International Association of Forensic Toxicologists (TIAFT)/Deutsche Forschungsgemeinschaft (DFG) (1992) *Thin-Layer Chromatographic Rf Values of Toxicologically Relevant Substances on Standardized Systems*, 2nd edn, VCH, Weinheim.

Volf, K. (1998) Detection of 1,4-benzodiazepines and benzophenones with a secondary amino group by thermal conversion and diazotization and spraying with Bratton–Marshall reagent. *J Planar Chromatogr*, **11**, 132–6.

Whelpton, R. (1997) Neuroleptics, in *Analytical Toxicology for Clinical, Forensic and Pharmaceutical Chemists*, Vol 5 (eds H. Brandenberger and R.A.A. Maes). Walter de Gruyter, Berlin, pp. 451–470.

Wilson, J.F., Smith, B.L., Toseland, P.A., Watson, I.D., Williams, J., Thomson, A.H., Capps, N.E., Sweeney, G., Sandle, L.N., Steering Committee for the United Kingdom National External Quality Assessment Scheme for Drug Assays. (2001) A survey of extraction techniques for drugs of abuse in urine. *Forensic Sci Int*, **119**, 23–7.

Zingales, I. (1967) Systematic identification of psychotropic drugs by thin-layer chromatography. I. *J Chromatogr*, **31**, 405–19.

Zingales, I. (1968) Systematic identification of psychotropic drugs by thin layer chromatography. II. *J Chromatogr*, **34**, 44–51.

7 Gas Chromatography

7.1 Introduction

In analytical toxicology, gas chromatography (GC) has a number of advantages over other widely used techniques such as HPLC and immunoassay. Firstly, GC has a range of sensitive detectors, such as the 'universal' FID and the selective NPD, ECD and MS detectors, which can be used in parallel. Secondly, stable, high efficiency (capillary) GC columns are now widely available. Thirdly, GC is easy to interface with techniques giving direct information about compound identity such as electron-ionization MS (EI-MS) and Fourier-transform infrared spectrometry (FTIR) (Grob and Barry, 2004; Rood, 2006). A compendium of GC terms and techniques is available (Hinshaw, 2002).

With GC, as with HPLC, qualitative and quantitative information can often be obtained in the same analysis provided that appropriate calibration and QC procedures are followed. Temperature programming in GC is analogous to gradient elution in HPLC, but is much simpler to perform and permits the analysis of compounds of different volatilities in one analysis. Moreover, the return to starting conditions is easy and the interdependence of M_r, retention time and column temperature is valuable in aiding peak assignment in STA. In addition, GC retention data are reproducible between different days, columns, instruments, operators, centres and so on (Box 7.1).

Box 7.1 Advantages of GC in analytical toxicology

- Can inject aqueous mixtures in some applications such as ethanol, although headspace almost as easy and applicable to wide range of volatiles
- Easy to use SPME/thermal desorption
- Range of stable, efficient capillary columns
- Sensitive universal (FID) and selective (NPD, ECD, MS) detectors
- Qualitative and quantitative
  - Wide-bore capillary columns used with injection liners facilitate quantitative work
  - Inter-dependence of elution time, column temperature, and M_r valuable in qualitative work
- Retention data reproducible between days/columns/operators/laboratories/countries
- Temperature programming and interfacing to MS or FTIR easy – need few column types because of high resolving power especially with GC-MS
- Can generate EI spectra from GC-MS
- Principal peak index valuable in compound identification
- Can be used for a wide range of gases and solvents

Disadvantages of GC include the requirement that the analyte or a derivative should be volatile and stable at the temperature required for the analysis (in practice below 400 °C or so). In addition,

Box 7.2 Limitations of GC in analytical toxicology

- Hardware/infrastructure expensive – need experienced operators, pure gas supplies, consumables, support from instrument manufacturer
- Septum injection – need frequent maintenance
- Sequential analysis – need autoinjectors
- Usually need some form of sample pretreatment
- Cannot adjust selectivity by modifying eluent (cf. HPLC)
- Cannot analyze very polar/high M_r compounds (cf. HPLC), although can sometimes derivatize
- Cannot analyze carbon monoxide directly by FID
- There is always the possibility of interference from coeluting compounds

substances with highly polar or ionizable functional groups may give poor performance (broad, tailing peaks), and some form of sample preparation is normally needed (Box 7.2). Moreover, some compounds that may be present in sample extracts may be thermally labile and the decomposition products may interfere in the analysis. Thus, in addition to the choice of the sample preparation procedure, the column and the chromatographic and detection conditions, due consideration must be given to other factors including sample collection and storage, choice of internal standard and quality assurance. This being said, GC remains the method of choice for gases and other volatiles such as ethanol and inhalational anaesthetics. GC is also widely used in the analysis of other compounds, in STA, and as an interface to MS, an especial advantage here being that EI spectra can be obtained (Section 10.4.3).

7.2 Instrumentation

Typically a GC system consists of a gas control unit, which supplies carrier gas to the column as well as gases, such as compressed air and hydrogen, to the detector, a sample injection system, an analytical column, and a detector with associated data acquisition/processing (Figure 7.1). The injector, column and detector ovens are normally heated independently, the detector generally being maintained at a higher temperature than the maximum temperature attained by the column oven to minimize the risk of fractionation or condensation of sample components in the detector. A similar consideration applies to the injector oven in isothermal operation. Some detectors, such

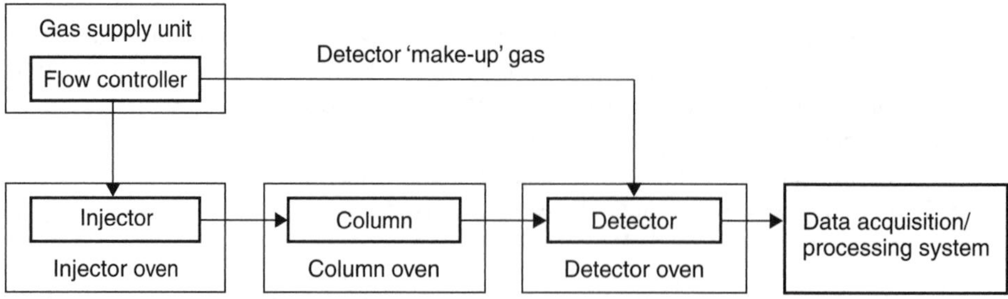

Figure 7.1 Block diagram of a gas chromatograph.

as the ECD (Section 7.2.2.2), require an additional 'make-up' gas flow in order to give optimum performance.

Gas chromatographs are usually purchased as a single unit, with gas supplies and data capture being separate. Increasingly instrument control is from a personal computer that also provides data capture facilities. Although commercially supplied cylinders are a convenient source of the gases needed in GC operation, alternatives are available for air (simple compressor), hydrogen (electrolytic hydrogen generator) and nitrogen (nitrogen generator based on a molecular sieve). However, regular monitoring and maintenance, and the use of appropriate filters to remove hydrocarbons, oxygen and moisture are mandatory. Filters should always be used with gas from cylinders.

7.2.1 Injectors and injection technique

Analytical toxicology is normally trace analysis and thus sensitivity is often limiting. Generally then, it is important that as much of a sample extract as possible is injected onto the column, other factors (the 'signal-to-noise ratio') being equal. In GC, sample capacity per injection is limited, but this does not matter greatly because GC detectors are very sensitive. However, in most GC methods the difficulties inherent in reproducibly injecting relatively small volumes of an extract necessitate the use of internal standardization. With packed columns sample injection is usually via a syringe through a silicone rubber septum in the injection port – it is important to use 'low-bleed' septa, especially with sensitive detectors such as the ECD. Use of a glass injection port liner that can be removed and cleaned minimizes the accumulation of non-volatile residues on the column, but on-column injection may be preferred if labile substances are to be analyzed. In such cases contaminated packing at the top of a column should be replaced with fresh material if efficiency is affected. Normally a minimal amount of solvent will be injected to reduce solvent effects at the detector. However, for gases and vapours much larger volumes can be injected via a gas tight syringe or a gas-sampling valve.

A variety of injection devices can be used with capillary columns and the terminology employed can be very confusing. Nevertheless, the importance of using an appropriate injection technique when working with such columns cannot be overemphasized and the subject is worthy of specialized study if the full capabilities of modern columns are to be realized (Grob, 2001). With relatively narrow-bore capillaries (<0.32 mm i.d.) some form of inlet splitter is normally needed to prevent overloading the column with injected material (Figure 7.2).

When using an inlet splitter, as with other GC injection devices, it is important that the portion of the injection solution passed to the column has the same composition as the rest of the solution. With wider bore capillaries (0.32 mm i.d. or greater) then (splitless) syringe injection either through a septum into a glass liner, or directly onto the column via a fused silica capillary guard column or retention gap (a 1–10 m length of deactivated tubing without stationary phase) is relatively simple and ensures deposition of the analytes on the column. If efficiency deteriorates, then removal of the first few cm or even an entire coil from the guard column often restores performance without altering analyte retention on the analytical column. A retention gap is sometimes needed to improve chromatographic performance when relatively large volumes (>2 μL) of solvent are injected, and of course can also serve as a guard column.

With on-column injection it is usual to inject at a temperature below the boiling point of the injection solvent. As the solvent evaporates in the stream of carrier gas, the analytes are concentrated in a small band and are then volatilized by increasing the column oven temperature.

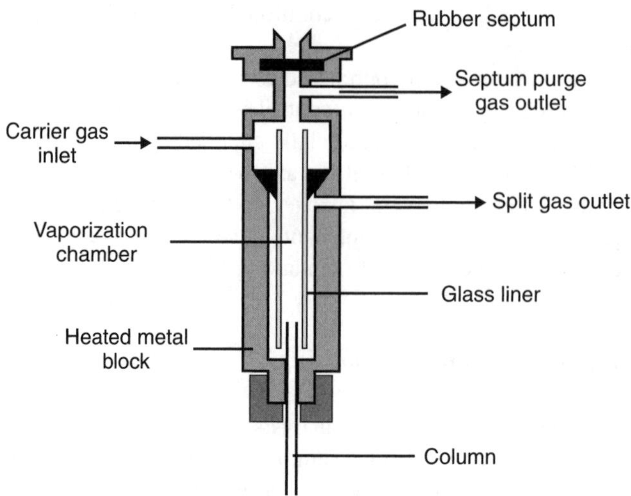

Figure 7.2 Split/splitless injector for GC.

7.2.1.1 *Cryofocusing/thermal desorption*

The great flexibility offered by the ability to use temperature programming in GC can be exploited in the cryofocusing or 'purge-and-trap' of volatile analytes, as discussed in Section 3.2.4. Thermal desorption and refocusing techniques are simple to use, easily automated, and applicable to a wide range of samples (Kolb and Ettre, 1997). A further major benefit of thermal desorption directly coupled to GC is enhanced sensitivity because all of the desorbable components are trapped at low temperature, either in a short column held in the GC oven filled with adsorbent such as Tenax-GC, or on the GC column itself, prior to flash vaporization by rapidly raising the oven temperature (Figure 7.3). Using a trap may be necessary if the matrix contains water. In this case the analytes are trapped on the adsorbent, and then the water is removed from the system by purging with nitrogen prior to heating to allow rapid transfer of the analytes to the GC column. Alternatively a Nafion trap can be used. A disadvantage is that thermal desorption techniques are not selective – all trapped volatiles are passed to the column unless removed by, for example, a nitrogen purge.

Cartridges filled with layers of different adsorbents can also be used. After an appropriate sampling time, trapped volatiles are flash-vaporized into a stream of carrier gas and carried onto the GC column. Alternatively, cartridges filled with activated charcoal can be used to 'trap' the volatiles, which are then extracted into a small volume of carbon disulfide prior to the analysis. This technique has been used widely in the analysis of volatile compounds in potable water samples, for example, but has not found wide application in the analysis of biological samples, mainly because of the difficulty in interpreting the results at concentrations below those which can be measured using ordinary headspace methods.

Sample delivery mechanisms may include headspace (HS) analysis, purging liquids by bubbling with nitrogen, pulse heating extraction for liquids or thermal desorption for solid matrices, or simply pumping ambient air through the cryogenic trap. In the main these cryofocusing and related techniques find application in occupational and environmental toxicology. However, fluorinated inhalation anaesthetics (halothane, enflurane, isoflurane, sevoflurane and desflurane), for example,

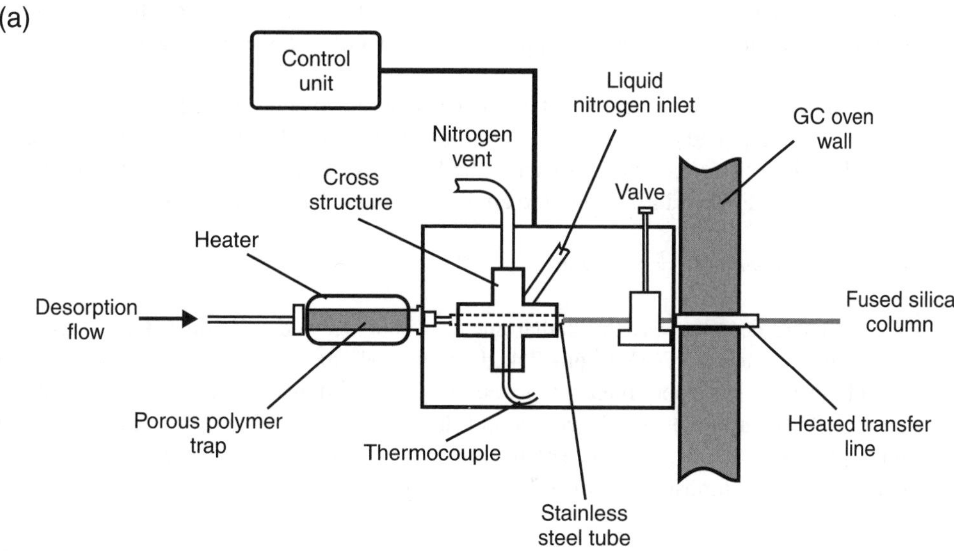

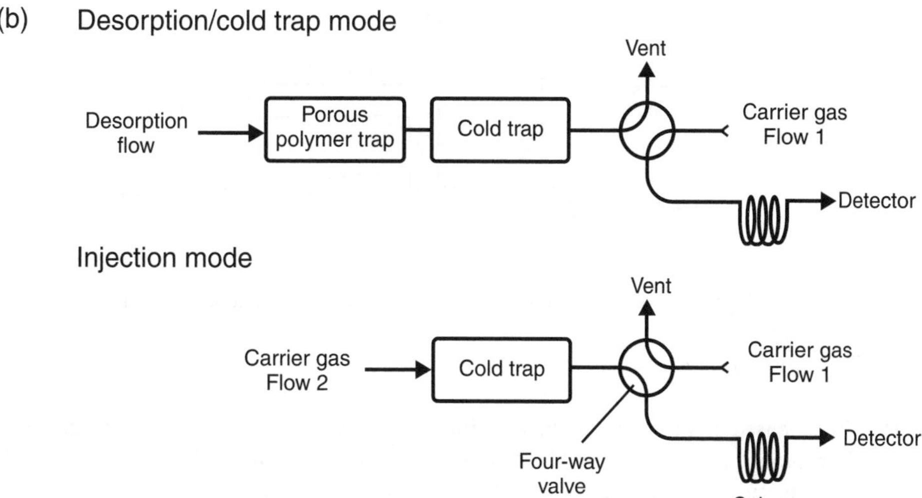

Figure 7.3 Schematic of (a) GC thermal-desorption unit (b) modes of use.

have been assayed in biological samples, principally GC with HS, purge-and-trap, or pulse heating extraction (Pihlainen and Ojanperä, 1998).

7.2.2 *Detectors for GC*

Of the four commonly used detectors, MS has assumed such importance that it is discussed in detail in Chapter 10. GC detectors are, in effect, destructive. Although the ECD is in essence non-destructive and thus can, in theory, be used in series with the FID or NPD, the high flow rate needed with ECD makes this impractical. Thus, post-column splitter systems are usually employed if more then one detector is to be used simultaneously, the relative detector response offering an additional parameter to aid compound identification even though this is rarely used in practice.

The flame photometric detector (FPD) can give a selective response for phosphorus- or sulfur-containing compounds, while the photoionization detector (PID) gives a higher signal-to-noise ratio than the FID for compounds such as barbiturates. However, nowadays the thermal conductivity detector, FPD and PID find little application in analytical toxicology.

7.2.2.1 *Thermal-conductivity detection*

The first GC detector was the katharometer or hot-wire detector, also known as the thermal conductivity detector (TCD; Figure 7.4). It is relatively insensitive (typically not better than 1 ng on column) and has a poor linear range, but may be useful in analytical toxicology for permanent gases and for carbon monoxide, compounds that do not respond on FID, ECD or MS. It is based on the principle that the resistance of an electrically heated filament placed in the column effluent changes as the composition of the effluent changes.

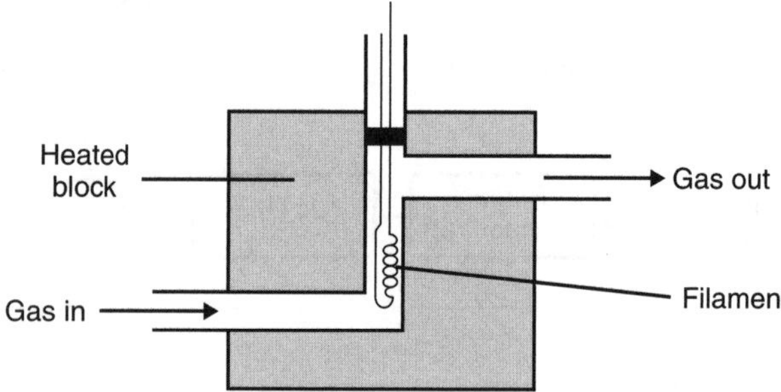

Figure 7.4 Schematic diagram of a GC thermal-conductivity detector.

7.2.2.2 *Flame-ionization detection*

The FID is based on the principle that the ionization produced as the column effluent passes into a hydrogen/air flame changes as the effluent composition changes (Figure 7.5). The change in ionization is monitored as a change in the standing current maintained by applying a potential difference between the flame tip and a collector electrode. The high sensitivity (10^{-11}–10^{-12} g s^{-1} methane, typically 0.1–10 ng on column), good stability (low background current) and wide linear range (up to 10^6) of this detector, and the large number of compounds (all organic molecules containing C–H bonds except formate) showing a response, have ensured that it remains widely used. The magnitude of the response is roughly proportional to the number of carbon atoms present in an analyte, although this is reduced if oxygen or nitrogen are also present in the molecule. Carbon monoxide does not respond. The lack of response to water enables aqueous injections to be performed when measuring ethanol and similar compounds (Section 7.6.2.2). On the other hand, the risk of interference is high, especially when analyzing samples obtained postmortem and thus selective detectors, notably MS, are favoured. There may be some build-up of siliceous deposits if used with packed columns or silicon-containing derivatizing reagents,

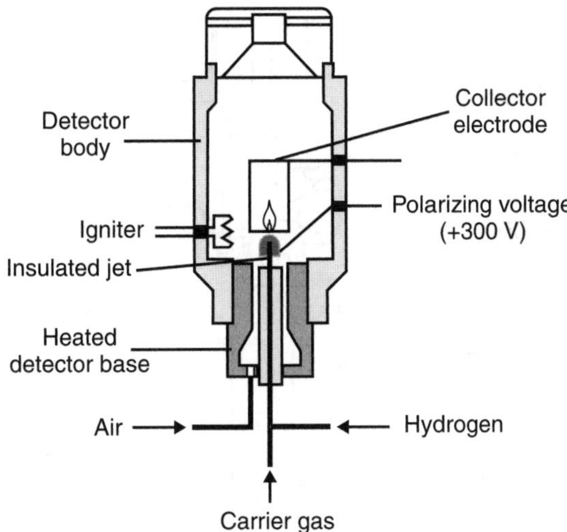

Figure 7.5 Schematic diagram of a GC flame-ionization detector.

necessitating periodic cleaning. The FID should normally be used in parallel with an ECD or NPD if retention measurements are to be based on Kovats retention indices (Section 5.3). A minimum operating temperature of 125 °C is recommended to prevent the condensation of water vapour.

7.2.2.3 Nitrogen–phosphorus detection

The NPD is based on the FID, but differs in that it contains a rubidium or possibly caesium silicate (glass) bead situated in a heater coil a little distance from the hydrogen flame. This introduces alkali metal vapour into the flame, which can facilitate a selective response to compounds containing phosphorus or C–N bonds via adjustment of the flame gases. Because many drugs and other poisons contain C–N bonds and many injection solvents and potential interferences do not, this detector (also known as the thermionic detector or the alkali flame-ionization detector, AFID; Figure 7.6) has been widely used in the nitrogen-selective mode (nitrogen:carbon response ratios about 5000:1). The increased sensitivity to phosphorus is a potential disadvantage as phosphorus-containing plasticisers show a good response (phosphorus:nitrogen response ratios about 10:1 in N mode). In the phosphorus mode, grounding the polarizing potential (Figure 7.6) increases the selectivity (phosphorus:carbon response ratios about 50,000:1), which can be exploited in a few instances, notably in the analysis of organophosphorus pesticides and their metabolites, or by producing phosphorus-containing derivatives.

Highly chlorinated solvents such as dichloromethane are not recommended for use with the NPD because they tend to volatilize the alkali metal from the source. Butyl acetate has proved a valuable extraction and injection solvent when used both with this detector and with the ECD. Use of nitrile-containing stationary phases such as OV-225 (Section 7.3.1) may cause increased baseline 'noise' with the NPD in the 'N' mode.

Modern versions of the NPD use an electrically heated rubidium silicate source and are somewhat more stable than early versions when a rubidium-containing bead was in essence 'balanced'

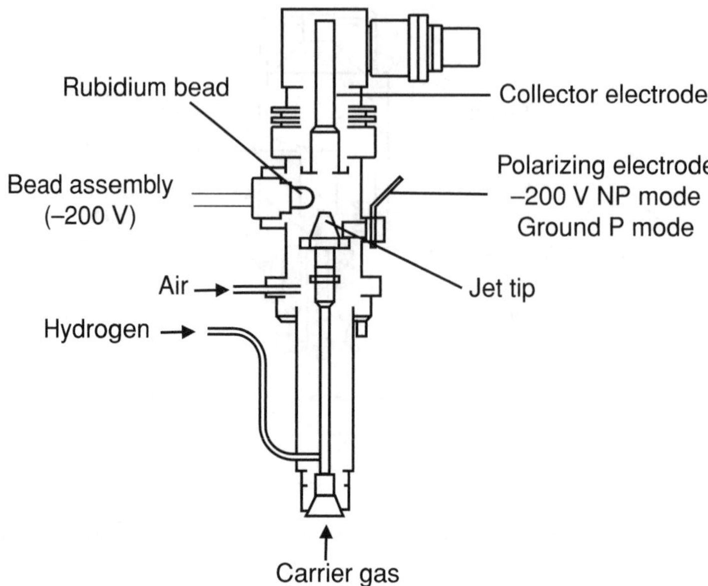

Figure 7.6 Schematic diagram of a nitrogen–phosphorus GC detector.

over an FID, but production and operation of this detector is still technically demanding. Correct adjustment of the position of the bead and of the detector gas flow rates are critical for successful operation. Bead life may vary from a few days to up to three months or so depending on bead quality and usage. As with the FID, stationary phase bleed from packed columns may necessitate periodic cleaning of the bead. Nevertheless, for some analyses including STA of basic drugs, the NPD detector may be a more sensitive option than MS.

7.2.2.4 Electron capture detection

The ECD (Figure 7.7) is based on the principle that electrons produced from a ^{63}Ni source or by plasma discharge (Section 7.2.2.5) are selectively 'captured' by certain analytes in the column effluent, thus causing a decrease in a standing current maintained between the source and a collector electrode. The detector can be operated either with a constant DC potential (DC mode), or with a pulsed potential (pulsed mode) applied across the electrodes. In DC mode a constant potential (a few V) is applied that is just sufficient to collect all electrons produced from the source and thus give a standing current. When an electron-capturing (electron deficient) molecule enters the detector, the molecule gains a negative charge. The mobility of the charged molecule is much smaller than that of the electrons, and the charged molecules are also more likely to be neutralized by collision with positive ions. Thus the standing current falls dramatically. In pulsed mode, the period of the pulsed potential is set such that relatively few negatively charged analyte molecules reach the anode, but the more mobile electrons do. When the pulse is 'off' the electrons re-establish equilibrium with the detector gas and overall a constant current is maintained. The operating variables are pulse duration, frequency and amplitude, and the output is a voltage dependent on the modulation frequency.

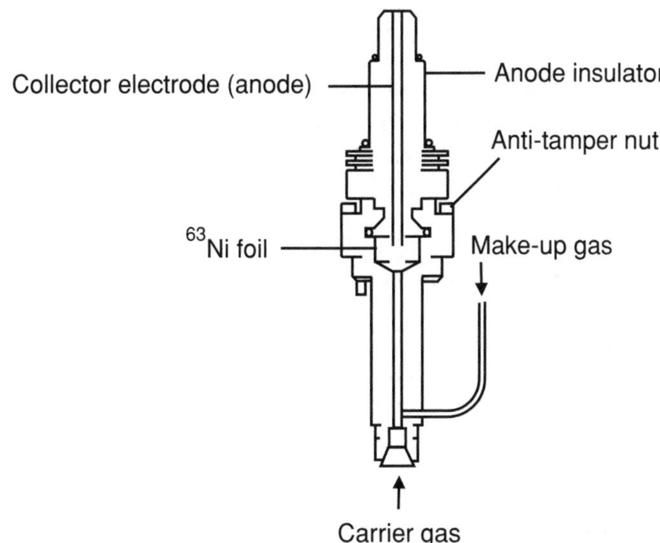

Figure 7.7 Schematic diagram of an electron capture GC detector.

The ECD is one of the most sensitive GC detectors available and shows an enhanced and selective response (about 10^3 g s^{-1} greater than the FID) to compounds containing a halogen, a nitro moiety and, to a lesser extent, to carbonyl-containing compounds. It is thus important for the analysis of halogenated solvents, pesticides and some halogen- or nitro-containing drugs, notably benzodiazepines. Derivatization with reagents such as heptafluorobutyric anhydride is widely used, not only to improve the volatility of suitable analytes, but also to exploit the high sensitivity and selectivity attainable. For some compounds the sensitivity of this detector easily exceeds that of MS detection – limits of sensitivity of a few fg are possible for organochlorine pesticides, for example.

Modern ECDs operated in the DC mode have linearity of response typically of the order of 10^4. Flow rates of 30–60 mL min^{-1} are necessary for the efficient operation of the ECD and thus a 'make-up' or 'purge' gas supply at the end of a capillary column is needed when using low carrier gas flow rates. Oxygen-free nitrogen normally provides a suitable make-up gas, but helium can be used with most modern detectors. Addition of a 'quench' gas such as methane, required with older models, is not normally necessary nowadays. Most manufacturers recommend operation close to the maximum operating temperature (300–350 °C) in order to minimize the risk of effluent deposits fouling the source.

The ECD suffers from a number of problems in routine use (Box 7.3). Most such detectors must not be used above 350 °C to avoid volatilization of the ^{63}Ni foil and it is a legal requirement to test routinely for radioactive leakage with a 'wipe test'. The ECD is particularly sensitive to impurities, especially oxygen and water vapour, in the carrier or purge gas, hence high-purity gas supplies are mandatory. Should a detector become contaminated, operation at the maximum operating temperature overnight with carrier and purge gas flow may be effective in restoring the response. A further, not insignificant, problem with this detector that has emerged in recent years is that of disposal of old instruments because of the presence of the radioactive source, hence the move to plasma discharge as the source of the electrons.

Box 7.3 Disadvantages of ^{63}Ni electron-capture detectors

- Limited number of analytes (must be electron deficient)
- Easily contaminated – with loss of standing current
- Large volume – requires make-up gas
- Limited linear range
- Safety issues
  - ^{63}Ni hard β-emitter
  - Maximum operating temperature generally 350 °C
- Legislative issues
  - Requires radioactivity 'wipe-test'
  - Difficult to dispose of old instruments

7.2.2.5 *Pulsed-discharge detection*

The pulsed-discharge detector (PDD, Figure 7.8), a relatively new development, can be operated in three modes: pulsed-discharge helium photoionization (He-PDPID), pulsed-discharge electron capture (PDECD), and helium-ionization emission (PDED) (Forsyth, 2004). The He-PDPID can detect permanent gases, volatile inorganic species and other compounds that give little or no response with the FID and has lower limits of detection [with minimum detectable quantities (MDQs) in the low picogram range] than can be achieved with TCD and a wide linear range (Roberge *et al.*, 2004). In the He-PDPID, helium is passed through a chamber where a glow discharge is generated and high-energy photons are produced. These pass through an aperture to another chamber where they ionize the gas or vapour species in the sample stream. The resulting electrons are directed by the bias electrode(s) to the collector electrode and are measured using a

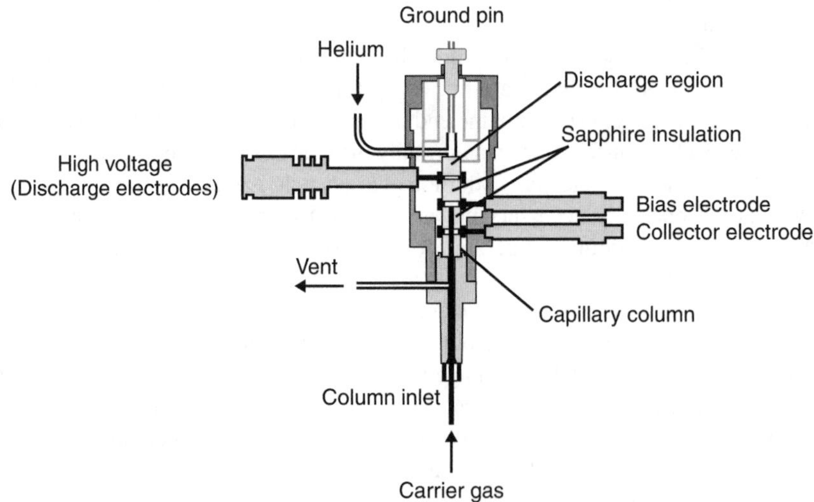

Figure 7.8 Schematic diagram of pulsed-discharge detector (PDD).

standard electrometer. When used as an ECD, a small percentage of an additional gas ('dopant') is added, which is ionized and provides electrons to produce a standing current. Several dopants have been tested – the best appear to be methane and xenon. The PDECD has a sensitivity similar to, or better than, ^{63}Ni ECD, but does not require licensing, wipe tests and other administrative or safety requirements. The PDED shows promise as an extremely selective and sensitive elemental detector.

7.2.2.6 *Flame-photometric detection*

The flame-photometric detector (FPD) exploits the chemiluminescence of sulfur or phosphorus when heated in a flame to give relatively specific detection for substances containing either element. The λ_{max} for emission of excited S_2 (the emitting species for sulfur) is about 394 nm, whilst excited HPO shows a doublet at 510–526 nm. An interference filter placed between the flame and the photomultiplier tube determines selectivity (Figure 7.9). This detector does not find wide application in analytical toxicology, but is used in pesticide development and other relatively specialized areas.

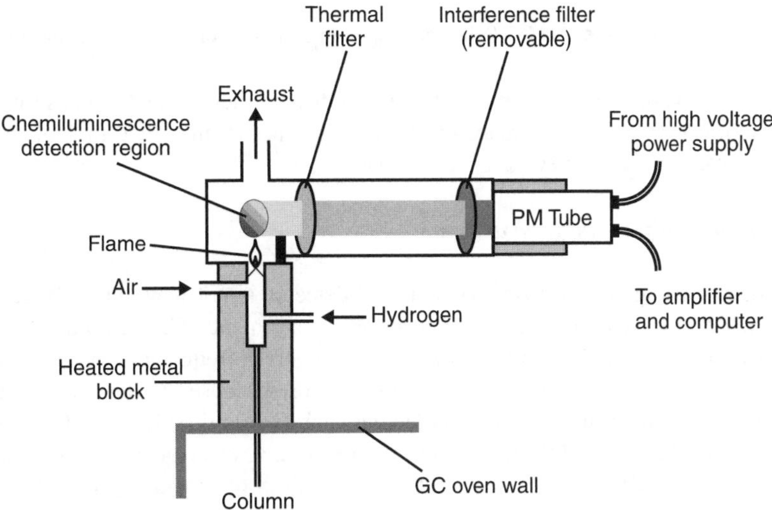

Figure 7.9 Schematic diagram of a GC flame-photometric detector.

7.2.2.7 *Atomic-emission detection*

The microwave-induced plasma-emission detector or atomic-emission detector (AED), once called a microwave plasma detector (MPD), can measure a number of elements, typically 15 or so, selectively and simultaneously in GC effluent. Analytes eluting from the GC column pass into a microwave plasma (or discharge) cavity where individual atoms are excited from ground state by the heat energy of the plasma. Each element emits light of a specific wavelength when decaying from the excited state and this can be measured via a photodiode array, usually in the range 170–800 nm (Figure 7.10). An associated computer can produce chromatograms made up

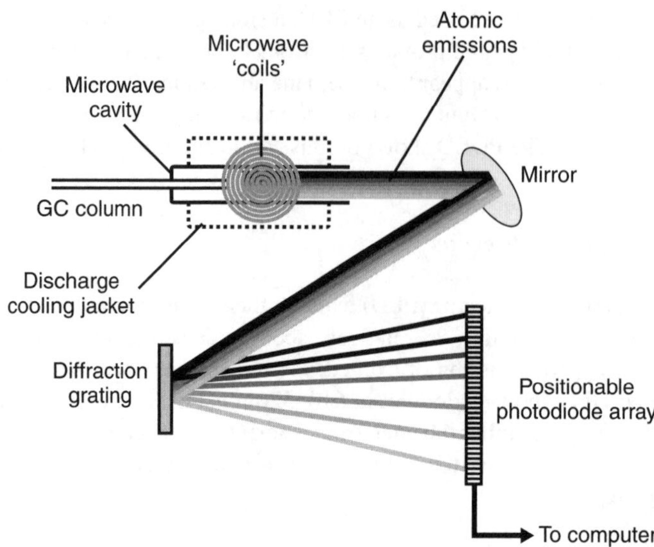

Figure 7.10 Schematic diagram of a GC atomic-emission detector.

of peaks from analytes that contain only a specified element. The AED has found application in conjunction with MS in pharmaceutical analysis if not in analytical toxicology *per se* (Hooker and Dezwaan, 1989; Laniewski *et al.*, 2004).

7.2.2.8 Fourier-transform infrared detection

The idea of using IR spectrometry as a GC detector is not new, but only became practicable with the development of Fourier transform (FT) instruments. FT converts data from an intensity versus time plot to an intensity (% transmission) versus frequency spectrum. GC-FTIR has not been widely applied in analytical toxicology largely because of cost and the fact that, with the exception of isomers and of simpler molecules such as volatile solvents, GC-MS often gives more information. However, FTIR detection is non-destructive and can be used in a quantitative as well as qualitative mode, although sensitivity is poor (limit of sensitivity at best 10 ng on column). GC with sequential FTIR/FID has been used to detect and measure volatiles in blood (Ojanperä *et al.*, 1998). Some other biomedical applications of GC-FTIR have been reviewed (Lacroix *et al.*, 1989).

7.3 Columns and column packings

Originally distinction was made between gas–solid chromatography (GSC) in which the stationary phase is an active solid and GLC in which the liquid stationary phase is coated onto an inert support. However, with the widespread use of capillary columns in which the stationary phase is chemically bonded onto the inner surface of the column, these distinctions are often less clear cut. As the mobile phase (carrier gas) has little effect on selectivity and only a small effect on efficiency, nitrogen is normally used with packed columns (flow rates 30–60 mL min^{-1}) and helium with capillaries (flow rates 1–10 mL min^{-1}). In all but the simplest systems it is vital to use oxygen-free carrier gas

Table 7.1 Capillary versus packed columns in GC: practical considerations.

	Packed	Capillary
Efficiency	Low (about 20,000 theoretical plates per column)	High (about 500,000 theoretical plates per column)
Peak symmetry	Poor for polar compounds	Good
Analysis time	Long	Short (for comparable resolution)
Temperature programming	Difficult (baseline drift)	Easy
Sample capacity	High	May be low
Sensitivity	Good (more sample injected)	Good (sharp peaks)
Quantitation	Yes	Direct injection
Interface to MS	Difficult (need splitter system)	Easy

as the presence of even very small amounts of oxygen can result in oxidation of certain stationary phases. A constrictor fitted to the detector end of the column limits back diffusion of air from an FID, for example. Some advantages to the use of capillary columns are summarized in Table 7.1.

Most columns, especially packed columns, require conditioning to remove volatile impurities before use. Proprietary columns should be supplied with full instructions. Otherwise, the column (not connected to the detector) should be purged with carrier gas to remove oxygen before heat is applied. The column should then be slowly brought to slightly above the intended maximum operating temperature either using a temperature programme, or in a series of steps, and held for about 12 h. The maximum temperature for the stationary phase recommended for isothermal operation should be taken into account.

Capillary columns should be conditioned according to the manufacturer's instructions and tested by injecting an appropriate mixture before use for sample analyses. If temperature programming is to be used, it is important that the column is not left at the maximum run temperature for a prolonged period. When conditioning columns it is obviously important to ensure that the GC oven temperature controller is working properly and that the carrier gas flow is adequate as there is no detector output to help indicate if anything is wrong.

7.3.1 *Packed columns*

Conventional borosilicate glass, stainless steel, or glass-lined stainless steel packed columns (0.5–4 m × 2–4 mm i.d.) still find application in some areas. Although less robust than steel, glass is less likely to adsorb or react with polar analytes. In addition, any settling of the packing and column contamination can be checked visually. Treatment of glass columns for about 8 h with a silanizing reagent such as trimethylchlorosilane (TMCS) (10 % v/v in toluene) and use of quartz or silanized glass wool to retain the packing may help minimize adsorptive effects. Excess reagent can be removed by rinsing with methanol and drying before the column is packed. A good source of information on these and other aspects of packed column GC remains the book by Supina (1974).

GSC packings are mainly used in the analysis of gases and solvents. Molecular sieve (synthetic zeolite) or silica gel packings are useful for the analysis of permanent gases and carbon monoxide.

The Chromosorb and Porapak series are cross linked divinylbenzene polystyrene copolymers with maximum operating temperatures of about 250 °C. Alcohols from methanol to pentanol can be separated on Porapak Q or Chromosorb 101. Tenax-GC is a porous polymer of 2,6-diphenyl-*p*-phenylene oxide and is used both as a stationary phase material and as a trap for volatiles prior to GC. Carbopaks B and C are graphitized carbon blacks having surface areas of 12 and 100 $m^2 g^{-1}$, respectively. The Carbopaks are usually used after modification with a light coating of a polar stationary phase such as Carbowax 20M, and can give good peak shapes and separations for alcohols and other volatiles. However, these materials are very friable and batch-to-batch variations in the peak shapes attained are common.

In packed column GLC the support should play little or no part in the separation. Calcined diatomaceous earth graded into appropriate size ranges (80–100 or 100–120 mesh) is widely used. Commercial brands include Chromosorb W and Chromosorb G (both Johns Manville) and Supelcoport (Supelco). Various deactivation procedures are employed, including acid washing to remove metallic impurities (denoted AW) and deactivation of surface silanols with hexamethyl-disilazane (denoted HMDS), dimethylchlorosilane (DMCS), or other silanizing reagents (Table 7.2). A light initial coating of a polar stationary phase applied before the primary phase may increase the apparent deactivation of the support, but may also influence the separations achieved. Pre-coating with potassium hydroxide (1–5 % w/w) has been widely used to improve the peak shapes given by strong bases such as the amfetamines. However, not all stationary phases are stable under strongly alkaline conditions.

Stationary phase loadings are normally expressed as percentage weight (or mass) of phase/weight (or mass) of support (% w/w or % m/m). Phase loading is not normally greater than 5 or 10 % (w/w). The phase is normally applied to the support dissolved in an appropriate solvent. After standing, the solvent is then removed either by gentle evaporation under vacuum in a rotary evaporator (taking care not to cause mechanical damage to the particles) or by filtration on a Büchner funnel followed by air drying. With the evaporation method the stationary phase may not be coated evenly. The filtration method gives a more uniform coating especially at lower phase loadings, but not all the stationary phase is adsorbed and some is lost at the filtration stage. Experience is needed to help assess the initial proportions of stationary phase, solvent and support to use. The stationary phase lost can be weighed by collecting the filtrate and evaporating to dryness.

There are several hundred GC stationary phases, but only a few – mainly those based on polysiloxane (often inaccurately called silicones) or polyethylene glycol – have found wide application in analytical toxicology. In the simplest of the polysiloxane phases, that is when R_{1-4} = methyl (Figure 7.10), the polymer is known as polydimethylsiloxane (PDMS) or simply methylpolysiloxane. Substitution of methyl moieties by other groups of increasing polarity alters the selectivity of the material. In the stationary phase, OV-17, for example, R_1 and R_2 = methyl, R_3 and R_4 = phenyl, and x and y both = 50 % (Figure 7.11). This phase can be referred to as 50 % dimethyl, 50 % diphenylpolysiloxane; 50 % methyl, 50 % phenylpolysiloxane; or even 50 % phenylpolysiloxane. In other phases, two different groups may be bonded to the same silicon atom. OV-225, for example, differs from OV-17 in that R_3 = cyanopropyl.

Polyethylene glycol (PEG) phases (Figure 7.11) are used either directly, or after substitution with alkyl moieties such as 2-nitroterephthalic acid as in the phase FFAP (free fatty-acid phase). Base-deactivated PEG phases (CAM, J&W) designed for amine analysis are also available. Although PEG phases have unique selectivities, they are less robust than the polysiloxanes and have lower operating temperature limits and a shorter life expectancy.

Table 7.2 Some stationary phases and stationary phase supports used in packed column GC (unless otherwise specified, intended mesh size is 80/100 or 100/120).

Material	Description	USP[a]
Chromosorb P-AW, Chromosorb W-AW	Diatomaceous earth, flux-calcined, acid-washed and water-washed until neutral, silanized	S1A
Chromosorb W-HP	Diatomaceous earth, base-washed, silanized	S1AB
Chromosorb P-AW DMCS	Crushed firebrick, calcined or burned with clay binder, acid-washed, silanized	S1C
Chromosorb W-NAW	Diatomaceous earth, untreated	S1NS
Gas Chrom 254, Chromosorb 101	Styrene-divinylbenzene copolymer (surface area <50 $m^2\ g^{-1}$; average pore diameter 0.3–0.4 μm)	S2
Super Q, Porapak Q	Ethylvinylbenzene-divinylbenzene copolymer (surface area 500–600 $m^2\ g^{-1}$; average pore diameter 7.5 nm)	S3
Porapak R	Styrene-divinylbenzene copolymer with aromatic -*O*- and -*N*-groups (surface area 400–600 $m^2\ g^{-1}$; average pore diameter 7.6 nm)	S4
Chromosorb T	High M_r tetrafluoroethylene polymer, 40/60 mesh	S5
Gas Chrom 220, Porapak P, Chromosorb 102	Styrene-divinylbenzene copolymers (surface area 250–350 $m^2\ g^{-1}$; average pore diameter 9.1 nm)	S6
Graphpac-GC	Graphitized carbon (surface area about 12 $m^2\ g^{-1}$)	S7
Porapak S	Copolymer of 4-vinylpyridine and styrene-divinylbenzene	S8
Tenax TA	Porous polymer of (2,6-diphenyl-*p*-phenylene oxide)	S9
3 % Carbowax 540 on Graphpac-GB, 80/120 or Carbograph 1, 80/120	Graphitized carbons (surface area of 100 $m^2\ g^{-1}$), modified with polyethylene glycol	S11
Graphpac-GB, Carbograph 1	Graphitized carbons (surface area 100 $m^2\ g^{-1}$)	S12

[a] US Pharmacopoeia Convention (1999) GC packing cross-reference.

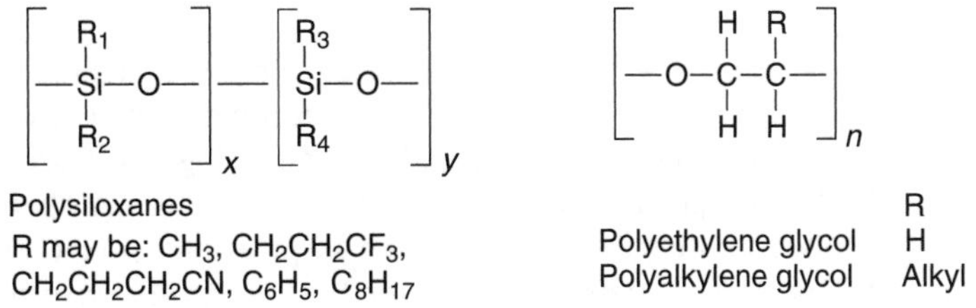

Polysiloxanes
R may be: CH_3, $CH_2CH_2CF_3$, $CH_2CH_2CH_2CN$, C_6H_5, C_8H_{17}

	R
Polyethylene glycol	H
Polyalkylene glycol	Alkyl

Figure 7.11 Structures of polysiloxane and polyethylene glycol stationary phases for GC (for explanation, see text).

As a general rule, retention on relatively non-polar phases such as PDMS is influenced mainly by analyte M_r, the dominant influence on volatility. However, hydrogen bonding and dipole interactions may also influence volatility and therefore GC separation on a given system. As a guide, polar compounds, such as many drugs and pesticides, show increased separation and give better peak shapes and hence greater sensitivity on more polar phases.

McReynolds (1970), building on the work of Rohrschneider, used the retention of benzene, butanol, 2-pentanone, nitropropane and pyridine to summarize the polarity of stationary phases. The sum of the difference of the retention indices (Section 5.3) of each of these compounds on the stationary phase under investigation as compared to the retention index of the compound on a standard non-polar phase (squalane) is used to derive the McReynolds constant for the phase. This gives a measure of the polarity of the phase, which can be used for classification purposes. The system is explained fully elsewhere (IUPAC, 1986; Supelco, 1997). In general, phases with McReynolds constants <100 are known as non-polar, 100–400 intermediate polarity and >400 polar. The McReynolds constants for some stationary phases and the US Pharmacopoeia (USP) Convention liquid-phase cross reference are given in Table 7.3. Note, however, that the 'polarity' of stationary phases, as reflected in the retention of polar as compared to non-polar analytes, increases with temperature.

Apiezon L is a hydrocarbon grease and has the advantage that, unlike the polymeric siloxane phases, it is stable if coated onto alkali-treated packing, although the maximum operating temperature (t_{max}) is then only 225 °C. SE-30, OV-1 and OV-101 (and also HP-1, DB-1, RTX-1 and SPB-1) are PDMS stationary phases that are broadly equivalent. Polar interactions are minimal with these phases and thus separations occur largely on the basis of M_r. However, a disadvantage is that peak tailing may occur with polar compounds. In general, peak tailing is reduced on more polar phases. SE-54, OV-7, OV-17 (and their broad equivalents HP-50 and RTX-50) and OV-225 are amongst the more polar polysiloxane-based phases available (Table 7.3). In general, the higher the polarity of polysiloxane phases, the lower the maximum operating temperature before column bleed becomes limiting.

Carbowax 20M (polyethylene glycol of average M_r 20,000) is a high-polarity phase, but has a relatively low t_{max}. However, as with Apiezon L, precoating the support with potassium hydroxide effectively minimizes peak tailing of strong bases and the t_{max} is unaffected. Polyester phases such as CHDMS (CDMS, cyclohexanedimethanolsuccinate) and also polyamides such as Poly A 103 have been advocated for specific separations of polar compounds such as barbiturates, but these approaches have been largely superseded with the widespread adoption of capillary column systems.

7.3.2 Capillary columns

Capillary columns were originally developed with the aim of resolving complex mixtures and can have great separating power. Glass and stainless steel have both been used to prepare such columns, but nowadays most applications in analytical toxicology use fused silica capillaries. An outer coating of polyimide protects the brittle silica giving mechanical strength and flexibility – the column may often be threaded through complex pipework to emerge, for example, at a detector inlet. As fused silica columns are inherently straight they are usually coiled and held in a metal support (cage). They should be handled with care because when they break they suddenly straighten and may cause injury. Typical dimensions are 0.05–0.53 mm i.d. and 10–50 m length. The polyimide coating imposes an upper operating temperature limit of 360 °C (400 °C for high-temperature polyimide coatings).

Table 7.3 Some GC stationary phases used in analytical toxicology.

Phase[a]	Type	Suggested operating temperature range (°C)	USP[b]	Sum of McReynolds constants
Squalane	Hydrocarbon	20/100	—	0
SPB-Octyl	50 % Dimethyl 50 % octylpolysiloxane	−60/280	—	51
Apiezon L	Hydrocarbon grease	50/300	—	143
SE-30	100 % Methylpolysiloxane gum	−60/300	G2	217
OV-1	100 % Methylpolysiloxane gum	−60/320	G2	222
OV-101	100 % Methylpolysiloxane oil	−60/320	G1	229
DB-1	100 % Methylpolysiloxane gum	−60/320	G1	229
SP-2100	100 % Methylpolysiloxane oil	−60/320	G1	229
Apiezon L/KOH[c]	2 % (w/w) Hydrocarbon grease + 1 % (w/w) potassium hydroxide	50/225	G44	301
SE-52	95 % Methyl, 5 % phenylpolysiloxane	−60/320	G27	334
HP-5	95 % Methyl, 5 % phenylpolysiloxane	30/310	G27	334
SE-54	94 % Methyl, 5 % phenyl, 1 % vinylpolysiloxane	−20/280	G36	337
OV-73	94.5 % Methyl, 5.5 % phenylpolysiloxane	−60/320	G27	401
RTX-1301	6 % Cyanopropylphenyl, 94 % methylpolysiloxane	−20/280	G43	592
OV-7	80 % Methyl, 20 % phenylpolysiloxane	−60/320	G32	592
OV-17	50 % Methyl, 50 % phenylpolysiloxane	30/310	G3	884
SP-2250	50 % Methyl, 50 % phenylpolysiloxane	30/310	G3	884
Poly A 103	Polyamide	50/275	G10	1072
OV-210	50 % Methyl, 50 % trifluoropropylpolysiloxane	40/230	G6	1520
OV-225	50 % Cyanopropylphenyl, 50 % methylpolysiloxane	40/230	G19	1813
CHDMS (Hi-EFF-8BP)	Cyclohexanedimethanol succinate	100/250	—	2017
Carbowax 20M	Polyethylene glycol (PEG) M_r 15 000	35/280	G16	2308
SP-1000	Terephthalic acid substituted PEG	60/200	G25	2409
FFAP	Polyethylene glycol-2-nitroterephthalic acid ester	50/250	G25	2546
DEGS (Hi-EFF-1BP)	Diethyleneglycol succinate polyester	20/200	G4	3504

[a]Key to suppliers: DB = J&W, OV = Ohio Valley, SP = Supleco, HP = Hewlett Packard (now Agilent), RTX = Restek (this list is not exhaustive)
[b]USP Convention (1999) liquid phase cross-reference
[c]Potassium hydroxide coated first

Column bleed and thus detector noise are reduced as compared to packed columns because: (i) the amount of stationary phase present is low and (ii) the phase is often cross-linked and chemically bonded to the wall of the capillary via surface silanols. Cross-linking is brought about via γ-irradiation, for example, which acts to initiate free radical reactions between adjacent stationary phase molecules. This has the effect of increasing the average M_r of the polymer, which in turn increases the three dimensional stability, viscosity and solvent resistance of the material. Modified polysiloxanes in which a phenyl moiety, for example (arylene-stabilized), is incorporated into the siloxane chain in place of some oxygen atoms (–O–Si(R_1R_2)–C_6H_5–Si(R_3R_4)–O–, cf. Figure 7.11) are also available. These phases give enhanced stability leading to a further decrease in column bleed, factors valuable in high sensitivity, high temperature analyses. The arylene substitution can be adjusted to maintain the same separation characteristics as the original, nonarylene-stabilized phenylsiloxane.

An important advantage of capillary columns is that thermal mass is low and thus temperature programming is facilitated. More importantly interfacing to MS is simplified because carrier-gas flow rates (and column bleed) are low. A further advantage with respect to packed columns is that when using polysiloxane stationary phases, accumulation of siliceous deposits on FID/NPD/ECD is minimized. On the other hand, one complication is that GC detectors other than the MS require higher flow rates than normally used with capillaries and thus a 'make-up' gas supply is needed.

PLOT (porous-layer open tubular), SCOT (support-coated open tubular) and WCOT (wall-coated open tubular) are terms used to describe different types of capillary column. In PLOT columns, a fine (10 μm or less) layer of stationary phase particles, for example activated alumina, is bonded to the internal surface of the capillary. Columns of this type, which are very retentive, are usually reserved for the analysis of very volatile substances such as permanent gases. They are not recommended for use with MS detection because of their tendency to shed particles of stationary phase. In SCOT columns a liquid stationary phase is coated onto particles such as barium chloride microcrystals that are bonded onto the inner surface of the capillary, whilst in WCOT columns a liquid stationary phase is coated, or more usually, chemically bonded onto the capillary wall directly. Although glass capillary WCOT columns can be prepared in-house it is much more convenient to buy pre-tested columns. The stationary phases used are, in general, those developed for packed columns (Table 7.3), but better column deactivation and more uniform phase coating mean that good peak shapes are often obtained even when using low-polarity phases. Moreover, when used with temperature programming most phases are stable when used 20–50 °C above their packed-column temperature limit for short periods.

In general, efficiency is increased with narrower column diameters/thinner film thicknesses, but sample capacity is greater with wider bores/thicker films, and thus the risk of column overloading, usually manifest as 'fronting' (front-tailing, 'shark's fin peak'), is minimized. Columns of 0.32 mm i.d. and 30–50 m length with a moderate phase loading (0.2–0.3 μm film thickness) and used with an appropriate injection system are suitable for many purposes in analytical toxicology such as STA (Section 7.6.1).

Columns of 0.53 mm i.d. and 15–30 m length with relatively high phase loading (1–5 μm film thickness) have similar capacity and efficiency to packed columns at carrier-gas flow rates of 8–20 mL min^{-1}, but with the advantages of better peak shapes, reproducibility and applicability in temperature programming. When used at these flow rates, detector make-up gas may not be required. Alternatively, these columns can be used at lower flow rates (1–2 mL min^{-1}) that are more suitable for interfacing with MS (Box 7.4). Columns with a relatively high film thickness (5 μm

Box 7.4 GC Capillary Column Selection

- Longer columns (25 or 50 m typical) give increased efficiency, but longer analysis times
- Narrower columns give more plates per metre, but other factors also important
  - 0.15 mm i.d. high efficiency with low carrier flow, but low capacity (split injection). Good for GC-MS
  - 0.32 mm i.d. reasonable efficiency and sample capacity (on-column injection). Good for STA and high sensitivity work
  - 0.53 mm i.d. gives efficiency and sample capacity at least equal to packed columns, but less tailing with polar compounds and temperature programming easier
- Film thickness (0.1–5 μm, often 0.25 μm) also important – influences both retention and sample capacity
- Cross-linked (immobilized) phases relatively stable. Large volume sample injections and column washing to remove non-volatile contaminants can be employed
- High efficiency means that fewer stationary phases needed. Commonly used polysiloxanes are:
  - OV-1/SE-30/O-101, HP-5, OV-17, OV-225
- Other phases used include Carbowax 20M, FFAP

film) can generally tolerate the injection of up to 1 μL of water and such columns may often be substituted directly for packed columns in a given method. Several manufacturers have supplied 'conversion kits' to enable these 'wide-bore' capillaries to be used in instruments configured for use with conventional packed columns. Some companies also produce capillary GC method optimization/simulation software for use with personal computers.

7.3.3 Multidimensional GC

In multidimensional GC (MD-GC) the sample is dispersed in different time dimensions. When a sample is separated using two dissimilar columns the total peak capacity will be the product of the individual column peak capacities. For maximum gain in resolution, separation in the individual time dimensions should be totally uncorrelated or orthogonal. Although this is rarely achieved in practice, significant gains in selectivity can be obtained without total orthogonality (Marriott et al., 2004).

In two-dimensional (2D) GC with heartcutting ('GC-GC') a fraction from the first retention axis is transferred for separation on the second retention axis. In contrast, in comprehensive 2D-GC ('GC_GC') fractions of the entire sample, after being separated on the first column, are passed on for analysis on the second column.

The earliest work in 2D-GC used valves to connect two columns and to fractionate the eluent from the first column. Deans (1968) introduced a valveless switching system by employing pneumatic pressure balancing. By using controlled carrier gas flows to the injection port and the column intersection point, fractions could be diverted to the second column or vented to waste (Figure 7.12). This design introduced the challenge of stabilizing carrier-gas pressure in the injector and at the midpoint connector, a challenge that has now been overcome by use of modern electronic mass flow controllers and electronic proportional back-pressure regulators with pressure sensing.

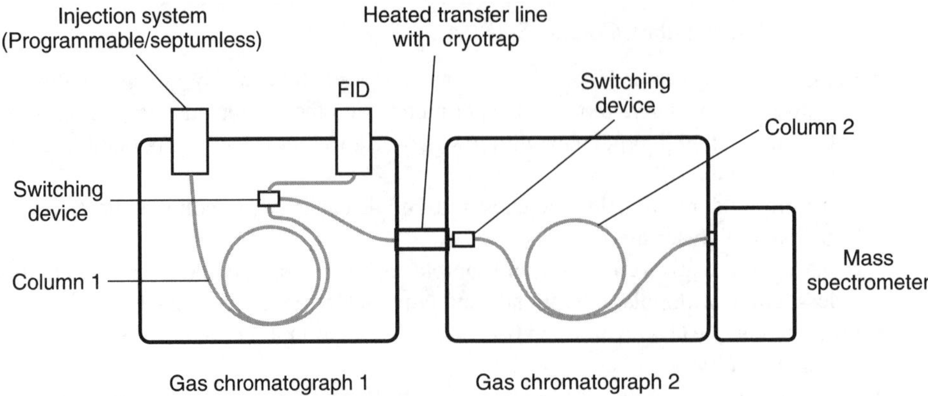

Figure 7.12 Schematic diagram of 2D-GC.

A more recent variant of 2D-GC uses two directly coupled columns and employs a longitudinally modulated cryogenic trap located between the columns (Marriott *et al.*, 2000). Compared with normal MD-GC, where the primary column effluent is temporarily diverted either to a monitor detector or to the second column, all of the effluent from the first column can in theory be passed to the second column. Alternatively, modulation of the trap allows target zones to be selected for enhanced separation on the second column. Because data are presented in a time-response format, quantitation and report generation are essentially the same as with any GC method. Advantages may also include sensitivity enhancement.

MD-GC can be a powerful analytical tool (Dalluge et al., 2003). Although toxicological applications are scant at present, some publications have shown the potential of this approach (Kueh *et al.*, 2003).

7.4 Derivatization for GC

Analytes with polar functional groups, particularly phenols, alcohols and carboxylic acids, sometimes give inefficient, tailing peaks on GC, leading to poor selectivity and sensitivity. Most aromatic and aliphatic hydroxyls form silyl ethers when treated with trimethylsilyl (TMS) reagents at room temperature or, in the case of sterically hindered compounds, on heating. HMDS has been widely used, but will not react with hindered hydroxyl groups. *N,O*-bis(trimethylsilyl)acetamide (BSA) is an alternative, but *N,O*-bis(trimethylsilyltrifluoro)acetamide (BSTFA) containing 1 % (v/v) trimethylchlorosilane (TMCS) is a more powerful reagent. *N*-Trimethylsilylimidazole (TSIM) is particularly effective at silanizing hindered hydroxyl groups, but will not react with amines. Because the silanizing reagents and the reaction products are easily hydrolyzed, water must be excluded. However, *tert*-butyldimethylsilylimidazole forms ethers that are relatively stable to hydrolysis. One disadvantage to the use of silylated derivatives is accumulation of decomposition products at the GC detector, leading to loss of sensitivity and necessitating regular cleaning. In this regard, BSTFA is said to be preferred if FID is used, as the combustion products are volatile and said not to foul the detector.

Carboxylic acids are usually analyzed as methyl or butyl esters. Methylation can be achieved by heating (60–100 °C, 30 min) with 14 % (w/v) boron trifluoride in methanol. After evaporating most of the solvent and adding water, the methylated derivative can be extracted into an organic solvent such as hexane. Diazomethane may also be used as a methylating reagent and reacts rapidly *in vitro* at room temperature; excess reagent may be simply removed by evaporation. However, care is needed because diazomethane is carcinogenic, highly toxic and potentially explosive. In addition, diazomethane will react slowly with phenols and this may be disadvantageous because mixtures of partially methylated products may result.

On-column methylation of barbiturates, hydantoins and some carboxylic acids can be achieved by injecting the sample mixed with 0.2 mol L^{-1} trimethylanilinium hydroxide in methanol. However, 'ghost' peaks may arise on subsequent injection of derivatizing reagent if incompletely methylated material has accumulated on the column.

Commonly used reactions before GC analysis of primary and secondary amines, and of phenolic hydroxyls, include acylation using reagents such as acetic and propionic anhydrides, or their acid chlorides. With paracetamol, extractive acetylation of paracetamol and *N*-butyryl-*p*-aminophenol (internal standard) can be performed using acetic anhydride and *N*-methylimidazole as catalyst (Huggett *et al.*, 1981). A specimen together with an IQC sample could be analyzed, in duplicate, within 20 min. The limit of accurate measurement of the assay was 10 mg L^{-1}.

Primary amines react with aldehydes or ketones such as acetone, benzaldehyde and cyclohexanone to give imines, whilst secondary amines produce enamines. Hydrazines and hydroxylamines react with carbonyl compounds to give hydrazones and oximes, respectively. Imine formation was exemplified by the reaction between acetone and amfetamine (Figure 1.2).

The reaction of boronic acids such as phenylboronic acid with 1,2- and 1,3-diols to form cyclic boronates is unaffected by the presence of water and has been used in the measurement of ethylene glycol in plasma, urine or dialysis fluid by packed column GC-FID (Flanagan *et al.*, 1987). 1,3-Propanediol was used as the reactive internal standard. The limit of accurate measurement was at least 0.1 g L^{-1}, and the linear range extended up to 5 g L^{-1}. Ethylene glycol is very polar and is especially difficult to measure using packed column GC; 2,3-butanediol and 1,2-propanediol have been mistaken for ethylene glycol in other GC assays.

Derivatization can also be used in headspace GC (Section 3.2.4) to increase sensitivity and improve chromatographic performance for specific compounds. Some acids, alcohols and amines may be adsorbed, not only on the analytical column, but also in the injection port, giving rise to tailing peaks and a poor detector response. In addition, they may be very soluble in the matrix, causing very poor partitioning into the headspace. Derivatization can improve volatility, as well as reducing the potential for adsorption onto the components of the GC system.

Common derivatization reactions that may be performed using the sample vial as the reaction vessel are esterification, acetylation and alkylation. However, although derivatization may improve chromatographic performance and volatility, the derivatization reaction itself may introduce other problems. The derivatization reagent, as well as any by-products from the reaction(s), may be volatile and partition into the headspace along with the derivatized analyte(s), and may elute at similar retention times to the compound(s) of interest. The derivatization reaction may also need to be performed at an elevated temperature, and in such cases the pressure inside the sample vial may exceed the pressure handling capabilities of the vial or the septum. Specially designed caps are available that allow excess pressure to be vented during derivatization reactions.

7.4.1 *Electron-capturing derivatives*

There are a number of halogenated silylating reagents available, for example bromomethyldimethylchlorosilane (BMDMCS) and chloromethyldimethylchlorosilane (CMDMCS), which may be used to impart electron-capturing properties to the derivative, but mainly perfluorinated

Table 7.4 Dervatizing reagents for GC-ECD.

Compound	Uses	Notes
Perfluorinated anhydrides		
Trifluoroacetic anhydride (TFAA) Pentafluoropropionic anhydride (PFPAA) Heptafluorobutyric anhydride (HFBAA)	React with alcohols, amines and phenols. Produce stable and highly volatile derivatives. Suitable for FID and ECD. Retention times increase with increasing M_r	Remove acid by-product (stream of nitrogen) before injection on to column. Bases (e.g. triethylamine) can be added as a proton receptor and thus promote the reaction
Perfluoracylimidazoles		
Trifluoroacetylimidazole (TFAI) Pentafluoropropanylimidazole (PFBI) Heptafluorobutyrylimidazole (HFBI)	Work best with amines and hydroxy compounds. Better choice for making ECD derivatives	React under mild conditions. By-product (imidazole) not acidic – less harmful to column. Reagents water sensitive – react violently
N-Methyl-bis(trifluoroacetamide) (MBTFA)	Reacts with primary and secondary amines. Reacts slowly with hydroxyl groups and thiols	Reacts under mild conditions. Reaction by-product (*N*-methyltrifluoroacetamide) is neutral, stable and volatile, and will not interfere in the analysis
Pentafluorobenzoyl chloride (PFBC)	Most reactive with phenols. Reacts with alcohols and secondary amines	Secondary amines will react with this compound
Pentafluorobenzyl bromide (PFBB)	Reacts with carboxylic acids, phenols and thiols	Strong lachrymator – use fume hood

acylating and alkylating reagents are employed (Table 7.4). A drawback to using electron-capturing derivatives is that biological extracts often contain many compounds with reactive functional groups and so the resulting chromatograms may contain additional peaks, which may interfere.

7.5 Chiral separations

HPLC has been widely applied in this area, but GC has also been used (Zief and Crane, 1988; Pasutto, 1992; König, 1998). HPLC has the advantage that the resolved analytes are more easily recovered from the column eluate and preparative-scale systems are common, whilst GC detectors are, in general, more sensitive. Modified lipophilic cyclodextrin (CD) derivatives have proved superior to all other previously used chiral stationary phases (CSPs) for capillary GC and have an almost unlimited range of application.

If the analyte contains a reactive functional group, then derivatization using an optically pure reagent and analysis of the resulting diastereoisomers on an achiral GC column may prove adequate (Srinivas *et al.*, 1995). A range of derivatizing reagents has been used including, for amines, (*S*)-(–)-*N*-trifluoroacetylprolyl chloride. Alternatively, analysis using a CSP may be employed. A number of chiral GC phases have been described including amides, diamides, dipeptides and polysiloxanes with chiral substituents (Liu and Ku, 1983), but the low thermal stability of these phases precluded their widespread use. However, modification of phenylmethylpolysiloxanes by introducing L-valine-*tert*-butylamide or L-valine-*S*-α-phenylethylamide moieties can give chiral phases with temperature stability as high as 230 °C. PDMS phases modified with (*S*)-(−)-*tert*-leucine derivatives have also been used (Abe *et al.*, 2000). In addition, a copolymeric stationary phase consisting of a chiral component, (1R-*trans*)-*N*, *N*′-1,2-cyclohexylenebisbenzamide, and an PDMS phase with good temperature stability has been developed (Juvanc *et al.*, 2002).

CDs are cyclic, chiral, torus-shaped molecules consisting of six or more D(+)-glucose moieties linked via α-(1–4) glycoside bonds. α-CDs contain six, β-CDs, seven and γ-CDs, eight amylose units (see Section 8.6.6 for further discussion of CDs). Peralkylated α-, β-, or γ-CDs can be used in fused silica capillaries dissolved in polysiloxanes to give chiral GC columns (König, 1998). Without the CD additive the columns are achiral. Enantiomers of polar compounds such as alcohols, diols and carboxylic acids can be resolved underivatized. Moreover, racemic alkanes and cycloalkanes can be separated using CD-modified phases. Enantiomers of halogenated anaesthetics (enflurane, halothane, isoflurane), have been resolved using such systems, (Meinwald *et al.*, 1991), as have enantiomers of ketamine and norketamine (Williams and Wainer, 2002). These columns will often separate positional isomers as well as enantiomers. The useable temperature range is about 30–240 °C.

A practical application of the use of capillary GC in the analysis of enantiomers is the measurement of amfetamine, metamfetamine, MDA, MDMA and MDEA enantiomers in plasma samples from clinical toxicology and from driving under the influence of drugs cases (Peters *et al.*, 2003a).

7.6 Applications of gas chromatography in analytical toxicology

The refinement of capillary column technology and the introduction of 'bench-top' GC-MS systems have led to renewed appreciation of the value of GC in analytical toxicology. In developing GC methods, knowledge of the M_r and structure of the compound of interest is important even if a published method is to be followed. Information on any co-formulated or co-administered drugs may also be valuable. Manufacturers may sometimes be able to give details of assay methods, current literature and potential internal standards for particular compounds.

7.6.1 Systematic toxicological analysis

The role of capillary GC in STA has been discussed extensively (Maurer, 1992; Peters *et al.*, 2003b), and more recently has been compared with HPLC-DAD, HPLC-MS and other techniques (Polettini, 1999; Maurer, 2004).

Packed column GC retention data for many drugs and other poisons on a range of stationary phases of differing polarities are highly correlated (Moffat *et al.*, 1974) due to the inter-relationship between M_r, volatility and retention that is dominant in GC. It was thus concluded that there was little to be gained from the use of more than one column in STA, and it was suggested that a low-polarity phase such as SE-30/OV-1/OV-101 (Table 7.3) should be used for this purpose as most of the compounds studied were eluted. Subsequent experience with modern high-efficiency capillary GC columns has borne out the suggestion that a single low- or moderate-polarity column with temperature programming is satisfactory for most purposes if used with a selective sample preparation procedure and some form of selective detection, ideally MS. PDMS phases such as

SE-30/OV-1/OV-101 now have the further advantage that retention index data (Section 5.3) for a large number of drugs and other compounds of interest generated on packed columns are available (de Zeeuw *et al.*, 1992). Such data are generally directly transferable to capillary columns.

The recent introduction of 'retention time locking' (RTL) for capillary GC (Agilent) has prompted a re-evaluation of the use of retention index data in STA. The long-term precision of three retention parameters, the absolute retention time, the relative retention time related to dibenzepin, and a retention index based on the alkylfluoroaniline series, were studied with 14 basic drugs on HP-5 (Agilent; Table 7.3) and DB-17 (J&W – equivalent to OV-17, Table 7.3) columns with and without the use of RTL. Using the constant flow mode, RTL gave better reproducibility with all three retention parameters compared to the non-RTL method on both columns. RTL offered a significant advantage within a single instrument method, not only between methods, with RSDs of <0.1 % in relative retention time (Rasanen *et al.*, 2003). RTL has also been employed in pesticides residue analysis (Cook *et al.*, 1999).

HP-5 has been used in STA for many years. A 25 m × 0.32 mm i.d. cross-linked fused silica capillary has been employed when screening for basic drugs in urine (Figure 7.13) and a further

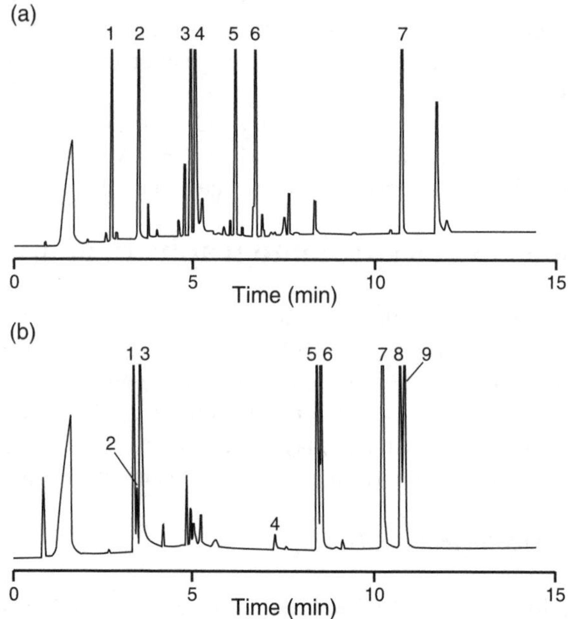

Figure 7.13 Capillary GC of drugs of abuse in urine (Caldwell and Challenger, 1989). Column: 25 m × 0.32 mm i.d. HP-5 (0.52 μm film). Oven temperatures: injector 250 °C, detector 300 °C. Column temperature: 90 °C (0.5 min), then to 250 °C at 40 °C min^{-1}, then to 310 °C at 5 °C min^{-1}. Carrier gas: helium (about 3.5 mL min^{-1}). Detector: NPD (N mode). Sample preparation: Box 7.5. Injection: Split/splitless (splitless mode), 2 μL extract. (a) Analysis of a sample from a patient prescribed methadone, but who was also taking pethidine and phentermine. Peaks: 1 = phentermine, 2 = nicotine, 3 = pethidine, 4 = norpethidine, 5 = EDDP (methadone metabolite), 6 = methadone, 7 = prazepam (IS). (b) Analysis of a sample from a patient taking a cough medicine containing codeine, ephedrine and promethazine. Peaks: 1 = phenylpropanolamine, 2 = nicotine, 3 = ephedrine, 4 = promethazine, 5 = codeine, 6 = norcodeine, 8 = prazepam (IS) – peaks 7 and 9 were described as promethazine metabolites, without further detail.

Box 7.5 Liquid–liquid extraction of basic drugs from urine prior to GC

- Add 0.25 mL aqueous sodium hydroxide (1 mol L^{-1}) and 0.5 mL internal standard solution (5 mg L^{-1} prazepam in butyl acetate) to 1 mL urine in a 4.5 mL polypropylene tube (Sarstedt)
- Vortex-mix (30 s) and centrifuge (3000 rpm, 10 min; bench-top centrifuge)
- Transfer 0.3 mL of extract to 1 mL autosampler vials with crimped-on rubber tops (Chromacol). Inject 2 μL

application of HP-5 has been in the analysis of organochlorine pesticides. DB-5 (J&W) is said to be directly equivalent to HP-5. SP-2250 (equivalent to DB-17 and OV-17, Table 7.3) has been used isothermally in the measurement of basic drugs (Dawling *et al.*, 1990a) including cocaine (Dawling *et al.*, 1990b).

7.6.2 *Quantitative analysis of drugs and other poisons*

As noted above, quantitative analyses can be performed using capillary GC provided that the injection is performed appropriately. The use of capillary GC-NPD to measure nicotine and cotinine in plasma, urine and saliva is illustrated in Figure 7.14. A liquid–liquid microextraction procedure was used (Box 7.6). Antifoam and phenol red were added to inhibit emulsion formation and to facilitate extract removal, respectively. The LLoD for both nicotine and cotinine was

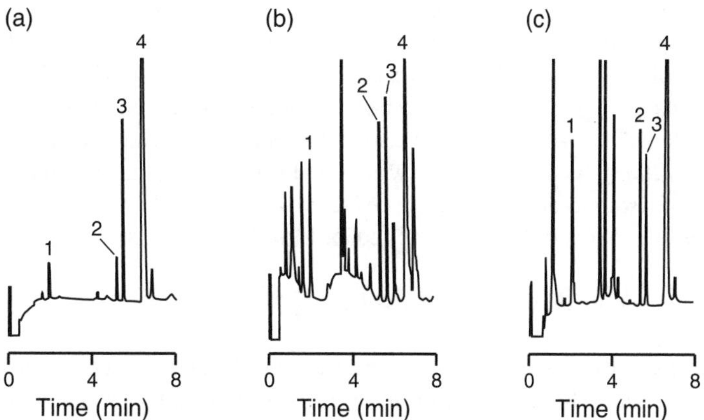

Figure 7.14 GC of nicotine and cotinine in (a) human plasma, (b) urine and (c) saliva (Reproduced from Feyerabend, C. and Russell, M.A.H. (1990) A rapid gas-liquid chromatographic method for the determination of cotinine and nicotine in biological fluids. J Pharm Pharmacol, **42**, 450–2, with permission from RPS Publishing). Column: 7 m × 0.32 mm i.d. HP-FFAP (0.52 μm film). Oven temperatures: Injector 170 °C; Detector 300 °C. Column temperature: 70 °C to 115 °C at 400 °C min^{-1} and hold (1.5 min), then to 200 °C at 400 °C min^{-1} and hold (2.25 min), then to 210 °C at 400 °C min^{-1} and hold (2.75 min). Carrier gas: helium, head pressure 105 kPa (15 psi). Detector: NPD. Detector gases: helium (make-up) 13 mL min^{-1}; hydrogen 3.3 mL min^{-1}; air 13 mL min^{-1}. Extraction procedure: Box 7.6. Injections: 2 μL sample extracts. Nicotine and cotinine concentrations (μg L^{-1}): plasma 23 and 27, urine 73 and 100, saliva 115 and 130, respectively. Peaks: 1 = nicotine, 2 = cotinine, 3 = 3-methylcotinine (internal standard), 4 = caffeine.

Box 7.6 Liquid–liquid microextraction of basic drugs prior to GC

- Add sample or standard (100 μL) to 60 × 5 mm i.d. glass tube (Dreyer tube)
- Add 100 μL internal standard solution (117 μg L^{-1} 5-methylcotinine), 300 μL aqueous sodium hydroxide (5 mol L^{-1}), 20 μL antifoam/phenol red mixture [5 % (v/v) Dow Corning antifoam RD emulsion, 200 mg L^{-1} phenol red] and 50 μL 1,2-dichloroethane
- Vortex-mix (1 min) and centrifuge (11,000 *g*, 2 min)
- Inject 2 μL of the 1,2-dichloroethane extract

0.1 μg L^{-1} (100 μL sample) allowing the method to be used to measure these compounds in samples from smokers and from non-smokers. The extraction time was 3 min, and by using multi-pipetting and vortexing systems 250 samples could be processed daily. The average RSD for nicotine (range 1–100 μg L^{-1}) was 3.9 % and for cotinine (range 1–1000 μg L^{-1}) 2.2 %.

7.6.2.1 *Measurement of carbon monoxide and cyanide*

GC offers advantages over spectrophotometric methods for carbon monoxide analysis (Section 4.3.3.2), especially if badly decomposed postmortem blood or tissues are to be analyzed (Mayes, 1993). However, because the sensitivity for carbon monoxide by FID is very poor, either TCD or analyte reduction with hydrogen on a heated Ni catalyst to produce methane before FID have had to be used. This latter procedure introduces an additional step, and requires non-standard apparatus. With the development of the helium discharge ionization detector (He-PDPID) it is possible to measure carbon monoxide directly with good sensitivity (Section 7.2.2.5). Helium is generally used as the carrier gas and as the ionized species. Sample manipulation, assay calibration, calculation of % COHb, and so on, are the same as with the use of TCD (Mayes, 1993).

Blood cyanide concentrations have been measured by GC-NPD using acetonitrile as internal standard after addition of phosphoric acid to the sample in a headspace vial (Mayes, 1993; Calafat and Stanfill, 2002). Assay calibration was by addition of potassium cyanide solution to alkalinized human blood. Isotope-dilution headspace GC-MS using $K^{13}C^{15}N$ or C^2H_3CN as internal standard has also been employed (Dumas *et al.*, 2005; Murphy *et al.*, 2006).

7.6.2.2 *Measurement of ethanol and other volatiles*

The earliest GC-FID method for the measurement of blood ethanol involved simple dilution of the sample (whole blood, plasma or urine) (50 μL) with internal standard solution (0.16 g L^{-1} aqueous propanol, 500 μL) followed by vortex-mixing (10 s) and direct injection of the resulting mixture onto a column packed with a molecular sieve such as Chromosorb 102 (Table 7.2). Other volatiles such as methanol, 2-propanol and acetone, were resolved and could be measured if required. Modified carbon black (Carbopak, Section 7.3.) materials can also be used (Figure 7.15).

Nowadays, static headspace sampling combined with temperature-programmed GC on a PDMS capillary column and dual detection (ECD/FID) can be used to screen for not only ethanol, methanol and 2-propanol, but also for a wide range of other volatile compounds in biological fluids (Figure 7.16). Alternatively a split injection/dual column (PDMS-PEG) system can be used (Sharp, 2001). It is at first sight surprising that many volatile compounds are relatively stable in

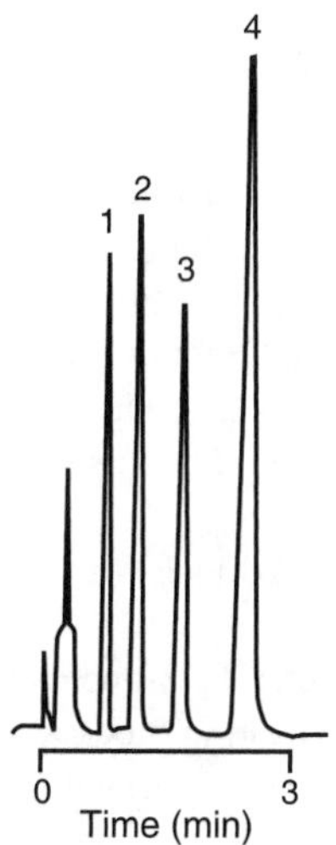

Figure 7.15 GC-FID of blood acetone, ethanol and 2-propanol. Column: 1 m × 2 mm i.d. glass packed with 0.2 % (w/w) Carbowax 1500 on Carbopak C (80–100 mesh). Column oven temperature: 100 °C. Carrier gas (nitrogen) flow: 30 mL min^{-1}. Detection: FID. Injection: 1 μL aqueous solution of ethanol (1), acetone (2), 2-propanol (3) (all 0.08 g L^{-1}) and propanol (4) (0.16 g L^{-1}).

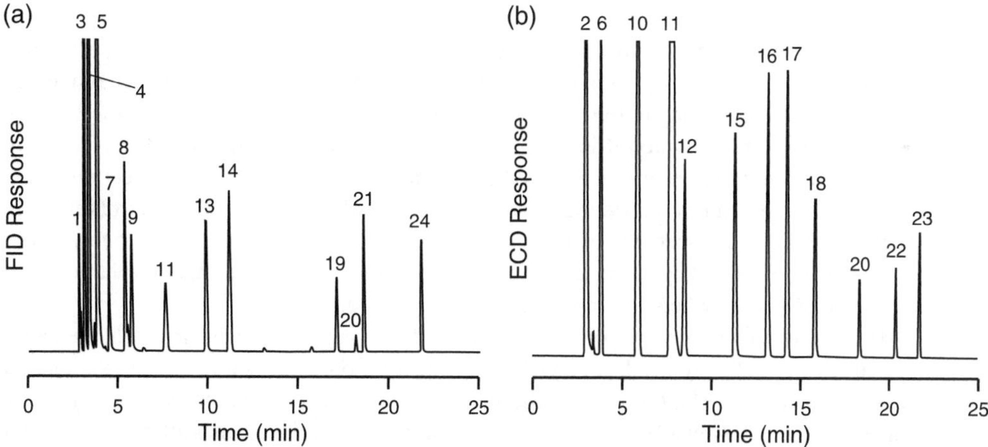

Figure 7.16 Screening for volatile substances using headspace capillary GC: (a) FID (b) ECD. Reproduced from Streete, P.J., Ruprah, M., Ramsey, J.D. and Flanagan, R.J. (1992) Detection and identification of volatile substances by headspace capillary gas chromatography to aid the diagnosis of acute poisoning. Analyst, 117, 1111–27, by permission of The Royal Society of Chemistry. GC Conditions: as Box 7.7. Injection: about 10 μL qualitative standard mixture. Peaks: 1 = propane, 2 = dichlorodifluoromethane, 3 = dimethyl ether, 4 = isobutane, 5 = butane, 6 = bromochlorodifluoromethane, 7 = ethanol, 8 = acetone, 9 = 2-propanol, 10 = fluorotrichloromethane, 11 = 1,1,2-trichlorotrifluoroethane, 12 = halothane, 13 = butanone, 14 = hexane, 15 = chloroform, 16 = 1,1,1-trichloroethane, 17 = carbon tetrachloride, 18 = trichloroethylene, 19 = 4-methyl-2-pentanone, 20 = 1,1,2-trichloroethane (internal standard), 21 = toluene, 22 = tetrachloroethylene, 23 = 2,2,2-trichloroethanol, 24 = ethylbenzene (internal standard).

Box 7.7 HS-GC procedure for screening for volatile compounds.

Apparatus and Reagents

- Column: 60 m × 0.53 mm i.d. fused silica. Stationary phase: SPB-1 (5 μm film). Carrier gas: Helium (8.6 mL min^{-1}). Oven temperature: 40 °C for 6 min, then to 80 °C at 5 °C min^{-1}, then to 200 °C at 10 °C min^{-1}. Injector temperature: 150 °C. Detector temperature: 275 °C. Detection: FID/ECD (split about 5:1). ECD purge: Nitrogen (about 35 mL min^{-1})
- Internal standard solution [25 mg L^{-1} ethylbenzene and 10 mg L^{-1} 1,1,2-trichloroethane in expired blood-bank whole blood:deionised water (1+24)]

Method

- Add internal standard solution (200 μL) to a 7 mL glass septum vial (Schubert, Portsmouth, UK)
- Seal the vial using a crimped-on PTFE-lined silicone disk
- Incubate the vial at 65 °C in a heating block and, after 15 min, withdraw a portion (100–300 μL) of headspace using a warmed (40 °C) gas-tight glass syringe and inject onto the column
- Subsequently add the sample (whole blood, plasma or serum) (200 μL) to the same vial using a 1.0 mL plastic disposable syringe and, after at least 15 min, take a second portion (100–300 μL) of headspace for analysis
- Remove the plunger from the gas-tight syringe and place the assembly on the heating block until the next injection to ensure evaporation of any remaining analyte(s)

blood if simple precautions are taken. The tube should be as full as possible and should only be opened when required for analysis, and then only when cold (4 °C). An anticoagulant (lithium heparin or EDTA) should be used, as in other methods in which whole blood is to be assayed. If the sample volume is limited it is advisable to select a container to match the volume of blood so that there is minimal headspace. Specimen storage between -5 and 4 °C is recommended (Gill *et al.*, 1988) and sodium fluoride (2 % w/v) should be added to minimize microbial metabolism (Section 2.3.1). In a suspected volatile substance abuse (VSA)-related fatality, analysis of tissues such as brain or lung may prove useful as relatively high concentrations of volatile compounds may be present even when nothing is detectable in blood.

Measurement of relatively non-volatile compounds such as ethanol and toluene can be performed isothermally by headspace GC after construction of a calibration graph. The value of attempting to measure blood concentrations of substances such as butane, which are gases at room temperature, is questionable and in most cases qualitative identification is all that is needed. In very general terms, blood concentrations of less volatile substances such as toluene of 10–20 mg L^{-1} and above can be associated with serious poisoning. In other words, the pharmacologically effective concentrations of volatile substances are similar to those of inhalational anaesthetics (Streete *et al.*, 1992), and are thus two or so orders of magnitude lower than those observed in poisoning with relatively water-soluble compounds such as ethanol.

In addition to HS-GC, purge-and-trap extraction coupled with GC and sequential FTIR/FID has been used to detect and measure volatile analytes in blood (5 mL) using 3-propanone as internal standard (Ojanperä, *et al.*, 1998). A PoraPLOT Q capillary column was used. FTIR identification limits ranged from 0.01 mg L^{-1} for ethyl acetate, butanone and sevoflurane to 24 mg L^{-1} for methanol, and generally allowed the detection of exposure to volatiles at concentrations below those associated with acute toxicity. Using FID the method was suitable for quantitative work. The throughput was five samples per day, the purge-and-trap stage being the limiting factor.

The alkyl nitrites that are abused by inhalation [isobutyl nitrite, isopentyl ('amyl') nitrite] are a special case in that: (i) they are extremely unstable *in vivo* and are rapidly hydrolyzed to the corresponding alcohols and (ii) usually also contain other isomers (butyl nitrite, pentyl nitrite). Any products submitted for analysis will usually contain the corresponding alcohols as well as the organic nitrites. Propanols and butanols may also arise, for example by microbial action from normal blood constituents *in vitro* and thus caution is needed in the interpretation of results in cases where these compounds are detected.

7.7 Summary

There is no doubt that GC is the system of choice for the analysis of solvents and other volatiles, in STA, and as an interface with MS (Box 7.8). The high efficiency and stability of modern capillary columns together with other features of GC, such as the ease of temperature programming and the reproducibility and transferability of retention data, are features that other chromatographic systems cannot match. The drawbacks of GC are the need for dedicated instruments, experienced operators and appropriate laboratory infrastructure. There is also the requirement to perform solvent extraction or some similar method of sample preparation and the restriction on the M_r of analytes to those that are stable and volatile, either derivatized or underivatized at the oven temperature required for the analysis (in practice up to M_r 750 or so). The great advantages of LC, especially when combined with MS, are that these latter restrictions are minimized. But all is not necessarily straightforward even with LC and LC-MS, as discussed in Chapters 8 and 10, respectively.

Box 7.8 Summary of the use of GC in analytical toxicology

- Modern fused silica capillaries offer unrivalled efficiency and stability, especially when used with temperature programming
  - MS interface easy (use in EI or CI mode)
- In general, use HS-GC for volatiles with 60 m × 0.56 mm i.d., 5 μm film, OV-1 or equivalent capillary column with split FID/ECD or MS detection
  - For ethanol, methanol direct injection often adequate
- For STE use 30 m × 0.35 mm i.d. chemically bonded fused silica capillaries (HP-5 or equivalent) with temperature programming and split FID/NPD or EI-MS detection
  - Take care to minimize losses of volatiles such as amfetamine during sample preparation
  - May need to derivatize polar analytes such as morphine
  - Split/splitless in splitless mode best for quantitative work
- ECD has unrivalled sensitivity/selectivity for appropriate analytes (may need to derivatize)
  - Modern detectors good linear range
  - Difficulty in disposing of old instruments

References

Abe, I., Terada, K. and Nakahara, T. (2000) Enantiomer separation of pharmaceuticals by capillary gas chromatography with novel chiral stationary phases. *Biomed Chromatogr*, **14**, 125–9.

Calafat, A.M. and Stanfill, S.B. (2002) Rapid quantitation of cyanide in whole blood by automated headspace gas chromatography. *J Chromatogr B*, **772**, 131–7

Caldwell, R. and Challenger, H. (1989) A capillary column gas-chromatographic method for the identification of drugs of abuse in urine samples. *Ann Clin Biochem*, **26**, 430–43.

Cook, J., Engel, M., Wylie, P. and Quimby, B. (1999) Multiresidue screening of pesticides in foods using retention time locking, GC-AED, database search, and GC/MS identification. *J AOAC Int*, **82**, 313–26.

Dawling, S., Ward, N., Essex, E.G. and Widdop, B. (1990a) Rapid measurement of basic drugs in blood applied to clinical and forensic toxicology. *Ann Clin Biochem*, **27**, 473–7.

Dawling, S., Essex, E.G., Ward, N. and Widdop, B. (1990b) Gas chromatographic measurement of cocaine in serum, plasma and whole blood. *Ann Clin Biochem*, **27**, 478–81.

Dalluge, J., Beens, J. and Brinkman, U.A. (2003) Comprehensive two-dimensional gas chromatography: a powerful and versatile analytical tool. *J Chromatogr A*, **1000**, 69–108

Deans, D.R. (1968) A new technique for heartcutting in gas chromatography. *Chromatographia*, **1**, 18–22.

Dumas, P., Gingras, G. and LeBlanc, A. (2005) Isotope dilution-mass spectrometry determination of blood cyanide by headspace gas chromatography. *J Anal Toxicol*, **29**, 71–5.

Flanagan, R.J., Dawling, S. and Buckley, B.M. (1987) Measurement of ethylene glycol (ethane-1,2-diol) in biological specimens using derivatisation and gas-liquid chromatography with flame ionisation detection. *Ann Clin Biochem*, **24**, 80–4.

Forsyth, D.S. (2004) Pulsed discharge detector: theory and applications. *J Chromatogr A*, **1050**, 63–8.

Gill, R., Hatchett, S.E., Osselton, M.D. Wilson, H.K., Ramsey, J.D. (1988) Sample handling and storage for the quantitative analysis of volatile compounds in blood: The determination of toluene by headspace gas chromatography. *J Anal Toxicol*, **12**, 141–6.

Grob, K. (2001) *Split and Splitless Injection for Quantitative Gas Chromatography: Concepts, Processes, Practical Guidelines, Sources of Error*, 4th edn, Wiley-VCH, Weinheim.

Grob, R.L. and Barry, E.F. (eds.) (2004) *Modern Practice of Gas Chromatography*, 4th edn, Wiley-Interscience, New York.

Hinshaw, J.V. (2002) A compendium of GC terms and techniques. *LCGC North America*, **20**, 1034–40 (http://www.lcgcmag.com/lcgc/data/articlestandard/lcgc/462002/38300/article.pdf, accessed 24 May 2006).

Hooker, D.B. and Dezwaan, J. (1989) Applications of a simultaneous atomic emission mass spectral gas chromatography detector in drug analysis. *J Pharm Biomed Anal*, **7**, 1591–7.

Huggett, A., Andrews, P. and Flanagan, R.J. (1981) Rapid micromethod for the measurement of paracetamol in blood plasma or serum using gas-liquid chromatography with flame-ionisation detection. *J Chromatogr*, **209**, 67–76.

IUPAC. (1986) Characteristics of liquid stationary phases and column evaluation for gas chromatography. *Pure Appl Chem*, **58**, 1291–306.

Juvanc, Z., Markides, K.E., Petersson, P., Johnson, D.F., Bradshaw, J.S. and Lee, M.L. (2002) Copolymeric (lR-trans)-N,N′-1,2-cyclohexylene-bisbenzamide oligodimethylsiloxane chiral stationary phase for gas chromatography. *J Chromatogr A*, **982**, 119–26.

Kolb, B. and Ettre, L.S. (1997) *Static Headspace-Gas Chromatography: Theory and Practice*, Wiley, London.

König, W.A. (1998) *Gas Chromatographic Enantiomer Separation with Modified Cyclodextrins*, Wiley, New York.

Kueh, A.J., Marriott, P.J., Wynne, P.M. and Vine, J.H. (2003) Application of comprehensive two-dimensional gas chromatography to drugs analysis in doping control. *J Chromatogr A*, **1000**, 109–24.

Lacroix, B., Huvenne, J.P. and Deveaux, M. (1989) Gas chromatography with Fourier transform infrared spectrometry for biomedical applications. *J Chromatogr*, **492**, 109–36.

Laniewski, K., Vagero, M., Forsberg, E. Forngren T, Hagman G. (2004) Complementary use of gas chromatography-mass spectrometry, gas chromatography-atomic emission detection and nuclear magnetic resonance for identification of pharmaceutically related impurities of unknown structures. *J Chromatogr A*, **1027**, 93–102.

Liu, R.H. and Ku, W.W. (1983) Chiral stationary phases for the gas-liquid chromatographic separation of enantiomers. *J Chromatogr*, **271**, 309–23.

Marriott, P.J., Haglund, P. and Ong, R.C. (2003) A review of environmental toxicant analysis by using multidimensional gas chromatography and comprehensive GC. *Clin Chim Acta*, **328**, 1–19.

Marriott, P.J., Massil, T. and Hugel, H. (2004) Molecular structure retention relationships in comprehensive two-dimensional gas chromatography. *J Sep Sci*, **27**, 1273–84.

Marriott, P.J., Ong, R.C., Kinghorn, R.M. and Morrison, P.D. (2000) Time-resolved cryogenic modulation for targeted multidimensional capillary gas chromatography analysis. *J Chromatogr A*, **892**, 15–28.

Maurer, H.H. (2004) Position of chromatographic techniques in screening for detection of drugs or poisons in clinical and forensic toxicology and/or doping control. *Clin Chem Lab Med*, **42**, 1310–24.

Maurer, H.H. (1992) Systematic toxicological analysis of drugs and their metabolites by gas chromatography-mass spectrometry. *J Chromatogr*, **580**, 3–41.

Mayes, R.W. (1993) ACP Broadsheet No 142: November 1993. Measurement of carbon monoxide and cyanide in blood. *J Clin Pathol*, **46**, 982–8.

McReynolds, W.O. (1970) Characterization of some liquid phases. *J Chromatogr Sci*, **8**, 685–91.

Meinwald, J., Thompson, W.R., Pearson, D.L., König, W.A., Runge, T. and Francke, W. (1991) Inhalational anesthetics stereochemistry: optical resolution of halothane, enflurane, and isoflurane. *Science*, **251**, 560–1.

Moffat, A.C., Stead, A.H. and Smalldon, K.W. (1974) Optimum use of paper, thin-layer and gas-liquid chromatography for the identification of basic drugs. 3. Gas-liquid chromatography. *J Chromatogr*, **90**, 19–33.

Murphy, K.E., Schantz, M.M., Butler, T.A., Benner, B.A. Jr, Wood, L.J. and Turk, G.C. (2006) Determination of cyanide in blood by isotope-dilution gas chromatography-mass spectrometry. *Clin Chem*, **52**, 458–67.

Ojanperä, I., Pihlainen, K. and Vuori, E. (1998) Identification limits for volatile organic compounds in the blood by purge-and-trap GC-FTIR. *J Anal Toxicol*, **22**, 290–5.

Pasutto, F.M. (1992) Mirror images: the analysis of pharmaceutical enantiomers. *J Clin Pharmacol*, **32**, 917–24.

Peters, F.T., Samyn, N., Wahl, M., Kraemer, T., De Boeck, G. and Maurer, H.H. (2003a) Concentrations and ratios of amphetamine, methamphetamine, MDA, MDMA, and MDEA enantiomers determined in plasma samples from clinical toxicology and driving under the influence of drugs cases by GC-NICI-MS. *J Anal Toxicol*, **27**, 552–9.

Peters, F.T., Schaefer, S., Staack, R.F., Kraemer, T. and Maurer, H.H. (2003b) Screening for and validated quantification of amphetamines and of amphetamine- and piperazine-derived designer drugs in human blood plasma by gas chromatography/mass spectrometry. *J Mass Spectrom*, **38**, 659–76.

Pihlainen, K. and Ojanperä, I. (1998) Analytical toxicology of fluorinated inhalation anaesthetics. *Forensic Sci Int*, **97**, 117–33.

Polettini, A. (1999) Systematic toxicological analysis of drugs and poisons in biosamples by hyphenated chromatographic and spectroscopic techniques. *J Chromatogr B*, **733**, 47–63.

Rasanen, I., Kontinen, I., Nokua, J., Ojanperä, I. and Vuori, E. (2003) Precise gas chromatography with retention time locking in comprehensive toxicological screening for drugs in blood. *J Chromatogr B*, **788**, 243–50.

Roberge, M.T., Finley, J.W., Lukaski, H.C. and Borgerding, A.J. (2004) Evaluation of the pulsed discharge helium ionization detector for the analysis of hydrogen and methane in breath. *J Chromatogr A*, **1027**, 19–23.

Rood, D. (2006) *The Troubleshooting and Maintenance Guide for Gas Chromatographers*, 4th edn, Wiley, New York.

Sharp, M.E. (2001) A comprehensive screen for volatile organic compounds in biological fluids. *J Anal Toxicol*, **25**, 631–6.

Srinivas, N.R., Shyu, W.C. and Barbhaiya, R.H. (1995) Gas chromatographic determination of enantiomers as diastereomers following precolumn derivatization and applications to pharmacokinetic studies: a review. *Biomed Chromatogr*, **9**, 1–9.

Supelco. (1997) *The retention index system in gas chromatography: McReynolds constants*. Supelco, Bellefonte, PA, (http://www.sigmaaldrich.com/Graphics/Supelco/objects/7800/7741.pdf, accessed 24 March 2006).

Supina, W.R. (1974) *The Packed Column in Gas Chromatography*, Supelco, Bellefonte, PA.

US Pharmacopoeia Convention. (1999) *USP 24—NF-19: US Pharmacopoeia and National Formulary*, USP Edition 24, NF Edition 19, United States Pharmacopoeia Convention, Rockville, MD.

Williams, M.L. and Wainer, I.W. (2002) Role of chiral chromatography in therapeutic drug monitoring and in clinical and forensic toxicology. *Ther Drug Monit*, **24**, 290–6.

de Zeeuw, R.A., Franke, J.P., Maurer, H.H. and Pfleger, K. (1992) *Gas-Chromatographic Retention Indices of Toxicologically Relevant Substances on Packed or Capillary Columns with Dimethylsilicone Stationary Phases*, 3rd edn, VHC, Weinheim.

Zief, M. and Crane, L.J. (1988) *Chromatographic Chiral Separations*, Dekker, New York.

8 High-Performance Liquid Chromatography

8.1 Introduction

High-performance liquid chromatography (HPLC) is well suited to the analysis of hydrophilic, thermally labile and/or high M_r compounds. In the analysis of drugs and other poisons HPLC has additional practical advantages of flexibility, generally low running costs, a range of selective detectors, which can usually be linked in series, and ease of automation (Box 8.1). These properties can often be exploited to facilitate the analysis of several compounds (e.g. drug and metabolites) at once. Major uses of HPLC include pharmacokinetic and metabolic studies, the measurement of plasma concentrations of drugs given in therapy (TDM), and in monitoring exposure to toxic chemicals.

Box 8.1 Advantages of HPLC in analytical toxicology

- Rapid, efficient separations – can control selectivity by adjusting eluent
- Analyse high M_r and/or polar molecules that are not amenable to GC
- Sensitive, selective, non-destructive detectors (UV, fluorescence) that can be used in series
- Ability to perform post-column reactions
- Can interface to MS-ESI, APCI, and so on – very selective detection for targeted analytes
- Can inject aqueous samples (but selectivity reduced compared with e.g. solvent extracts)
- Running costs low – can recirculate isocratic eluent
- Portable equipment – normally no gas supplies, high-temperature ovens, and so on
- Relatively easy to automate
- Silica-based packings are robust – can be used with polar and nonpolar eluents

Generally HPLC of drugs and other poisons, and their metabolites, is performed isocratically (constant eluent composition), although gradient elution (changing eluent composition during an analysis) may be needed if a range of compounds of widely different polarities is to be measured or a rapid analysis time is required. Operation at ambient temperature is often used for simplicity and to minimize dissolution of the column packing. However, HPLC is not without its difficulties. Both sample volume and detector sensitivity may be limiting and potential interferences are legion. Accumulation of sample components on the column and/or of particulate material on the top of columns may also contribute to the loss of efficiency usually observed in routine use. Some further limitations are summarized in Box 8.2.

Whilst analyte thermal stability is not as important as in GC, many compounds (especially metabolites) are unstable in biological samples or if subjected to extremes of pH, for example

Box 8.2 Limitations of HPLC in analytical toxicology

- Hardware/infrastructure expensive; need experienced operators, pure solvent supplies, consumables support
- HPLC resolves relatively few peaks in useful range ($k = 2$–10, compare GC with temperature programming)
- Elution independent of M_r (cf. GC)
- Gradient elution comparable to temperature programming in GC, but more problems (baseline drift, re-equilibration)
- Only ED, fluorescence and MS detectors give sensitivity comparable to GC detectors, but relatively selective
- No universal detector comparable to FID in sensitivity
- Can screen with DAD, but with LC-MS cannot use EI to generate principal peaks index
- Efficiency influenced by many extraneous factors (injection volume, injection solvent, 'dead' volumes in injectors/tubing/detector, detector sampling rate, void formation, blockage of frits, etc.)
- Always the possibility of co-elution

during sample preparation. Thus, as well as the choice of the column/eluent combination, due consideration must be given to the collection and storage of the sample (Chapter 2), the choice of sample preparation procedure (Chapter 3) and the detection conditions. Selection of a suitable internal standard is also an important step.

Operationally the big difference between GC and HPLC is the nature of the mobile phase. Liquids are not as compressible as gases. This means that in HPLC, smaller particle-size packings can be used giving high efficiencies with relatively short columns at ambient temperature. In addition, the composition of the eluent can be altered to control retention and selectivity giving a very high degree of control over a given separation.

8.2 HPLC: general considerations

A typical HPLC system consists of an eluent reservoir, a high-pressure pump (or pumps in the case of high-pressure mixing gradient systems) with flow controller, a sample injection system, stainless steel guard and analytical columns packed with stationary-phase material, a detector, usually with a low volume flow cell, and a data capture system (Figure 8.1). The most commonly used detectors in analytical toxicology are UV/visible, including diode array (DAD), fluorescence, electrochemical (ED) and mass spectrometric (MS).

Isocratic elution is clearly the simplest mode of HPLC and is analogous to isothermal operation in GC. However, if the eluent composition (organic solvent content, pH or ionic strength) is made to change in a pre-determined way, either by using a second solvent reservoir and pump, and a mixing chamber (Figure 8.1), or by use of two solvent reservoirs, a mixing chamber and one pump (low-pressure mixing), the technique is called gradient elution (cf. temperature programming in GC). Ternary gradients are also possible. The advantage is, of course, that later eluting peaks are sharpened. Instruments can be programmed to return to starting conditions, but this may take longer than simply cooling a GC oven. Baseline drift during gradient elution with UV detection can be minimized if the absorptivity of the components of the gradient at the detection wavelength

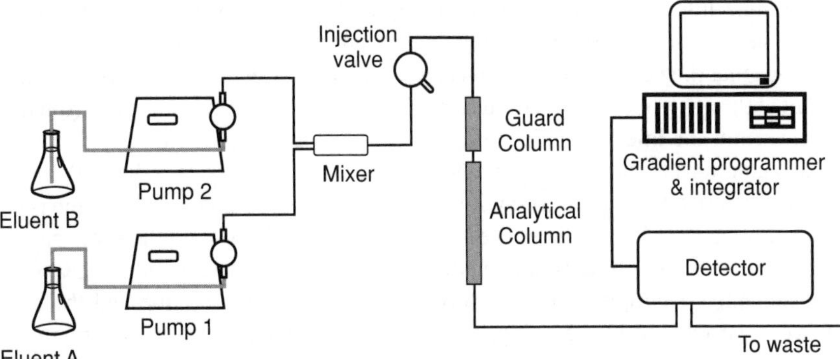

Figure 8.1 Schematic diagram of a gradient HPLC system with high-pressure mixing.

is balanced beforehand. An alternative, use of an eluent splitter to divert a portion of the eluent to the reference cell in a double-beam detector, is rarely used.

8.2.1 The column

As with other forms of elution chromatography, the HPLC column is the heart of the system. In order to maintain a column in good working order it is important to use it carefully, especially with regard to the integrity of the column bed. The higher efficiencies (Section 5.1.2) that are possible with HPLC as opposed to conventional LC are due to the introduction of packing materials with particle sizes typically in the range 3–10 μm. The back-pressures that are generated require packing materials that can stand such forces, and, as silica can be used for long periods at pressures up to 28,000 kPa (4000 psi) without physical damage, it is the most commonly used HPLC packing material. Furthermore, the surface silanol groups can be exploited to form a large range of chemically modified phases. These are discussed in more detail below (Section 8.4.2).

To withstand the operating pressures used, the column material is most commonly packed in stainless steel tubing with porous frits (typical pore size 0.5–2 μm, depending on the size of the packing material), at each end. Many manufacturers have also developed cartridge systems using stainless steel, glass or polyetheretherketone (PEEK, Figure 8.2) columns in a stainless steel or aluminium enclosure. If the inlet frit becomes partially blocked with particles not previously removed from the eluent and sample extracts, not only will the column back-pressure increase, but also column efficiency may be adversely affected as the sample will not be dispersed evenly onto the top of the column.

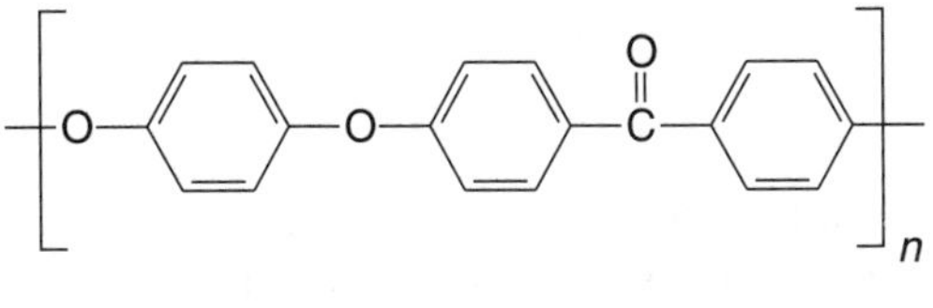

Figure 8.2 Structure of polyoxy-1,4-phenyleneoxy-1,4-phenylenecarbonyl-1,4-phenylene (polyetheretherketone, PEEK).

An in-line filter placed between the pump and the injector to trap particles from the piston seal can be beneficial, particularly if the outlet valve does not have its own filter. Filters have the advantage that they can be quickly and easily replaced. A guard column is a small column placed in front of the analytical or preparative column to protect it from impurities from the sample, wear particles from the injection valve and so on. It should be approximately 5 % of the size of the working column and should be changed frequently. Use of a short (say 50 mm) column of narrower bore than the analytical column (say 2 mm i.d. if a 4.6 mm i.d. analytical column is being used) will help minimize peak broadening. The guard column should contain the same packing material as the analytical column. A sacrificial guard column packed with relatively large particle size unmodified silica, for example, may be inserted in-line between the pump and the injection valve to trap wear particles from the pump, to act as a pulse dampener and to saturate the eluent with silica.

Columns should not be used with eluents that may, for example, dissolve the packing, or with samples or sample extracts that may introduce particles or cause precipitation of solid material in the system. It is important to ensure that organic solvents are only used with solvents with which they are miscible, and this should be borne in mind when changing eluents. Unmodified silica columns are normally supplied filled with hexane as they are usually used for normal-phase chromatography (Section 8.5.1). However, if used with methanol or methanol:buffer eluents, the hexane will have to be removed. Purging with MTBE or ethyl acetate, for example, which are miscible with hexane, and then with methanol, before using methanol:water ensures safe removal of the hexane. Isopropanol can be used instead of ethyl acetate, thereby obviating the need for the methanol step, but is more viscous and hence tends to generate high back-pressures.

Use of too high an inlet pressure may crush silica packings, causing void formation and, in the case of chemically modified silica, revealing untreated surfaces with consequent impairment of selectivity and/or efficiency. Even at normal operating pressures there is a tendency for voids to form at the top of columns during use, although this seems to be not as common with spherical particles as it used to be with older, irregularly shaped silica packings. The presence of a void will cause peak broadening, but the column may still give acceptable results. However, sudden changes in pressure or dropping a column may cause channels to form in the packing with resultant loss of performance, which may be manifest as poor peak shapes or even double peaks.

Columns are supplied with an indication of the direction in which they were packed and it is usual to maintain the flow direction in use. However, at least one manufacturer supplies columns that are designed to be used with the flow in either direction, the instruction being to reverse the flow regularly to reduce the risk of clogging of the frits. Remember not to connect to the detector until flow has been established for several column volumes to minimize the risk of particles being swept into the detector flow cell. Generally, it is possible to reverse the direction of flow of HPLC columns, particularly with columns packed with spherical particles, without appreciable deterioration in performance. However, it is better to stop the inlet frit becoming blocked in the first place. Monolithic silica columns (Section 8.4.2.4) should only be used in the indicated flow direction.

8.2.1.1 Column oven

It is generally accepted for reversed-phase separations that a column temperature of 35–40 °C improves efficiency and reproducibility. In general, a change in HPLC column temperature of 1 °C will change a retention time by 1–2 %. Thus, a column temperature change of 5–10 °C during a day is likely to be a source of significant variation in retention. Therefore, the use of column temperature control in HPLC not only increases column efficiency, but also improves retention time reproducibility and selectivity control. Baseline drift, which may be observed with flow-sensitive detectors such as ED, can be eliminated by careful temperature control. Column ovens with high-velocity circulating air fans analogous to GC ovens can be used. Alternatively contact heating via aluminium blocks or by circulating fluid from a constant-temperature bath has been employed. Ideally, the column temperature should be controlled within ±0.2 °C to eliminate peak drift. Peltier effect ovens provide both heating and cooling within precise limits. If an oven is not available, the column should not be placed where it might be subject to excessive changes in temperature, for example near an open window or in direct sunlight for part of the day.

In the case of the immunosuppressant ciclosporin, HPLC operation at 60 °C or so is necessary as at lower temperatures this drug exists as different conformers, each of which has a slightly different retention time on reversed-phase HPLC giving rise to a very broad peak (Hasumi *et al.*, 1994).

8.2.2 The eluent

Use of pure water is important, particularly with gradient elution when impurities may be retained and concentrated on the column at the start of the analysis only to elute as peaks as the proportion of the organic component in the eluent is increased. Water purified by reverse osmosis may be suitable, particularly for isocratic separations. Deionized water is usually suitable for gradient systems, although contaminants may arise from impurities in the ion-exchange resins. Thus, for high-sensitivity gradient work, use of certified HPLC quality water is advisable. Buffer salts and other reagents should be of the highest available purity. Generally, organic solvents should be HPLC grade, although this usually only indicates that the solvent has an appropriately low UV cut-off. Acetonitrile is available in several grades, a high purity grade being for use down to approximately 205 nm.

It is important that HPLC eluent preparation is reproducible if consistent separations and retention times are to be achieved. Solvent volumes should be measured in appropriate apparatus and liquids mixed carefully with cooling if necessary. Measuring cylinders may be suitable for large volumes, whereas for small volumes pipetting may be more appropriate.

Eluent pH may be controlled by: (i) using a buffer of known pH, (ii) adjusting the pH of the eluent just prior to making up to final volume, or (iii) adding a known volume of an acid or base such as trifluoroacetic acid or triethylamine to a set volume of eluent. It should be noted that adding organic solvent to a buffer alters the hydrogen ion activity, such that the pH of the mixture differs from the buffer pH (Tindall, 2002a, 2002b, 2003). The difference between the buffer pH and the 'apparent pH' (pH*) of the organic solvent–buffer mixture increases as more organic solvent is added. The terms pH and pH* are often used indiscriminately, and it is not always possible from published reports to deduce how an eluent was prepared.

Confusing pH or pH* can result in failure to reproduce a separation and if the pH* is too high, the life of a silica-based column may be drastically shortened. Measuring accurate pH* in eluents with a high proportion of organic modifier requires a 'sleeve junction' pH electrode. This has a relatively large opening through which aqueous saturated potassium chloride can be brought into contact with the test mixture without precipitating potassium chloride in the opening. Consequently, the pH of the buffer is normally adjusted before mixing with organic solvent, which is acceptable provided the resulting eluent is suitable and reproducible. However, changing the volume of buffer added can have a marked effect on pH*.

Any particles should be removed from the eluent before use by filtration through a 0.45 μm membrane. Filtration to 0.2 μm is recommended when porous graphite ED electrodes are being used. Some methods filter the buffer before it is mixed with organic solvent or even before the pH is adjusted. As a precipitate may form on pH adjustment or addition of solvent, it may be better to filter the eluent just prior to use, providing this does not change the eluent composition significantly.

The final stage in eluent preparation is 'degassing'. Dissolved gas should be removed to prevent bubble formation in the pump head (cavitation), which affects flow control, and in the detector flow cell, which affects baseline stability. Removal of oxygen from the eluent may also be important if low-wavelength UV or reductive ED is to be used. For some applications vacuum filtration may suffice. If additional degassing is desired after filtration, the vacuum pump can be left on and a stopper placed over the top of the flask containing the filtered eluent. Further eluent degassing will be indicated by the formation of small bubbles. Bubbling (sparging) with helium (balloon gas may be cheaper than helium *per se*) is the most effective way of degassing and a slow constant stream of helium should be maintained to prevent baseline drift due to oxygen redissolving in the eluent if working at 200–215 nm. However, both vacuum degassing and helium sparging may alter eluent composition by removing more volatile eluent components selectively. Placing the eluent reservoir in an ultrasonic bath for a few minutes is an alternative. In-line degassing units, which apply a vacuum to the eluent as it flows through semiporous fluoropolymer tubing, are becoming increasingly popular.

8.2.3 The pump

The pump has the role of delivering the eluent at a constant and reproducible flow rate, independent of column back-pressure. Although positive displacement (syringe) pumps were manufactured, nowadays usually eluent flow is induced by the reciprocating action of a sapphire piston and hence the pump is fitted with inlet and outlet check valves to maintain the flow in the required direction. During the filling stroke the inlet check valve opens as eluent is drawn into the pump head (Figure 8.3). On the displacement stroke the increased pressure closes the inlet check valve and opens the outlet valve so that high-pressure eluent is pumped in the required direction. An outlet valve filter prevents pump seal debris from entering the valve where it might lodge in a ball valve, causing the valve to malfunction. A flexible, but pressure-resistant seal (often graphite filled PTFE) prevents leakage of eluent around the piston.

The pump components that have contact with the eluent(s) have to be of inert materials, typically glass, ceramic and stainless steel. To reduce contamination of the eluent with iron the system may be pacified or constructed using titanium or PEEK. Pacification of iron involves removal of the column and any other sensitive components and flushing (30 min each at 1 mL min^{-1}) the pump, injection valve and connecting stainless steel tubing with: (i) filtered, deionized water, (ii) nitric acid (6 mol L^{-1}), (iii) glacial acetic acid and (iv) filtered, deionized water until the effluent pH is 4 or greater. Step (ii) can be replaced with aqueous EDTA (0.5 mol L^{-1}, pH 8) followed by deionized water before the glacial acetic acid step. Addition of EDTA to the eluent may have the same effect, however. Pacification may be helpful with high sensitivity ED, but is rarely required with other detection systems, unless the analytes are iron sensitive.

Routine maintenance should include checking that the valves are functioning correctly and that the piston seal is not leaking. When working with aqueous buffer systems, rinse/wash the piston frequently with water to remove crystallized buffer salts and thus prolong the life of both the

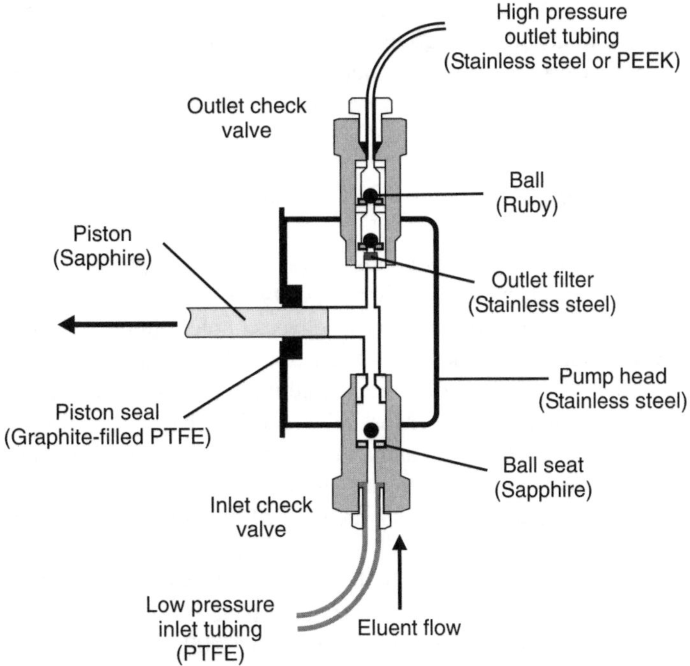

Figure 8.3 HPLC pump head during a filling stroke.

piston and the piston seal. The buffer should be flushed from the pump head using a compatible solvent before the pump is switched off. When changing a piston seal, any debris adhering to the piston may be removed by gentle polishing with toothpaste, for example. A scored (damaged) piston usually has to be replaced.

The filtered and degassed eluent should be placed into the eluent reservoir. The inlet tubing to the pump should be fitted with a filter to prevent particles that may have accidentally entered the reservoir passing through to the pump. Some pumps require the reservoir to be positioned higher than the pump to provide a static head pressure and this arrangement is useful when the pump is being primed. The reservoir may be pressurized, but this is not normally necessary. The eluent flow rates used are typically 1–2 mL min^{-1} with 2–5 mm i.d. columns containing 3–5 μm average particle size packings. Most pumps have an upper operational pressure limit of 34,500 kPa (5000 psi). Some pumps can be used with microbore or capillary HPLC columns (flow rates in the μL min^{-1} range).

Fluctuations in pressure (pulsations) when delivering the eluent are undesirable because of possible effects on detector baseline stability, hence HPLC pumps are designed to minimize such fluctuations (<1 % or better short-term variation). In older systems pulse dampeners consisting of, typically, a coil of capillary tubing, were inserted between the pump and the injection system in order to help minimize pump pulsation. Use of a sacrificial guard column (Section 8.2.1) has a similar effect. Alternative pulse dampeners are of the membrane type. One such dampener has a PTFE membrane with a reservoir of heptane to absorb the pulsations, but the membrane can rupture, releasing heptane into the eluent. The manufacturers may include a blue dye in the heptane to indicate if this happened. Some pumps include membrane pulse dampeners under the casing, and it is advisable to know of their existence should they fail. A more robust design uses a cast iron block to absorb the pulsations. Dual piston pumps, working either in parallel or in series, are a further way to minimize fluctuations in flow. A malfunctioning check valve on the parallel pump design can lead to major problems, particularly with detectors that are sensitive to flow-rate changes.

8.2.4 Sample introduction

Nowadays sample introduction is almost always via a sample injection loop operated either manually or electromechanically. There are two types of HPLC sample valve: internal loop and external loop. Internal loops are designed to deliver sample volumes of less than 1 μL, and external loops can deliver sample volumes from a few μL to 20 mL or more. External loops use narrow bore tubing as the volume measure, and the internal type has a small hole or a narrow space in a disk, which gives the fixed volume. Most valves work by having a stationary body, often made of stainless steel, known as the stator, and a moving part, the rotor, which must be made of a softer material such as Vespel or PEEK to ensure a leak-free seal between the two surfaces. If necessary, valves made of biocompatible materials such as PEEK or titanium rather than stainless steel may be used.

HPLC sample injection valves are built to withstand pressures up to 69,000 kPa (10,000 psi), but are often supplied, factory-set, to withstand only 42,000 kPa (6000 psi) because, in general, operating pressures are only up to about 21,000 kPa (3000 psi). Adjusting the valve to operate at pressures above 28,000 kPa (4000 psi) will increase the wear and tear on the rotor.

A diagram of a six-port HPLC sample injection valve is shown in Figure 8.4. In the load position the eluent flow passes via a slot on the rotor from the pump (port 2) to the column (port 3). The ends of the sample loop are connected to ports 1 and 4. The sample loop is charged by a syringe through the injection port, which is aligned so the end of the needle abuts directly against one end

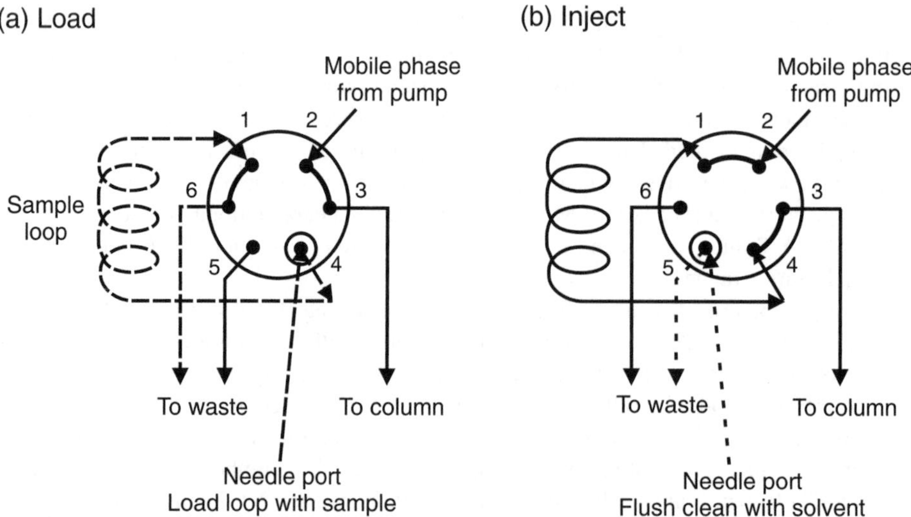

Figure 8.4 Six-port HPLC injection valve showing (a) load and (b) inject positions.

of the sample loop (port 4). Thus, it is important to ensure the appropriate needle is used. Fluid in the loop is discharged via a slot in the rotor, which, in this position, connects port 1 to waste (port 6). On rotating the valve to the inject position the sample loop is placed between the column and the pump supplying eluent by connecting ports (1) and (2) and ports (3) and (4). The eluent then sweeps the sample, in the opposite direction to the one in which it was loaded, onto the column. The injection port is now over the second waste port (5) allowing the injection port to be rinsed. After sampling, the rotor is returned to the loading position, the external sample tube washed with solvent and the sample loop filled with the next sample for analysis. As the sample loop can be any length or diameter, and may even include a short column on which analyte can be concentrated, several configurations are possible. For maximum precision the sample loop should be partially filled, no more than 50 %, with a carefully controlled volume of sample or be overfilled (>2 loop volumes).

Automated sample injectors are widely used, typically with high precision motorized syringes for accuracy and precision. There is a risk of cross contamination (carry-over) in such systems, particularly where there are wide differences in analyte concentration between one sample and the next, which must be guarded against.

The possible effect of the injection solvent on chromatographic efficiency must be considered. In many published methods, the analyte is taken up for injection in HPLC eluent. However, if the injection volume is too large this will contribute to band broadening. Preferably, samples should be injected in a solvent of lower 'eluting power' than the eluent. This allows for concentration of the analyte(s) on the top of the precolumn or column. For example, in reversed-phase systems, using an injection solvent containing less organic modifier than the eluent allows injection of larger volumes with no detrimental effect on efficiency and peak shape. Conversely, the use of an injection solvent with a significantly higher proportion of organic solvent than that of the eluent will lead to the analysis beginning well before the arrival of the eluent, unless a very small injection volume is used which, of course, limits sensitivity.

8.2.5 System operation

An accurate record of key column performance parameters such as system back-pressure, analyte retention time and LLoD, and baseline noise and drift should be kept. Any deterioration in performance will then soon become apparent and preventative maintenance can be undertaken.

In reversed-phase HPLC, and other situations where aqueous buffers are used, it is prudent to rinse a column periodically with deionized water to reduce the risk of precipitation of inorganic salts, or with organic solvent to remove retained lipophilic material. Organic solvents should not be used until inorganic salt residues have been removed. Typically a column that is being used with an aqueous buffer as part of the eluent should be flushed sequentially with deionized water, methanol or acetonitrile, and deionized water before equilibration with fresh eluent. Similar flushing should be carried out before a column is stored, the final flush being with methanol or acetonitrile:water (1 + 1) to prevent bacterial growth.

If column performance is still poor after column reconditioning, then the fault may be due to dissolving or compression of the column inlet bed. To repair any column inlet, it is necessary to remove the fitting. Before proceeding with column repair on a reversed-phase column, it should be conditioned to methanol:water (1 + 1). This will solvate the C_8 or C_{18} phase with a moderately lipophilic mixture and so prevent the packing from expanding out of the column after the fitting has been removed. The fitting or frit closure should be removed from the inlet end of the column and the packing examined. It should be off-white.

If the packing is coloured, it may indicate that corrosion of steel components of the HPLC system has occurred, or that impurities present in the sample extracts have accumulated on the top of the column. If the inlet bed has some depression, then it has to be made level to prevent peak broadening or, possibly, peak splitting. The void should be filled with suitable packing material (5 or 10 μm column packing added as a dense slurry with a spatula and levelled off, or 30–40 μm glass beads). Attempting to use smaller sizes of packing material is rarely successful. A new frit should always be used in the fitting or closure when the column is reassembled.

Pump problems may be manifest as erratic retention times, noisy baselines and/or spikes on the baseline due to pressure variations. Pump leakage usually occurs at inlet or outlet fittings, or at the piston seal. Salt crystals on a fitting usually indicate a leak. Problems attributed to solvents used as eluents include: (i) fouling of the pump and/or clogging of the check valves from particles, (ii) low sensitivity due to contaminated solvent and (iii) drifting baselines and/or retention times due to impurities, for example in the eluent buffers.

Eluent filtration before use should be routine. If there is a danger of precipitation of buffer salts, the eluent can be filtered after the buffer and organic modified have been mixed, although care must be taken to ensure that this does not change the composition of the eluent – a 1 % change in organic content could change retention by 5–10 % in a reversed-phase system. This will be more apparent when the eluent contains only a small proportion of a relatively volatile solvent, or if volatile buffer components such as ammonia or trifluoroacetic acid are used.

8.3 Detection in HPLC

The ideal HPLC detector would have: (i) high sensitivity, (ii) universal or specific response as required, (iii) wide linear dynamic range, (iv) minimal band broadening and (v) a stable response with temperature or flow-rate change. Currently, no detector fulfils all of these requirements.

HPLC detectors are of two types: bulk property or selective. The former measure a change in a physical property of the eluent such as refractive index (RI), conductivity or light scattering that is affected by the presence of an analyte. However, such detectors generally do not have sufficient sensitivity/selectivity for use in analytical toxicology, and thus detectors that exploit chemical or physical properties of an analyte, namely UV/visible absorption (including diode-array and scanning instruments), fluorescence, electrochemical oxidation or reduction and MS are used almost exclusively (Table 8.1). UV/visible absorption and fluorescence detectors are non-destructive and are relatively insensitive to small fluctuations in flow rate.

Further options include post-column collection or modification of the analyte to improve sensitivity or selectivity. Whatever the detection system, it is important that an appropriate low dead volume detection cell and an adequate signal sampling rate or time constant (usually 0.2 s or less) are used to ensure that the high separating power of modern microparticulate packings at the commonly used flow rates of 1–2 mL min^{-1} is not lost.

As HPLC does not have a simple, sensitive, universal detector analogous to the FID in GC, MS has become widely used. Although interfacing MS and GC is relatively simple, in HPLC interferences from the eluent must be eliminated. In order to overcome this problem, many interfaces have been developed for different purposes. The interfacing techniques available for HPLC-MS are discussed further in Section 10.5.

Table 8.1 Detection in HPLC

Principle	Variant	Option
Spectroscopic	Atomic absorption spectrometry (AAS)[a]	
	Chemiluminescence (CLD)	
	Fluorescence	Variable excitation and emission wavelength (combinations of filter and monochromator)
	Fourier-transform infrared (FTIR)	
	Mass spectrometric (MS)	Electrospray ionization (ESI) Atmospheric pressure chemical ionization (APCI) Positive or negative selected ion monitoring (SIM) Inductively coupled plasma (ICP)
	Nuclear magnetic resonance (NMR)	
	Phosphorescence	
	UV/Visible	Fixed wavelength Dual wavelength Multiwavelength scanning (filter or monochromator) Diode array
Electrochemical (ED)	Amperometric	Continuous or pulsed
	Coulometric	
	Conductivity	
Immunoassay (IA)[a]		
Bulk property	Refractive index (RI)	
	Evaporative light scattering (ELSD)	
	Chemiluminescent nitrogen (CLND)	
	Chiral	Optical rotation (OR) Circular dichroism (CD)
Radioactivity		

[a] Off-line detection.

Combining HPLC with atomic spectroscopic methods can give information on the atomic composition of the analyte. An example is HPLC-ICP-MS (Section 11.4.4). This is a more powerful technique than HPLC-AAS (atomic absorption spectrometry) because ICP has high sensitivity and can detect many elements simultaneously.

Detector problems can be electrical, optical or mechanical. When a problem is encountered, turn the pump off – if there are still spikes on the baseline the problem is probably electrical. The presence of gas bubbles in the detector cell may also cause spiking on the baseline – thorough eluent degassing should cure the problem by removing dissolved gas and thus preventing formation of new bubbles. Detector cell bubble formation can also be reduced by using a flow restrictor to maintain a higher pressure post-column, given that the detector cell(s) can withstand the pressure.

8.3.1 UV/visible absorption detection

Provided that the analyte has sufficient absorbance in the range 200–600 nm, the sensitivity of UV/visible absorption detectors can be less than a few ng injected if there is no contribution to baseline noise from small changes in eluent flow rate and temperature. UV/visible detectors use either a low-pressure mercury lamp as the light source, which is particularly useful for monitoring at 254 nm (an intense emission line), or, more usually, a deuterium discharge lamp (200–400 nm) and a tungsten lamp (400–700 nm). The most popular commercially available flow cells for UV/visible detectors have cell volumes of 8 μL or less and a path length of 10 mm (Figure 8.5), but for microbore or capillary HPLC a lower cell volume (0.1–0.3 μL) must be used so as not to compromise the increased sensitivity generated by the low-volume column.

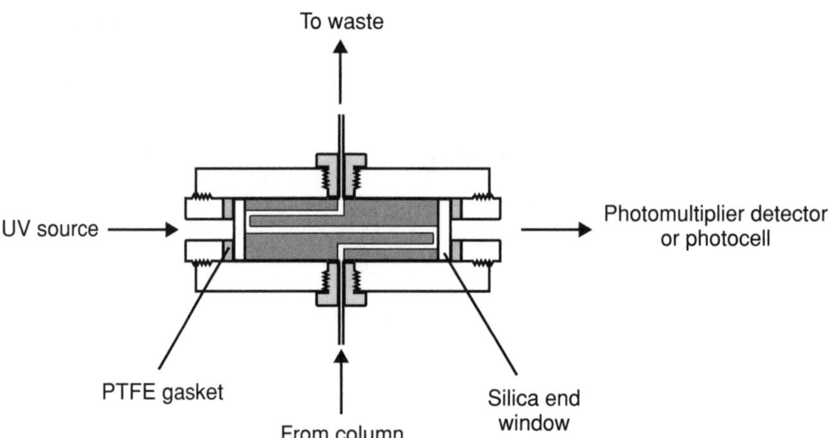

Figure 8.5 Schematic diagram of UV/visible detector flow cell for HPLC.

Use of relatively nonselective detection wavelengths (200–220 nm) can pose problems if high-sensitivity work with biological extracts is contemplated; many reports advocating such wavelengths for particular analytes make no mention of potential interferences. Sensitivity may also be reduced when working near the UV cut-off of the eluent. The presence of oxygen dissolved in the eluent may also limit sensitivity at low wavelengths (200–210 nm).

Multiwavelength detectors, that is detectors capable of generating absorbance data for more than one narrow band of wavelengths at once, have proved useful in STA (Section 8.8.3). These instruments employ either a fixed grating/DAD (Figure 4.3) combination or a high-speed scanning monochromator in conjunction with an 8–13 μL flow cell. A data collection rate of one spectrum per second is normally optimal in conventional HPLC. A three-dimensional plot (absorbance against wavelength and time) is obtained, which can be used in both qualitative and quantitative work. Reference to a specific wavelength can be used to reduce baseline drift during gradient work.

Several protocols have been developed to help assess analyte identification and peak homogeneity using such instruments, including absorbance ratio data and peak purity parameters (White, 1988; Naish *et al.*, 1989). If the ratio of two different wavelengths is not constant across a peak then the peak is not homogeneous. Although advocated for barbiturate identification (White, 1980), absorbance ratioing is not ideal as the absorbance of these compounds at 220–254 nm is greatly influenced by small changes in eluent pH.

8.3.2 Fluorescence detection

Fluorescence is when a molecule absorbs light and emits light at a longer (less energetic) wavelength during decay from the excited state (Section 4.4.1). Generally mercury or xenon light sources are used; fluorescence detectors fitted with a deuterium lamp are no longer available commercially. Fluorescence detection can be very sensitive (typically a 10–100-fold increase as compared to typical UV/visible detectors), but it is selective and only a few analytes of toxicological interest fluorescence unless derivatized. The response obtained is dependent on both the energy of the incident light and the optical configuration of the detector. Some analytes, for example quinine, only fluoresce under acidic conditions. In such cases post-column adjustment of the eluent pH is easily accomplished and repeating the analysis without post-column flow gives an additional identification parameter.

Fluorescence detectors can be double monochromator, double filter, or a combination of monochromator (excitation) and filter (emission) instruments (Section 4.4.3). The more expensive scanning instruments use a beam splitter to divert 50 % of the incident light to a reference photocell. The discrete excitation and emission wavelengths obtainable with double monochromator instruments provide maximum selectivity for the target analyte. However, as compounds with similar chemical structures, including metabolites, frequently have different fluorescence spectra, this may not always be an advantage. With some instruments, it is possible to programme excitation and emission wavelength changes during an analysis, but this presupposes that the analytes, their retention times and their fluorescence spectra are known. The use of wide bandpass filters allows more compounds to be detected. One manufacturer has developed a compact, inexpensive detector with excitation (330–365 nm) and emission (440–530 nm) filters chosen specially to measure fluorescent OPA derivatives.

Detector cells vary in their design. Some are simply a narrow-bore quartz tube with inlet and outlet connections at the bottom and the top, the excitation and emission light paths being at right angles to the tube and each other. With others, the incident light is directed down the length of the tube and the emitted light detected at right angles. This arrangement has been used in combined UV/fluorescence detectors (Figure 8.6).

8.3.3 Chemiluminescence detection

Chemiluminescence detection (CLD) is potentially more sensitive than fluorescence detection as it does not require irradiation of the flow cell, and so the problem of light scattering with consequent reduction in signal-to-noise ratio is avoided. The instrumentation is simpler and

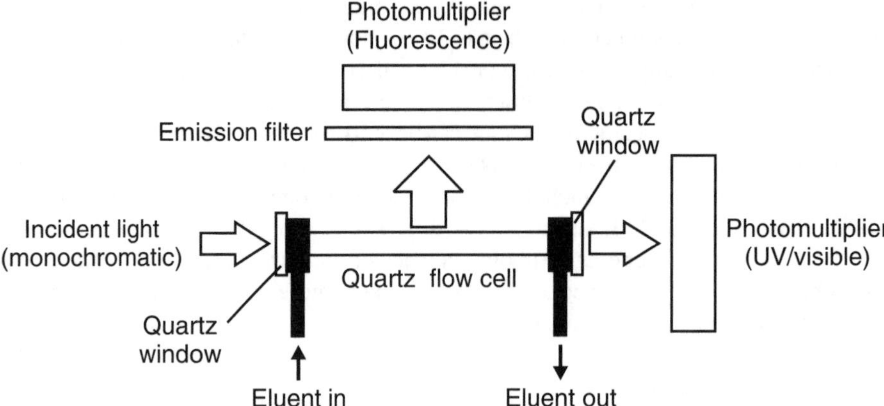

Figure 8.6 Schematic diagram of a combined fluorescence/UV detector cell for HPLC.

CLD has a wider linear dynamic range than conventional fluorescence detection. The reactants are usually introduced post-column (Section 8.3.10), or the chemiluminescence may be photo-induced. CLD has found use in environmental studies, for example measurement of nitrogen oxides, but has not been widely adopted in analytical toxicology. HPLC-CLD of opioids using potassium permanganate oxidation has been described (Abbott *et al.*, 1987). A similar approach has been used in the CE-CLD of morphine, 6-MAM and diamorphine; inclusion of β-cyclodextrin in the eluent increased the luminescence signal (Gong *et al.*, 1999). Gómez-Taylor *et al.* (2006) have investigated the photochemiluminescence of 97 drugs, which may awaken interest in CLD. The use of CLD in HPLC and CE analysis has been reviewed (Li *et al.*, 2003).

8.3.4 Electrochemical detection

There are two basic types of electrochemical detector for HPLC: (i) amperometric or coulometric and (ii) electrical conductivity. The former are reaction detectors, that is the analytes are either oxidized or reduced at the surface of the working electrode, and it is these that have found most application in analytical toxicology. To maintain the electric circuit a second (counter) electrode is required and this may be of the same material as the working electrode or simply the stainless steel of the outlet tubing. A two-electrode system would be too unstable for practical use and a three-electrode circuit is normally employed. The third electrode, frequently silver/silver chloride, having a known potential, is used as a reference by which the potential applied to the working electrode can be kept stable via feedback circuitry (Figure 8.7). Detection can be enhanced by derivatization of suitable analytes to produce electroactive species.

The most widely used working electrodes are carbon (glassy carbon, carbon paste, porous graphite) as this material can be used for both oxidation and reduction and is not easily poisoned by the reaction products. Platinum, silver, gold, or gold amalgam may be used for more specialized applications. Amperometric detectors are used with glassy carbon working electrodes in a wall jet or thin-layer arrangement (Figure 8.8). These give adequate sensitivity for many compounds; carbon paste electrodes may be more sensitive, but maintaining satisfactory performance is more difficult. Porous graphite flow-through cells offer the possibility of coulometric detection, that is detection involving near quantitative conversion, and such cells are available in a dual-electrode

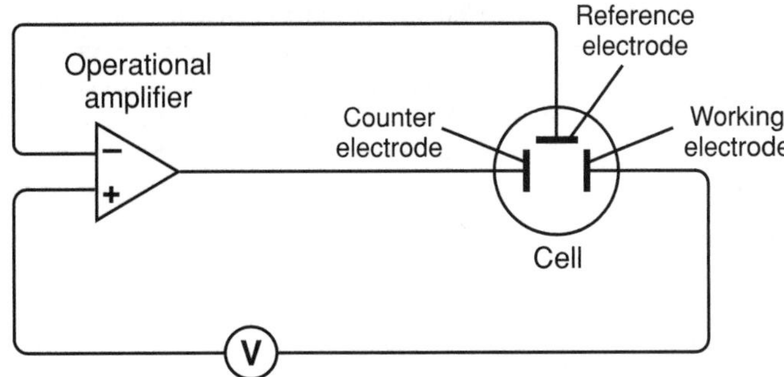

Figure 8.7 Schematic diagram of a three-electrode electrochemical detector circuit.

configuration (ESA Coulochem). Although relatively expensive, such instruments can give an enhanced signal-to-noise ratio for analytes such as physostigmine and may be more flexible in routine use than wall-jet assemblies (Flanagan *et al.*, 2005). Porous electrodes need meticulous care to avoid blockage, and should always be used with graphite filters.

Electrochemical reduction has been used to detect drugs, such as some benzodiazepines and disulfides but, in general, reductive detection is more difficult than oxidation, particularly at high negative potentials because dissolved oxygen must be excluded. Electrochemical oxidation, on the other hand, has been widely used to detect basic drugs, especially morphine and related alkaloids. The sensitivity achieved is often superior to that found with UV detection. Very high sensitivity is not normally required, but relatively high applied potentials may be needed to detect some basic drugs (Table 8.2). Thus, the signal-to-noise ratio is similar to that encountered in, for example, the analysis of easily oxidized species that require lower applied potentials such as catecholamines in plasma. ED is not widely used for acidic or neutral compounds such as salicylate or paracetamol as these drugs, although easily oxidized, are often present at relatively high concentrations and UV detection is usually adequate.

A great drawback to the use of ED in HPLC is electrode deactivation. Loss of response is normally accompanied by a decreased standing current, while excessive noise due to, for example,

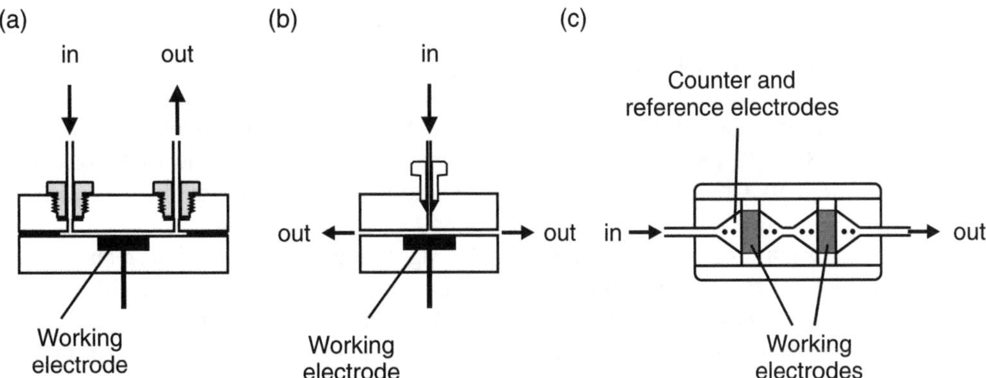

Figure 8.8 Simplified designs of HPLC-ED cells: (a) thin-layer cell, (b) wall-jet cell, (c) porous flow-through cell.

Table 8.2 Suggested optimal ED conditions for certain functional groups

Functional group	Optimum oxidation potential (V) against Ag/AgCl	Approximate detection limit (ng on-column)	Typical standing current (nA)
Phenol, aromatic amine	0.7	0.1	2–5
Phenothiazine sulfur	0.8	0.1	5–10
Imidazoyl nitrogen, indole	0.9	0.2	10–20
Tertiary aliphatic amine	1.0	0.5	50–100
Secondary aliphatic amine	1.2	2	200–500
Primary aliphatic amine	1.6	20	>2000
Pyridyl nitrogen, quaternary ammonium compound	>1.6	—	—

Column: 125 × 5 mm i.d. Spherisorb S5W silica. Eluent: methanolic ammonium perchlorate (10 mmol L^{-1}) + 1 mL L^{-1} methanolic sodium hydroxide (0.1 mol L^{-1}), pH* 6.7. Flow rate: 2.0 mL min^{-1}. Working electrode: V25 grade glassy carbon (Le Carbone); reference electrode: Ag/AgCl (Jane *et al.*, 1985). Maximum useable applied potential for biological extracts normally + 1.2 V

a contaminated eluent is usually paralleled by an increase in the standing current. The ED is very sensitive to flow-rate changes, so that pump pulsations are usually very apparent and even small changes in ambient temperature can lead to marked baseline drift. Thus, temperature control is important if high-sensitivity operation is contemplated. Other factors being equal, UV/visible and fluorescence detection are more robust and, therefore, preferred.

8.3.5 *Chemiluminescent nitrogen detection*

The CLND is a gas-phase detector that exploits the reaction of nitric oxide and ozone to produce excited nitrogen dioxide with subsequent emission of light:

$$NO + O_3 \rightarrow [NO_2]^* \rightarrow NO_2 + O_2 + h\nu$$

The nitric oxide is produced by heating nitrogen-containing compounds at 1050 °C in an oxygen atmosphere and, after separation of other pyrolysis products, it is mixed with ozone under reduced pressure (about 2.7 kPa) in a reaction chamber in front of a photomultiplier (Taylor *et al.*, 1998). A PEEK needle valve was used to control the flow of gas to the reaction chamber (Figure 8.9). Obviously, CLND cannot be used with nitrogen-containing eluents (solvent or buffer).

The sensitivity of CLND is said to be 0.1 pg of nitrogen, with a linear range $>10^4$. For a compound such as imipramine ($M_r = 280$), which conveniently contains 10 % nitrogen by weight, this equates to 1 pg of drug. As this measurement was performed using flow injection analysis (FIA) giving very sharp peaks, chromatographic sensitivity is probably in the region 1–2 pg s^{-1} for imipramine. The reproducibly of response to equimolar amounts of nitrogen, irrespective of the analyte, means that this detector has been used to measure nitrogen-containing compounds without primary standards, such as screening combinatorial libraries, metabolic studies (Taylor *et al.*, 2002) and in screening drug seizures (Laks *et al.*, 2004).

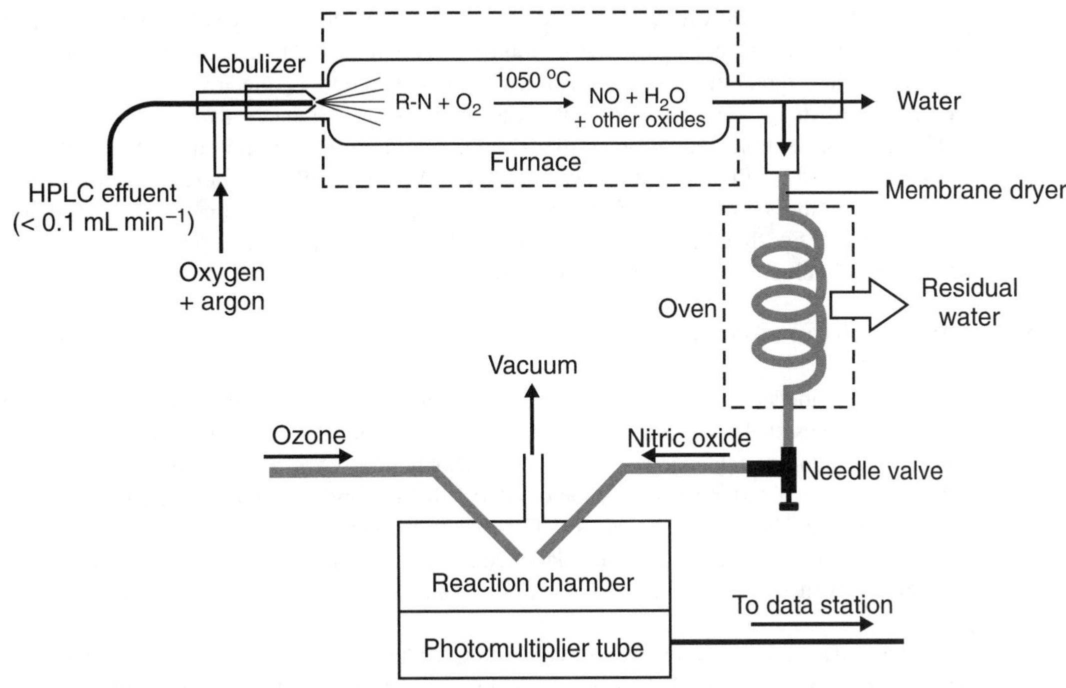

Figure 8.9 Schematic diagram of a chemiluminescent nitrogen detector. Reproduced from Taylor, E.W., Quian, M.G. and Dollinger, G.D. (1998) Simultaneous on-line characterization of small organic molecules derived from combinatorial libraries for identity, quantity and purity by reversed-phase HPLC with chemiluminescent nitrogen, UV, and mass spectrometric detection. Anal Chem, **70**, 3339–47, by permission of the American Chemical Society.

8.3.6 Evaporative light scattering detection

In the ELSD, effluent from the HPLC column is converted into a fine mist by a nebulizer and the solvent evaporated to leave a stream of solute particles onto which a beam of light is directed. An optical detector, set at an appropriate angle, measures the scattered light. As detection is not dependent any spectral properties, the ELSD is often referred to as a universal detector, but clearly it cannot be used with volatile analytes. Like CLND described above, the ELSD should be useful for measuring compounds that are poorly characterized, such as products from combinatorial libraries and drug metabolites. The results from ELSD may be influenced by HPLC eluent composition, for example during gradient programming, but probably the biggest drawback is its lack of sensitivity. When Fries *et al.*, (2005) evaluated an ELSD for *in vitro* metabolism studies, the matrix was found to interfere, and even with less complex, 'contrived' samples, the LoD was 1–10 mg L^{-1}. ELSD may prove useful for quantifying endogenous compounds (biomarkers), which are relevant to analytical toxicology (Aradottir and Olsson, 2005).

8.3.7 Charged aerosol detection

The charged aerosol detector (CAD) is a relatively new form of HPLC detection (Corona Plus, ESA). As with the ELSD, the eluent is passed though a nebulizer and the solvent evaporated at ambient temperature. However, in the CAD the particles are charged by collision with a stream of postively charged nitrogen gas. High-mobility ions are removed by a negatively charged, low-voltage trap. The remaining analyte ions transfer their charge to a collector (filter-electrometer), the transferred charge being proportional to the analyte mass (Gamache *et al.*, 2005).

The CAD has similar limitations to the ELSD, including those imposed by the volatility of the analyte and the HPLC eluent. However, the CAD is claimed to be more stable with regard to gradient elution, less influenced by differences in aerosol particle diameters, and at least 10 times more sensitive than ELSD. It is too early to tell whether CAD will be developed to the point where it is suitable for routine use in analytical toxicology.

8.3.8 Radioactivity detection

Radioactively labelled analytes can be monitored via a flow cell using various radioactivity (RA) detection systems such as a sodium iodide scintillation counter, a Geiger–Muller counter or a liquid scintillation counter, depending on the radionuclide being used. Scintillation produces light via matrix ionization caused by the emitted radiation, and the light is amplified via a photomultiplier tube. Usually, either ^{3}H- or ^{14}C-labelled substance will be used so that the labelled material is as similar as possible to the unlabelled compound. Other suitable radionuclides include ^{32}P and ^{35}S, if these elements are present in the compound being studied, but generally ^{14}C is preferred.

The basic design has features in common with CL detectors. In order to enhance sensitivity, a large-volume flow cell is needed, but too large a volume causes band broadening and thus loss of sensitivity. To some extent this can be minimized by constructing the cell as a coil of tubing placed in front of the photomultiplier. However, having to integrate the counts, that is counting at discrete time intervals, reduces resolution and clearly to achieve high sensitivity the counting interval will need to be increased, resulting in further loss of resolution. Tritium, being a 'soft' β-emitter (Box 12.2) has to be counted in solution, as the low energy electrons will not penetrate the detector cell walls. The added stream of liquid scintillator cocktail contributes to peak broadening and increases the volume of low-level radioactive waste for disposal.

Despite the limitations associated with RA detectors (Box 8.3), they are useful if large numbers of samples are to be assayed. However for small-scale studies, collecting fractions of eluent and counting off-line is a viable, if not preferable, alternative. For targeted analytes, the non-labelled compounds can be added beforehand so that these can be detected, say, by UV detection and individual peaks collected for RA counting.

Box 8.3 Practical considerations in the use of radioactivity detectors

- Not particularly sensitive
- Cannot recirculate the HPLC eluent
- ^{3}H-labelled analytes require addition of liquid scintillant
- Large volumes of low-level radioactive waste
- Band spreading leads to poor resolution and loss of sensitivity
- Integration of the signal produces a histogram
- Incompatible with high-efficiency HPLC separations

8.3.9 *Chiral detection*

Chiral detectors are important in the analysis and purification of enantiomers. Several are available commercially and are based on either optical rotation of the plane of polarized light (OR detectors) or by measuring the difference between absorption of right and left-handed circularly polarized light, in other words circular dichroism (CD detectors). CD detectors are generally 10–100 times more sensitive than OR detectors and are suitable for compounds that absorb in the region 220–420 nm. Because there is an absorption signal as well as a CD signal, CD detectors can be used to measure an excess of one or other enantiomer without performing a chiral separation and, by measuring specific rotation of separated peaks, can give proper peak assignment. OR detectors, on the other hand, can be used to detect enantiomers that do not possess chromophores. The useful detection limit for typical pharmaceuticals ranges from 1 to 100 μg on column, depending primarily on the specific rotation exhibited by the analyte in the eluent and secondarily on retention time and peak shape.

Laser CDs are said to be more sensitive than conventional instruments, but as yet do not seem to have the sensitivity required for monitoring chiral compounds in biological fluids. This coupled with the cost of these instruments makes it seem unlikely that they will be adopted widely in analytical toxicology.

8.3.10 *Post-column modification*

By comparison with GC carrier gases, HPLC eluents are essentially non-compressible, and so HPLC also offers a range of options for post-column analyte modification, including post-column pH adjustment, hydrolysis, derivatization and photolysis. Such modifications are normally employed to facilitate or enhance fluorescence as fluorescence detectors are relatively insensitive to flow-rate changes. Post-column derivatization or pH manipulation may not be practicable with ED as this detector is especially sensitive to perturbations of flow rate.

To minimize band broadening during post-column reagent addition, a specially designed system is needed. The column effluent is mixed with reagent solution provided by an auxiliary pump using a T-piece or, preferably, a Y-connector, which gives better mixing (Figure 8.10). The reaction coil can be constructed of narrow-bore stainless steel or PTFE tubing, or other suitable material, which can be placed in an oven or heating block if the reaction requires an elevated temperature. The reaction time is influenced by the geometry of the reaction coil and the flow rates of the eluent and reagent streams. For example, if the effluent and the reagent flow rate are both 1 mL min^{-1}, the length (L) of 0.2 mm i.d. tubing needed to give a residence time in the coil of 1 min is 63.7 m [volume of a cylinder $\pi r^2 L$, hence $L = 2/(\pi r^2) = 63.7$ m].

Doubling the i.d. of the reaction coil will increase the residence time fourfold, but this is likely to reduce the system efficiency. Reducing the flow rate of either the eluent or the reagent stream will increase the residence time and, provided sufficient reagent can be delivered, reducing the reagent flow rate to a minimum will maximize sensitivity.

Subjecting compounds to high-energy light often yields highly fluorescent products. Advantages of photolysis over post-column derivatization are that a second pump is not required, the sample is not diluted and, because there is less flow-rate disturbance, it is amenable to both fluorescence and ED (Lihl *et al.*, 1996). Thin-walled PTFE tubing, crocheted to form a flat mat, may be used for the reaction chamber, as this gives a large surface area for irradiation.

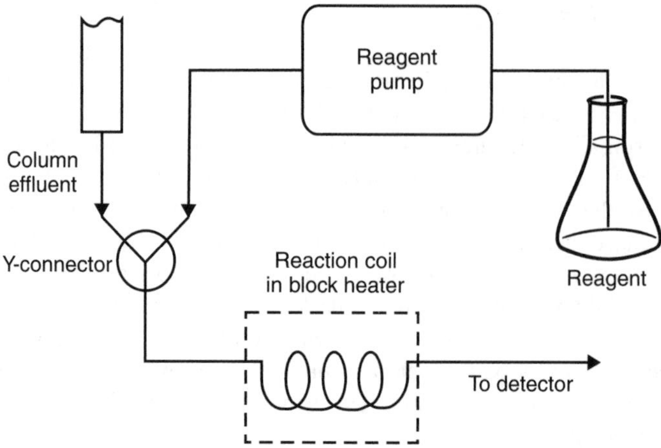

Figure 8.10 Arrangement for post-column reagent addition in HPLC.

Post-column coulometric electrochemical oxidation of analytes to fluorescent products is an alternative to photolysis; this approach has been used to measure plasma amoxicillin, for example (Mascher and Kikuta, 1990).

8.3.11 Immunoassay detection

Immunoassay techniques (Chapter 12) can sometimes be useful for off-line work. The column effluent is collected in small (about 1 mL) fractions, each fraction is assayed and a chromatogram constructed by plotting assay result against fraction number (Daldrup *et al.*, 1985). The procedure is lengthy, but very high sensitivity/selectivity can be achieved. Moreover, time may be saved as relatively simple sample preparation procedures may often be used, as detection is so selective. HPLC with immunoassay detection has been used for cannabinoids, opiates, lysergide and cardioactive glycosides in biological fluids, amongst other compounds.

8.4 Columns and column packings

In analytical toxicology conventional stainless-steel columns (100 or 125 $\times$ 4.6 or 5 mm i.d.) packed with 5 μm materials are often preferred because at flow-rates of 1.5–2.0 mL min^{-1} retention, analysis time, selectivity and sensitivity can usually be adjusted within acceptable limits. Longer (150 or 250 mm) columns, smaller (3 μm, or less) materials or even microbore (1–2 mm i.d.) columns (Table 8.3) may give enhanced efficiency, hence sensitivity, in specific applications; 10 μm materials give lower back-pressures, but otherwise have few advantages and are rarely used.

It should be remembered that the smaller the particle size of the packing, the greater the care needed in sample and eluent preparation to prolong column life because particles and other impurities can collect more rapidly on the smaller particle column causing back-pressure, selectivity and resolution problems. Reduced analysis time is commonly achieved by using shorter columns and higher flow rates. Shorter columns require smaller particles in order to maintain efficiency. However, the increased column back-pressure when using smaller silica particles limits column length.

Table 8.3 Broad classification of HPLC column types

Type	Internal diameter (mm)	Length (mm)	Normal particle size (μm)	Eluent flow-rate (mL min^{-1})
Preparative	100	600	10–50	>5
Semi-preparative	25	250	5–20	2–5
Normal	3.9–5	50–250	5–10	1–2
Minibore	2.1–3.9	50–200	3–5	0.5–1.5
Microbore	1.0–2.1	50–150	3–5	0.1–0.5
Capillary	0.05–1.0	50–250	2–5	0.0002–0.1

8.4.1 Column configuration

Low dead volume (LDV) tubing and fittings are used to connect the column to the injection device and to the detector. Smaller bore columns can give greater sensitivity and use less solvent, but if the system dead volume or the detector cell volume are too high the extra-column band broadening will make the smaller bore column appear to perform poorly. Inert plastic (PEEK) finger-tight male and stainless steel female fittings are normally employed. The plastic fittings do not lock onto the connecting tubing in the same way as stainless steel fittings and slide to fit any type of column fitting correctly.

If conventional male end-fittings are used then plastic (PTFE, Vespel, PEEK) ferrules should be used even at the inlet end of the column as this minimizes the risk of physical damage to the stainless-steel fittings and tubing from over-tightening, and allows adjustment of the tubing as with the finger-tight fittings. Stainless steel or, increasingly, PEEK tubing may be used for pre-column connections. Use of narrow bore (0.15 mm i.d.) PTFE tubing for post-column connections is not only convenient, but also minimizes band broadening, particularly when several detectors are connected in series.

8.4.2 Column packings

Silica gel [silicon dioxide, $(SiO_2)_n$, xH_2O] is the most widely used column matrix as it is cheap, relatively stable, inert and has good mechanical strength. Moreover, it is available with various average particle sizes (3–80 μm), pore sizes (4–33 nm) and surface areas (100–800 m^2 g^{-1}) and can be chemically modified via surface silanol moieties (Si–OH) to give a large range of stationary phases (Section 8.4.2.1). Other oxides (e.g. aluminium, magnesium or zirconium oxides) or cross-linked polymers (styrene/divinylbenzene, polyacrylamide, etc.) are used on occasion as they have good mechanical strength. Sometimes the matrix alone can be used as the stationary phase, for example silica is used in adsorption chromatography (Section 8.5.1) and styrene/divinylbenzene is used in reversed-phase HPLC (Section 8.5.2).

The use of good mechanical strength packings makes it possible to pack small particle size material into columns using high pressure to obtain high-efficiency columns. Furthermore, the risk of the column packing collapsing in use, which would lead to high back-pressures, void formation and loss of efficiency, is minimized. Large pore polymer supports [pore size >50 nm

(500 Å)], which are often used for size-exclusion chromatography, SEC) do not have much cross linking, hence they must be handled and stored carefully. Similarly, large pore silicas (>50 nm, which are used as the support in 'protein' columns) have thinner supporting walls and also need to be handled with care.

The silicas available to each manufacturer, whether made in-house or purchased, may differ in pore size, surface area, pore volume and surface silanol pH. Silica gel is made by precipitation from silicate or silica sol solutions in the pH range 4–10, and the surface pH reflects the pH of the starting mixture. Most manufacturers try to make the surface pH as close to 7 as possible to minimize interactions with acidic or basic analytes. Silica gel can be made with different pore size and surface area depending on its intended use.

The classical silica gel used in adsorption chromatography (Section 8.5.1) has a pore size of 6–8 nm and a surface area of 300–500 $m^2\ g^{-1}$. Because pore size may decrease after chemically bonding a stationary phase, some manufacturers use larger pore size material (10–12 nm) for their base silica to allow adequate access to the bonded phase. For separating analytes with $M_r > 2000$ such as peptides and proteins, 30, 50 or 100 nm, or even larger pore diameter base silicas may be used. However, as pore diameter increases, surface area decreases, and this can limit sample capacity.

Many silica gel packings will decompose if used outside of the pH range 2–8.5. In an attempt to solve this problem, packings have been prepared using highly purified silica and special bonding procedures with stable silanes. There has also been interest in macroporous cross-linked polystyrene-divinylbenzene (PSDVB) copolymer packings which are chemically inert and stable over a wide pH range. Such materials have been advocated for some applications including the analysis of the anticonvulsant vigibatrin and related compounds (Cope and Davidson, 1987). However, in general, PSDVB columns are less efficient than those packed with siliceous materials; other problems are a tendency to instability at high pressures and an eluent-dependent tendency to swell or shrink.

Porous graphitic carbon (PGC), commercially available as Hypercarb (Shandon), has also been suggested as an HPLC packing (Knox *et al.*, 1986). Although chemically inert and with good mechanical strength, PGC columns are prone to void formation. In addition, peak shape is dependent, to an extent, on the nature of the eluent, and efficiency and peak shape both decrease very rapidly with increasing k. Trifluoroacetic acid is a useful modifier in the analysis of both acidic and basic compounds (Lim, 1991), but there is, as yet, no simple way of adjusting selectivity.

8.4.2.1 Chemical modification of silica

Usually silica gel is acid washed, perhaps using nitric acid and/or hydrochloric acid, to remove metal ions. After washing with distilled water to bring the silica to a neutral pH, the material is dried overnight at about 120 °C. For chemical modification, the silica is then placed in a flask with dried toluene or other suitable solvent and a mono-, di-, or trifunctional silane reagent added. The functional (leaving) group can be a halide, or a methoxy or ethoxy moiety. Generally, only one of the functional groups bonds to form a Si–O–Si bond. Less often two of the functional groups react to form adjacent Si–O–Si bonds. The remaining functional groups on each reagent molecule hydrolyze to form silanols after the initial reaction has been completed. These additional silanols formed from hydrolysis of di- and tri-functional reagents can cross link with one another near the surface of the silica support, and thus bonded phases made with any di- or tri-functional reagent are termed 'polymeric' phases. By analogy, any packing made with a monofunctional silane reagent is referred to as a 'monomeric' phase.

Monomeric and polymeric phases often have different selectivities. A monofunctional silane reagent can only bond to the surface silanols and any excess is removed by treatment with water:

$$2\left[\begin{array}{c} R' \\ | \\ R\text{—}Si\text{—}X \\ | \\ R' \end{array}\right] + H_2O \longrightarrow \begin{array}{c} R' \quad\quad R' \\ | \quad\quad\quad | \\ R\text{—}Si\text{—}O\text{—}Si\text{—}R \\ | \quad\quad\quad | \\ R' \quad\quad R' \end{array} + 2HX$$

where R = C_{18}, C_8, cyanopropyl, and so on, and R′ = a shorter alkyl group (methyl, ethyl or propyl). This smaller alkyl group gives less steric bulk to the reagent, hence promoting more complete coverage of surface silanols.

Depending on the reaction conditions, which may include adding water to the reaction mixture, and post-reaction treatment, a range of stationary phase concentrations can be achieved at the silica surface. The surface carbon loading (%) can be ascertained by routine carbon and hydrogen analysis; nitrogen analysis may be helpful for amino- and nitrile-containing phases, as well as some nitrogen-containing chiral packings. However, surface coverage or bonding density will depend on the surface area of the silica as clearly columns of the same carbon loading, but different surface areas, will have different bonding densities. Furthermore, if the silica has a wide distribution of pore size, the surface coverage may be uneven and calculation of bonding density based on carbon loading and surface area may be misleading. Each manufacturer attempts to keep the carbon loading or bonded phase coverage numbers as constant as possible to give batch-to-batch reproducibility. Each number varies, of course, with the chain length or number of aromatic rings on the silane used; the longer the aliphatic chain or the greater the number of aromatic rings, the more weight added for each molecule bonded. The carbon loading is usually in the range 5–25 %, the surface coverage ranging from 1.19 to 3.8 μmol m^{-2}. It is not possible to react every surface silanol because of steric hindrance, which is more apparent when bonding higher M_r silanes.

If a packing is made from the same reagent and base silica, then, generally, the higher the carbon loading, the more retentive it will be for any given eluent/analyte combination. The column with the lower carbon loading will have a greater proportion of residual (unreacted) silanols. It is possible to take bonded material and derivatize residual silanols by using a reagent such as TMCS or HMDS. This process is called ‘end-capping’. Because the M_r of these reagents is small, they do not add much to the carbon loading (only 0.1–0.2 % by weight carbon) compared to the initial bonded phase. Of course, not all bonded phases can be end-capped – the above reagents react with diol and amino moieties, for example.

Diol columns are prepared using a silane reagent with the structure Si–O–CHOH–CH_2OH. Nitrile and amino phases are most often made from dimethylcyanopropylsilane and dimethylaminopropylsilane, respectively. Many ion-exchange materials are manufactured via multistep reactions. Thus, after a phenylpropyl (Si–CH_2–CH_2–CH_2–C_6H_5) moiety has been attached to the silica surface, sulfonation of the ring gives a strong cation exchange (SCX) material. Similarly, a bonded secondary amine can be quaternized to form a strong anion exchange (SAX) phase.

8.4.2.2 Bonded-phase selection

Important parameters to be considered when deciding which stationary phases to evaluate for a new application include: (i) the particle shape (spherical or irregular), (ii) the pore size and surface area, (iii) the type of bonding (monomeric or polymeric phase), (iv) the chain length bonded, (v) the carbon loading (or, if available from the manufacturer, the μmol m^{-2} coverage of the bonded phase) and (vi) whether the phase is end-capped. Other parameters that might not be available except by direct measurement are the pH of the silica surface and the percentage coverage of the surface silanols.

The suitability of a column for a particular separation should be tested as part of method development. A different column of the same type from the same manufacturer should give the same results, but the same type of column from a different manufacturer may give different results. Such columns may need to be optimized as regards the eluent composition in order to bring the separation and analyte retention times to approximately the same range as the column used initially. The elution order on packings with residual silanols may be pH dependent. The flow rate (linear velocity) will need to be adjusted if the diameters of the columns tested are different.

8.4.2.3 Stability of silica packings

The manufacturer's instructions regarding proper HPLC column storage and use should always be followed. The one big disadvantage of silica gel is that it is soluble in water and, to an extent, in methanol. This is true whether it is chemically modified or unmodified, although, in theory, bonded -phase material should dissolve at a slower rate. The higher the eluent ionic strength and the higher the pH, the more rapid will be the dissolution of the packing. In general, silica based packings should not be used above pH 7.5–8.0. Mixing buffer solutions or ion-pairing reagents with organic solvents can raise pH*, possibly to values >8 (Section 8.2.2).

Stationary phases bonded via Si–O–Si bonds are generally stable, except under strongly acidic conditions (<pH 2) as noted above. However, amine-modified packings are particularly reactive (e.g. with aldehydes and ketones) and should be used with care. Other silica packings produced by *in situ* bonding or coated-polymer methods are even more stable in the presence of water at extremes of pH, but may have other limitations as regards compatibility with organic solvents.

8.4.2.4 Monolithic columns

In GC the use of thin-film open-tubular capillary columns with low thermal capacity and an inert carrier gas with low viscosity offers rapid mass transfer between phases and low back-pressures. This gives rise to very efficient separations and, especially when combined with temperature programming, gives a powerful analytical technique. Whilst HPLC has many complementary features to GC, disadvantages to conventional HPLC are slow mass transfer between phases and limitations caused by column back-pressure. Moreover, slow mass transfer within and between phases limits the practicality of open-tubular columns.

In a conventional packed HPLC column, eluent must flow through pores in the packing particles or through the voids between particles. In an attempt to avoid some of the problems associated with packed columns by providing an accessible flow path within the column, a variety of alternative matrices have been studied. Of the 'continuous media' developed, monolithic columns have been studied extensively and HPLC columns based on a porous silica 'monolith' (Nakanishi and Soga, 1991) have been marketed (Chromolith, Merck).

Monolithic columns consist of a single, rigid or semi-rigid, macroporous (2 μm or so average pore size) rod. Like other continuous media, monolithic columns can allow higher flow rates than conventional packed columns at reasonable column back-pressures. Analyte capacity is usually provided by smaller pores within the monolith structure. Two types of monolithic columns have been developed for chromatography: organic polymers based on polymethacrylates, polystyrenes, or polyacrylamides, and inorganic polymers based on silicates. Of these, the silicate monolith columns that are now available with a range of bonded phases (http://www.merck.de/servlet/PB/menu/1229750/index.html, accessed 4 August 2007) offer enhanced possibilities for fast HPLC analysis (Smith and McNair, 2003; Bugey and Staub, 2004; Samanidou *et al.*, 2004), although there remain concerns as to the reproducibility with which monolith columns can be manufactured.

8.4.2.5 *Hybrid particle columns*

Hybrid particle technology (HPT, Waters) combines the properties of silica and organic polymers to produce columns with increased mechanical strength. Ultra-performance liquid chromatography (UPLC) exploits second generation HPT packings based on bridged ethylsiloxane/silica hybrids (BEH) made by reacting tetraethoxysilane and bis(triethoxysilane)ethane (4 + 1) (Figure 8.11). As BEH packing can withstand packing pressures of up to 172,000 kPa (25,000 psi) and operating pressures up to 103,500 kPa (15,000 psi), UPLC utilizes 1.7 μm particles and narrow-bore columns to give increased efficiency (Figure 8.12), resulting in increased sensitivity and reduced analysis times (Wyndham *et al.*, 2003; Lurie, 2005). These columns are designed for use

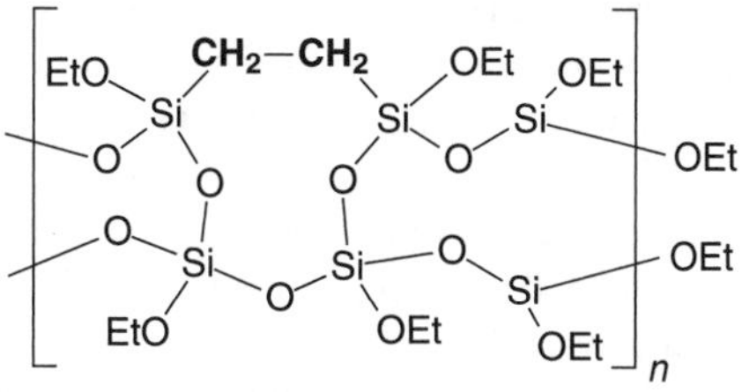

Polyethoxysiloxane

Figure 8.11 Example of a bridged ethylsiloxane/silica hybrid HPLC packing.

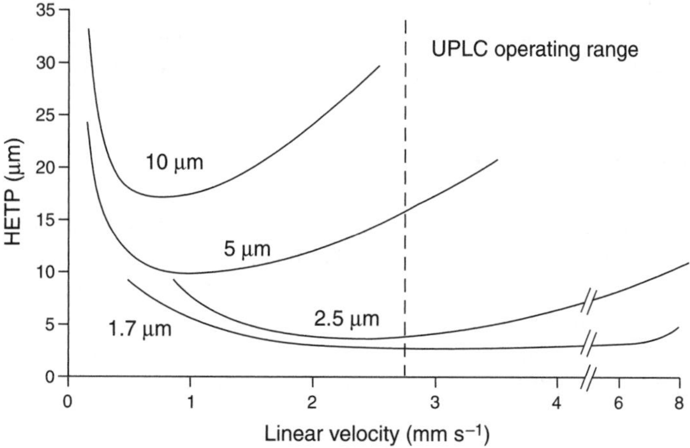

Figure 8.12 Effect of particle size and linear velocity on HETP.

with the Waters Acquity UPLC and similar systems, which are designed to cope with the higher pressures and increased efficiencies attainable.

8.5 Modes of HPLC

Available modes of HPLC include adsorption, liquid–liquid partition, ion-exchange, ion-pair, size-exclusion and affinity chromatography. Each mode allows for different combinations of molecular interactions between the analyte and the stationary and mobile phases, and this gives the opportunity for almost infinite adjustment of selectivity. Partitioning or ion-exchange of an analyte between a stationary phase chemically bonded to an inert support material and the mobile phase is commonly used. In practice, several different mechanisms may operate simultaneously.

8.5.1 Normal-phase chromatography

Juxtaposition of a non-polar mobile phase and a polar stationary phase is called 'normal-phase' chromatography. This is exemplified by adsorption chromatography using silica gel as the stationary phase, analytes being retained by adsorption onto the polar silanols at the silica surface. Typical eluents in this mode are alkanes such as heptane, or alcohol-modified alkanes. Alcohols form a bilayer at the silica surface and such layers can be modified with miscible ions. A systematic method for choosing a suitable HPLC solvent in such cases is: (i) reject those with unsuitable physicochemical properties (boiling point, viscosity, UV absorption, etc.), (ii) from the remaining solvents select those that give appropriate retention (e.g. $2 \leq k \leq 10$) and (iii) perform the final choice on the basis of suitable separation (α) of the components of interest. Solvent polarity can be estimated by solvent strength. Solvent strength can be calculated by the additivity value of each solvent if a mixture is used. It is possible to recreate HPLC detector-compatible eluents from classical TLC systems (Snyder and Kirkland, 1979). As the eluent does not contain water, it is necessary to present the analyte in a non-aqueous medium after, for example, LLE or SPE.

Other adsorbents used in normal-phase HPLC either directly or after chemical modification include alumina (Al_2O_3) and zirconia (ZrO_2), but neither material is widely used in toxicological analyses. Polystyrene gel packings have also been studied. Here, retention is attributed to interaction between the analyte and π-electrons in the aromatic moiety, and although retention is said to be normal phase when heptane is used as the mobile phase, retention is said to be 'reversed-phase' if methanol is used. This also applies if aminopropyl- or cyanopropyl-modified silica is used as the stationary phase.

8.5.2 Reversed-phase chromatography

'Reversed-phase' HPLC uses a polar eluent and a non-polar stationary phase. Typically the stationary phase is silica modified by the addition of octadecyl (ODS, C_{18}), octyl (C_8), ethyl (C_2), methyl (C_1), or phenylpropyl (phenyl) silyl moieties. There are many different types of phases depending on the conditions of preparation and synthesis; even C_{18} phases differ when obtained from different manufacturers (Section 8.4.2.1). Although reversed-phase is the most popular separation mode in HPLC, the retention mechanism is often unclear, especially as far as polar analytes are concerned. Partition, adsorption and ionic mechanisms may contribute to the retention of different analytes under the same eluent conditions, and vice versa. This is reflected in the fact that the effect on retention of altering the proportion of water in the eluent may not be predictable.

Optimization strategies for reversed-phase HPLC on organosilane-modified silica columns have been described (Naish-Chamberlain and Lynch, 1990). While all very well for acidic and neutral compounds, such strategies may not be applicable to basic drugs, particularly when ionic interactions with the stationary phase predominate. There are four major inter-molecular interactions between solute and solvent molecules in liquid chromatography: dispersion, dipole, hydrogen bonding and dielectric. Dispersion interactions are the attraction between each pair of adjacent molecules. Strong dipole interactions occur when both sample and solvent have permanent dipole moments that are aligned. Strong hydrogen-bonding interactions occur between proton donors and proton acceptors. Dielectric interactions favour the dissolution of ionic molecules in polar solvents. The total interaction of the solvent and sample is the sum of the four interactions. The total interaction for a sample or solvent molecule in all four ways is known as the 'polarity' of the molecule. Polar solvents dissolve polar molecules and, for normal-phase partition chromatography, solvent strength increases with solvent polarity, whereas solvent strength decreases with increasing polarity in reversed-phase systems (Snyder and Kirkland, 1979).

Typical eluent solvents in reversed-phase HPLC are methanol and/or acetonitrile modified by adding water to adjust retention and selectivity. Adding water increases the polarity of the eluent, which generally increases analyte retention. Addition of water to methanol also markedly increases viscosity and thus column back-pressure, other factors being equal. Water:acetonitrile mixtures tend to be less viscous than water:methanol mixtures and this, together with the lower UV cut-off of far UV-grade acetonitrile, are reasons acetonitrile may be preferred. To separate compounds that are not soluble in methanol, acetonitrile or tetrahydrofuran (THF) and water, dichloromethane:methanol, or THF:methanol mixtures can be used as eluent, but use of such mixtures is uncommon.

Restricted access media (RAM) is a general term applied to reversed-phase HPLC packings with large pore-size that aim to elute proteins and other high M_r compounds whilst retaining and separating hydrophobic compounds such as drugs (Section 3.2.1). Internal surface reversed-phase (ISRP) chromatography was the first such column/eluent combination to be proposed (Hagestam and Pinkerton, 1985; Cook and Pinkerton, 1986). However, the eluent pH has to be maintained at about 7 and a water-rich eluent has to be used to minimize the risk of protein precipitation. Although some publications have resulted (Rainbow *et al.,* 1989; Rainbow *et al.,* 1990), the system is not widely used. Semi-permeable surface (SPS) and hydrophobic shielded phase (Hisep) columns are further variants on this theme and have been evaluated in some applications (Haque and Stewart, 1999).

8.5.3 Ion-exchange chromatography

Ion-exchange chromatography is a process whereby an ionized analyte is attracted by ionic (electrostatic) forces to species of opposite charge on the stationary phase. When the analyte has a positive charge and the stationary phase a negative charge the technique is called cation-exchange chromatography, and when the charges are reversed it is known as anion-exchange chromatography. Eluents used in conventional ion-exchange chromatography generally consist of aqueous solutions of a suitable salt or mixture of salts, with sometimes a small amount of an organic solvent being added. The salt mixture may itself be a buffer, or a separate buffer can be used. The prime component of the eluent is the counter ion, which has the role of eluting sample components from the column in a reasonable time. In cation-exchange chromatography, the presence of other positively charged ions in the eluent decreases analyte retention. Generally solute ions are retained in the following order:

$$Ba^{2+} < Ca^{2+} < Cu^{2+} < Zn^{2+} < Mg^{2+} < K^{+} < NH^{4+} < Na^{+} < H^{+} < Li^{+}$$

As a result, if K^+ is used as the counter ion instead of Na^+, analyte elution is more rapid. In anion-exchange chromatography, the retention order is as follows:

$$\text{Citrate} < SO_4^{2-} < \text{oxalate} < NO_3^- < Br^- < SCN^- < Cl^- < CH_3COO^- < OH^- < F^-$$

Because the eluent is a solution containing various inorganic salts, retention can be changed by different kinds of salts, their concentration and the eluent pH.

Nowadays two main types of matrix are used, silica gel and polymeric organic materials. Of these, silica gel is used in HPLC. Most silica-based ion-exchange materials are strong anion- or strong cation-exchange in nature (SAX or SCX, respectively).

HPLC of inorganic and organic anions such as chloride, nitrate, oxalate and sulfate, is often known as ion chromatography (IC). This is a specialized subset of ion-exchange chromatography in which a suppressor column is used to exchange the counter ion of the analyte ion after the separation column and, therefore, enhance the conductivity detection of the analyte ions. Anion measurements are sometimes necessary in analytical toxicology. Bromide ion, for example, is a metabolite of the anaesthetic halothane and indeed bromide salts have been used as anticonvulsants. Although often thought of as requiring a dedicated instrument, conventional HPLC columns and equipment can be used for IC (Wheals, 1987).

8.5.4 Ion-pair chromatography

Although ionic compounds can be separated by ion-exchange chromatography, the ion-exchange mode may not give the desired selectivity. Instead, ion-pair chromatography can be used with, for example, a C_{18}-modified silica column. According to the simplest explanation, the analyte ion is paired with an ion of opposite charge added to the eluent to give a neutral species (ion-pair) that can be separated by reversed-phase partition. Typical ion-pairing agents include alkane sulfonates and alkane quaternary ammonium compounds. Because the analytes and counter ions are ionic and thus hydrophilic, typical eluent solvents are methanol or acetonitrile–aqueous buffer mixtures.

8.5.5 Size-exclusion chromatography

Size-exclusion chromatography (SEC) is based on the effective shape and size of molecules in solution. It is also called gel-permeation chromatography (GPC) if used with organic solvents, or gel-filtration chromatography (GFC) if used with aqueous solvents. The stationary phases are porous particles with closely controlled pore size. Unlike other modes of HPLC, in SEC there should be no interaction between the analyte and the surface of the stationary phase – if necessary silica packings can be treated with TMCS to minimize adsorption effects. When analytes are larger than the pore size, they elute in the void volume of the column (V_0). Molecules that are much smaller than the pore size are able to freely diffuse in the internal volume (V_i) of the packing and elute in a volume equal to the total volume of the column (V_t). Molecules of intermediate size will be able to penetrate some of the pores, and will elute in a volume between V_0 and V_t, depending on their size (Figure 8.13). SEC columns have defined working ranges and should be selected on the basis of the size of analytes being investigated. The columns are calibrated using polystyrenes of known M_r (hydrodynamic volume is approximately proportional to M_r). The choice of eluent is simple as only one solvent is required. Although SEC was applied originally mainly to polymer characterization, it has now many uses in other fields including the analysis of proteins and peptides.

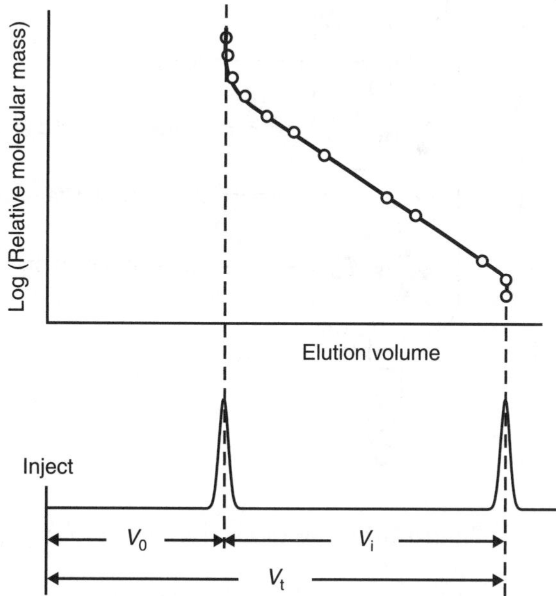

Figure 8.13 Typical calibration curve for size-exclusion chromatography.

8.5.6 *Affinity chromatography*

The retention mechanisms described above are, in the main, based on small differences in the adsorptivity, hydrophobicity or ionic interactions of analytes (Hage, 1999). The interactions are general and not specific for any given analyte, and therefore, various kinds of molecules can be separated. In contrast, affinity chromatography is based on a specific interaction such as an enzyme–substrate or antibody–antigen interaction. This kind of molecular recognition mechanism is an important interaction in HPLC separations, and the concept of affinity chromatography can extend to other separation problems such as isomeric and chiral separations.

To separate an antibody, for example, an immunoadsorbent has to be prepared. This consists of a solid matrix to which the antigen has been coupled (usually covalently). Agarose or Sephadex i.e. modified cellulose, or other polymers can be used as the matrix. When the sample is added, as long as the capacity of the column is not exceeded, those antibodies in the mixture that are specific for the antigen will bind (non-covalently) and be retained. Antibodies of other specificities and other serum proteins will pass through unimpeded (Figure 8.14). An elution reagent is passed through the column to release the antibodies from the immunoadsorbent. Buffers containing a high salt concentration and/or of low pH are often used to disrupt the antibody–antigen interaction.

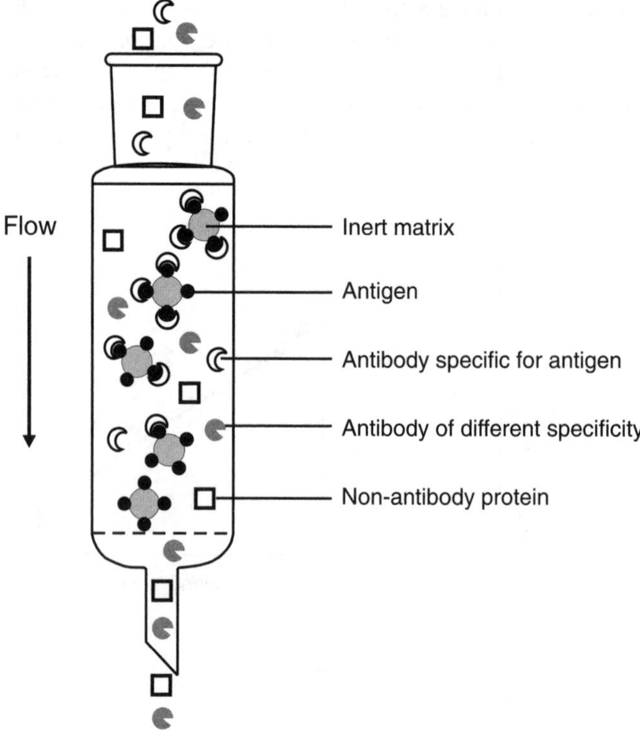

Figure 8.14 Schematic of affinity chromatography.

A denaturing agent, such as 8 mol L^{-1} urea, will also break the interaction by altering the configuration of the antigen-binding site of the antibody molecule. Another, gentler, approach is to elute with a soluble form of the antigen. These compete with the immunoadsorbent for the antigen-binding sites of the antibodies and release the antibodies from the matrix. The eluate is then dialyzed against, for example, buffered sodium chloride solution to remove the elution reagent.

8.5.7 Semipreparative and preparative chromatography

Preparative chromatography is rarely used in analytical toxicology, but could be used to obtain reference materials such as metabolites and internal standards if these are not available from alternative sources. When the isolation of relatively large quantities of an analyte is required (more than 1 g), larger bore (10–150 mm i.d.) and longer length columns are needed. Above about 25 mm i.d., columns containing 3–5 μm packings not only become increasingly expensive because of the large amount of packing needed, but also difficult to pack because high flow-rate pumps are needed to deliver large volumes of slurry. In practice, preparative columns are always heavily overloaded to generate high product throughput, but this simultaneously decreases efficiency. Thus, these large columns (i.e. i.d. $>$25 mm) are often packed with larger particle size media (e.g. 25–40 or 40–63 μm) to minimize costs. Although efficiency is less, as long as the selectivity is the same as that used for the initial separation on an analytical column, the compound(s) of interest can usually be resolved, although the analysis time may be longer (McDonald and Bidlingmeyer, 1987).

8.6 Chiral separations

Chiral HPLC is more versatile than chiral GC because it can be used with non-volatile compounds and is generally the method of choice for enantiomer separations, and for semipreparative and even preparative work. HPLC can provide fast and reproducible analyses and allows on-line detection and quantitation of both mass and optical rotation of enantiomers, if appropriate detectors are used. Problems in routine use are the very high cost of certain columns and the fact that packings designed for enantiomer separations may not separate metabolites; a second, conventional HPLC column may thus be needed in series with the CSP column.

Chiral separation is a complex field (Chen *et al.*, 2005). The approaches adopted include the use of CSPs, the use of chiral additives in conventional reversed-phase systems (Section 8.6.8) and also the formation of chiral derivatives (Section 8.7.3). Direct chiral separations using CSPs are more widely used and are often more predictable than those using chiral eluent additives. Some 100 HPLC CSPs have been marketed. However, no single CSP has the ability to separate all classes of racemic compounds.

Some CSPs are manufactured by immobilizing single enantiomers onto the column matrix. Enantiomer resolution relies on the formation of transient diastereoisomers on the surface of the column packing. The enantiomer that forms the most stable diastereoisomer will be most strongly retained. The forces involved are very weak (typically a free energy of interaction difference of only 0.03 kJ mol^{-1} between a pair of enantiomers and the stationary phase will lead to resolution) and require careful optimization by adjustment of the eluent and the column temperature to maximize selectivity.

To achieve discrimination between enantiomers there must be three points of interaction of one enantiomer with the CSP, at least one of which is stereochemically dependent (Pirkle and House, 1979) (Figure 8.15). The intermolecular forces involved in chiral recognition are polar/ionic interactions, π–π interactions, hydrophobic effects and hydrogen bonding. These can be augmented

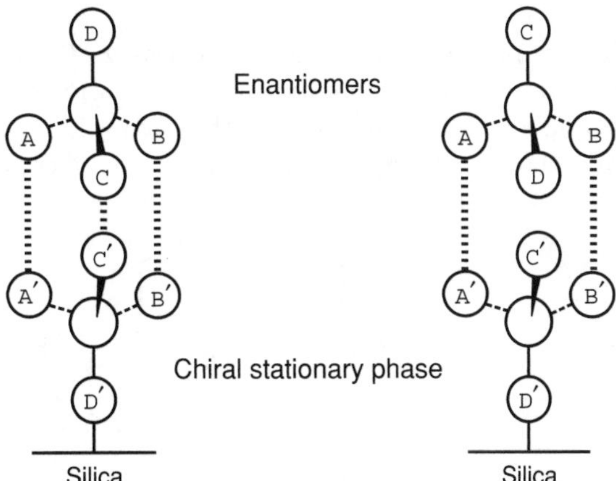

Figure 8.15 Representation of chiral recognition. The structure with three-point interaction (*left*) is retained more strongly than its enantiomer with only two (*right*).

by the formation of inclusion complexes and binding to specific sites such as peptide or receptor sites in complex phases. These inter-molecular forces may be manipulated by choosing suitable eluents – polar interactions, for example, may be controlled by adjusting eluent pH. The effect of temperature is important in chiral HPLC. Lower temperatures increase chiral recognition, but also reduce the speed of mass transfer between phases giving broader peaks. Thus, there is often an optimum temperature for a separation and this gives another factor to exploit in method development.

The type of column required to separate a particular class of enantiomers is often highly specific, making the choice of column for a given separation appear complex. Furthermore, the high cost of chiral columns may limit the number of CSPs available. Fortunately, study of the structure of chiral phases and visualizing the potential interactions with the analyte can narrow down the choice of column markedly. The key to developing a successful enantioseparation of a particular class of racemates on a given system is the understanding of the possible chiral recognition mechanisms that may come into play. The nature of the analyte or derivatized analyte, the stationary phase and the eluent must be taken into consideration. The most effective CSPs are those based on the macrocyclic glycopeptides and the cyclodextrins.

8.6.1 Chiral stationary phases

8.6.1.1 Amylose and cellulose polymers

These CSPs are derivatized amylose (starch) or cellulose polymers coated onto a silica support. The cellulose phases available include microcrystalline triacetate- (MCTA-), tribenzoate-, trisphenyl-, carbamate- and tris(3,5-dimethylphenylcarbamate)-cellulose. These phases are used with hexane:ethanol, hexane:2-propanol, or even 100 % ethanol eluents. Chlorinated solvents are avoided because they could remove the cellulose from the silica support. Amylose CSPs show different selectivity to cellulose CSPs, but are less robust. Starch is water-soluble, hence eluents must not contain water.

8.6.1.2 Crown ethers

Crown ether CSPs are ring structures based on repeating $-OCH_2CH_2-$ units. The name 18-crown-6 (1,4,7,10,13,16-hexaoxacyclooctadecane, Figure 8.16; $M_r = 264$) indicates that there are 18 atoms in a ring, 6 of which are oxygen. The interior of the cavity is hydrophilic, whilst the exterior is hydrophobic. Varying the size of the crown ether varies the cavity size. Crown

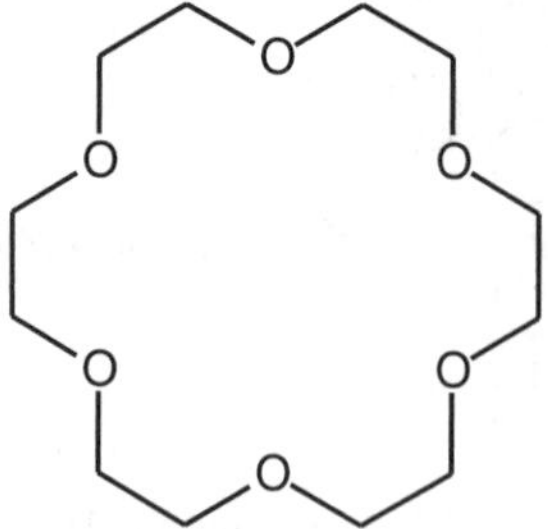

Figure 8.16 Molecular structure of 18-crown-6.

ethers are carcinogens, but can be immobilized to form HPLC stationary phases that complex with protonated primary amines. Acidic eluents must be used of course. The most commonly used crown ether CSP, 18-crown-6, is available commercially (Crownpak, Daicel). Crownpak columns give very efficient separations and are available in both (+) and (−) forms, allowing reversal of elution order which can be helpful, for example if trace analysis of one enantiomer is to be performed in the presence of an excess of the other enantiomer.

8.6.1.3 Cyclodextrins

Cyclodextrins bonded to 5 μm silica particles are popular CSPs. They can be used with methanol:aqueous buffer mixtures and with nonaqueous eluents. Good sample capacity means they are useful in preparative work. Cyclodextrins are produced by partial degradation of starch followed by the enzymatic coupling of the D-(+)-glucose units via α-(1-4)glycoside linkages into crystalline, homogeneous, toroidal structures of different molecular size. α-Cyclodextrin (cyclohexamylose, Figure 8.17; M_r 973), β-cyclodextrin (cycloheptamylose; M_r 1135) and γ-cyclodextrin (cyclooctamylose; M_r 1297) contain 6, 7 and 8 glucose units, respectively. β-Cyclodextrin, for example, has 35 chiral centres.

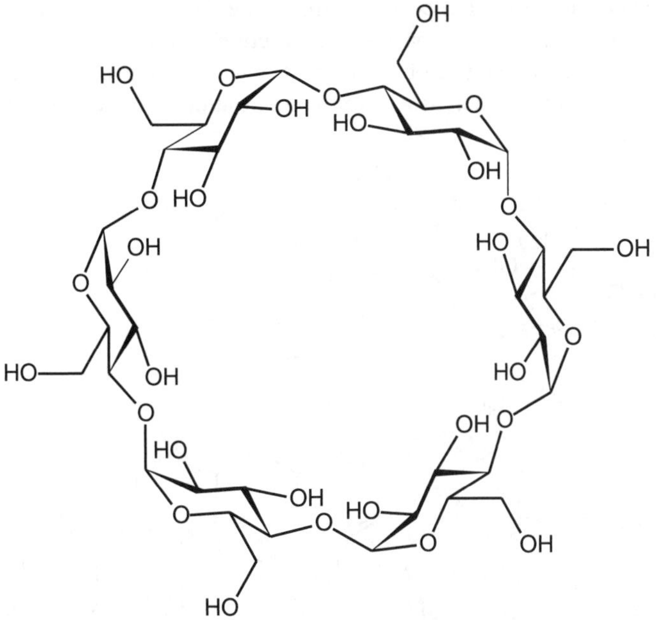

Figure 8.17 Molecular structure of α-cyclodextrin.

The mouth of the torus-shaped cyclodextrin molecule has a larger circumference than the base and is linked to secondary hydroxyl groups of the C_2 and C_3 atoms of each amylose unit. The primary hydroxyl moieties are located at the base of the torus on the C_6 atoms and these groups are used to anchor the cyclodextrin to the silica matrix. The size of the cyclodextrin cavity increases with increasing number of amylose units in the molecule. When α-, β-, or γ-cyclodextrins are derivatized, the hydroxyl moieties on the 2-positions react first. However, the derivative remains size-selective and interactions are determined by the size of the analytes as well as the functional groups present.

8.6.1.4 *Ligand-exchange chromatography*

Chiral ligand-exchange chromatography (CLEC) can resolve enantiomers via the formation of diastereomeric metal complexes (Kurganov, 2001). The method is used primarily in amino acid analysis. A chiral amino acid–copper complex is bound to silica or to a polymeric stationary phase, and copper ions are included in the eluent. Amino acids may then be separated via formation of diastereomeric copper complexes. Water stabilizes the complexes, and steric factors then determine which of the two complexes is more stable – in one instance a water molecule is usually sterically hindered from co-ordinating to the copper. Other transition metals have been used in CLE. The optimum temperature for such work is about 50 °C.

8.6.1.5 *Macrocyclic glycopeptides*

Macrocyclic antibiotics such as rifamycin, teicoplanin or vancomycin chemically bonded to silica gel can be used as CSPs (Armstrong *et al.*, 1994). Vancomycin (Chirobiotic V; $M_r = 1449$; Figure 8.18) contains 18 chiral centres surrounding three 'pockets' or 'cavities' (A–C) bridged by five aromatic rings. It has a basket-like structure with a single flexible carbohydrate 'flap'. Furthermore, it contains a range of functionalities useful for enantiomeric separations [hydrogen bonding, hydrophobic pockets (inclusion complexation), aromatic moieties (π–π interaction)

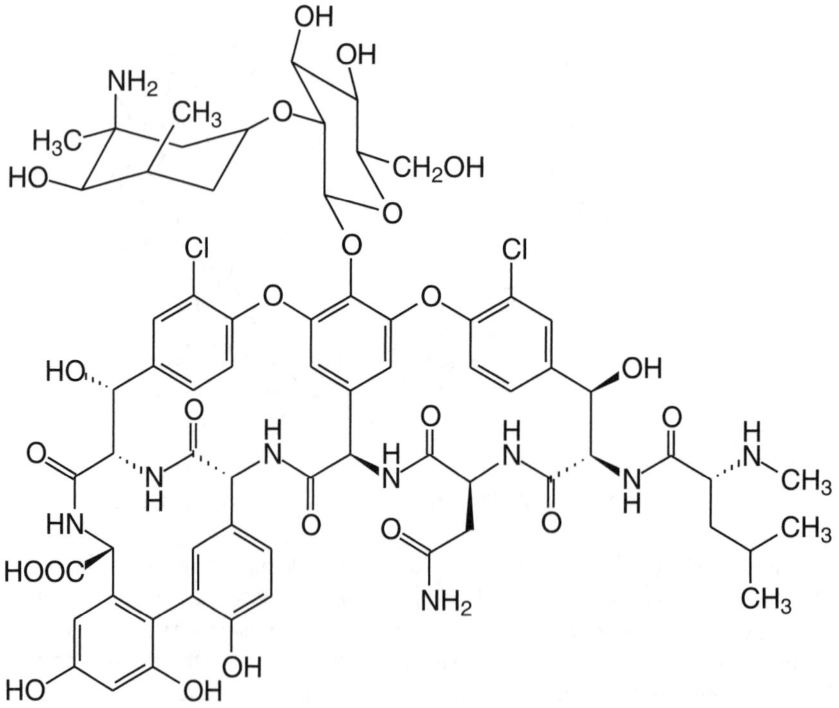

Figure 8.18 Molecular structure of vancomycin.

and ionizable groups]. Vancomycin CSPs are stable, have a relatively high sample capacity, and can be used with nonaqueous or aqueous eluents in the pH range 4.0–7.0. Ammonium nitrate, triethylammonium acetate and sodium citrate buffers have all been used with these materials.

Teicoplanin (Chirobiotic T; $M_r = 1885$) has 20 chiral centres, three carbohydrate moieties and four fused rings surrounding four 'pockets'. Teicoplanin CSPs are said to complement vancomycin phases, and can be used with similar eluents. Teicoplanin CSPs are stable over the pH range 3.8–6.5, although can be used outside these limits for short periods. Unusually, this CSP gives longer retention and increased resolution as the proportion of organic modifier in the eluent is increased. Avoparcin (Chirobiotic A), a 1:4 mixture of α- and β-avoparcin ($M_r = 1909$ and 1944, respectively), has also been used as a CSP. The avoparcin aglycone contains three connected semirigid macrocyclic rings (one 12-membered, two 16-membered) that form a possible inclusion pocket. This glycopeptide contains seven aromatic rings with four phenol moieties, four carbohydrate chains, 16 hydroxyls, one carboxylic acid, two primary amines, one secondary amine, six amide linkages, one or two chlorine atoms for α- and β-avoparcin, respectively and 32 chiral centres.

8.6.1.6 Pirkle brush-type phases

These CSPs consists of low M_r chiral substances bonded to silica gel (Pirkle and House, 1979). Each bonded moiety has a limited number of chiral centres, but as the groups involved are small there can be a large number bonded to the silica surface. It follows that there is a relatively high probability of the analytes interacting with a chiral centre. An advantage of brush-type CSPs is that, because the interacting stationary phase moiety is small and the analytes are not strongly retained, chiral selectivity is the dominant factor influencing retention.

There are two main types of Pirkle CSP, π-acceptors and π-donors. The most common π-acceptor phase is *N*-(3,5-dinitrobenzoyl)-phenylglycine bonded to propylamino-modified silica. These CSPs are capable of separating a large range of compounds that possess a π-donor aromatic group. This may be introduced by derivatization with naphthoyl chloride or other appropriate reagent (Section 8.7.3). A further advantage of brush-type CSPs is that they are easily synthesized. Urea-linked CSPs can be prepared by injecting the appropriate chiral isocyanate onto aminopropyl-modified silica columns (Buckley and Whelpton, 1992). Spherisorb Chiral 2 (*N*-1-napthylethyl-*N*′-propylurea) was prepared by injection of isocyanate, whereas the Chiral 1 (*N*-1-phenylethyl-*N*′-propylurea) was prepared in the more conventional manner by reaction of the appropriate silane with silica. It is not always easy to predict how even structurally related compounds will behave on these phases. Amfetamine (as the 3,5-dinitrobenzamide derivative), for example, was resolved on Chiral 2, but not Chiral 1 [Figure 8.19(a)]. In contrast, the norfenfluramine derivative was partially resolved on Chiral 1, but not at all on Chiral 2 [Figure 8.19(b)]. Tranylcypromine, again as the 3,5-dinitrobenzamide, was resolved on both stationary phases [Figure 8.19(c)].

Pirkle brush-type CSPs are readily available commercially at reasonable cost (Regis, Sumitomo). The columns can give good selectivity and hence relatively high sample capacity. Disadvantages are that the columns will only separate aromatic compounds and thus (achiral) derivatization may be needed to aid resolution.

A CSP derived from dimethyl *N*-3,5-dinitrobenzoyl-α-amino-2,2-dimethyl-4-pentenylphosphonate covalently bound to 5 μm mercaptopropyl-modified silica (Burke-II phase) has been used for the analysis of enantiomers of underivatized β-blockers (Pirkle and Burke, 1991). Lowering the temperature reduced the retention of the less retained enantiomer and increased

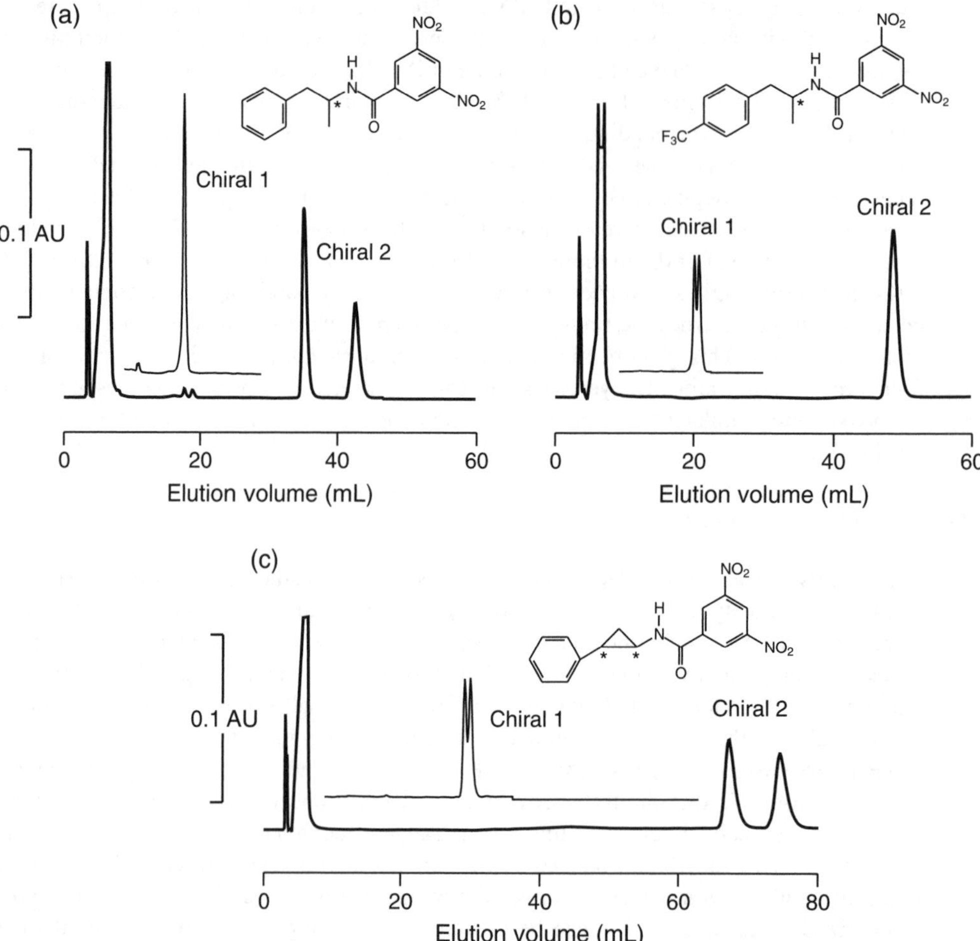

Figure 8.19 Analysis of 3,5-dinitrobenzamide derivatives of: (a) amfetamine, (b) norfenfluramine and (c) tranylcypromine. Columns: 250 × 4.6 mm i.d. Spherisorb 5 Chiral 1 and Chiral 2. Eluent: hexane:dichloroethane:ethanol (50 + 10 + 1). Flow rate: 1 mL min $^{-1}$. Detection: UV, 254 nm. Adapted by permission of The Royal Society of Chemistry from Whelpton and Buckley (1992b).

the retention of the more retained enantiomer without appreciable band broadening. The separation between enantiomers (α) depends on the difference in the Gibbs free energy (ΔG) for the interactions with the CSP:

$$\text{In}(\alpha) = \frac{(\Delta G)_R - (\Delta G)_S}{RT} \tag{8.1}$$

where R is the gas constant and T the absolute temperature. If the interaction is enthalpy driven then lowering the temperature increases α, but predicting the effect of temperature change is not easy. Some separations have been shown to have an optimum temperature.

New brush-type CSPs continue to be marketed. A brush-type CSP 'mixed' (π-donor, π-acceptor) derived from (*R*)-*N*-(3,5-dinitrobenzoyl)allylglycine-2,6-dimethylanilide, for example, has been described (Uzunov *et al.*, 1995). Quinidine carbamate CSPs, have been used to separate enantiomeric acids via ionic interaction with the immobilized basic moiety. If a quinine- rather than a quinidine-CSP was used, the elution order of the enantiomeric pair studied was reversed (Lammerhofer *et al.*, 2004).

8.6.1.7 Protein-based phases

CSPs based on human serum albumin (HSA) or α_1-acid glycoprotein (AAG) have been described (Fitos *et al.*, 1992). Other protein CSPs include bovine serum albumin and egg proteins (ovomucoid). For use as CSPs they are normally immobilized on spherical 5 μm silica. Eluent additives such as octylamine reduce overall retention and increase chiral selectivity. HSA-based columns are especially useful for the resolution of acidic, zwitterionic and nonprotolytic compounds. Enantiomers of acidic compounds can be resolved without derivatization. AAG may be more suitable for basic compounds. Racemates that do not bind to plasma protein will not be resolved with these columns.

8.6.2 Chiral eluent additives

Chiral eluent additives (CEAs) have some advantages and some disadvantages as compared with CSPs (Box 8.4). CEAs are of several types: ion pair, inclusion complex, ligand exchange and protein interaction. Chiral ion-pairing agents may be used with unmodified silica, although the loading capacity is very low. Enantiomers are resolved as diastereomeric ion pairs. Propranolol enantiomers, for example, have been resolved using tartaric acid as the counter ion. Other counter ions used in this way include quinine, *N*-benzoxycarbonylglycyl-L-proline and (+)-10-camphorsulfonic acid.

Box 8.4 Advantages and disadvantages of eluent additives in chiral HPLC

Advantages:

- Can use standard (less expensive) columns
- High loading capacities are possible
- Solute character may be modified (e.g. by ion pairing)
- Wide range of additives available
- Scaling up for semipreparative or preparative work not limited by column availability

Disadvantages:

- May have to remove the chiral selector after the analysis
- Separations relatively difficult to develop
- Can be expensive on a large scale without recycling additive(s)
- Crown ethers too toxic for use as CEAs

Inclusion complexes may be formed by adding cyclodextrin to aqueous eluents. Resolution can sometimes be predicted from results on cyclodextrin CSPs, although the elution order is often reversed. Acidic or basic extraction can be used to remove cyclodextrins from the column effluent.

The second approach is to use SPE on a C_{18} column. After loading, the SPE column is washed with methanol:water (1 + 9) to remove the cyclodextrin and any buffer salts, and the analyte is then eluted with methanol.

Ligand chromatography can be carried out by using a chiral selector such as *N*-alkyl-L-hydroxyproline with copper(II) acetate in the eluent on a C_{18} column. The chiral selector is strongly adsorbed onto the stationary phase – if water-rich eluents are used there is very little column bleed hence the CEA may be omitted from the eluent for the actual analysis.

Both bovine serum albumin and AAG have been used as CEAs. Bovine serum albumin may be added to an aqueous eluent at a concentration of around 3 g L^{-1} and used with C_{18}, CN, or diol columns. Although inexpensive, it has a UV cut-off at 340 nm that limits its use. AAG has the advantage of being UV transparent, but, being a human blood product, it is expensive. As with protein stationary phases the preparative capacity is very low.

Finally, columns packed with PGC (Section 8.4.2) can be used to separate stereoisomers after absorptive modification of the surface with a chiral reagent or by adding chiral eluent modifiers (Dolphin, 1989).

8.7 Derivatives for HPLC

Except for facilitating chiral separations, derivatization in HPLC is usually performed to enhance sensitivity/selectivity and reagents are available for imparting UV, fluorescent and electrochemical properties to the product (Imai, 1987; Lingeman and Underberg, 1990). The need for a derivatizable functional group on the analyte is self-evident. Many reagents can be used in either pre- or post-column mode. As in GC there is always the option of pre-column derivatization (Section 3.7), but this: (i) alters the chromatographic characteristics of the analyte, (ii) may also derivatize unwanted sample components and (iii) may make choice of internal standard more difficult. For post-column work, the ideal reaction would go to completion in a few seconds, but provided the reaction is first order with respect to analyte and zero order with respect to reactants, the signal will be directly proportional to the analyte concentration. However, it is crucial to ensure constant reaction times. Thus, reagent consumption is higher than with pre-column derivatization as a large excess must be maintained.

The use of derivatization to increase UV absorptivity is not widely used in analytical toxicology and related fields. Many of the available reagents were developed for use with early single-wavelength detectors. Consequently, several are based on *p*-nitrobenzoyl or *p*-nitrophenyl moieties that have absorption maxima (λ_{max})at about 254 nm, a strong emission line in mercury lamps. Carbonyl compounds can be treated with 2,4-dinitrophenylhydrazine, giving hydrazones with λ_{max} 360 nm, $\varepsilon = \sim 20{,}000$ L cm^{-1} mol^{-1} and at 254 nm $\varepsilon = \sim 10{,}000$ L cm^{-1} mol^{-1}.

8.7.1 Fluorescent derivatives

Fluorescent derivatives offer enhanced sensitivity as compared to UV detection. Primary amines may be derivatized using *o*-phthaldialdehyde (OPA) in the presence of a thiol (2-mercaptoethanol, *N*-acetylcysteine, 3-mercaptopropionate, *tert*-butylthiol) or sodium sulfite and by other methods developed in the first instance for amino acid analysis. 2-Mercaptoethanol has the disadvantage that the derivatives are unstable, but if the reaction is carried out in a suitable autosampler, the reaction time can be carefully controlled.

Fluorescence detection is relatively insensitive to changes in eluent flow rate hence an alternative approach is post-column derivatization, which has the advantage that the analyte can be analyzed directly (Section 8.3.10). The aim is to achieve rapid, selective derivative formation in the presence of the eluent with minimal reagent 'background'. OPA is a particularly good reagent for post-column procedures because it is 'silent', that is to say it has no native fluorescence.

8.7.2 *Electroactive derivatives*

Formation of electroactive derivatives for HPLC is uncommon, especially as ED is very sensitive to flow-rate changes hence post-column derivatization is not normally an option. OPA derivatives are electroactive and ED may be useful especially with analytes derivatized at more than one site that do not fluoresce due to quenching. The derivative formed by reaction of the γ-amino acid baclofen with OPA/*tert*-butyl thiol has been analyzed by HPLC-ED. 2,3-Naphthalenedialdehyde (NDA) in the presence of cyanide ion is an alternative to OPA (Lunte *et al.*, 1989; Kline and Matuszewski, 1992). Other examples of electroactive derivatives are largely of research interest and not currently used in analytical toxicology and related areas. In general, the derivatizing reagents needed have to be custom synthesized.

8.7.3 *Chiral derivatives*

Chiral derivatization is always performed pre-column as the object is separation and not enhanced detection (Toyo'oka, 2002; Kato *et al.*, 2001; Kraml *et al.*, 2005). Requirements for an ideal chiral derivatization reagent are summarized in Box 8.5. Clearly chiral derivatization would be the last resort in semipreparative or preparative work.

Box 8.5 Requirements for an ideal chiral derivatization reagent

- High chemical and optical purity
- Economic to purchase
- Stable (no racemization during storage)
- Rapid derivatization under mild reaction conditions to give a single reaction product
- Negligible racemization during derivatization
- Reaction rates and yields essentially the same for each enantiomer
- Resulting diastereoisomers have adequate detector response

Separations on π-donor brush-type (Pirkle) CSPs require the analyte to contain a π-acceptor such as an aromatic nitro moiety. Readily available reagents include 3,5-dinitrobenzoyl chloride, 4-nitrophenyl isocyanate, 2,4-dinitrophenylhydrazine and 4-nitrophenacyl bromide, which may be chosen on the basis of the functional group (amine, alcohol, carboxylic acid, etc.) to be derivatized (Whelpton and Buckley, 1992a). Carboxylic acids may be converted to acid chlorides and reacted with 2,4-dinitroaniline. Amide, urea and carbamate derivatives are the most suitable as these provide sites for hydrogen bonding and dipole stacking, a second point of interaction. The third interaction is the spatial arrangement of groups around the asymmetric centre (Figure 8.20). Ester derivatives are not usually as successful as amides. Derivatization with a chiral reagent to

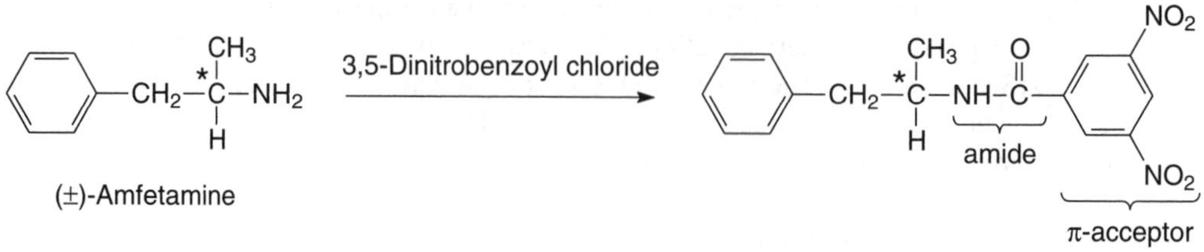

Figure 8.20 Derivatization of racemic amfetamine to provide points of interaction with a Pirkle-type π-donor column.

form a diastereoisomer and subsequent analysis on a conventional, achiral stationary phase is a further option.

The principal reactions used in chiral derivatization of amino acids and amino alcohols are based on the formation of amides, carbamates, urea and isourea. Diastereoisomeric amide reactions are also widely used for the resolution of primary amines. For UV/visible detection, derivatives that absorb in the visible region are, of course, preferable as this gives enhanced selectivity and hence sensitivity. Several chiral fluorigenic agents are also available.

8.8 Use of HPLC in analytical toxicology

8.8.1 Acidic and neutral compounds

Generally acidic and neutral compounds are analyzed conveniently by reversed-phase chromatography using cyanopropyl-, phenylpropyl-, C_8-, C_{18}-, or similarly modified silica packings with aqueous methanol or acetonitrile eluents (Box 8.6). Ionic interactions with the silica matrix, itself weakly acidic, are relatively unimportant and the effect of altering the ratio of the organic component of the eluent to the aqueous component has by far the greatest impact and is generally predictable, more water giving increased retention and *vice versa*. Similar separations can often be achieved using phenyl-modified packings as compared to ODS-modified materials, but retention of later eluting components may not increase as rapidly as with ODS. This can result in shorter analysis times and in greater sensitivity as peak broadening may be reduced. On the whole, therapeutic doses, and hence plasma concentrations, tend to be higher (w/w) with acidic and neutral drugs than with many bases, thus simplifying detection. However, this is not helpful with drugs such as barbiturates, which can only be detected by low-wavelength UV unless derivatized.

Box 8.6 Considerations in the HPLC of acidic and neutral compounds

- Use on bonded-phase silica packing with aqueous methanol or acetonitrile eluent
- Need to buffer eluent in pH range 2–8 to give stable retention for ionizable compounds (higher pHs may encourage hydrolysis of packing)
- Can adjust retention/selectivity by:
  - Choice of bonded phase (and of manufacturer)
  - Eluent solvent composition (generally increasing water content = increased retention)
  - With ionic compounds can adjust pH, ionic strength, add counter-ions, ion-pairing agents, etc.

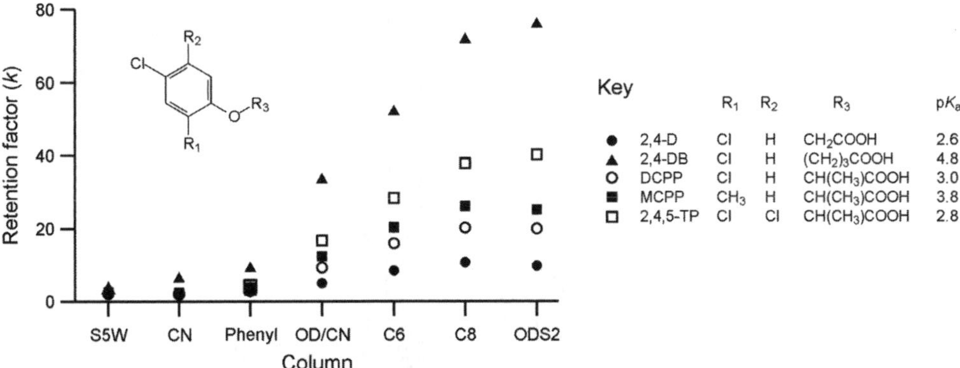

Figure 8.21 Effect of the nature of the column packing on the HPLC of some chlorophenoxy acid herbicides. Column: 100 × 4.6 mm i.d. 5 μm Waters Spherisorb [S5 MM2 (50 % ODS, 50 % cyanopropyl)]. Flow rate: 1.8 mL min^{-1}. Detection: UV, 280 nm. Temperature: about 20 °C. Injection: 25 μL aqueous solution of 2,4-D, DCPP, MCPP, 2,3,5-TP and 2,4-DB (50 mg L^{-1} each analyte). Eluent: aqueous potassium dihydrogen orthophosphate (50 mmol L^{-1}, pH 3.5):acetonitrile (75+25) [Key: 2,4-D = 2,4-dichlorophenoxyacetic acid, 2,4-DB = 4-(2,4-dichlorophenoxy)butyric acid, DCPP = 2-(2,4-dichlorophenoxy)propionic acid, MCPP = 2-(4-chloro-2-methylphenoxy)propionic acid, 2,4,5-TP = 2-(2,4,5-trichlorophenoxy)propionic acid)].

The analysis of some chlorophenoxy herbicides on modified and unmodified silica packings at constant eluent composition is illustrated in Figure 8.21. The elution sequence was the same in each case. S5 C_6, S5 C_8 and S5 ODS2 (C_{18}) columns gave essentially similar retention and peak shapes, whilst a 'mixed-mode' column (S5 OD/CN, 50 % C_{18}-, 50 % cyanopropyl-modified, a material designed for use with both acids and bases) gave more moderate retention and good peak shapes. Retention was much reduced on the phenyl- and nitrile-modified materials and was even less on unmodifed silica (S5W). Nevertheless, the phenyl column gave the best compromise between peak shape, analysis time and resolution (Figure 8.22).

In general, anticonvulsants, benzodiazepines, caffeine and other xanthines and very polar amphoteric analytes such as catecholamines, being acidic, neutral or weakly basic drugs, are analyzed on reversed-phase systems. In the case of the xanthines, addition of THF to the eluent has been advocated to ensure the separation of the caffeine metabolite paraxanthine and theophylline.

8.8.2 Basic drugs and quaternary ammonium compounds

Aspects of the HPLC of basic drugs are summarized in Box 8.7. Bonded-phase/aqueous methanol or acetonitrile eluent systems have been widely used in the HPLC of basic drugs. However, as well as buffer salts, various additives (pairing or counter ions such as alkylsulfonates, alkylamines or quaternary ammonium compounds) are often needed to give efficient performance. Moreover, in contrast to the behaviour of acidic/neutral compounds, the effect of altering the amount of water in the eluent on retention may not be predictable. This is probably because ionic interactions between the analyte and surface silanols are as strong, or stronger, than hydrophobic interactions with the bonded phase. Indeed, similar retention and elution sequences (and generally better peak shapes) to those obtained on a range of bonded phases can often be obtained on unmodified silica (Law, 1990).

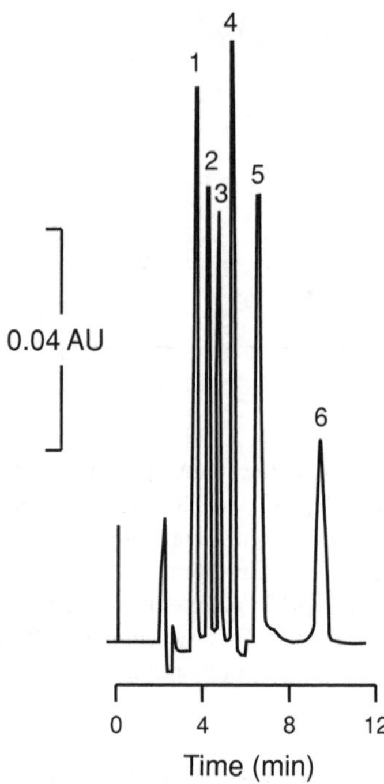

Figure 8.22 HPLC of some chlorophenoxy herbicides and ioxynil (4-hydroxy-3,5-di-iodobenzonitrile). Column: 250 mm Sphensorb S5 Phenyl. Eluent: potassium dihydrogen orthophosphate (50 mmol L^{-1}, pH 3.5):acetonitrile (3 + 1). Flow rate 1.8 mL min^{-1}. Detection: UV, 240 nm. Injection: 20 μL Tris buffer containing 0.10 g L^{-1} each compound (0.025 g L^{-1} ioxynil). Peaks 1: 2,4-D, 2: DCPP, 3: MCPP, 4: 2,4,5-TP, 5: ioxynil, 6: 2,4-DB.

The analysis of a test mixture of basic drugs on a range of packings using a typical reversed-phase eluent of constant composition is shown in Figure 8.23. The advantage of unmodifed silica as regards peak shape in the analysis of basic drugs is clear in this instance.

In practice, reversed-phase HPLC may permit the simultaneous analysis of basic drugs and neutral or even acidic drugs or metabolites (amides, glucuronides or sulfates). Factors which may

Box 8.7 HPLC of basic compounds

- Basic drugs (and quaternary ammonium compounds) often give poor peak shapes on bonded-phase silicas with aqueous methanol or acetonitrile eluents due to ionic interactions with surface silanols
- Can improve peak shapes by 'end-capping' or adding counter ions or other modifiers to the eluent
- Alternatively, can use unmodified silica in ion-exchange mode with methanol or aqueous methanol eluents of appropriate pH and ionic strength
- Microparticulate strong cation exchangers (SCX) can be used in the same way as unmodified silica and give excellent results with many compounds

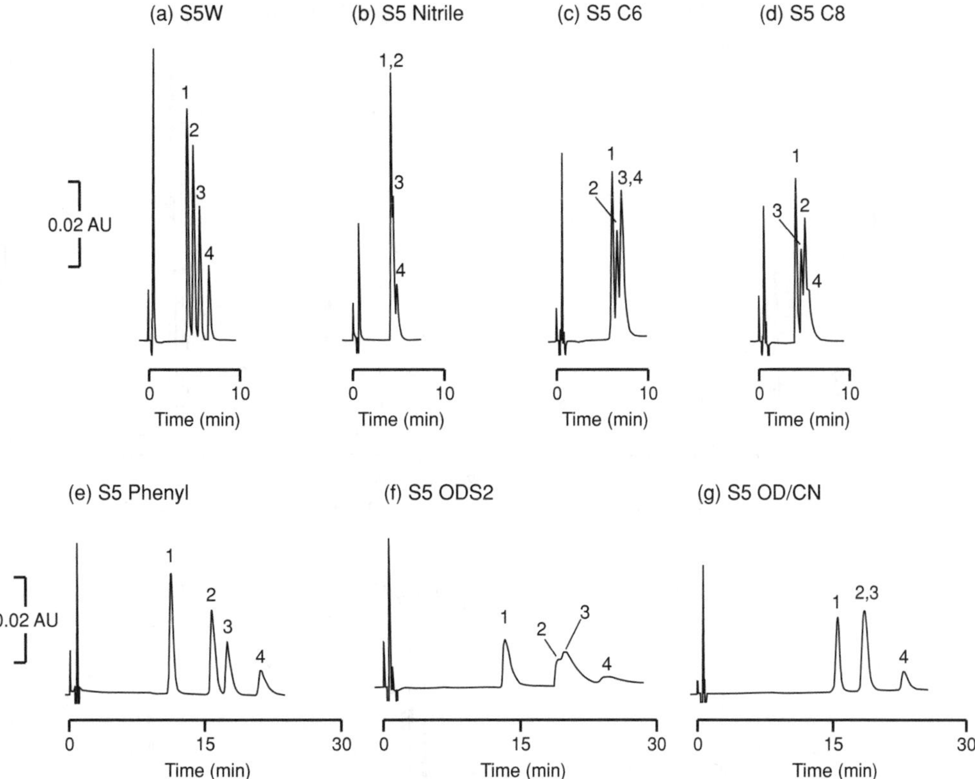

Figure 8.23 Effect of the column packing on the HPLC of some basic drugs using a reverse-phase eluent. Column: 100 × 4.6 mm i.d. 5 μm Waters Spherisorb (S5 OD/CN: 50 % ODS, 50 % cyanopropyl). Flow rate: 2.0 mL min^{-1}. Detection: UV, 215 nm. Temperature: about 22 °C. Injection: 20 μL of solution of nortriptyline (1), amitriptyline (2), imipramine (3) and methdilazine (4) (10 mg L^{-1} each compound in methanol). Eluent: aqueous potassium dihydrogen orthophosphate (10 mmol L^{-1}, pH 6.0):methanol: acetonitrile (30 + 10 + 60).

improve peak symmetry for bases are use of an eluent pH between 2.5 and 3.5, generally higher buffer concentrations, use of potassium instead of sodium salts and addition of amine modifiers such as triethylamine (30 to 50 mmol L^{-1}) or dimethyloctylamine (5 to 10 mmol L^{-1}). However, such additives greatly restrict the use of electrochemical oxidation detection. Many manufacturers now produce special 'base-deactivated' bonded-phase silicas designed to minimize interactions with surface silanols, and thus permit more predictable reversed-phase behaviour and give better peak shapes.

8.8.2.1 Nonaqueous ionic eluent systems

It has long been known that efficient performance can be obtained for many basic drugs on unmodified silica using methanol containing perchloric acid (0.01 or 0.02 % v/v, about 1 or 2 mmol L^{-1}) as eluent. Perchloric acid is a useful ionic modifier if a strongly acidic eluent is required. Methanolic solutions of perchlorates have UV cut-offs of about 200 nm. An ammonium perchlorate modified

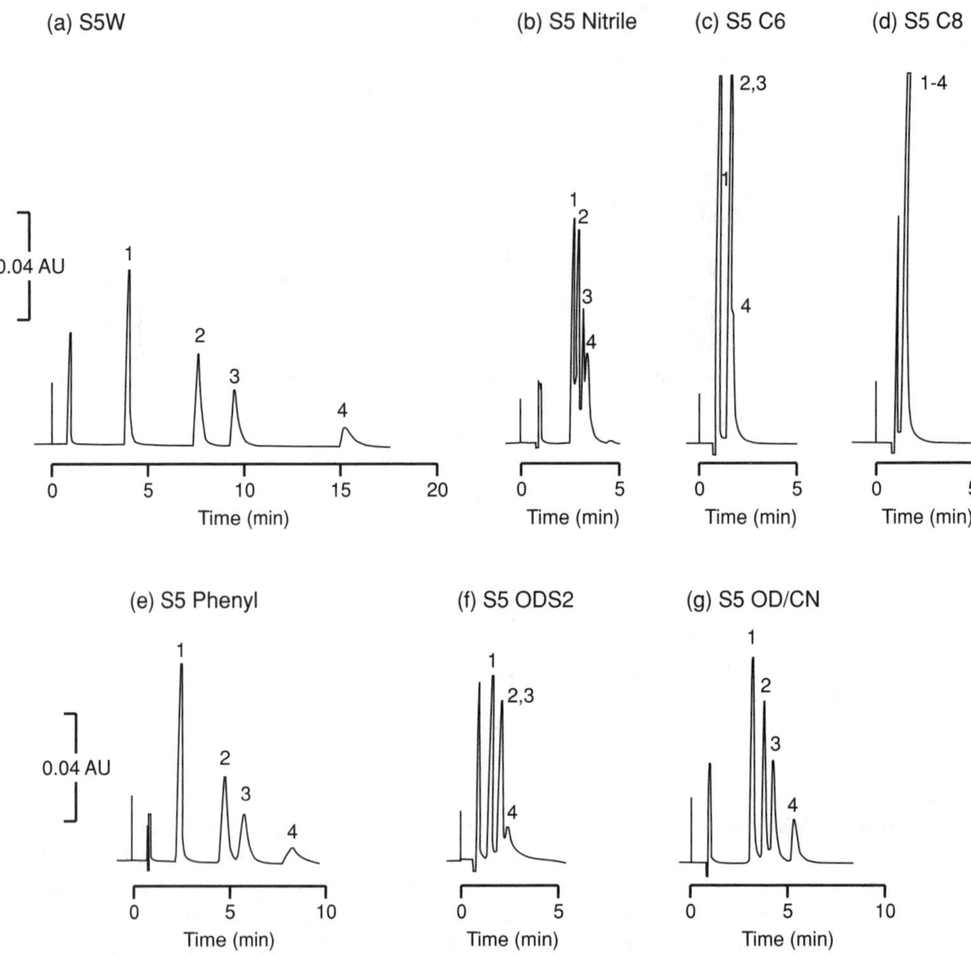

Figure 8.24 Effect of the nature of the column packing on the chromatography of some basic drugs using a nonaqueous ionic eluent. Eluent: methanolic ammonium perchlorate (5 mmol L^{-1}, pH* 6.7). Other chromatographic conditions and injection as for Figure 8.23.

eluent (10 mmol L^{-1}, pH* 6.7) balances retention, peak shape and electrochemical response for many analytes (Jane *et al.*, 1985).

The retention of protonated bases and quaternary ammonium compounds on unmodified silica using methanol or indeed aqueous methanol eluents is probably mediated largely by cation exchange with surface silanols (Figure 8.24). Eluent pH and ionic strength are thus major influences on retention (Flanagan and Jane, 1985). The peak shape of some analytes, notably alkaloids such as morphine, quinine and strychnine, on nonaqueous ionic eluent systems is also influenced by eluent pH, higher pH values usually giving better peaks. Silica column/nonaqueous ionic eluent systems are useful for the analysis of basic drugs in biological extracts as only protonated bases and quaternary ammonium compounds are retained. In addition, *N*-dealkylated, phenolic and other metabolites such as sulfoxides are usually resolved (Cashman and Yang, 1990).

Small changes in eluent solvent composition rarely alter retention even if water is added up to 20 % (v/v); adding more water enhances retention markedly for most analytes, but selectivity remains largely unaltered (Flanagan and Jane, 1985). However, penbutolol and 4-hydroxypenbutolol can be resolved using iso-octane:MTBE:methanol (37.5 + 37.5 + 25) at pH 6.7. Direct analysis of MTBE plasma extracts with recirculation of the eluent does not alter retention unless recirculation is prolonged (several hundred injections) when some decrease in retention may occur.

One problem with unmodified and indeed many modified silicas is that some compounds are poorly retained even if the eluent ionic strength is very low. Microparticulate strong cation exchangers, however, give good retention and peak shape for many basic drugs (Croes *et al.,* 1995; Figure 8.25). Practical advantages are that methanol is much less viscous than water/methanol

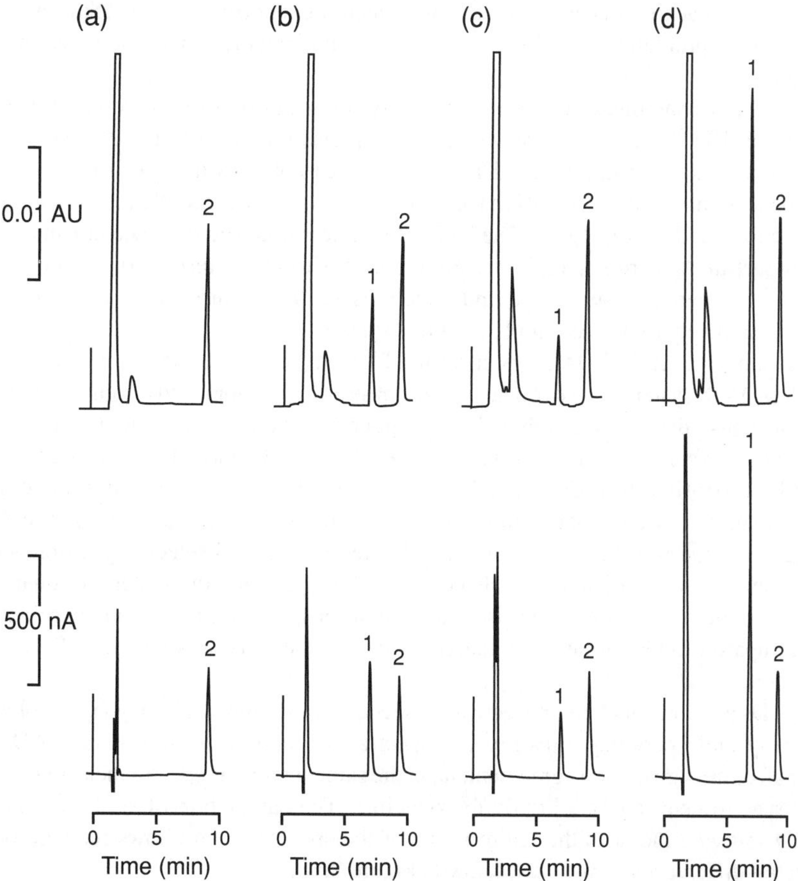

Figure 8.25 HPLC of lidocaine using Waters Spherisorb S5 SCX. Column: 250 × 5 mm i.d. Eluent: 20 mmol L^{-1} ammonium perchlorate in methanol (pH* 6.7). Flow rate: 2.0 mL min^{-1}. Detection: UV, 215 nm and ED, + 1.1 V against Ag/AgCl. Extraction: 200 μL sample + 50 μL aqueous Tris buffer (4 mol L^{-1}, pH 11) + 50 μL aqueous bupivacaine (internal standard, 10 mg L^{-1}) + 200 μl MTBE. Vortex-mix and centrifuge. Injection: 100 μL extract. Samples: (a) Newborn calf serum. (b) Standard in calf serum (1.0 mg L^{-1} lidocaine). (c) Equine serum (Gibco, Paisley, UK) found to contain 0.7 mg L^{-1} lidocaine. (d) Expired blood-bank human plasma found to contain 3.3 mg L^{-1} lidocaine. Peaks: 1 = lidocaine, 2 = bupivacaine.

mixtures and thus peak efficiency is high, and column back-pressures are relatively low. In addition, using methanol as the eluent solvent minimizes the risks of silica dissolution and of air-bubble formation; eluent degassing is not normally needed. A further advantage of nonaqueous ionic eluent systems is that relatively large volume solvent extracts can be injected directly in a 'noneluting' solvent with no loss of efficiency.

8.8.3 Systematic toxicological analysis

Normally toxicology screening is performed using immunoassays and GC-MS. However TLC and capillary GC with specific detectors are still used in some toxicology laboratories. Problems in using HPLC for unknown screening are: (i) variation in retention data due to, for example, variation in packing materials or lack of temperature control, (ii) the lack of a universal detector and (iii) the problems normally associated with gradient elution as compared to GC temperature programming.

Silica column/nonaqueous ionic eluent systems can be used to screen for basic drugs; UV (254 nm):ED (+1.2 V against Ag/AgCl) response ratios give an extra parameter to aid peak assignment (Jane *et al.*, 1985). Unmodified silica columns have also been used with aqueous methanol or acetonitrile eluents in basic drug analyses, most recently together with on-line sample extraction (Kelly *et al.*, 1989). The REMEDI system used standardized eluents and columns with UV spectral data from a rapid-scanning detector to characterize drugs and some metabolites. While commercially successful and widely used, the recent upsurge in affordable LC-MS and HPLC-DAD equipment is supplanting this system.

A number of HPLC-DAD systems for STA have been described (Lai *et al.*, 1997). A gradient HPLC-DAD system based on a mixed-phase (C_{18}/cyanopropyl) column with retention data and response data for more than 250 compounds including basic, acidic and neutral drugs and metabolites, some of which were not amenable to GC and TLC, has been described (Elliott and Hale, 1997, 1998). Compounds were identified using both retention indices (calculated by interpolation between a series of reference drug markers, separate scales for acidic and for basic drugs) and spectral data. The column had potentially useful selectivity differences compared to C_{18} columns. The database was also useful in helping adapt the system for quantitative analyses of drugs and metabolites under isocratic conditions, exploiting the inherent flexibility of HPLC as compared to GC in targeted analyses. Sample capacity was also an advantage in analyzing powders.

DADs are now much improved with respect to wavelength accuracy and resolution, sensitivity, linearity and operating software. UV spectra measured with up-to-date DADs from different manufacturers give good agreement, have the same quality as spectra measured by a conventional UV spectrometer and are highly reproducible. The calculation of similarity parameters by the DAD software includes the entire range of the spectrum and allows recognition of very small spectral differences (Pragst *et al.*, 2001; Figure 8.26).

In a systematic study of 2682 toxicologically relevant substances, retention times relative to an internal standard and spectra in the range 195–380 nm were recorded using a 5 μm Lichrospher-RP8ec, 250 × 4 mm i.d. column with an acetonitrile:aqueous phosphate buffer (0.1 mol L^{-1}, pH 2.3) mixture in two different proportions – Eluent A: 37 + 63 [internal standard: 5-(4-methylphenyl)-5-phenylhydantoin] and Eluent B: 63 + 37 (internal standard: 4-phenylbenzophenone). There were 1650 different chromophores or chromophore combinations.

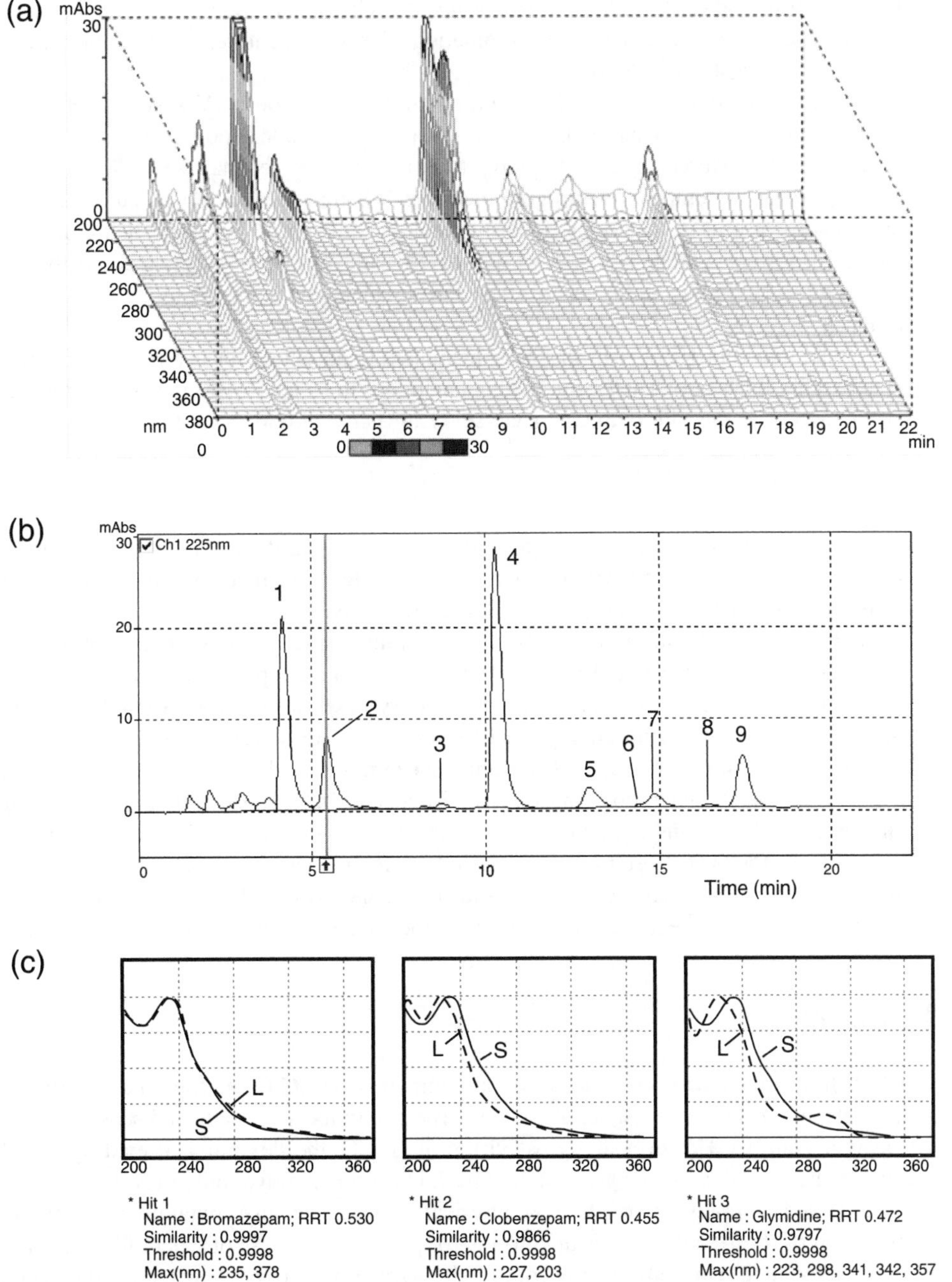

Figure 8.26 HPLC-DAD of a dichloromethane extract (pH 9) of a blood sample from a poisoned patient. (a) 3-D chromatogram. (b) Chromatogram (225 nm). Compounds (mg L^{-1}) identified by UV spectrum and retention time: Peaks: 1 = tramadol (1.2), 2 = bromazepam (0.15), 3 = oxazepam (0.01), 4 = nordazepam (0.51), 5 = nortriptyline (0.07), 6 = nortrimipramine (0.01), 7 = amitriptyline (0.08), 8 = trimipramine (0.02) and 9 = diazepam (0.12). (c) Result of the library search of peak 2 (wavelength nm). A similarity index above 0.9990 is a very strong evidence for identical spectra. S = sample spectrum, L = library spectra.

In all, 1619 substances (60.4 %) were identified unambiguously by spectrum alone; the identification rate increased to 84.2 % by the combination of spectrum and relative retention time (Herzler *et al.*, 2003; Pragst *et al.*, 2004).

Therefore, modern HPLC-DAD in combination with a good UV spectral database and defined sample preparation and HPLC retention data is a valuable supplementary technique in STA (Maurer, 2000). Moreover, sample preparation may be simpler than with GC, although the resolving power of temperature programmed GC-MS may be lacking (Maurer, 2004). HPLC-DAD is also useful in metabolite identification, because in many cases such compounds have the same or very similar UV spectra compared with the parent substance and their retention times on reversed-phase columns are altered in a manner characteristic of the expected metabolic product.

Finally, although HPLC-MS has been evaluated to an extent for use in STA (Saint-Marcoux *et al.*, 2003), the lack of a detection system analogous to EI in GC-MS means that for the present, HPLC-MS is best used for targeted analyses (Dams *et al.*, 2000; Maurer, 2004).

8.8.4 Chiral analyses

Although chiral methodology is of primary interest in drug development, the possibility of inversion *in vivo* is real. Ibuprofen is normally marketed as a racemate, but a proportion of inactive R-(−)-ibuprofen undergoes inversion to the active enantiomer *in vivo* via the production of an unstable metabolite (Evans, 2001). Citalopram, a selective serotonin reuptake inhibitor, is also prescribed as a racemate. Enantioselective HPLC of citalopram and *N*-desmethylcitalopram (norcitalopram) in femoral blood from 53 autopsy cases revealed an S:R ratio for citalopram of 0.67 ± 0.25 and for norcitalopram of 0.68 ± 0.20 (Holmgren *et al.*, 2004). The S:R ratios increased with increasing citalopram concentration and were also associated with high parent drug:metabolite ratios. In contrast, Rochat *et al.*, (1995) found the ratios of S-(+)-citalopram and S-(+)-norcitalopram to their inactive enantiomers were only some 0.35 ± 0.05. A further factor is that escitalopram (S-citalopram) is available in its own right and is used at doses 50 % or so of those of citalopram itself, the doses, of course, assuming little selective inversion to the active enantiomer from the inactive enantiomer when the racemate was given.

8.9 Summary

HPLC has supplanted UV/visible spectrophotometry and GC for measurement of many specific analytes and metabolites, especially when harnessed to the selectivity and sensitivity offered by MS (Chapter 10). However, the great virtue of HPLC, the flexibility offered by the ability to modify the eluent, is also its greatest drawback in routine use. Not only may the column packing be dissolved, compressed, or otherwise physically altered, but also contaminants may accumulate at the top of (or in) columns, leading to loss of performance. Use of eluent filtration and of pre- or guard-columns can help prolong column life. Use of modern packing technology can facilitate on-line analysis of plasma for targeted analytes, although HPLC still has a more limited role in STA than GC (Box 8.8). Basic limitations are that there is no sensitive 'universal' HPLC detector analogous to the FID in GC and secondly that there is no parallel in HPLC-MS to EI in GC-MS.

Box 8.8 Summary of the use of HPLC in analytical toxicology

- Must consider selectivity from sample work up (LLE often best) and detector as well as the column
- 5 μm packings often adequate if used with appropriate column/eluent combination
- Routinely expect 3000–4000 theoretical plates per 125 mm column at 1–2 mL min^{-1} eluent flow with clinical sample extracts
- 3 μm packings or microbore/capillary systems can give higher efficiencies, but may also give higher back-pressures, longer analysis times, reduced solute capacity and require modifications to injectors/detectors
- Use of methanolic or largely methanolic eluents in the HPLC of basic drugs (SCX-modified silica packings) gives good peak shapes and facilitates direct analysis of solvent extracts
- Use of DAD can be helpful in STA, but in general temperature-programmed GC preferred

References

Abbott, R.W., Townshend, A. and Gill, R. (1987) Determination of morphine in body fluids by high-performance liquid chromatography with chemiluminescence detection. *Analyst*, **112**, 397–406.

Armstrong, D.W., Tang, Y., Chen, S. Zhou Y., Bagwill C. and Chen J-R. (1994) Macrocyclic antibiotics as a new class of chiral selectors for liquid chromatography. *Anal Chem*, **66**, 1473–84.

Aradottir, S. and Olsson, B.L. (2005) Methodological modifications on quantification of phosphatidylethanol in blood from humans abusing alcohol, using high-performance liquid chromatography and evaporative light scattering detection. *BMC Biochem*, **6**, 18 (electronic journal: http://www.pubmedcentral.nih.gov/articlerender.fcgi?artid=1249557, accessed 17 July 2006).

Binder, S.R., Regalia, M., Biaggi-McEachern, M. and Mazhar, M. (1989) Automated liquid chromatographic analysis of drugs in urine by on-line sample cleanup and isocratic multi-column separation. *J Chromatogr*, **473**, 325–41.

Buckley, D.G. and Whelpton, R. (1992) Direct injection of chiral isocyanates, not marketed, onto aminopropyl columns to render them chiral, in *Bioanalytical Approaches for Drugs, Including Anti-Asthmatics and Metabolites*, Vol. 22 (eds E. Reid and I.D. Wilson), Royal Society of Chemistry, Cambridge, pp. 317–320.

Bugey, A. and Staub, C. (2004) Rapid analysis of benzodiazepines in whole blood by high-performance liquid chromatography: use of a monolithic column. *J Pharm Biomed Anal*, **35**, 555–62.

Cashman, J.R. and Yang Z.-C. (1990) Analysis of amine metabolites by high-performance liquid chromatography on silica gel with a non-aqueous ionic eluent. *J Chromatogr*, **532**, 405–10.

Chen, J., Korfmacher, W.A. and Hsieh, Y. (2005) Chiral liquid chromatography-tandem mass spectrometric methods for stereoisomeric pharmaceutical determinations. *J Chromatogr B*, **820**, 1–8.

Cope, M.J. and Davidson, I.E. (1987) Use of macroporous polymeric high-performance liquid chromatographic columns in pharmaceutical analysis. *Analyst*, **112**, 417–21.

Cook, S.E. and Pinkerton, T.C. (1986) Characterization of internal surface reversed-phase silica supports for liquid chromatography. *J Chromatogr*, **368**, 233–48.

Croes, K., McCarthy, P.T. and Flanagan, R.J. (1995) HPLC of basic drugs and quaternary ammonium compounds on microparticulate strong cation-exchange materials using methanolic or aqueous methanol eluents containing an ionic modifier. *J Chromatogr A*, **693**, 289–306.

Daldrup, T., Michalke, P. and Szathmary, S. (1985) HPLC in forensic chemistry, in *Practice of High Performance Liquid Chromatography* (ed. H. Engelhardt), Springer, New York, pp. 241–285.

Dams, R., Lambert, W.E., Clauwaert, K.M. and De Leenheer, A.P. (2000) Comparison of phenyl-type columns in the development of a fast liquid chromatographic system for eighteen opiates commonly found in forensic toxicology. *J Chromatogr A*, **896**, 311–9.

Dolphin, J. (1989) Enantiomer separations. *Lab Practice*, **38**, 71–3.

Elliott, S.P. and Hale, K.A. (1998) Applications of an HPLC-DAD drug-screening system based on retention indices and UV spectra. *J Anal Toxicol*, **22**, 279–89.

Elliott, S.P. and Hale, K.A. (1997) Development of a high-performance liquid chromatography retention index scale for toxicological drug screening. *J Chromatogr B*, **694**, 99–114.

Evans, A.M. (2001) Comparative pharmacology of *S*(+)-ibuprofen and (*RS*)-ibuprofen. *Clin Rheumatol*, **20** (Suppl 1), S9–14.

Fitos, I., Visy, J., Simonyi, M. and Hermansson, J. (1992) Chiral high-performance liquid chromatographic separations of vinca alkaloid analogues on alpha 1-acid glycoprotein and human serum albumin columns. *J Chromatogr*, **609**, 163–71.

Flanagan, R.J. and Jane, I. (1985) High-performance liquid chromatographic analysis of basic drugs on silica columns using non-aqueous ionic eluents. I. Factors influencing retention, peak shape and detector response. *J Chromatogr*, **323**, 173–89.

Flanagan, R.J., Perrett, D. and Whelpton, R. (2005) *Electrochemical Detection in the HPLC of Drugs and Poisons*, Royal Society of Chemistry, Cambridge.

Fries, H.E., Evans, C.A. and Ward, K.W. (2005) Evaluation of evaporative light-scattering detection for metabolite quantification without authentic analytical standards or radiolabel. *J Chromatogr B*, **819**, 339–44.

Gamache, P.H., McCarthy, R.S., Freeto, S.M. Asa D.J., Woodcock M.J., Laws K., and Cole R.O. (2005) HPLC analysis of nonvolatile analytes using charged aerosol detection. *LCGC North America*, February.

Gómez-Taylor, B., Palomeque, M., Garcia Mateo, J.V. and Martinez Calatayud, J. (2006) Photoinduced chemiluminescence of pharmaceuticals. *J Pharm Biomed Anal*, **41**, 347–57.

Gong, Z., Zhang, Y., Zhang, H. and Cheng, J. (1999) Capillary electrophoresis separation and permanganate chemiluminescence on-line detection of some alkaloids with β-cyclodextrin as an additive. *J Chromatogr A*, **855**, 329–35.

Hasumi, H., Nishikawa, T. and Ohtani, H. (1994) Effect of temperature on molecular structure of cyclosporin A. *Biochem Mol Biol Int*, **34**, 505–11.

Hage, D.S. (1999) Affinity chromatography: a review of clinical applications. *Clin Chem*, **45**, 593–615.

Hagestam, I.H. and Pinkerton, T.C. (1985) Internal surface reversed phase silica supports for liquid chromatography. *Anal Chem*, **57**, 1757–63.

Haque, A. and Stewart, J.T. (1999) Direct injection HPLC analysis of some non-steroidal anti-inflammatory drugs on restricted access media columns. *Biomed Chromatogr*, **13**, 51–6.

Herzler, M., Herre, S. and Pragst, F. (2003) Selectivity of substance identification by HPLC-DAD in toxicological analysis using a UV spectra library of 2682 compounds. *J Anal Toxicol*, **27**, 233–42.

Holmgren, P., Carlsson, B., Zackrisson, A.L., Lindblom, B., Dahl, M.L., Scordo, M.G., Druid, H. and Ahlner, J. (2004) Enantioselective analysis of citalopram and its metabolites in postmortem blood and genotyping for CYD2D6 and CYP2C19. *J Anal Toxicol*, **28**, 94–104.

Imai, K. (1987) Derivatization in liquid chromatography, in *Advances in Chromatography* (eds J.C. Giddings, E. Grushka and P.R. Brown), Dekker, New York, pp. 215–245.

Jane, I., McKinnon, A. and Flanagan, R.J. (1985) High-performance liquid chromatographic analysis of basic drugs on silica columns using nonaqueous ionic eluents. II. Application of UV, fluorescence and electrochemical oxidation detection. *J Chromatogr*, **323**, 191–225.

Kato, M., Fukushima, T., Shimba, N., Shimada, I., Kawakami, Y. and Imai, K. (2001) A study of chiral recognition for NBD-derivatives on a Pirkle-type chiral stationary phase. *Biomed Chromatogr*, **15**, 227–34.

Kelly, M.T., Smyth, M.R. and Dadgar, D. (1989) Analysis of basic drugs by high-performance liquid chromatography using on-line solid-phase extraction and an unmodified silica column. *Analyst*, **114**, 1377–80.

Kline, W.F. and Matuszewski, B.K. (1992) Improved determination of the bisphosphonate alendronate in human plasma and urine by automated precolumn derivatization and high-performance liquid chromatography with fluorescence and electrochemical detection. *J Chromatogr*, **583**, 183–93.

Knox, J.H., Kaur, B. and Millward, G.R. (1986) Structure and performance of porous graphitic carbon in liquid chromatography. *J Chromatogr*, **352**, 3–25.

Kraml, C.M., Zhou, D., Byrne, N. and McConnell, O. (2005) Enhanced chromatographic resolution of amine enantiomers as carbobenzyloxy derivatives in high-performance liquid chromatography and supercritical fluid chromatography. *J Chromatogr A*, **1100**, 108–15.

Kurganov, A. (2001) Chiral chromatographic separations based on ligand exchange. *J Chromatogr A*, **906**, 51–71.

Lai, C.K., Lee, T., Au, K.M. and Chan, A.Y. (1997) Uniform solid-phase extraction procedure for toxicological drug screening in serum and urine by HPLC with photodiode-array detection. *Clin Chem*, **43**, 312–25.

Laks, S., Pelander, A., Vuori, E., Ali-Tolppa, E., Sippola, E. and Ojanpera, I. (2004) Analysis of street drugs in seized material without primary reference standards. *Anal Chem*, **76**, 7375–9.

Lammerhofer, M., Gyllenhaal, O. and Lindner, W. (2004) HPLC enantiomer separation of a chiral 1,4-dihydropyridine monocarboxylic acid. *J Pharm Biomed Anal*, **35**, 259–66.

Law, B. (1990) Use of silica with reversed-phase type eluents for the analysis of basic drugs and metabolites. *Trends Anal Chem*, **9**, 31–6.

Li, F., Zhang, C., Guo, X. and Feng, W. (2003) Chemiluminescence detection in HPLC and CE for pharmaceutical and biomedical analysis. *Biomed Chromatogr*, **17**, 96–105.

Lihl, S., Rehorek, A. and Petz, M. (1996) High-performance liquid chromatographic determination of penicillins by means of automated solid-phase extraction and photochemical degradation with electrochemical detection. *J Chromatogr A*, **729**, 229–35.

Lim, C.K. (1991) Porous graphitic carbon in biomedical applications, in *Advances in Chromatography* (eds J.C. Giddings, E. Grushka and P.R. Brown), Dekker, New York, **32:** pp. 1–19.

Lingeman, H. and Underberg, W.J.M. (eds) (1990) *Detection Orientated Derivatization Techniques in Liquid Chromatography*, Dekker, New York.

Lunte, S.M., Mohabbat, T., Wong, O.S. and Kuwana, T. (1989) Determination of desmosine, isodesmosine, and other amino acids by liquid chromatography with electrochemical detection following precolumn derivatization with naphthalenedialdehyde/cyanide. *Anal Biochem*, **178**, 202–7.

Lurie, I.S. (2005) High-performance liquid chromatography of seized drugs at elevated pressure with 1.7 μm hybrid C18 stationary phase columns. *J Chromatogr A*, **1100**, 168–75.

Mascher, H. and Kikuta, C. (1990) Determination of amoxicillin in plasma by high-performance liquid chromatography with fluorescence detection after online oxidation. *J Chromatogr*, **506**, 417–21.

Maurer, H.H. (2004) Position of chromatographic techniques in screening for detection of drugs or poisons in clinical and forensic toxicology and/or doping control. *Clin Chem Lab Med*, **42**, 1310–24.

Maurer, H.H. (2000) Screening procedures for simultaneous detection of several drug classes used for high throughput toxicological analyses and doping control. A review. *Comb Chem High Throughput Screen*, **3**, 467–80.

McDonald, P.D. and Bidlingmeyer, B.A. (1987) Strategies for successful preparative liquid chromatography, in *Preparative Liquid Chromatography* (ed. B.A. Bidlingmeyer), Elsevier, Amsterdam, pp. 1–103.

Naish, P.J., Lynch, R.J. and Blaffert, T. (1989) A study of the practical limitations of principal components analysis and the labelling of unresolved HPLC peaks. *Chromatographia*, **27**, 343–58.

Naish-Chamberlain, P.J. and Lynch, R.J. (1990) Evaluation of a complete HPLC solvent optimization system involving piecewise quadratic modelling. *Chromatographia*, **29**, 79–89.

Nakanishi, K. and Soga, N. (1991) Phase separation in gelling silica-organic polymer solution: systems containing poly(sodium styrenesulfonate). *J Am Ceram Soc*, **74**, 2518–30.

Pirkle, W.H. and Burke, J.A. (1991) Chiral stationary phase designed for beta-blockers. *J Chromatogr*, **557**, 173–85.

Pirkle, W.H. and House, D.W. (1979) Chiral high pressure liquid chromatographic stationary phases 1. Separation of the enantiomers of sulfoxides, amines, amino acids, alcohols, hydroxy acids, lactones, and mercaptans. *J Org Chem*, **44**, 1957–60.

Pragst, F., Herzler, M. and Erxleben B.-T. (2004) Systematic toxicological analysis by high-performance liquid chromatography with diode array detection (HPLC-DAD). *Clin Chem Lab Med*, **42**, 1325–40.

Pragst, F., Herzler, M., Herre, S., Erxleben, B-T. and Rothe, M. (2001) *UV-Spectra of Toxic Compounds. Database of Photodiode Array UV Spectra of Illegal and Therapeutic Drugs, Pesticides, Ecotoxic Substances and Other Poisons*. Handbook and CD. Dieter Helm, Heppenheim.

Rainbow, S.J., Dawson, C.M. and Tickner, T.R. (1990) Direct serum injection high-performance liquid chromatographic method for the simultaneous determination of phenobarbital, carbamazepine and phenytoin. *J Chromatogr*, **527**, 389–96.

Rainbow, S.J., Dawson, C.M. and Tickner, T.R. (1989) Non-extraction HPLC method for the simultaneous measurement of theophylline and caffeine in human serum. *Ann Clin Biochem*, **26**, 527–32.

Rochat, B., Amey, M. and Baumann, P. (1995) Analysis of enantiomers of citalopram and its demethylated metabolites in plasma of depressive patients using chiral reverse-phase liquid chromatography. *Ther Drug Monit*, **17**, 273–9.

Saint-Marcoux, F., Lachatre, G. and Marquet, P. (2003) Evaluation of an improved general unknown screening procedure using liquid chromatography-electrospray-mass spectrometry by comparison with gas chromatography and high-performance liquid-chromatography—diode array detection. *J Am Soc Mass Spectrom*, **14**, 14–22.

Samanidou, V.F., Ioannou, A.S. and Papadoyannis, I.N. (2004) The use of a monolithic column to improve the simultaneous determination of four cephalosporin antibiotics in pharmaceuticals and body fluids by HPLC after solid phase extraction—a comparison with a conventional reversed-phase silica-based column. *J Chromatogr B*, **809**, 175–82.

Smith, J.H. and McNair, H.M. (2003) Fast HPLC with a silica-based monolithic ODS column. *J Chromatogr Sci*, **41**, 209–14.

Snyder, L.R. and Kirkland, J.J. (1979) *Introduction to Modern Liquid Chromatography*, 2nd edn, Wiley, New York, chapters 2 and 6.

Taylor, E.W., Jia, W., Bush, M. and Dollinger, G.D. (2002) Accelerating the drug optimization process: identification, structure elucidation, and quantification of in vivo metabolites using stable isotopes with LC/MS^n and the chemiluminescent nitrogen detector. *Anal Chem*, **74**, 3232–8.

Tindall, G.W. (2002a) Mobile-phase buffers, Part I. The interpretation of pH in partially aqueous mobile phases. *LCGC North America*, **20**, 1028–32.

Tindall, G.W. (2002b) Mobile-phase buffers, Part II. Buffer selection and capacity. *LCGC North America*, **20**, 1114–8.

Tindall, G.W. (2003) Mobile-phase buffers, Part III. Preparation of buffers. *LCGC North America*, **21**, 28–32.

Toyo'oka, T. (2002) Resolution of chiral drugs by liquid chromatography based upon diastereomer formation with chiral derivatization reagents. *J Biochem Biophys Methods*, **54**, 25–56.

Uzunov, D.P., Zivkovich, I., Pirkle, W.H., Costa, E. and Guidotti, A. (1995) Enantiomeric resolution with a new chiral stationary phase of 7-chloro-3-methyl-3,4-dihydro-2*H*-1,2,4-benzothiadiazine *S*,*S*-dioxide, a cognition-enhancing benzothiadiazine derivative. *J Pharm Sci*, **84**, 937–42.

Wheals, B.B. (1987) Ion chromatography of inorganic anions on a dynamically modified polystyrene-divinylbenzene packing material and its application to anion screening by sequential ultraviolet absorbance and electrochemical detection. *J Chromatogr*, **402**, 115–26.

Whelpton, R. and Buckley, D.G. (1992a) Resolution of compounds on urea-bonded chiral HPLC columns—effect of derivatization, in *Bioanalytical Approaches for Drugs, Including Anti-Asthmatics and Metabolites*, Vol. 22 (eds E. Reid and I.D. Wilson), Royal Society of Chemistry, Cambridge, pp. 311–315.

Whelpton, R. and Buckley, D. (1992b) Comparison of Chiral 1 and Chiral 2 high-performance liquid chromatographic columns for the resolution of some drug enantiomers. *Anal Proc*, **29**, 249–51.

White, P.C. (1988) Comparison of absorbance ratios and peak purity parameters for the discrimination of solutes using high-performance liquid chromatography with multi-wavelength detection. *Analyst*, **113**, 1625–9.

White, P.C. (1980) Use of dual-wavelength UV detection in high-performance liquid chromatography for the identification of barbiturates. *J Chromatogr*, **200**, 271–6.

Wyndham, K.D., O'Gara, J.E., Walter, T.H. Glose KH, Lawrence NL, Alden BA, Izzo GS, Hudalla CJ, Iraneta PC. (2003) Characterization and evaluation of C_{18} HPLC stationary phases based on ethyl-bridged hybrid organic/inorganic particles. *Anal Chem*, **75**, 6781–8.

9 Capillary Electrophoretic Techniques

9.1 Introduction

Of the many electrophoretic techniques available, the most widely used in analytical toxicology and related areas are capillary electrophoresis (CE, also called capillary zone electrophoresis—CZE), micellar electrokinetic (capillary) chromatography (MEKC or MECC) and capillary electro(kinetic) chromatography (CEC). CE, sometimes referred to as free solution capillary electrophoresis (FSCE) to distinguish it from MEKC, refers to the separation of ions in free solution based on different velocities in an electric field. Selectivity is primarily due to differences in mass and charge – small charged molecules migrate more quickly than larger ones carrying the same charge. The technique depends heavily on the availability of high-quality fused silica capillary tubing and extremely sensitive detectors.

In CE a fused-silica capillary, usually 25–75 μm i.d. (i.e. narrower than normally used in capillary GC – Section 7.3.2) and coated with polyimide to provide strength, is placed between two reservoirs of a buffer solution, that is referred to as the background electrolyte (BGE). In nonaqueous CE (NACE), the aqueous buffer is replaced by a background electrolyte in (a mixture of) organic solvents. The capillary is filled with buffer and platinum electrodes are placed in both reservoirs so that a potential difference of around 5–30 kV can be applied across the capillary, in turn creating a current of 150–300 μA. With spectrophotometric detection, the peaks are detected on-column, a small area of polyimide being removed to provide a window for the light path to the detector (Figure 9.1). The preparation and care of a capillary is summarized in Box 9.1.

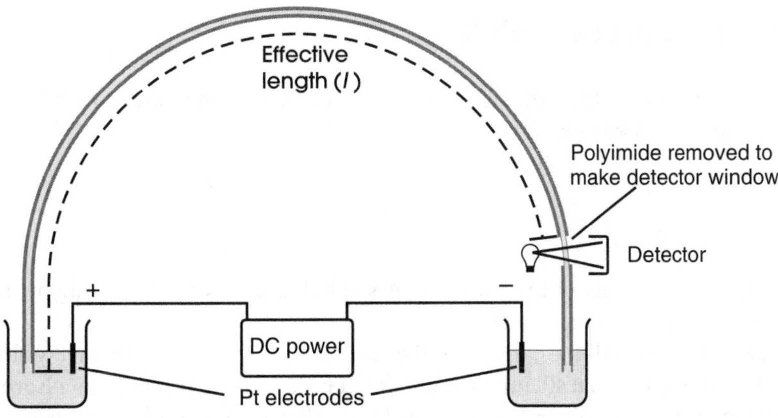

Figure 9.1 Schematic diagram of the arrangement for capillary electrophoresis.

Box 9.1 Preparation, conditioning and storage of capillaries for electrophoresis

Cutting to length
- Cut to length with ceramic cutter – pull lengthways to break capillary
- Burn off 1 mm of polyimide coating at each end[a]
- Check for clean cut with aid of a microscope

Dectector window
- Measure appropriate distance to detector
- Burn off small area of coating[a]
- Carefully wipe off burnt coating with methanol soaked tissue
- Inspect with microscope to ensure clear light path

Conditioning
- Flush with aqueous sodium hydroxide (1 mol L^{-1}, 30 min)
- Flush with deionized water (30 min)
- Equilibrate with BGE (20 min)

Storage
- Flush with sodium hydroxide and water as above
- Rinse with methanol
- Dry with a stream of nitrogen
- Store in a safe place – in a cartridge if possible

[a]Use a small flame such as a match held under a piece of aluminium sheet with a small hole drilled in it, or an electrically heated wire

The length of the capillary from the anode reservoir to the detector window is known as the effective length (l) and is usually a few centimetres shorter than the total length (L), which is typically 15–100 cm. The applied field, E (V cm^{-1}), is the applied potential difference (V) divided by L. The same basic apparatus and detectors are used for MEKC and CEC.

9.2 Electrophoretic mobility

Ions are separated on the basis of size and charge. The electrophoretic mobility (μ_e) of an ion of radius r carrying a charge, q, is:

$$\mu_e = \frac{q}{6\pi\eta r} \tag{9.1}$$

where η is the viscosity of the buffer. Thus small, highly-charged ions migrate faster than larger ions carrying less charge.

Under alkaline conditions, the walls of a fused silica capillary are negatively charged because of ionization of the residual silanols and possibly because of the presence of adsorbed ions. Cations migrate to the wall, forming a layer of ions that is relatively immobile (Stern layer). There is a more diffuse layer of solvated ions that extends into the bulk liquid (Goüy layer). The potential between the Stern layer and the interface with the diffused double layer, is known as the zeta potential, ζ. It is possible to calculate the thickness of the double layer which, depending on the conditions, is of the order of 1–10 nm (Knox and Grant, 1987). When an electric field is applied, hydrated cations in the diffuse layer are attracted towards the cathode (negative potential) resulting in shearing in the region of the double layer. However the bulk of the liquid in this sheath is carried towards the cathode. This is known as the electro-osmotic flow (EOF) and its velocity is unaffected by the diameter of the capillary, provided that it is about 20 times greater than the thickness of the double layer.

The EOF increases the rate of migration of cations and transports nonionized molecules and even anions (provided the EOF is strong enough to overcome electrophoretic migration of anions to the anode) to the cathode. Thus, cations and anions move in the same direction and can be measured as they flow past the detector (Figure 9.2). An important feature of the EOF is that, because the forces responsible for the flow are uniformly distributed across the capillary, the flow is uniform and the flow profile is perpendicular to the capillary wall (plug flow), unlike the parabolic flow profile in pressure-driven systems such as GC and HPLC. This leads to less zone dispersion and, in theory, much greater efficiencies than those achievable by normal chromatographic techniques. The advantage of nonlaminar flow is exploited in CEC (Section 9.9.1).

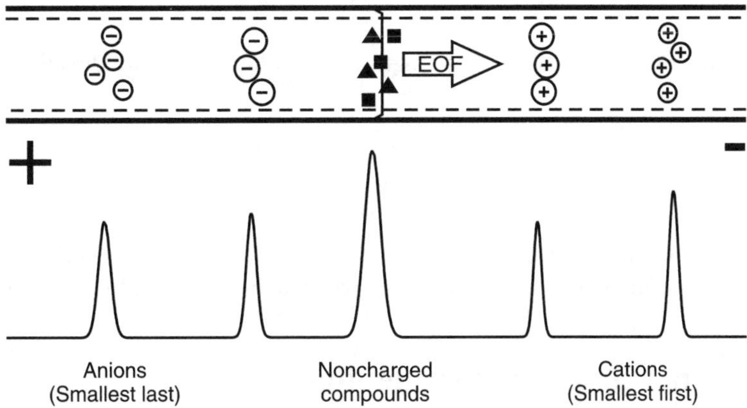

Figure 9.2 Diagram showing how electro-osmotic flow carries analytes to the cathode.

The mobility of the EOF is determined to a large extent by the zeta potential:

$$\mu_{EOF} = \frac{\varepsilon\zeta}{\eta} \tag{9.2}$$

where ε is the dielectric constant of the BGE. The zeta potential, and hence the EOF, increases as the pH of the BGE is increased because this increases the number of ionized silanols on the capillary wall. The velocity of the EOF is proportional to the applied electric field, E. Other influences on EOF include temperature (via changed viscosity of the BGE), the proportion of any organic modifier, the concentration of any added surfactant and changes to the nature of the capillary wall. It is important to be able to control the EOF: too high a flow rate and the analytes will not be resolved, too low a flow rate and cations may be adsorbed onto the negatively charged wall of the capillary.

Addition of cationic surfactants at concentrations below their critical micellar concentration (CMC) can reverse the direction of the EOF. A layer of surfactant is attracted to the negative charge on the capillary wall and a second layer forms (as a result of interactions of the hydrophobic tails) with the positive charges facing the bulk fluid. Hence, the effective ionization is now positive rather than negative.

Increasing the ionic strength of the BGE decreases ζ and hence decreases the EOF. This can improve resolution because migration times are prolonged. It also increases the pH stability of the BGE, which may be important as the pH is altered during electrophoresis. Consequently, it is recommended that the BGE is renewed frequently, for example every five analyses, as pH is generally one of the major influences on migration time. Furthermore, this practice ensures that the liquid levels in the vials are equal, as even a small degree of siphoning can have a marked effect on efficiency. The BGE should be filtered and degassed before use – because of the small volumes involved, filtration can be performed using syringes fitted with membrane filters (0.2 μm) and degassing by sonication is usually adequate.

From the above discussion, it is clear that the migration time of an analyte will be a function of the electrophoretic mobility and the EOF. The apparent mobility, μ_a, is given by:

$$\mu_a = \mu_e + \mu_{EOF} \tag{9.3}$$

and is proportional to the effective length of the column (l) and indirectly proportional to the applied field:

$$\mu_a = \frac{l}{Et} = \frac{lL}{Vt} \tag{9.4}$$

where t is the migration time (s). Thus, the units of mobility are cm^2 V^{-1} s^{-1}. Nonionized molecules will migrate with the EOF and so cannot be separated by CE. Anions have negative mobility.

9.3 Efficiency and zone broadening

The efficiency of a capillary electrophoretic separation can be expressed in terms of the number of theoretical plates, $N = 1/\sigma^2$ (Section 5.2.2). The main determinant of zone broadening is longitudinal diffusion because radial diffusion is negligible as a result of the plug-flow profile. The variance due to longitudinal diffusion is given by:

$$\sigma^2 = 2D_m t \tag{9.5}$$

where D_m is the diffusion coefficient of the analyte. Large molecules, such as peptides and proteins, have small values of D_m and usually produce sharp peaks on CE. Substituting Equation (9.4) for t gives:

$$N = \frac{\mu_e V l}{2D_m L} \tag{9.6}$$

Clearly, from Equation (9.6), high applied potentials result in high efficiency and therefore improved resolution. As with chromatography, the more quickly a compound migrates, the sharper the peak will be. Because N is directly proportional to V, applied potentials are not normally reduced as a means of reducing the EOF. There are a number of practical reasons why N is less than predicted by Equation (9.6), as discussed below.

9.3.1 *Joule heating*

It is important not only to understand the role of Joule heating, but also to minimize its effects. The heat generated by the electric current, which is proportional to the power (current × voltage), creates temperature gradients that cause local changes in viscosity and hence band broadening. Use of narrow-bore capillaries and temperature regulation (forced air or liquid cooling) can reduce these effects, but very narrow capillaries will limit sample loading and the effective light path length if using spectrophotometric detection. Measuring current as a function of applied potential is used to check the working potential range of a method (Ohm's Law plot). This will be linear whilst the contribution from Joule heating is acceptable.

9.3.2 *Electrodispersion*

A difference between the conductivity of the sample zone and the conductivity of the BGE causes peak distortion. For a cation, when the conductivity in the sample zone is high, the leading edge of the sample zone is diffuse, with a high concentration of the ions at the trailing interface. This leads to tailing peaks. Conversely, fronting occurs when the sample zone is of lower conductivity than the BGE. The reverse is true of species of the opposite charge. When the sample and BGE have similar conductivities then the peaks are more symmetrical. Neutral species are unaffected. These effects may be observed when a sample containing a range of species with different mobilities is injected (Figure 9.3). Although such distortions always occur with ions, the consequences may be small compared to other dispersive effects.

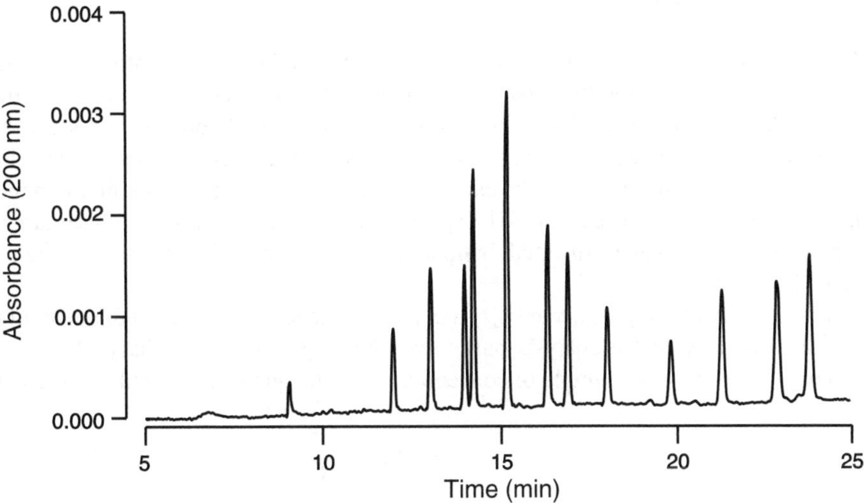

Figure 9.3 Electropherogram of substance P and 12 of its fragments showing the effect of electrodispersion. Early peaks are prone to tailing whist later peaks are prone to fronting. Conditions: Fused silica capillary (57 cm × 75 μm i.d.); BGE: phosphate buffer (50 mmol L^{-1}, pH 2.44); hydrodynamic injection (10 s).

9.3.3 Adsorption of analyte onto the capillary wall

Interaction of the analyte with the capillary wall may lead to either peak tailing or complete loss of analyte. Proteins and peptides are frequently adsorbed by electrostatic attraction to the negatively charged capillary wall and by hydrophobic interactions. Adsorption of compounds will affect the zeta potential and hence the EOF, leading to changes in migration times. Thus, injection of biological samples containing protein may result in marked changes in analyte migration times, even if the effect on the analyte is negligible when aqueous standards are injected. It may be necessary to flush the capillary with aqueous sodium hydroxide (say, 0.1 mol L^{-1}, 10 min), and then to re-equilibrate with BGE (possibly 10 min) between injections. This, of course, can add considerably to the cycle time.

9.4 Sample injection

In capillary electrophoretic techniques samples are introduced via either hydrodynamic or electrokinetic injection. In either case, too large a volume of sample injected (‘plug length’) will have a deleterious effect on efficiency, just as it does in chromatography. The contribution of the plug length to the total variance is:

$$\sigma_{\text{inj}}^2 = \frac{w_{\text{inj}}^2}{12} \tag{9.7}$$

where w_{inj} is the injection plug length. Ideally the variance due to the injection should be less than that due to longitudinal diffusion [Equation (9.5)]. If possible the injection plug length should not exceed 1–2 % of the length of the column. However, if sensitivity is a problem then it may be necessary to inject a larger amount of sample.

9.4.1 Hydrodynamic injection

Hydrodynamic injection is the most widely used method of sample introduction because it has the advantage that sample composition is unaltered by the injection. A small volume (a few nL) of sample can be introduced into the end of the capillary by hydrostatic pressure, which can be applied via: (i) positive pressure at the injection end, (ii) vacuum at the exit end or (iii) raising the sample vial above the height of the exit vial to cause siphoning. For the first two methods either the sample vial or exit vial needs to be sealed; this is usually achieved with septum caps through which the gas lines can be inserted. If siphoning is used, the difference in the heights of the vials should be 5–10 cm.

The quantity of sample injected (Q) will be influenced by the internal diameter of the capillary (d), the viscosity of the analysis buffer (η), the length of the capillary (L) and the pressure differential (ΔP). For a sample of concentration, c, the quantity injected in t seconds is:

$$Q = \Delta P \frac{\pi d^4}{128 \eta L} ct \tag{9.8}$$

The pressure differential for siphoning is given by:

$$\Delta P = \Delta h g \rho \tag{9.9}$$

where Δh is the difference in height, ρ is the density of the buffer and g is the gravitational constant. To obtain the highest reproducibility of injection, t should not be too short, yet compatible with not reducing efficiency by injection of too large a sample, and the viscosity of the BGE should be maintained by good temperature control. The viscosity of the sample solution has little effect because of the very small volumes injected.

9.4.2 Electrokinetic injection

By placing the injection end of the capillary into the sample vial and applying a potential across the capillary (usually 3–5 times less than the running potential), a volume of sample will be drawn into the capillary by the EOF. However ions will also be drawn in (or excluded) as a result of electrophoretic mobilities. The quantity of an analyte injected is given by:

$$Q = (\mu_{\text{EOF}} + \mu_e)\pi r^2 Ect \tag{9.10}$$

where the applied field, $E = V/L$. It is clear from Equation (9.10) that the composition of the fluid injected will differ from that of the original sample, there being an increase in the more mobile cations and a decrease in the concentration of anions, for which μ_e is negative. Furthermore, the composition of the residual sample will be changed – in some circumstances repeat injections from the same sample vial are clearly different, reflecting the changing sample composition with each injection.

9.4.3 Sample 'stacking'

When the conductivity of the sample buffer is less than that of the BGE, the ions migrate faster in the sample zone and become concentrated at the interface with the analysis buffer, a phenomenon commonly referred to as 'stacking'. Stacking is most easily achieved by injecting the sample (either hydrodynamically or electrokinetically) dissolved in water. If this is not feasible and the sample and BGE have similar conductivities, then a small volume (a few nL) of water may be injected before the sample. Greater than 10-fold increases in the amounts injected can be achieved. A potential problem to be considered when using sample stacking is that most of the voltage drop occurs in the stacking zone, resulting in the production of a significant amount of heat with the attendant risk of destroying thermally labile analytes.

9.5 Detection

Detection systems for CE and related techniques are spectrophotometric (UV and fluorescence), electrochemical and mass spectrometric. The optical systems normally detect the analytes on-column. Using silica capillaries allows wavelengths as low as 190 nm to be used. An obvious problem with using the capillary is that the path length is limited to the internal diameter of the capillary unless bubble cell capillaries are used. In these the capillary i.d. has been increased (about threefold) in the detector region. UV detectors, which are particularly suited to CE because of their relatively small size and rapid response times (Section 4.3.2) are usually fixed wavelength (a deuterium lamp with interference filters) or diode array.

Of the fluorescence detectors, laser-induced fluorescence (LIF) detectors offer greatly enhanced sensitivity, but the excitation wavelengths available depend on the types of laser employed.

Table 9.1 Examples of gas lasers and appropriate derivatizing reagents

Laser	Excitation (nm)	Derivative
Helium Cadmium	325	Dansyl choride
		o-Phthaldialdehyde (OPA)
Helium Cadmium	442	Naphthalene-2,3-dicarboxaldehyde (NDA)
		3-(4-Carboxybenzoyl)quinoline-2-carboxaldehyde (CBQCA)
Argon ion	488	Fluorescein isothiocyanate (FITC)
		5-(4,6-Dichlorotriazinyl)aminofluorescein (DTAF)
		3-(2-Furoyl)quinoline-2-carboxaldehyde (FQ)
		7-Nitrobenz-2-oxa-1,3-diazole (NBD)
Argon ion	514	
Helium Neon	543 or 594	
Krypton	568	

Although the wavelength range available is being extended by the introduction of diode-based lasers, it is still often necessary to derivatize the analyte to produce a suitable fluorophore (Table 9.1). In the case of argon ion lasers, derivatives of fluorescein isothiocyanate (FITC) are suitable, but tertiary amines have to be *N*-dealkylated, for example with 1-chloroethyl chloroformate (CECF, Figure 9.4), before they can be derivatized. Obviously, secondary and tertiary amines can

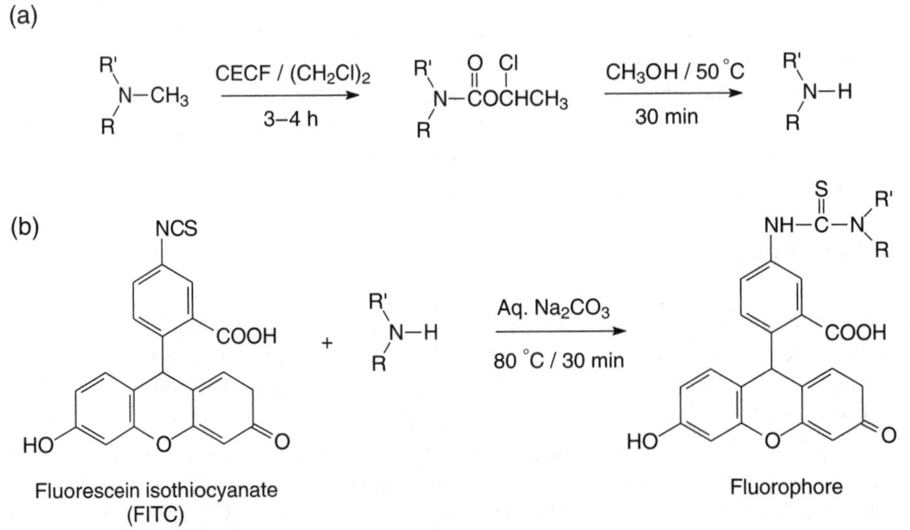

Figure 9.4 Multistage preparation of fluorescent derivatives of tertiary amines (after Alnajjar *et al.*, 2004).

be reacted with FITC without prior dealkylation – amfetamine, for example, reacts directly (Páez *et al.*, 1996).

Synthesis of FITC produces several isomers and the most abundant, and the one that is generally used, is known as FITC isomer I or 5-FTIC. With FITC derivatization, sensitivities of 50–100 ng L^{-1} for morphine, codeine and 6-monoacetylmorphine can be achieved using LIF detection with excitation and emission at 488 and 520 nm, respectively (Alnajjar *et al.,* 2004), compared to sensitivities of 250–300 μg L^{-1} attainable using native fluorescence (excitation 245 nm, emission 320 nm).

Caslavska and Thormann (2004) obtained high sensitivities using a xenon–mercury lamp in a specially designed fluorescence detector. This approach has the advantage over LIF in that the excitation wavelength is adjustable and compounds can be detected using native fluorescence, thus obviating the need for derivatization.

The small cell volumes that can be achieved with ED make use of this detection mode with CE attractive (Wallingford and Ewing, 1987). Rather than develop stand-alone detectors for CE, the trend has been towards separations 'on a chip' with combined electrophoretic separation and ED (Vandaveer *et al.*, 2004). The potential for miniaturization and rapid analysis times makes CE-ED on a chip of interest in clinical chemistry and possibly point of care testing (POCT) for drugs of abuse (e.g. Dolnik and Liu, 2005 – see Chapter 13).

The low flow rates used in CE make MS detection an obvious choice and several commercial interfaces are available. Most are based on ESI, which is sensitive and can generate ions for a wide range of species. MS, particularly MS-MS, can provide information on the identity of peaks and SIM can be used to resolve overlapping peaks if need be. If volatile buffers are used, the coupling of CE with MS is relatively straightforward, but the presence of involatile surfactants makes coupling MEKC and MS more difficult, as surfactants may cause severe ion suppression (Section 10.5.7). Quadrupole and ion-trap instruments have been successfully interfaced to CE. MDA isomers, for example, have been measured using ESI and MS-MS (Gaus *et al.*, 1996; Ramseier *et al.*, 2000). In both cases a volatile BGE was used for CE-MS. Chip-based approaches have also been described (Deng *et al.*, 2001).

9.6 Reproducibility of migration time

A criticism of CE and related techniques is the problem of the reproducibility of migration time. Small changes in EOF have a considerably larger effect on the migration time of late migrating compounds than compounds migrating close to the EOF. For example, from basic electrophoretic equations it has been calculated that when the t_{EOF} in a CE or MEKC system changes from 3.0 to 3.1 min, the migration time of a solute first migrating at 4.0 min will increase to 4.2 min (5 % increase), whilst a compound migrating at 20.0 min will have a migration time of 25.5 min (27.5 % increase) (Hilhorst *et al.*, 2001). Thus, care must be taken to ensure minimal changes in EOF, for example by regular cleaning and conditioning of the capillary and frequently replenishing the BGE to maintain the correct operating pH. The use of dynamically coated capillaries may also be beneficial (Boone *et al.*, 2001). However, these procedures alone may not be sufficient if migration times are be used to aid analyte identification in STA, and migration times may be reported relative to an internal standard used as a migration time marker. Better results may be obtained using more than one marker and some integration software packages allow electrophoretograms (and chromatograms) to be scaled to two retention time markers. The use of mobility or migration time ratios has been proposed in order to obtain more consistent results. The use of markers to calculate migration indices has also been proposed (Hyötyläinen *et al.,* 1996), but this approach is likely to remain 'in house' rather than be adopted universally as retention indexes have been in GC. Use of corrected migration values, although improving between-laboratory migration time reproducibility in MEKC, had no effect on migration time reproducibility in CE (Boone *et al.*, 2002).

Other ways of investigating peak identity include standard addition, that is adding an appropriate amount of a known compound and repeating the analysis to see whether the added compound co-migrates with the compound under investigation. Another way is to change the separation conditions. Changing the BGE is relatively simple in CE and this approach is particularly useful for amphoteric substances such as amino acids and peptides. By, for example, increasing the buffer pH, ionization of amines will be suppressed whilst the degree of ionization of acidic groups will be increased. Thus, such changes will result in altered migration times and migration order, which may give valuable information as to the nature of an unknown compound. The use of detection systems such as DAD and MS, in particular, circumvents some of these problems, as identification is not based solely on migration time.

9.7 Applications of capillary electrophoresis

Because CE separates analytes on the basis of mass and charge, it is considered an orthogonal technique to chromatography. *Inter alia*, it is very unlikely that two analytes will have identical chromatographic retention and electrophoretic migration times and thus use of both techniques gives good discriminating power. This approach was used to confirm of the presence of a substance P (SP) fragment, SP(1–7) in an HPLC peak. Collecting the peak corresponding to the retention time of SP(1–7) and subjecting the fraction to CE, confirmed the presence of the SP fragment, amongst various unknown endogenous compounds (Figure 9.5). Fragments SP(1–4) and SP(2–11) were added to the fraction collected as migration time markers.

9.8 Micellar electrokinetic capillary chromatography

The inability of CE to separate neutral molecules has been addressed, in part, by the use of MEKC (Terabe *et al.*, 1984). This is one of the most widely used modes of CE. Micelles are formed when surfactants (detergents) are added to solutions above their critical micellar concentration (CMC) and it is the interactions between the micelles and the neutral analytes that effects the separation. The surfactants are usually ionized so that the migration can be controlled by the electric field and/or the EOF (Mukerjee and Mysels, 1971; Myers, 1988).

Although negatively charged micelles migrate towards the anode, the EOF is usually strong enough to sweep them towards the detector. Neutral species partition in and out of the micelles and the greater the interaction of the analyte with the micelle, the longer the migration time. When the analyte is not in the micelle it is carried along by the EOF (Figure 9.6).

If a neutral analyte does not partition into the micelles then it will not be resolved from the EOF, but on the other hand if it is highly bound to the micelles it will migrate with them. For a neutral analyte partitioning into the micelle, the retention factor, k, is:

$$k = \frac{\text{amount of analyte in micelle}}{\text{amount of analyte in buffer}} \tag{9.11}$$

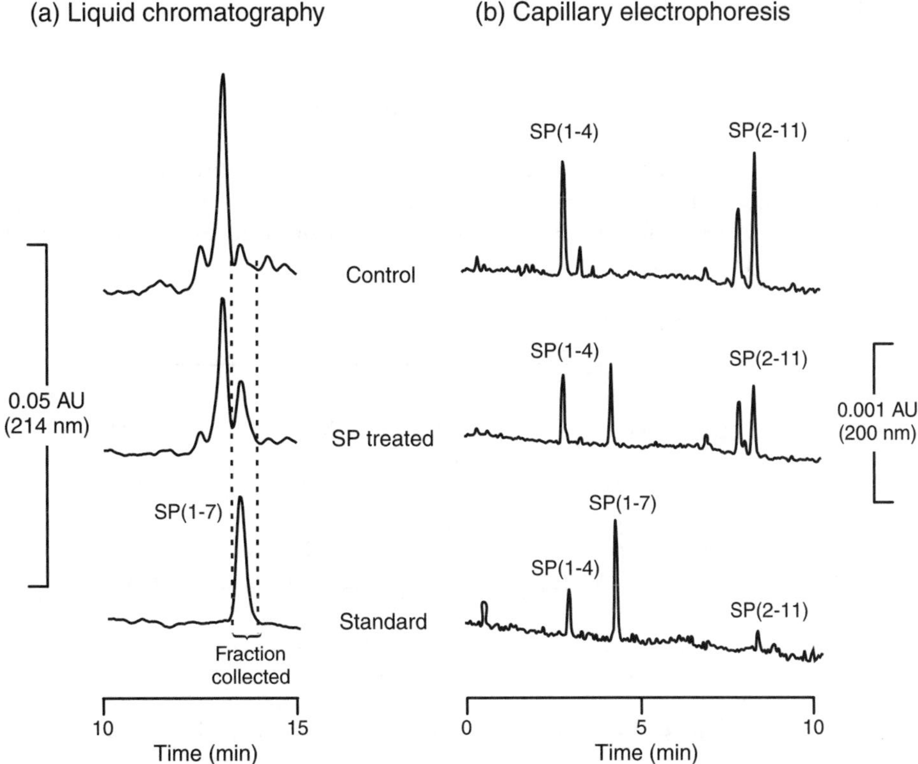

Figure 9.5 Separation of substance P fragment SP(1–7) by gradient HPLC and confirmation by CE after SP treatment of rats. Reprinted from Whelpton, R., Michael-Titus, A.T., Stephens, S.M., Yau, K.W. and Fengas, D. (1998) Identification of substance P metabolites using a combination of reversed-phase high-performance liquid chromatography and capillary electrophoresis. J Chromatogr B, **716**, 95–106, with permission from Elsevier.

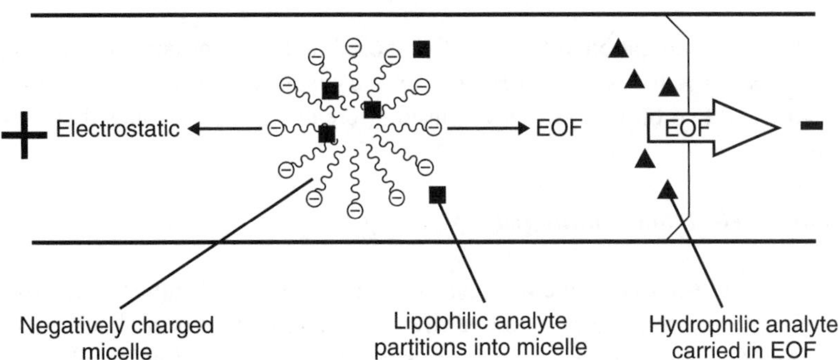

Figure 9.6 Principle of micellar electrokinetic capillary chromatography.

and the migration time t_R is:

$$t_R = \frac{1 + k}{1 + (t_0/t_{mc})k} \tag{9.12}$$

where t_0 is the migration time of an analyte that does not interact with the micelle (i.e. the migration time of the EOF) and t_{mc} is the migration time of the micelle. Thus, there is a region (or time window) in the electropherogram, between the EOF and the micelles, in which the analytes must be separated (Figure 9.7).

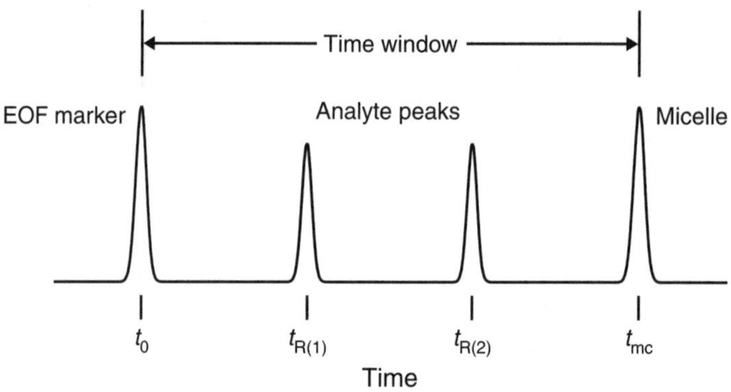

Figure 9.7 Elution time window for nonionized species in MEKC.

The high efficiencies obtainable with MEKC result in high peak capacities (Section 5.2.5) hence multicomponent mixtures can be analyzed (Figure 9.8). MEKC is also very versatile as cationic, anionic or neutral surfactants may be used, and partitioning can also be adjusted by the inclusion of organic modifiers. However, MEKC separations are, as the name implies, chromatographic and MEKC should not be considered as orthogonal to chromatography.

9.9 Other capillary electrokinetic modes

Other modes of separation related to CE include CEC, capillary gel electrophoresis (CGE), capillary isoelectric focusing (CIEF) and capillary isotachophoresis (CITP) (Watzig and Gunter, 2003). In CITP two buffer systems are used to create separated zones that move with the same velocity.

9.9.1 Capillary electrochromatography

CEC is a hybrid separation method that uses an electric field rather than hydraulic pressure to propel the eluent through a packed bed. Because there is little or no back-pressure it is possible to use small diameter particles, which when coupled with the plug-like flow profile, means that very high efficiencies can be attained. The A and C terms in the van Deemter equation (Section 5.2.3) are relatively small in CEC because the constant velocity of eluent between the particles means that the eddy diffusion (A) term is negligible and the ability to use small particles reduces

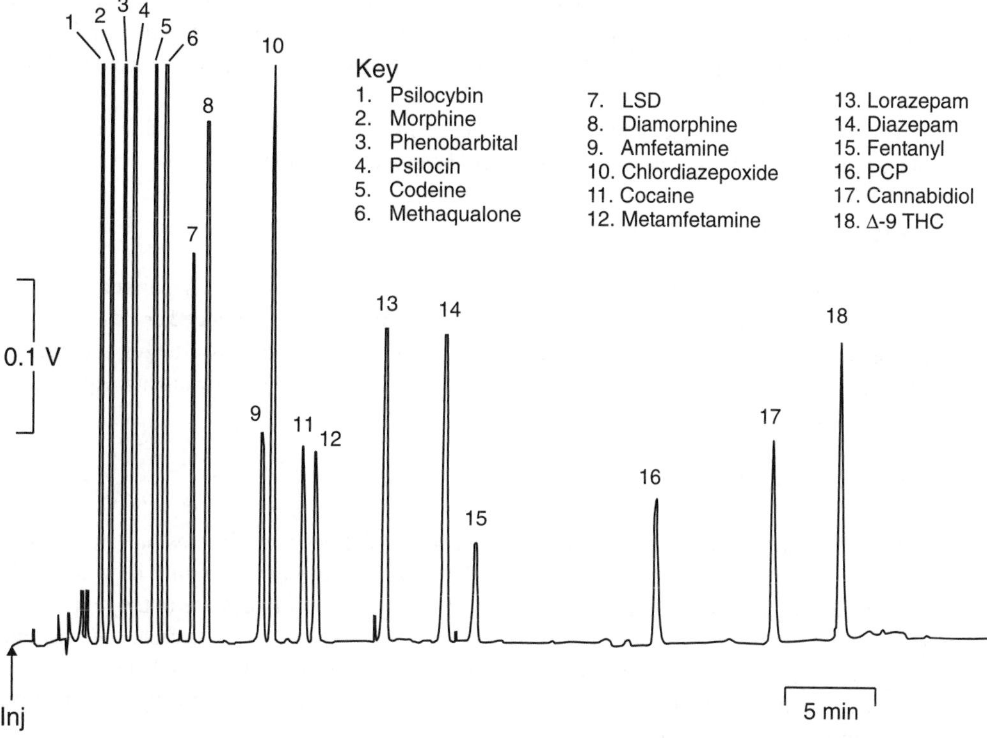

Figure 9.8 Micellar electrokinetic capillary chromatography of some illicit drugs. Conditions: Fused silica capillary (25 cm × 50 μm i.d.). BGE: 85 mmol L^{-1} SDS, 8.5 mmol L^{-1} phosphate, 8.5 mmol L^{-1} borate:acetonitrile (85+15), pH 8.5. Temperature: 40 °C. Potential 20 kV. Redrawn from Weinberger, R. and Lurie, I.S. (1991) Micellar electrokinetic capillary chromatography of illicit drug substances. Anal Chem, **63**, 823–7, by permission of the Americal Chemical Society.

the C term, so that efficiencies similar to those in CE are achievable. Typical particles sizes are 1–3 μm. Particle sizes >0.5 μm and very dilute aqueous buffers (<0.01 mol L^{-1}) should be used (Knox and Grant, 1987). Capillaries are packed using a high-pressure pump (62 MPa). A retaining frit is made from silica and a slurry of the stationary phase pumped into the capillary. Once packed, a second retaining frit is burnt in place and excess packing removed by reversing the direction of flow. A detection window is formed just beyond the retaining frit. This technique is not easy to learn, but packed capillaries are now available from several manufacturers in a range of dimensions and with a range of packings. The problem of bubble formation in the region of the outlet frit, due to degassing of the BGE, has been reduced by pressurizing the column.

As with HPLC, C_{18}-modified silica has been widely used for CEC. However, in CEC an important contribution to the EOF arises from negatively charged residual silanols on the packing, a contribution that is reduced when using C_{18} phases. Thus, the EOF is strongly dependent on pH, and this effectively makes rapid separations under acidic conditions difficult. Because of this drawback, there has been interest in many other types of bonded phase for use in CEC, including ion-exchange phases, particularly where a combination of ion-exchange and C_{18}-modified packing, so-called 'mixed mode' phases, is used. SCX/alkyl chain mixed-mode phases have been shown to produce an EOF that is stable over a wide pH range due to the SCX moieties retaining a negative charge even at low pH. Mixed-mode phases have been used with chiral selectors to resolve enantiomers (Section 9.10.1). The EOF can be reversed using SAX-based CEC columns (Byrne *et al.*, 2001).

The preparation and use of monolithic columns in CEC has been reviewed (Svec *et al.*, 2000). SCX functional groups and phases have been developed for these columns.

9.9.2 Capillary gel electrophoresis

In CGE the capillary is filled with a polymer mixture and molecules are separated on the basis of size. The cross-linked polymers, such as polyacrylamide, may be produced *in situ* in the capillary and as a consequence remain in the capillary. CGE is particularly useful for separating DNA fragments because the mass-to-charge ratios are very similar and so they are not separated by free solution CE.

9.9.3 Capillary isoelectric focusing

For CIEF, amphoteric buffers and analytes, such as peptides and proteins, are pumped into the capillary. When the electric field is applied, the buffer molecules move until they are uncharged, creating a pH gradient in the capillary. The analytes migrate until they are in the region where the pH is equal to the pI of the species. This 'focusing' of the peptides to discrete regions, indicated by there being no longer any current, can separate molecules with pI values that vary by as little as 0.005. The contents of the capillary can be 'swept' past the detector by the application of pressure to the anodic end.

9.10 CE techniques in analytical toxicology

It was not until the 1990s that a number of techniques based on electrokinetic phenomena, including CE and MEKC, began to be applied to analytical toxicology (Hilhorst *et al.*, 2001; Thormann, 2002). However, the application of the technique in this area remains limited despite further publications (Hempel, 2003; Lin *et al.*, 2003; Smyth and Brooks, 2004; Tagliaro and Bortolotti, 2006).

The role of capillary electrophoretic methods in STA has been reviewed (Boone and Ensing, 2003). The poor reproducibility and low sample capacity, and hence poor sensitivity, of current methods remain major limitations, as do the high cost of CE instruments and the physical fragility of many systems. However, orthogonality to chromatography and the requirement for very small sample size are potential advantages in the longer term. Application to hair analysis for drugs of abuse may be such an example (Gottardo *et al.*, 2007).

9.11 Chiral separations

Chiral separations have been described for CE, MEKC and CEC (Fanali *et al.*, 2001; Bonato, 2003). The approaches adopted parallel those described for HPLC (Section 8.6) and include the use of chiral additives, chiral surfactants such as bile acids (for MEKC) and, in the case of CEC, the use of additives or CSPs.

In CE, neutral chiral selectors have been used with the pH values of the BGE adjusted to ensure that the analyte was ionized: low pH for bases, high pH for acids. Modified cyclodextrins have been used widely, as have crown ethers, proteins and other additives used in chiral HPLC. Baclofen enantiomers have been resolved after reaction with NDA using α-cyclodextrin as chiral selector (Chiang *et al.*, 2001). Native and recombinant chicken AGPs (20 μmol L^{-1}) have been used to resolve racemic eperisone, pindolol and tolperisone; the native protein was preferred, but even then efficiency was poor with broad, tailing peaks (Matsunaga *et al.*, 2003). Vancomycin and teicoplanin have been assessed as chiral selectors in CE (Bednar *et al.*, 2001). Pirkle-type compounds have been used as additives, for example (R)-DNBPG has been used to resolve the sulfoxide metabolites of albendazole (Thormann *et al.*, 2002). An advantage of CE is that, because only small volumes of BGE are required, the amounts of the relatively expensive chiral additives needed are also small.

Otsuka *et al.*, (2000) used silica-gel based 5 and 3 μm particles coated with cellulose tris(3,5-dimethylphenylcarbamate) with a number of test racemates including enilconazole, homatropine, ibuprofen, pindolol, propranolol, verapamil and warfarin. Not all pairs of enantiomers, notably those of propranolol, could be resolved. Vancomycin, incorporated into diol-modified silica as a CSP for CEC columns, has been used to separate enantiomers of venlafaxine and *O*-desmethylvenlafaxine (Figure 9.9). The method was applied to plasma to which the analytes were added.

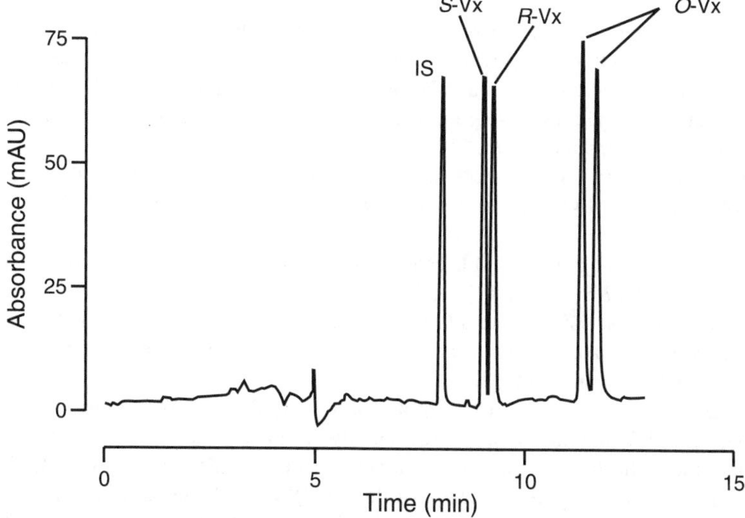

Figure 9.9 Enantiomeric separation of venlafaxine (Vx) and O-desmethylvenlafaxine (O-Vx) (2.5 g L^{-1} of each racemate). Capillary: 75 mm i.d. 35 cm total length, packed with 11 cm diol silica, 23 cm vancomycin CSP silica, and 1 cm diol silica; Eluent: ammonium acetate (100 mmol L^{-1}, pH 6.0):water:acetonitrile (5 + 5 + 90); Applied voltage, 27 kV; Temperature 20 °C; Sample injection: 12 kV (10 s) followed by an eluent plug (12 bar, 0.2 min); Internal standard (IS) imipramine (2.5 g L^{-1}). Reprinted from Fanali, S., Rudaz, S., Veuthey, J.L. and Desiderio, C. (2001) Use of vancomycin silica stationary phase in packed capillary electrochromatography. II. Enantiomer separation of venlafaxine and O-desmethylvenlafaxine in human plasma. J Chromatogr A, **919,** 195–203, with permission from Elsevier.

Lämmerhofer *et al.* (2001) have described the synthesis of monolithic CEC columns incorporating chiral functional groups, and more recently chiral monoliths with SCX groups have been developed (Preinerstorfer *et al.*, 2005).

The role of CE, CEC and MEKC in the resolution of enantiomers of SSRIs (fluoxetine, citalopram, paroxetine, sertraline); norepinephrine reuptake inhibitors (reboxetine); serotonin and norepinephrine reuptake inhibitors (venlafaxine, milnacipran, duloxetine); and noradrenergic and specific serotonergic antidepressants (mirtazapine) has been reviewed (Mandrioli and Raggi, 2006).

9.12 Summary

Despite the promise of unparalleled efficiency because flow is electrically rather than pressure driven, the practical problems caused by the need to limit and control heat production limit column diameter, hence sample capacity and thus sensitivity. Moreover the analytical system is not robust and CEC columns are very difficult to pack reproducibly. Therefore, whilst achieving prominence in specialized areas, the promise of CE remains largely unfulfilled as regards analytical toxicology. In contrast, recent developments in HPLC-MS, especially, are having a major impact, as discussed in the next chapter.

References

Alnajjar, A., Butcher, J.A. and McCord, B. (2004) Determination of multiple drugs of abuse in human urine using capillary electrophoresis with fluorescence detection. *Electrophoresis*, **25**, 1592–600.

Bednar, P., Aturki, Z., Stransky, Z. and Fanali, S. (2001) Chiral analysis of UV nonabsorbing compounds by capillary electrophoresis using macrocyclic antibiotics: 1. Separation of aspartic and glutamic acid enantiomers. *Electrophoresis*, **22**, 2129–35.

Bonato, P.S. (2003) Recent advances in the determination of enantiomeric drugs and their metabolites in biological fluids by capillary electrophoresis-mediated microanalysis. *Electrophoresis*, **24**, 4078–94.

Boone, C.M. and Ensing, K. (2003) Is capillary electrophoresis a method of choice for systematic toxicological analysis? *Clin Chem Lab Med*, **41**, 773–81.

Boone, C.M., Jonkers, E.Z., Franke, J.P., de Zeeuw, R.A. and Ensing, K. (2001) Dynamically coated capillaries improve the identification power of capillary zone electrophoresis for basic drugs in toxicological analysis. *J Chromatogr A*, **927**, 203–10.

Boone, C.M., Manetto, G., Tagliaro, F., Waterval, J.C., Underberg, W.J., Franke. J.P., de Zeeuw, R.A. and Ensing, K. (2002) Interlaboratory reproducibility of mobility parameters in capillary electrophoresis for substance identification in systematic toxicological analysis. *Electrophoresis*, **23**, 67–73.

Byrne, C.D., Smith, N.W., Dearie, H.S., Moffatt, F., Wren, S.A. and Evans, K.P. (2001) Influence of the unpacked section on the chromatographic performance of duplex strong anion-exchange columns in capillary electrochromatography. *J Chromatogr A*, **927**, 169–77.

Caslavska, J. and Thormann, W. (2004) Monitoring of drugs and metabolites in body fluids by capillary electrophoresis with XeHg lamp-based and laser-induced fluorescence detection. *Electrophoresis*, **25**, 1623–31.

Chiang, M.T., Chang, S.Y. and Whang, C.W. (2001) Chiral analysis of baclofen by α-cyclodextrin-modified capillary electrophoresis and laser-induced fluorescence detection. *Electrophoresis*, **22**, 123–7.

Deng, Y.Z., Zhang, N.W. and Henion, J. (2001) Chip-based quantitative capillary electrophoresis/mass spectrometry determination of drugs in human plasma. *Anal Chem*, **73**, 1432–9.

Dolnik, V. and Liu, S. (2005) Applications of capillary electrophoresis on microchip. *J Sep Sci*, **28**, 1994–2009.

Fanali, S., Catarcini, P., Blaschke, G. and Chankvetadze, B. (2001) Enantioseparations by capillary electrochromatography. *Electrophoresis*, **22**, 3131–51.

Gaus, H.-J., Gögüs, Z.Z., Schmeer, K., Behnke, B., Kovar, K.A. and Bayer, E. (1996) Separation and identification of designer drugs with capillary electrophoresis and on-line connection with ionspray mass spectrometry. *J Chromatogr A*, **735**, 221–6.

Gottardo, R., Bortolotti, F., De Paoli, G., Pascal, J.P., Mikšík I. and Tagliaro, F. (2007) Hair analysis for illicit drugs by using capillary zone electrophoresis-electrospray ionization-ion trap mass spectrometry. *J Chromatogr A*, **1159**, 185–9.

Hempel, G. (2003) Biomedical applications of capillary electrophoresis. *Clin Chem Lab Med*, **41**, 720–3.

Hilhorst, M.J., Somsen, G.W. and de Jong, G.J. (2001) Capillary electrokinetic separation techniques for profiling of drugs and related products. *Electrophoresis*, **22**, 2542–64.

Hyötyläinen, T., Sirén, H. and Riekkola, M.-L. (1996) Determination of morphine analogues, caffeine and amphetamine in biological fluids by capillary electrophoresis with the marker technique. *J Chromatogr A*, **735**, 439–47.

Knox, J.H. and Grant, I.H. (1987) Miniaturization in pressure and electroendosmotically driven liquid-chromatography – Some theoretical considerations. *Chromatographia*, **24**, 135–43.

Lämmerhofer, M., Peters, E.C., Yu, C., Svec F. and Fréchet J.M. (2000) Chiral monolithic columns for enantioselective capillary electrochromatography prepared by copolymerization of a monomer with quinidine functionality. 1. Optimization of polymerization conditions, porous properties, and chemistry of the stationary phase. *Anal Chem*, **72**, 4614–22.

Lin, C.C., Li, Y.T. and Chen, S.H. (2003) Recent progress in pharmacokinetic applications of capillary electrophoresis. *Electrophoresis*, **24**, 4106–15.

Mandrioli, R. and Raggi, M.A. (2006) Advances in the enantioseparation of second-generation antidepressant drugs by electrodriven methods. *Electrophoresis*, **27**, 213–21.

Matsunaga, H., Sadakane, Y. and Haginaka, J. (2003) Separation of basic drug enantiomers by capillary electrophoresis using chicken α_1-acid glycoprotein: insight into chiral recognition mechanism. *Electrophoresis*, **24**, 2442–7.

Mukerjee, P. and Mysels, K.J. (1971) *Critical Micelle Concentrations of Surfactant Systems*. National Standard Reference Data Series NSRDS-NBS 36. National Bureau of Standards, Washington DC.

Myers, D. (1988) *Surfactant Science and Technology*, VCH, New York.

Otsuka, K., Mikami, C. and Terabe, S. (2000) Enantiomer separations by capillary electrochromatography using chiral stationary phases. *J Chromatogr A*, **887**, 457–63.

Páez, X., Rada, P., Tucci, S., Rodríguez, N. and Hernández, L. (1996) Capillary electrophoresis-laser-induced fluorescence detection of amphetamine in the brain. *J Chromatogr A*, **735**, 263–9.

Preinerstorfer, B., Lindner, W. and Lämmerhofer, M. (2005) Polymethacrylate-type monoliths functionalized with chiral amino phosphonic acid-derived strong cation exchange moieties for enantioselective nonaqueous capillary electrochromatography and investigation of the chemical composition of the monolithic polymer. *Electrophoresis*, **26**, 2005–18.

Ramseier, A., Siethoff, C., Caslavska, J. and Thormann, W. (2000) Confirmation testing of amphetamines and designer drugs in human urine by capillary electrophoresis-ion trap mass spectrometry. *Electrophoresis*, **21**, 380–7.

Smyth, W.F. and Brooks, P. (2004) A critical evaluation of high performance liquid chromatography-electrospray ionisation-mass spectrometry and capillary electrophoresis-electrospray-mass spectrometry for the detection and determination of small molecules of significance in clinical and forensic science. *Electrophoresis*, **25**, 1413–46

Svec, F., Peters, E.C., Sýkora, D. and Fréchet, J.M. (2000) Design of the monolithic polymers used in capillary electrochromatography columns. *J Chromatogr A*, **887**, 3–29.

Tagliaro, F. and Bortolotti, F. (2006) Recent advances in the applications of CE to forensic sciences (2001–2004). *Electrophoresis*, **27**, 231–43.

Terabe, S., Otsuka, K., Ichikawa, K., Tsuchiya, A. and Ando, T. (1984) Electrokinetic separations with micellar solutions and open-tubular capillaries. *Anal Chem*, **56**, 111–3.

Thormann, W. (2002) Progress of capillary electrophoresis in therapeutic drug monitoring and clinical and forensic toxicology. *Ther Drug Monit*, **24**, 222–31.

Thormann, W., Prost, F. and Prochazkova, A. (2002) Capillary electrophoresis with (*R*)-(−)-*N*-(3,5-dinitrobenzoyl)-α-phenylglycine as chiral selector for separation of albendazole sulfoxide enantiomers and their analysis in human plasma. *J Pharm Biomed Anal*, **27**, 555–67.

Vandaveer, W.R., Pasas-Farmer, S.A., Fischer, D.J., Frankenfeld, C.N. and Lunte, S.M. (2004) Recent developments in electrochemical detection for microchip capillary electrophoresis. *Electrophoresis*, **25**, 3528–49.

Wallingford, R.A. and Ewing, A.G. (1987) Capillary zone electrophoresis with electrochemical detection. *Anal Chem*, **59**, 1762–6.

Watzig, H. and Gunter, S. (2003) Capillary electrophoresis-a high performance analytical separation technique. *Clin Chem Lab Med*, **41**, 724–38.

10 Mass Spectrometry

10.1 Introduction

Mass spectrometry (MS) is concerned with the vapour-phase separation of ionized atomic or molecular species according to their mass-to-charge ratio (*m*/*z*) (Downard, 2004; Watson, 1997). When an analyte is ionized, a characteristic ion, representing the intact atom or molecule and/or a group of ions of different masses that represent fragments of the ionized species, are formed. These ions are separated by manipulation of magnetic and/or electrostatic fields in a high vacuum (typically 10^{-5} Pa), and the plot of their relative abundance versus the *m*/*z* of each ion constitutes a mass spectrum. The mass resolution available is instrument-dependent – ideally resolution should be less than 1 atomic mass unit (u). Inductively coupled plasma-MS (ICP-MS) is a specialized branch of MS used in the analysis of trace elements and toxic metals and is discussed further in Chapter 11.

MS can give more information about an analyte using less sample than other techniques. Identification of a molecule from its mass spectrum is much easier than with other types of spectral information, especially with the use of modern computerized databases. MS is also the most reliable technique for accurate mass measurement. Disadvantages of MS compared to some other techniques are that the sample taken for analysis is consumed and the investments in capital equipment and operator training are both relatively high. Although with modern bench-top instruments and associated software MS operation is now relatively simple poor quality spectra can be misleading and may give rise to inaccurate results (Barwick *et al.*, 2007).

Accurate mass measurements can be used to aid compound identification, and molecular fragmentation patterns ('fingerprints') can be used to identify analytes either empirically, or by comparison with published data. Alternatively, only ions with a particular *m*/*z* of interest can be monitored (selected ion monitoring, SIM). Controlled fragmentation where ions issuing from one mass analyzer and subjected to further fragmentation/separation in subsequent analyzer(s) (MS-MS and MS^n) can also be used in both qualitative and quantitative work. The whole field is developing rapidly and reference to manufacturers' and current scientific literature is needed to keep abreast of events.

Direct sample introduction can be performed with relatively simple matrices, such as exhaled air (respiratory MS), but complex mixtures such as plasma and urine are best analyzed via 'hyphenated' techniques, notably GC-MS and LC-MS (Box 10.1). Some form of sample preparation is, of course, usually needed, although sample handling can sometimes be simplified as compared with conventional GC and HPLC procedures (Van Bocxlaer *et al.*, 2000). A fundamental restriction with MS, however, is that it is achiral, hence chiral separation by GC, HPLC, or CE has to be performed before MS in order to detect and measure enantiomers (Cody, 1992; Chen *et al.*, 2005; Speranza *et al.*, 2005).

Box 10.1 Mass spectrometry in analytical toxicology

Qualitative work

- GC-MS more than LC-MS valuable in STA (analyte identification/metabolite characterization)
  - Accurate mass measurement (instrument dependent)
  - Fragmentation patterns ('fingerprints')
  - Databases of GC retention data and principal ion abundance

Quantitative work

- GC-MS and LC-MS can give sensitive/selective measurement especially if selected ion monitoring (SIM) and/or controlled fragmentation (MS-MS or MS^n) are used
  - Ideally need stable isotope internal standard (increased assay cost)
  - Sensitivity may not be as great as, for example, GC-ECD or HPLC with fluorescence detection
  - GC-MS restricted by thermal stability of analyte or derivative
  - Co-eluting components may interfere in LC-MS ('ion suppression')

General disadvantages/restrictions

- May be high capital, running and operator training costs (instrument dependent)
- Sample consumed
- Achiral

10.1.1 Historical development

The work of Sir Joseph J. Thomson (1856–1940) and Francis W. Aston (1877–1945) at the Cavendish Laboratories in Cambridge led to the development of MS – both received Nobel Prizes, Thompson (Physics, 1906) for the discovery of the electron and Aston (Chemistry, 1922) for his work with the 'mass spectrograph'. About this time, Arthur J. Dempster (1856–1950) at the University of Chicago developed a magnetic sector instrument with direction focusing. He also developed the first electron ionization (EI, often incorrectly called electron impact) source, which ionizes volatilized analytes with a beam of electrons from a hot wire filament. Alfred O.C. Nier (1911–1994) at the University of Minnesota carried on Aston's studies of the isotopic composition of elements and isolated ^{40}K and ^{235}U, amongst other major contributions that included miniature instruments used in space exploration.

The first practical EI mass spectrometers (sector instruments) appeared in the early 1940s. In 1948, Vickers in Manchester produced the MS-2, the first commercial MS. This had limited resolution and a maximum mass range of around 300 u. At this time, the measurement of ion time-of-flight (ToF) as a mass analysis technique was also discovered – this is based on the principle that ions with different m/z values travel at different speeds when the same accelerating voltage is applied.

In the early 1950s the fragmentation pathways of ions were beginning to be understood, but instruments still had limited mass range. In 1953 the quadrupole mass analyzer was patented, although it was work by Finnigan MAT in the 1980s that made the quadrupole ion trap (QIT) MS the versatile instrument it is today. Hans G. Dehmelt (1922–) and Wolfgang Paul (1913–1993) shared the 1989 Nobel Prize in Physics for the development of the quadrupole analyzer and the ion-trap technology with which it has become associated. Quadrupole instruments and ToF systems, together with advances in laboratory computers, have lead to the relatively affordable 'bench-top' MS systems that are now in widespread use.

Chromatographic and electrophoretic methods, particularly capillary GC and CE, have proved ideal sample introduction methods for MS because of the low flow rates used. More recently interfaces for HPLC have been developed that not only cope with the presence of relatively large amounts of HPLC eluent, but also permit the ionization of large molecules such as proteins (Abian, 1999). John B. Fenn (1917–) and Koichi Tanaka (1959–) shared the 2002 Nobel Prize in Chemistry for their work with electrospray ionization (ESI) and soft laser desorption (SLD) techniques, respectively. A related methodology, matrix-assisted laser desorption ionization (MALDI), was introduced in 1988 by Karas and Hillenkamp – this employs short pulses of laser light focused on the sample adsorbed on a matrix that absorbs UV light. Other ionization techniques such as surface enhanced laser desorption ionization (SELDI) and direct ionization on silicon (DIOS), coupled with mass spectrometers that are not limited by a maximum m/z value (ToF instruments), have also extended the scope of MS.

10.2 Instrumentation

The basic components of a mass spectrometer are an inlet device, an ion source and an ion selection system (mass analyzer), both under vacuum, a detector (typically some kind of electron multiplier) and a data system. Sample introduction to the source may be made with a probe inserted via a vacuum interlock, but in this case the only analyte separation that can be achieved is by heating the probe slowly to vaporize lower boiling point analytes first. Moreover, probe injection is not easily automated. Hence, more usually, samples are introduced after a chromatographic or other separation step, as discussed above.

The ions from the inlet device enter the instrument via a focusing slit. The kinetic energy imparted by motion through an electric field gives the particles inertia dependent on their mass, and the analyzer steers the particles to a detector on the basis of their m/z values by varying an electric and/or magnetic field. The smallest ions carrying the highest charge move most rapidly. The analyzer can be set to select a narrow range of m/z values, or to scan through a range of values to record the ions present.

There are several types of mass analyzer in addition to magnetic sector and ToF instruments (Box 10.2). Quadrupole mass analyzers and ion traps use electrical fields to selectively stabilize or

Box 10.2 Mass spectrometry: Types of mass analyzers

- Sector or double focusing (magnetic/electrostatic)
  - Often higher resolution than quadrupole instruments
  - Upper m/z limit
- Time-of-flight (ToF)
  - No upper m/z limit
- Quadrupole
  - Inherently higher sensitivity than sector instruments
  - Link in series to perform controlled fragmentation (MS-MS)
- Quadrupole ion trap (QIT)
  - Perform MS^n on single instrument
- Fourier-transform ion cyclotron resonance (FTICR)

destabilize ions falling within a narrow range of m/z values. Fourier-transform ion cyclotron resonance mass spectrometry (FTICR-MS) measures mass by detecting the image current produced by ions spinning in a magnetic field.

10.2.1 Sector instruments

Sector 'double-focusing' instruments consist of a combination of an electrostatic focusing device ('E' sector) and a large electromagnetic sector ('B' sector) – different manufacturers use different arrangements (Figure 10.1).

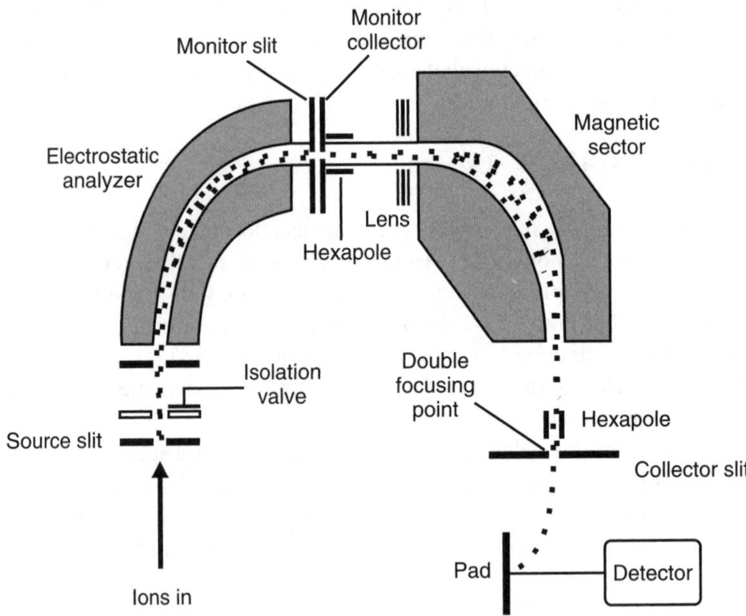

Figure 10.1 Schematic diagram of a double-sector mass spectrometer.

Ions from the source are accelerated by applying an appropriate voltage, and enter the electrostatic analyzer ('E sector'). Because of the physical nature of the ionization process, ions of the same m/z value are often produced with different energies. If uncorrected, this markedly reduces the resolution and mass accuracy attainable. The electrostatic analyzer focuses ions of the same m/z into a more coherent path through to the magnetic sector ('B sector'), which consists of a broad flight tube through a variable magnet that gives an angle of deflection of typically 60–120° to the ions according to Equation (10.1):

$$m/z = B^2r^2/2V \tag{10.1}$$

where B is the magnetic field strength, r is the radius of path through the magnet and V is the accelerating velocity. Changing the magnetic field strength focuses ions of differing m/z at the 'double-focusing point'. A typical sweep time for the magnetic field across a mass range of

50–800 u would be 1 s, but faster scan speeds are needed if high-resolution chromatography is used to introduce analytes to the source. After the ions have been separated, they enter a detector, the amplified output from which is sent to a computer that records the data, and converts the electrical impulses into formats suitable for display.

10.2.2 Quadrupole instruments

As its name implies, a quadrupole mass analyzer consists of four parallel rods (Figure 10.2) that have fixed direct current (DC) and alternating radio frequency (RF) potentials applied to them (variants of the system use from one to eight devices or rods, but the name and principle of operation is the same). The mass range and resolution of the instrument is set by the length and diameter of the rods, larger diameter rods giving increased sensitivity, narrower or longer rods increasing the resolving power. In simple terms ions are focused and passed along the middle of the space between the rods. Their motion depends on the electric fields so that only ions with a given m/z will be in resonance and thus pass through to the detector. All other m/z values will be nonresonant and will hit the rods, and thus not be detected.

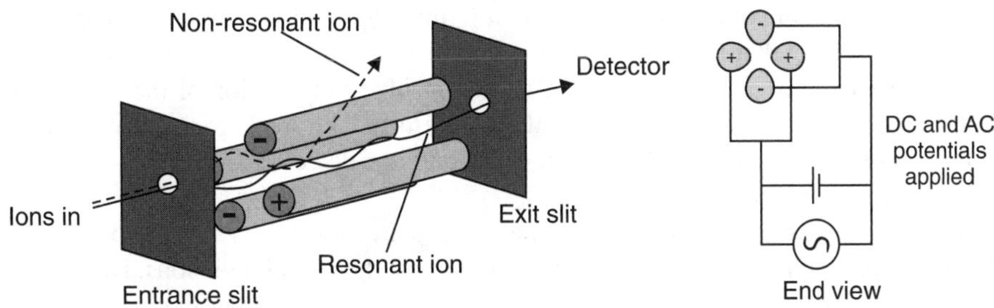

Figure 10.2 Schematic diagram of a quadrupole mass spectrometer.

Quadrupole analyzers do not require all the ions that enter the instrument to have the same kinetic energy and hence sensitivity is inherently higher than in magnetic sector instruments. The RF can be varied to bring ions of different m/z into focus on the detector at different times, usually from lower to higher m/z, and thus build up a mass spectrum. Alternatively, the applied RF can be used to select only ions with a particular m/z value. This process, commonly called SIM, allows for a longer time to be spent monitoring a single ion, and for also rapid switching between other selected ions. The result is increased sensitivity towards the selected analyte(s) and reduced noise, but of course the overall amount of MS data collected is much reduced, as other ions are not detected. Thus, SIM is used primarily for high sensitivity quantitative analysis, rather than for analyte identification.

10.2.3 Quadrupole ion-trap instruments

The ion trap consists of three hyperbolic electrodes: a central ring electrode, and entrance and exit end-cap electrodes (Figure 10.3). The ring electrode RF potential, a constant frequency/variable amplitude AC potential, produces a three-dimensional quadrupolar potential field in the trap. This captures incoming ions in a stable oscillating trajectory. The exact motion of the ions depends

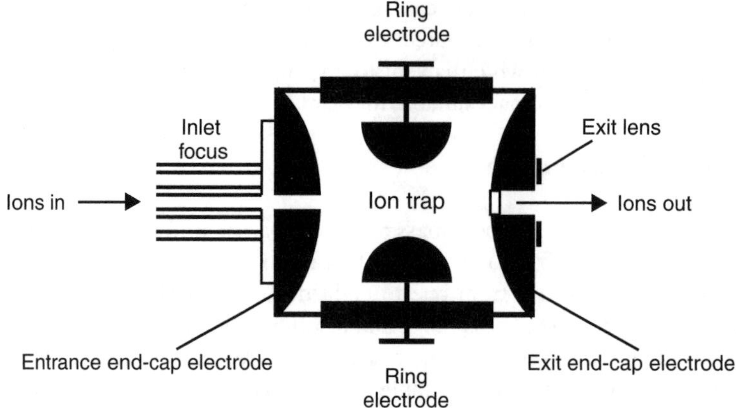

Figure 10.3 Schematic diagram of an ion-trap quadrupole mass spectrometer.

on the applied voltage and the individual m/z of the ions. The energy of the ions in the trap is quenched by helium at a pressure of 133 Pa (1 mm Hg) so that they focus near the centre of the trap, that is to say that their centrifugal energy is reduced. To detect the ions, the potentials are altered gradually to destabilize the ion motion resulting in ejection of the ions through the exit end cap, which also has a DC potential applied to it. The ions are usually ejected in order of increasing m/z. This stream of ions is focused onto the detector, an electron- or photo-multiplier tube, to produce the mass spectrum.

Because the distances ions travel in the ion trap are short, the vacuum does not need to be as high as in other types of MS instruments to minimize the risk of collisions. However, because the ions are confined, interactions with other ions are more likely than in conventional quadrupole instruments. In modern ion traps, because such ion–ion reactions can generate atypical spectra, not only is ionization performed outside the trap, but also the number of ions in the trap at any one time is controlled, thus reducing the incidence of ion–ion reactions.

10.2.4 Ion cyclotron resonance

FTICR-MS is analogous to Fourier-transform nuclear magnetic resonance (FT-NMR) spectroscopy. As the ions pass close to two opposed electrodes they induce a small alternating current in the circuit encompassing the electrodes. The amplitude and frequency of this current is proportional to the number of circulating ions and the cyclotron frequency of the ions, respectively. When the circulating ions are excited by a RF pulse of radiation the decay in the induced current is converted into a mass spectrum using Fourier transformation. This technique provides extremely high resolution and mass accuracy.

10.2.5 Controlled fragmentation (MS-MS)

The power of MS can be increased dramatically by linking mass analyzers in series (multistage MS). In some instruments, quadrupole analyzers are linked in series. In 'triple quads', for example, three such analyzers are used (Figure 10.4). These instruments usually generate pseudomolecular

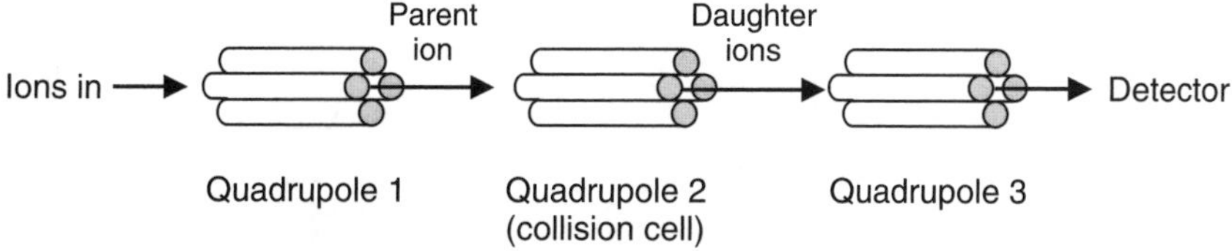

Figure 10.4 Schematic diagram of a triple-quadrupole mass spectrometer (tandem MS or MS-MS).

ions by chemical ionization (CI). The first quadrupole (Q1) is set to accelerate ions of a selected m/z into the second analyzer (Q2) where they are fragmented by high-energy collisions with helium or another collision gas. The selected ion passed through from Q1 is referred to as the precursor (or parent) ion. The ions formed from the fragmentation of the precursor ion are called product (or daughter) ions. In such cases the fragmentation produced is usually fairly reproducible. The third quadrupole (Q3) is used to scan the product ions, or selectively allow one or more such ions through to the detector. The fragmentation can be varied by changing the nature of the collision gas (argon, helium, nitrogen or xenon), the energy of ions emitted from Q1, temperature and pressure.

MS-MS or MS^n may be performed in ion-trap instruments. Firstly, all ions are ejected from the trap except for a precursor ion. A voltage is applied to the end-cap electrode, 180° from the field generated by the RF on the ring electrode. When the voltage applied to the end cap resonates with the energy of a particular m/z value, ions with that value are destabilized and fragment. The amount of energy used can be varied, thus varying the amount of fragmentation. Once fragmented, the product ions are scanned out of the trap to the detector. The nature of ion trapping/ejection makes a QIT especially suited to MS^n experiments in structure elucidation studies. The isolation and fragmentation steps can be repeated several times, the precise number of repeats only being limited by the trapping efficiency of the instrument. MS^5 experiments are routine, as is the coupling of liquid chromatography to perform LC-MS^n studies (Hopfgartner *et al.*, 2003).

MS-MS is a very powerful analytical tool. If the first ion isolated is the molecular ion, the likelihood of interference from other compounds (i.e. from co-eluting peaks if a chromatographic method of sample introduction is used) is much reduced. This increases the certainty of analyte identification and at the same time reduces dependence on the pre-MS stages of the analysis. MS^n methods also enhance the ability to elucidate molecular structures. Rather than seeing the total spectrum formed from the fragmentation of a molecule, individual ions can be isolated and their fragmentation also evaluated. Unfortunately, however, the fragmentation produced in MS-MS rarely reproduces the EI spectrum of a molecule making established spectral libraries of little help.

10.3 Presentation of mass spectral data

Mass spectra are constructed by plotting the intensity (equivalent to the number) of ions detected against their m/z value. It is usual for the ion intensities to be normalized as percentages of the intensity of the most abundant ion (base peak). With hyphenated techniques such as GC-MS, LC-MS and CE-MS, mass spectral data are collected throughout the separation and can be displayed as mass chromatograms (sometimes called mass fragmentograms). These are plots of scan number (analogous to retention time) versus either the intensity of a chosen m/z ratio (SIM), or the sum

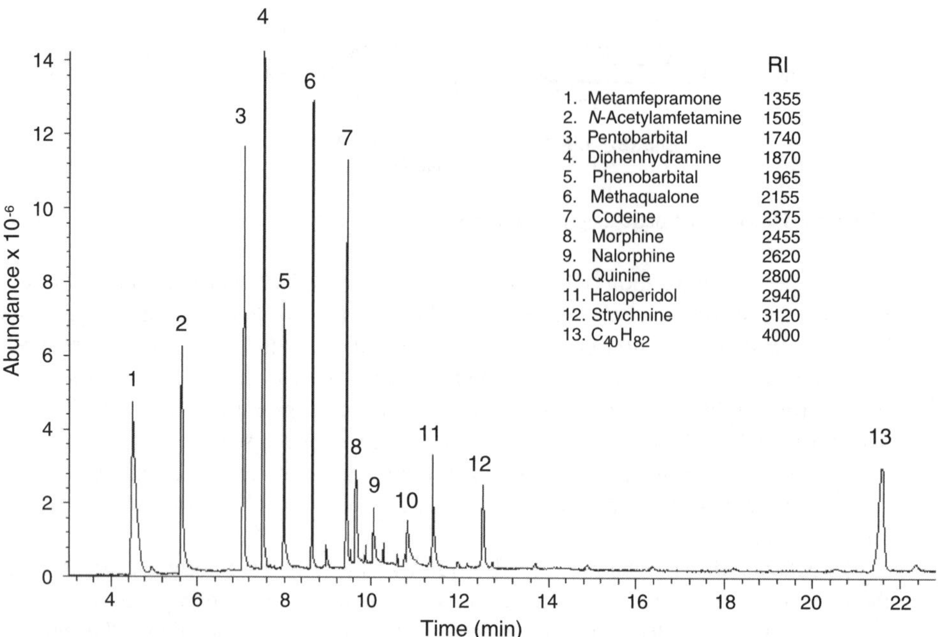

Figure 10.5 GC-MS reconstructed total ion chromatogram of a standard test solution. Injection: 1 mL of solution of each compound (50 mg L^{-1}) in methanol. Redrawn from from Pfleger *et al.*, (2007) with permission from John Wiley & Sons Ltd.

of total ion intensity (reconstructed total ion chromatograms, RTIC). Either type of plot can be used for quantitative measurements. RTIC plots often look like GC-FID chromatograms. In the example above (Figure 10.5) a standard solution of drugs covering the relevant retention time was used daily for checking instrument performance. The quality criteria were as follows: (i) all compounds should be resolved, (ii) the peaks should be sharp and of sufficient abundance and (iii) the relation of the peak areas of underivatized morphine to codeine should be at least 1:10.

When using SIM it is usual to monitor two or more ions, for example one characteristic of the analyte and another characteristic of the internal standard, in separate channels (Figure 10.6). This added dimension is a further advantage of GC-MS or LC-MS. The use of appropriate integration parameters such as threshold and peak width remains important, however.

10.4 Gas chromatography-mass spectrometry

Both EI and CI are used routinely in the analysis of relatively low M_r, volatile, thermally stable organic compounds, especially when coupled with GC, and both allow the same basic setup and source design. EI produces positively charged ions via removal of an electron, whilst CI can lead to the production of either positively or negatively charged species (Box 10.3). These techniques are, in effect, limited to the analysis of compounds up to M_r 1000 or so, although for GC, the practical mass limit is of the order of 750 u. A range of other desorption ionization techniques (ESI, etc.) are available and are discussed under LC-MS below (Section 10.5).

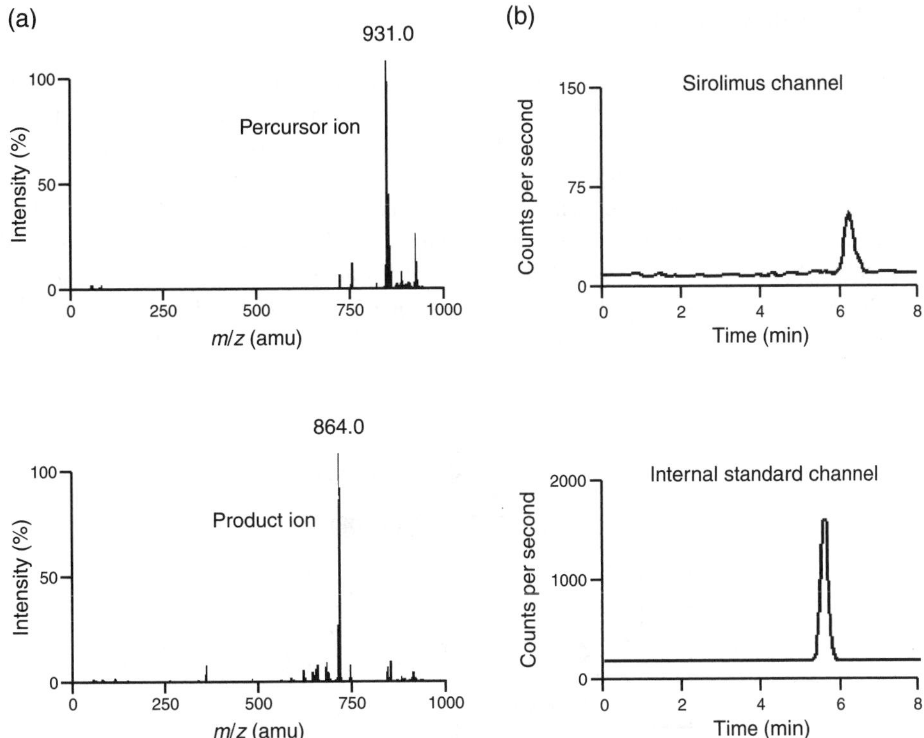

Figure 10.6 Analysis of whole blood sample from patient treated with sirolimus (rapamycin) by HPLC-MS-MS. Instrument: Sciex API2000. Post-column split 10:1. Source: heated ESI (300 °C). Nebulizer gas: air (positive ionization mode). Q1: ammonium adduct ions of sirolimus ($[MHNH_3]^+$, *m/z* 931.8 ± 1) and the internal standard nor-rapamycin ($[(M–CH_2)NH_3]^+$, *m/z* 917.5 ± 1) selected. Q2: collision chamber (nitrogen). Q3: Product ions (possibly sirolimus $[(M–CH_3OH–H_2O)NH_3]^+$, nor-rapamycin $[(M–CH_2–CH_3OH–H_2O)NH_3]^+$). Sirolimus concentration 4.1 μg L^{-1}. (a) ion scan of sirolimus precursor and product ions, (b) mass chromatograms at *m/z* 864.5 ± 0.5 u (sirolimus) and *m/z* 850.0 ± 0.5 u (IS). Redrawn from Holt *et al.*, (2000) with permission from The American Association for Clinical Chemistry; the composition of the product ions is as suggested by Wallemacq *et al.* (2003).

Box 10.3 Modes of ionization in GC-MS

Electron ionization (EI)
- 'Hard ionization'
- Positively charged ions, may include molecular ion ($M^{+\cdot}$, a radical cation)
- Fragmentation pattern (fingerprint) often produced

Chemical ionization (CI)
- 'Soft ionization'
- Requires 'reagent gas' (usually ammonia, isobutane or methane) in the ion chamber
- Pseudomolecular ion (MH^+) usually produced in positive ion (PICI or PCI) mode, with little fragmentation
- Negative ion (NICI or NCI) mode may yield intact molecular anions, but only applicable to electron-deficient analytes

10.4.1 Electron ionization

The vaporized analyte (or its decomposition products if probe injection of a thermally labile analyte has been attempted) first pass into an EI chamber (Figure 10.7). Here interaction occurs with a homogeneous beam of electrons produced normally from an electrically heated filament (rhenium or tungsten wire). Typically the electrons have an energy of 70 eV, much higher than the strength of the bonds within a typical analyte (approximately 10–20 eV). The beam is directed across the source chamber to ground. A fixed magnet is placed, with opposite poles slightly off-axis, across the chamber to create a spiral beam to increase the chance of interactions between the electrons and the vaporized analyte – even then typically only 0.1 % of the analyte is ionized. There are no actual collisions between analyte molecules or fragments and electrons.

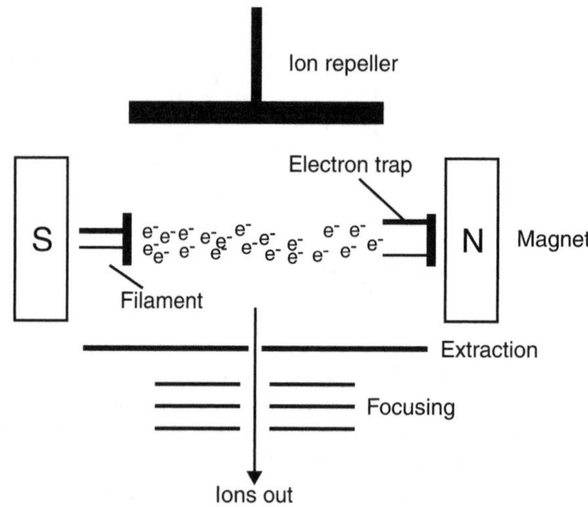

Figure 10.7 Schematic side view of an EI source.

The pathway leading to a charged molecule is initiated by loss of one electron from the analyte or by analyte decomposition. A molecule with one electron missing is called the molecular ion and is represented by $M^{+\cdot}$ (a radical cation). The peak resulting in the mass spectrum from this ion gives the molecular mass of the analyte. The ions are expelled from the source by application of a potential to a repeller electrode, which is maintained at the same charge as the ions.

Due to the large amount of energy imparted to the molecular ion it often fragments producing smaller ions with characteristic m/z values and relative abundances, the plot of which provides the fingerprint for that molecular structure. This information may be used to identify compounds of interest even in the absence of the molecular ion, and to help elucidate the structure of unknown components of mixtures. With an analyte AB, two processes that might occur are the direct result of energy transfer to the analyte, causing primary fragmentation [Reactions (10.1) and (10.2)]. This is the second cause (after thermal decomposition) of the presence of fragment ions in the spectrum. The third process is electron ejection from the analyte to create the energized radical ion.

This can then lose energy either through 'ion cooling' and stabilize to give the radical molecular ion, or by further fragmentation [Reaction (10.3)]. Some instruments allow the ionization voltage to be adjusted, a lower voltage giving less ionization and less fragmentation, which can be useful in analyte identification. Be this as it may, the relatively high degree of analyte fragmentation in EI spectra often results in the technique being termed 'hard' ionization.

$$AB + e^{\ominus\star} \longrightarrow A^{\oplus} + B^{\ominus} + e^{\ominus} \qquad \text{(Reaction 10.1)}$$

$$AB + e^{\ominus\star} \longrightarrow A^{\oplus} + B^{\star} + 2e^{\ominus} \qquad \text{(Reaction 10.2)}$$

$$\begin{array}{c} \text{Secondary fragmentation} \\ \uparrow \\ AB + e^{\ominus\star} \longrightarrow [AB^{\oplus\cdot\star}] + 2e^{\ominus} \\ \downarrow \\ AB^{\oplus\cdot}\ \text{Radical molecular ion} \end{array} \qquad \text{(Reaction 10.3)}$$

10.4.2 Chemical ionization

In positive ion CI (PCI, PICI), there is a reagent gas (usually ammonia, isobutane or methane) in the ion chamber as well as the analyte, and analyte ionization is by reaction with ionized reagent gas rather than by direct interaction with the electron beam as in EI. Reagent gas is continually introduced at a concentration sufficient to allow the desired reactions to proceed, but the vacuum is not as high as that usually employed in EI, in turn increasing the likelihood that an analyte molecule will collide with ionized reagent gas. Analyte ionization is most commonly the result of transfer of a proton from the ionized reagent gas to the analyte. A small positive charge is used to repel the ions out of the ionization chamber.

In PCI, although the initial reaction with the reagent gas is a high-energy process, ionization is caused by proton transfer and it is therefore a less energetic process than EI. Because less residual energy is possessed by the protonated molecules, fragmentation is greatly reduced, thus giving less information about the detailed structure of a molecule. However, it generally produces a pseudomolecular ion (MH^+) so the M_r of the analyte can be obtained, which is not always the case when using EI. Furthermore, MH^+ is usually a suitable ion for SIM and hence quantitative analyses. However, CI still requires the analyte to be present in the vapour phase, so thermal degradation can lead to varying degrees of analyte fragmentation. However, CI is generally a much 'softer' ionization method than EI, and until the development of desorption ionization methods, was the preferred way to analyze small, polar analytes such as many drugs and drug metabolites.

Ion formation in CI with methane as the reagent gas is illustrated in Reactions (10.4)–(10.6). In Reaction (10.4), methane is ionized by an electron beam in the same way as with EI. There may be several further reactions [Reactions (10.5) and (10.6)]. Reaction (10.7) shows ionized reagent gas reacting with unionized gas to form the carbanion (protonated methane). This step requires the reagent gas to be at a critical pressure – too low a pressure, and no ionization of the analyte will take place. Reaction (10.8) shows proton transfer from the carbanion to the analyte AB to form protonated analyte (ABH^+). If the pressure of the reagent gas is too high, then secondary reactions [Reactions (10.9) and (10.10)] may also occur, leading to formation of analyte adduct ions – these are seen as $M_{AB} + 15$ and $M_{AB} + 29$ *m/z* peaks in the spectrum (i.e. 14 and 28 *m/z* higher than the ABH^+).

$$CH_4 + e^{\ominus} \longrightarrow CH_4^{\oplus\cdot} + 2e^{\ominus} \quad \text{(Reaction 10.4)}$$

$$CH_4^{\oplus\cdot} \longrightarrow CH_3^{\oplus} + H^{\oplus} \quad \text{(Reaction 10.5)}$$

$$CH_3^{\oplus} + CH_4 \longrightarrow C_2H_5^{\oplus} + H_2 \quad \text{(Reaction 10.6)}$$

$$CH_4^{\oplus\cdot} + CH_4 \longrightarrow CH_5^{\oplus} + CH_3^{\cdot} \quad \text{(Reaction 10.7)}$$

$$M + CH_5^{\oplus} \longrightarrow CH_4 + MH^{\oplus} \quad \text{(Reaction 10.8)}$$

$$M + CH_3^{\oplus} \longrightarrow MCH_3^{\oplus} \quad \text{(Reaction 10.9)}$$

$$M + C_2H_5^{\oplus} \longrightarrow MC_2H_5^{\oplus} \quad \text{(Reaction 10.10)}$$

Ammonia is also a commonly used reagent gas and ionizes as shown in Reactions (10.11) and (10.12). However, subsequent reactions are generally dependent on the nature of the analyte. Amine-containing molecules often react as shown in Reaction (10.13), whilst nonbasic polar molecules usually form adducts. Non-polar analytes do not usually ionize, and thus the use of ammonia as reagent gas confers a degree of selectivity on the ionization process. Reagent gases may also sometimes be combined to optimize ionization of a given analyte.

$$NH_3 + e^{\ominus} \longrightarrow NH_3^{\oplus\cdot} + 2e^{\ominus} \quad \text{(Reaction 10.11)}$$

$$NH_3^{\oplus\cdot} + NH_3 \longrightarrow NH_4^{\oplus} + NH_2^{\cdot} \quad \text{(Reaction 10.12)}$$

$$NH_4^{\oplus} + RNH_2 \longrightarrow RNH_3^{\oplus} + NH_3 \quad \text{(Reaction 10.13)}$$

Negative ion CI (NICI or NCI) involves capture of a relatively low-energy electron such as those resulting from the ionization of methane [Reaction (10.1)], to generate a negatively charged ion. This process can yield intact molecular anions that are readily detected. Because the energy of the initiating electrons is low, high electron-affinity analytes, that is molecules containing a halogen atom, or a nitro- or carboxyl-moiety, are the best candidates for this work. In essence NCI-MS is a form of ECD and electron-capturing derivatives may be prepared as described in Section 7.4.1. Many biological molecules are not electron deficient and therefore are not ionized by this method. Hence the technique is often 100–1000 times more sensitive than PCI and so it is used widely in analytical toxicology (Maurer, 2002).

10.4.3 Application in analytical toxicology

In order for a compound to be analyzed by GC-MS it must be sufficiently volatile and thermally stable. Samples are usually analyzed after LLE, SPE or some other sample preparation process (Chapter 3). Some compounds may require prior derivatization to give reliable chromatography or to enhance the quality (selectivity) of the spectral information obtained. Trimethylsilyl (TMS) derivatives provide abundant ions of *m/z* $[M-15]^+$ due to the loss of a methyl group, for example (Section 10.6), whilst derivatization with reagents containing trifluoroacetyl, pentafluoropropionyl or heptafluorobutyryl groups yield derivatives with high electron affinity, thus facilitating NCI.

Quadrupoles or QITs are widely used, while the advent of the 'bench-top' MS in the form of the mass-selective detector (MSD; Agilent) and ion-trap detector (ITD; Thermo-Finnigan) brought GC-MS and even tandem GC-MS (GC-MS-MS) within the reach of many laboratories.

Capillary GC provides almost the ideal sample delivery system for MS because modern vacuum systems can easily cope with the relatively low flow rates used. The end of the column is simply passed through a heated transfer line to terminate at the entrance to the ion source. If the same GC conditions are used as in a non-MS method, then retention times will be shorter due to the vacuum at the MS end of the column.

If a packed or wide-bore capillary column (flow rate, 10–30 mL min^{-1}) system is used then a suitable interface is needed prior to entry to the ion source in order to maintain the necessary vacuum. The jet separator is one such system and relies on the fact that in passing the column effluent across a gap from a jet to a second jet, the relatively mobile carrier gas (usually helium) is largely removed, whilst the momentum of the heavier analyte molecules carries them across the gap (Webb, 1991). System maintenance may include cleaning or replacement of the injection port liner, shortening (10–20 cm) or even replacement of the GC column, and cleaning of the ion source. System performance must be tested using calibrators and 'blank' control samples. The use of 'blanks' is vital as analyte carry over is a major problem in trace analysis.

GC-MS can be used in the EI mode to produce definitive information as to peak identity either by providing complete spectra, or from the five or six principal peaks in the mass spectrum. Alternatively, the CI mode can be used to ascertain the M_r of the analyte, which is particularly useful in STA as a large number of potential candidate structures can be eliminated. Quantitative information can also be obtained if the instrument is used in the SIM mode if appropriate calibration standards are available (Peters *et al.*, 2003). GC-MS is especially valuable in the analysis of anaesthetic gases, solvents and other volatile analytes where LC-MS is clearly not feasible, and in screening for unknown compounds when GC temperature programming can be used in conjunction with EI spectral libraries to screen for the presence of a very wide range of compounds (Maurer, 2004; Smith, 2004).

An example of the use of GC-MS in compound identification is illustrated in Figure 10.8. The antidepressants citalopram and clomipramine have similar GC retention times (citalopram RI 2377, clomipramine RI 2388 on SE-30/OV-1) and NPD response, if not derivatized. Although the presence of metabolites may help differentiate these compounds, the EI spectra are clearly different (citalopram principal peaks at m/z 58, 238, 208, 42, 324 and 190 u, clomipramine principal peaks at 58, 85, 269, 268, 270, 271 and 242 u). Note, however, that the principal peak on both spectra is at m/z 58, arising from fragmentation to $(CH_3)_2NCH_2^+$ or other ions with this m/z.

Not only citalopram and clomipramine, but also many other basic drugs and metabolites give rise to the $(CH_3)_2NCH_2^+$ fragment or $C_2H_4NHCH_3^+$ (also m/z 58) hence it is generally of little use in compound identification. In addition, in the case of citalopram this fragment carries almost all of the ion current, making it very difficult to see the other ions, including the molecular ion (324 u). Another common fragment is m/z 44 corresponding to $CH_3NHCH_2^+$; this dominates the EI spectrum of another antidepressant, fluoxetine. These ions (m/z 44 and m/z 58) dominate the EI spectra of two important analytes, amfetamine and metamfetamine, respectively (Figure 10.9), and illustrate the need to derivatize these compounds to give reliable MS identification (Figure 10.10). Acetylation is another procedure commonly used for this purpose. Use of the STA procedure summarized in Box 3.14 can facilitate the detection and identification of a wide range of basic and neutral drugs and metabolites by GC-MS as their acetylated derivatives.

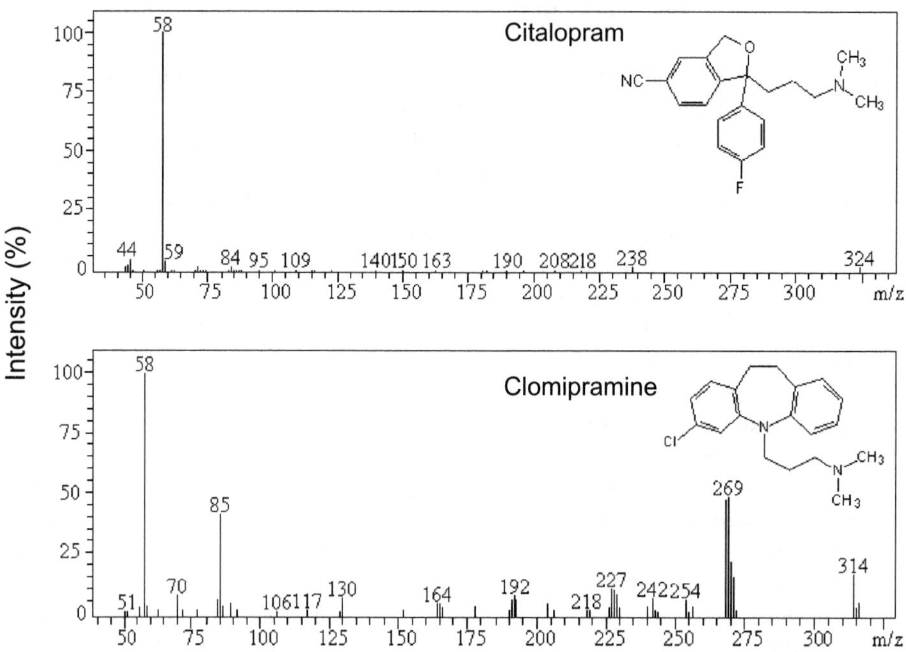

Figure 10.8 EI Mass spectra of citalopram (M_r 324.4) and clomipramine (M_r 314.9).

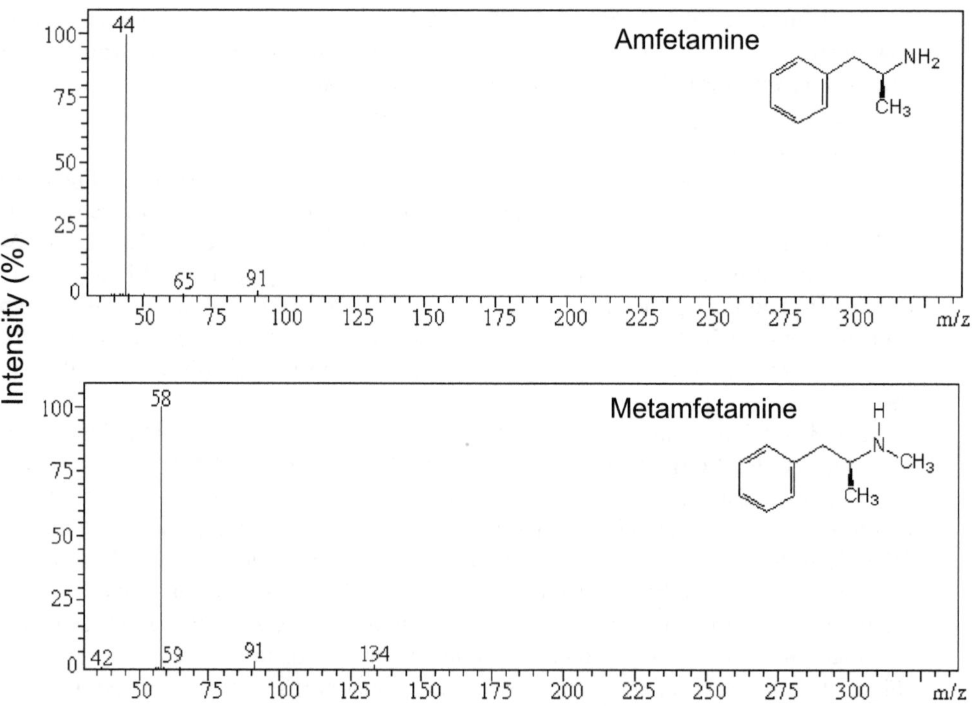

Figure 10.9 EI Mass spectra of amfetamine (M_r 135.2) and metamfetamine (M_r 149.2).

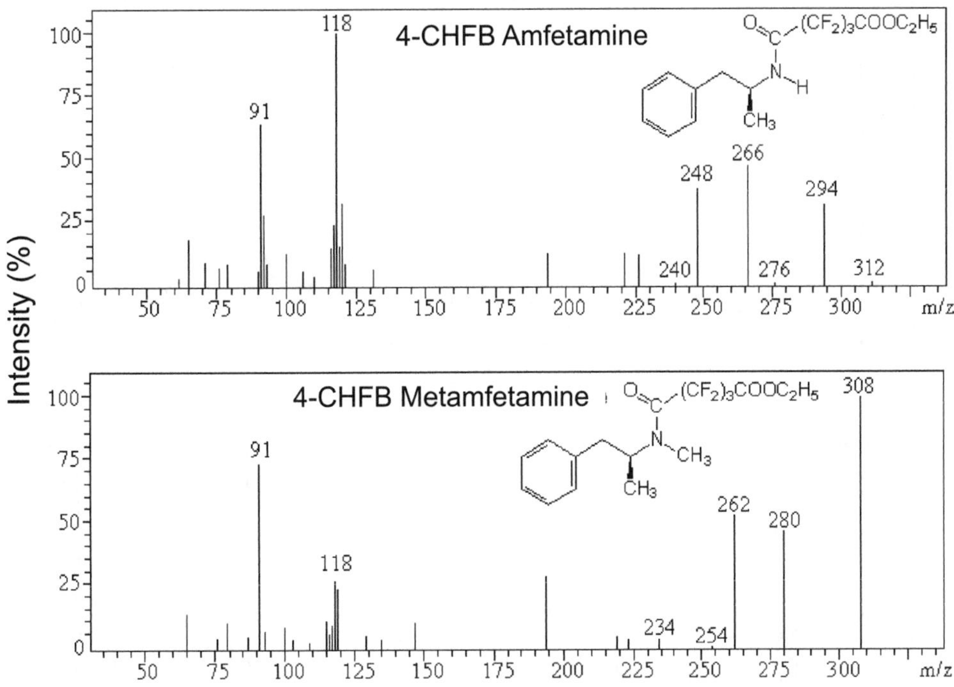

Figure 10.10 EI Mass spectra of the 4-carboxyhexafluorobutyryl (4-CHFB) derivatives of amfetamine (M_r 385) and metamfetamine (M_r 399).

Of course, many drugs, drug metabolites and other poisons can be detected if they are: (i) extracted under the conditions applied, (ii) volatile under the GC conditions used and (iii) if their mass spectra are in the reference libraries (e.g. Pfleger *et al.*, 2007; McLafferty and US Department of Commerce, 2005). In order to widen the scope of the screen, comprehensive urine analysis by full-scan GC-MS is recommended.

As an example of the use of GC-MS in STA, merged mass chromatograms of ions characteristic of amfetamine, MDA, MDMA and of the appropriate $^2H_5$-labelled internal standards are shown following analysis after SPE and derivatization with HFBA (Figure 10.11, top). Peaks 2, 10 and 17 indicate the possible presence of amfetamine, MDA and MDMA, respectively, and peaks 1, 3, 9, 16 and 18 indicate the internal standards. Merged mass chromatograms of target and two qualifier ions (additional ions characteristic of the particular analyte) of the three drugs and their respective internal standards are also shown (Figure 10.11, bottom). The presence of the respective target and qualifier ions in the peaks of these analytes and small deviations of their peak area ratios (target versus qualifier ions) from those obtained for reference substances (range from −7.4 to +2.2 %) were used to show the presence of the parent compounds in the sample.

A similar approach can be used to identify designer drugs derived from piperazine, such as benzylpiperazine (BZP), methylenedioxybenzylpiperazine (MDBP), trifluoromethylphenylpiperazine (TFMPP), *m*-chlorophenylpiperazine (mCPP) and *p*-methoxyphenylpiperazine (MeOPP). BZP in an authentic plasma sample was confirmed by the presence of respective target and

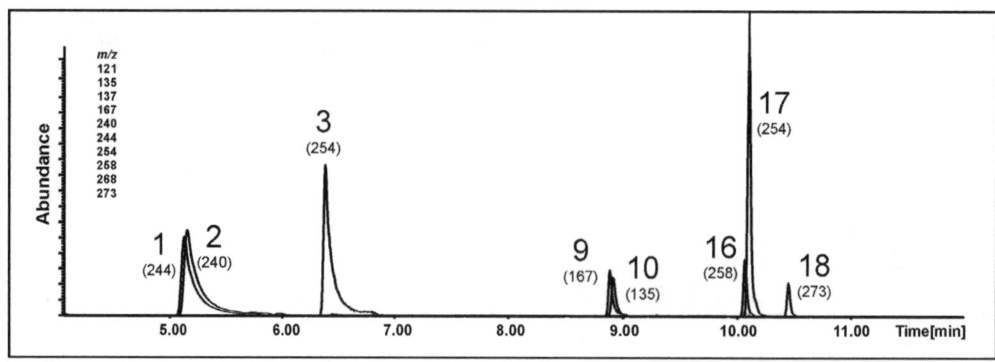

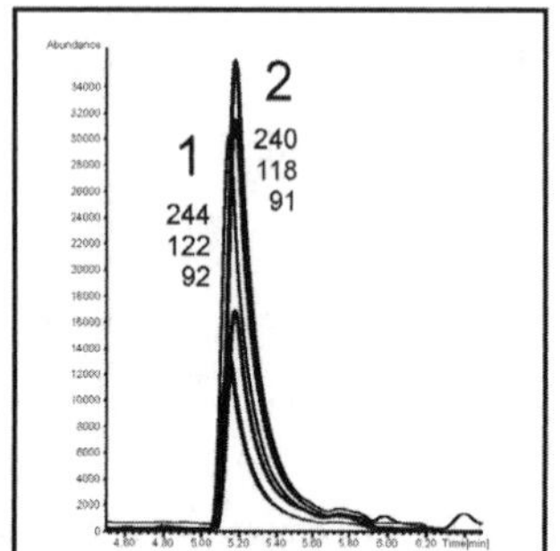

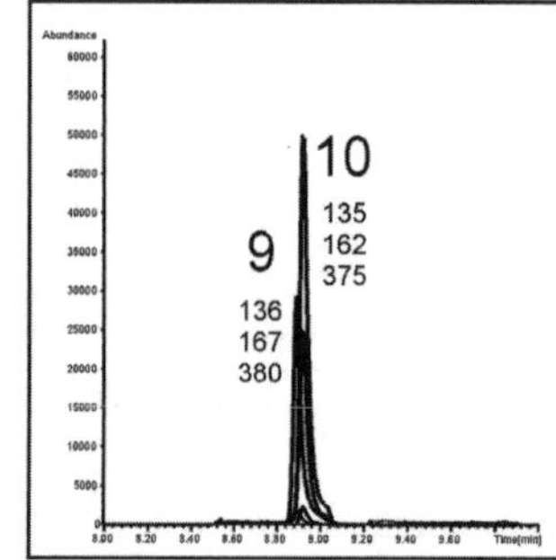

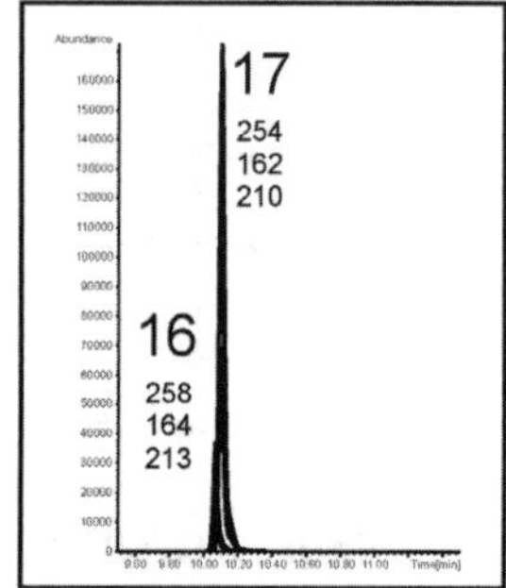

Figure 10.11 Top: Merged mass chromatograms of the given characteristic ions of HFBA derivatives of amfetamine, MDA, MDMA and internal standards obtained on analysis of a patient plasma sample. Bottom: Merged mass chromatograms of target and qualifier ions of amfetamine (left), MDA (middle) and MDMA (right). Peaks 2, 10 and 17: amphetamine, MDA and MDMA (111, 99 and 461 $\mu g\, L^{-1}$), respectively. Peaks 1, 3, 9 and 16: amfetamine-$^2H_5$, metamfetamine-$^2H_5$ MDA-$^2H_5$ and MDMA-$^2H_5$(all 100 $\mu g\, L^{-1}$), respectively. Reproduced from Peters *et al.* (2003), with permission from John Wiley & Sons Ltd.

qualifier ions and peak area ratios, as described above. In the absence of suitable ^{2}H-labelled internal standards, *p*-tolylpiperazine was used (Peters *et al.*, 2003).

As a further example, mass chromatography with the ions *m/z* 157, 161, 174, 200, 216 and 330 was used in the analysis of the designer drug 1-(3-trifluoromethylphenyl) piperazine (TFMPP) and its metabolites by GC-EI-MS after acid hydrolysis, extraction and acetylation (Staack *et al.*, 2003). The identities of the peaks in the mass chromatograms (Figure 10.12) were confirmed by computerized comparison of the underlying mass spectra with reference spectra. The selected ions *m/z* 174 and 200 were used for detection of TFMPP itself, and the ions *m/z* 216 and 330 were used for indicating the presence of its main metabolite, hydroxy-TFMPP. The ion *m/z* 174 was selected for monitoring*N*-(3-trifluoromethylphenyl)ethylenediamine and the ions *m/z* 161 and 157 were used to monitor the aniline metabolites 3-trifluoromethylaniline and hydroxy-3-trifluoromethylaniline, respectively (Figure 10.13). Screening for *N*-(hydroxy-3-trifluoromethylphenyl)ethylenediamine was not useful because it was only excreted in small amounts and could only be detected in urine after application of higher doses of TFMPP.

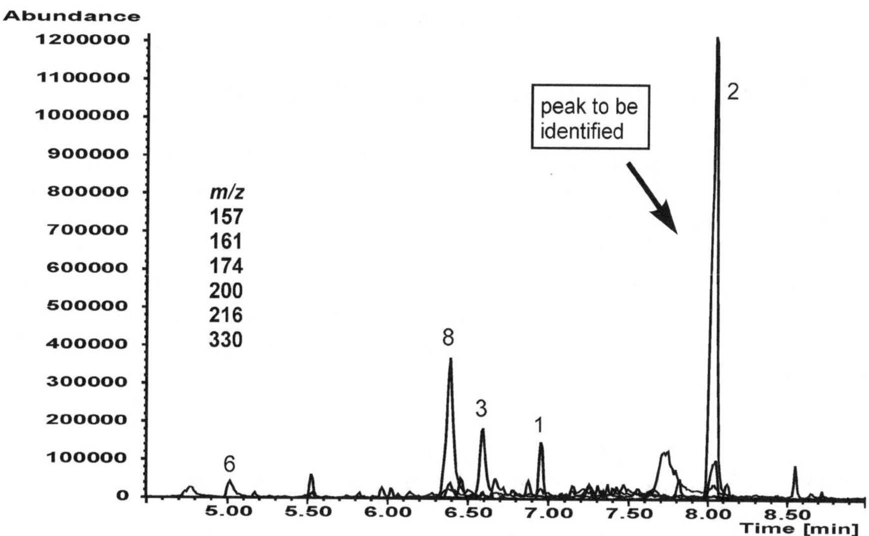

Figure 10.12 Reconstructed mass chromatograms of an acetylated extract of a rat urine sample collected over 24 h after a dose of 1 mg kg^{-1} body weight TFMPP. Reproduced from Staack *et al.* (2003) with permission from John Wiley & Sons Ltd.

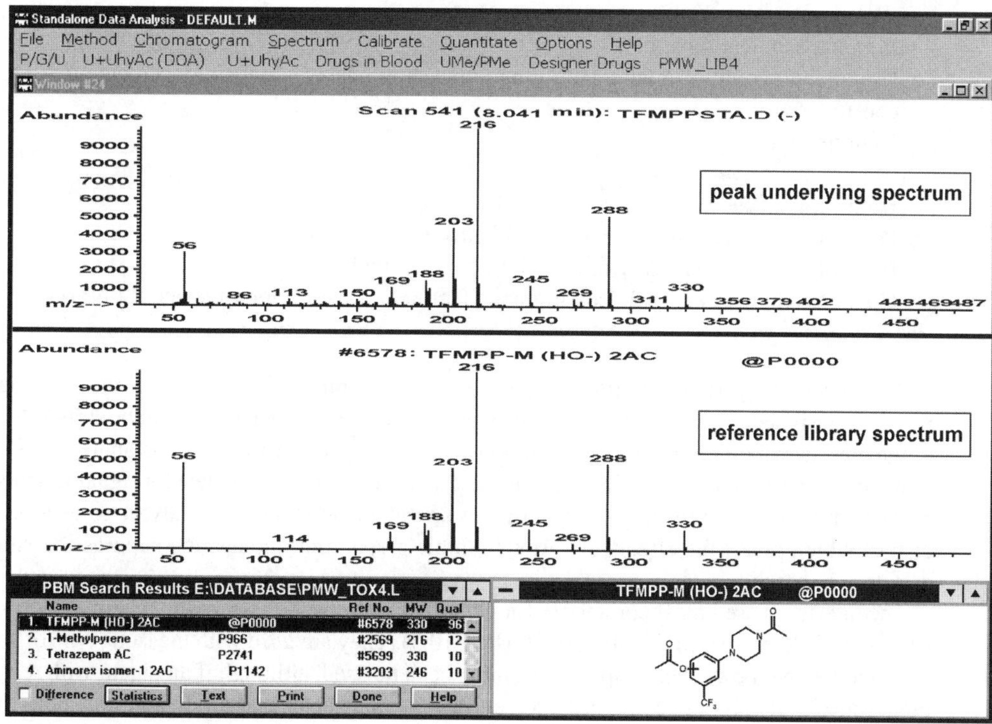

Figure 10.13 Mass spectrum underlying the marked peak (Figure 10.12), the reference spectrum, the structure and the library search 'hit list'. Reproduced from Pfleger *et al.* (2007) with permission from John Wiley & Sons Ltd.

10.5 Liquid chromatography-mass spectrometry

The primary advantage of LC-MS (HPLC-MS) over GC-MS is that it is capable of analyzing a much wider range of compounds. Polar and thermally labile/high M_r analytes may all be chromatographed using LC-MS without derivatization. A simple syringe pump may be used to deliver a solution of the analyte if LC is thought unnecessary, although it is usual to use at least a short column. Selectivity and hence sensitivity is often increased over HPLC-UV, with the result that sample volumes may be reduced and sample preparation simplified. HPLC analysis times may also be shortened, giving reduced eluent consumption and increased sample throughput (Box 10.4).

Box 10.4 Some advantages and potential disadvantages of LC-MS

Advantages

- Wide range of analytes
  - Not limited by sample volatility or thermal stability
- High selectivity/sensitivity
  - May require less sample than with conventional HPLC
- Wide linear range
- Minimal sample preparation
- Use of short columns/gradient elution to give short analysis times/reduced eluent consumption
- High flexibility

Potential disadvantages

- High capital cost
- Possibility of ion suppression or ion enhancement
- Possibility of interference from co-eluting compounds
- No unified ionization mode/conditions to facilitate construction of spectral libraries

Analytes eluting from the HPLC column have to be introduced to the MS via a specialized interface designed to cope with the presence of the eluent. Even then it may not be straightforward as the presence of co-eluting sample components may impair ionization of the analyte and/or any internal standard ('ion suppression') giving false quantitative results. To test for signal suppression, post-column analyte infusion (provided there are adequate stocks of analyte) gives a baseline against which to test the effects of various sample preparation modes, ion suppression showing as a drop in baseline.

The interfaces used most commonly for LC-MS are atmospheric-pressure chemical ionization (APCI) and electrospray ionization (ESI) (Box 10.5). They are used with the same type of source. In general eluent flow rates of up to 1 mL min^{-1} can be used, although if an eluent with relatively high water content is employed then a lower flow rate may be necessary. In addition, non-volatile salts and additives tend to deposit on the sampling cone resulting in a rapid decrease in instrument sensitivity. Thus, appropriate sample clean-up and the use of volatile eluents are required. Acetic, formic and trifluoroacetic acids may be used for acidic eluents and volatile (i.e. those that sublime) ammonium salts (ammonium acetate, carbonate, formate) for higher pH eluents. Additives such

Box 10.5 Some modes of ionization in LC-MS

Atmospheric pressure chemical ionization (APCI)

- Analogous to CI in GC except that
  - Corona discharge used as ionization source
  - Nebulizer gas and vaporized solvent molecules used instead of reagent gas
- Molecular species and adduct ions result, with minimal fragmentation

Atmospheric-pressure photoionization (APPI)

- Gas discharge lamp generates vacuum-UV photons that ionize nebulized effluent molecules
- Dopant (e.g. acetone, anisole) added to the auxiliary or sheath gas may aid indirect ionization

Electrospray ionization (ESI)

- Employs the same interface as APCI except that ionization source effluent spray needle maintained at + or − 2.5–4 kV against counter electrode
- Singly or multiply charged analyte molecules repelled towards a sampling cone – solvent evaporation assisted by nebulizer gas
- Usually little if any fragmentation

Ionspray ionization (ISI)

- As with ESI except that no external source of ionization used – limited to flows of a few $\mu L\ min^{-1}$

Flow fast-atom bombardment (FAB) ionization

- Very effective for thermally labile, polar, low volatility/high M_r compounds
- Solutes are dispersed in a matrix, for example glycerine, and bombarded with high velocity argon or xenon gas
- Eluent flow rates limited to approximately 10 $\mu L\ min^{-1}$

Particle-beam ionization

- Also called monodispersive aerosol generation for introduction of liquid chromatographic effluent (MAGIC)
- Similar to TSP – heater, nebulizer, and skimmer give increased analyte concentration
- Differs from TSP in that the solutes are completely desolvated at the interface by a nozzle and skimmer – especially useful for interfacing to CI and EI sources

Thermospray (TSP)

- Column effluent sprayed into the ion source as small droplets heated to approximately 300 °C evaporates forming ions
- If organic component in the eluent is high can be used in conventional CI mode with a supplemental filament, if low a glow discharge can be used to promote stable ionization

as those used for ion pairing in HPLC as or surfactants in MEKC can cause problems and are best avoided if possible.

All the modes of ionization used with LC are 'soft' methods, that is to say that there is at present no true equivalent to EI in GC-MS. This is a disadvantage in analytical toxicology because spectral libraries are largely based on EI spectra, which benefit from the fragmentation that occurs commonly with this ionization mode. However the use of controlled fragmentation techniques compensates for this to an extent (Maurer, 2005; Section 10.6).

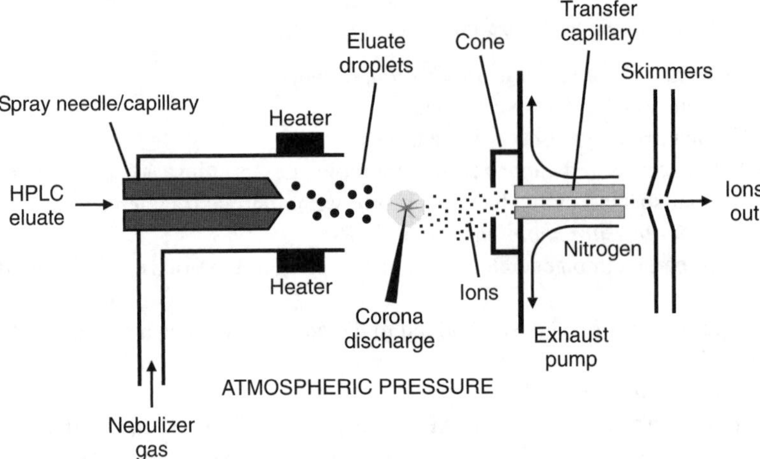

Figure 10.14 Schematic of the components of an APCI source.

10.5.1 *Atmospheric-pressure chemical ionization*

APCI is analogous to CI except that, as its name implies, the ionization occurs at atmospheric pressure. The primary applications of APCI are in the analysis of low mass, thermally stable compounds such as many xenobiotics. The interface consists of nebulizer, vaporizer, ionization source, exhaust and electrical source control (Figure 10.14), and is the same as that used in ESI except for the introduction of a corona discharge needle. The analyte solution (normally HPLC eluate) is introduced into a pneumatic nebulizer and desolvated in a heated quartz tube. The vaporized solutes reach a needle electrode and a high-voltage corona discharge (about 10 kV) ionizes eluent molecules. The corona discharge replaces the electron filament used in positive ion CI and produces primary N_2^+ and N_4^+ ions by EI. These primary ions collide with the vaporized solvent molecules to form secondary reactant ions such as H_3O^+ and $(H_2O)_nH^+$ (Figure 10.15). These

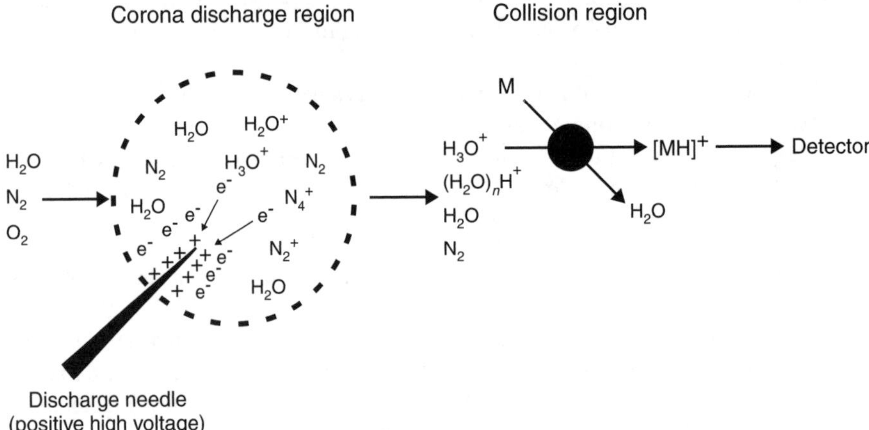

Figure 10.15 APCI: explanation of corona discharge (plasma) region and collision regions.

secondary reactant ions undergo repeated collisions with the analyte resulting in the formation of analyte ions.

The high frequency of collisions results in a high ionization efficiency and vaporization of the analyte ions. Once the ions are formed, they enter the pumping and focusing stage in much the same way as in ESI, although sensitivity tends to be less. Molecular species and adduct ions predominate in the spectra with little or no fragmentation. In order to collimate ions, send them effectively into the second small hole and then dissociate the cluster ions, the drift voltage ranging from 50 to 200 V is charged between the first and the second small holes. If this voltage is too high, the collision energy is increased and fragmentation is likely.

10.5.2 Atmospheric-pressure photoionization

APPI is a relatively new ionization technique for LC-MS. The APPI source is based on a gas discharge lamp that generates vacuum-ultraviolet (VUV) photons of 10–10.6 eV energy. The energy of the photons is normally greater than a first ionization potential (IP) of an analyte because many organic compounds have IPs in the range 7–10 eV. On the other hand, the IPs of the most common LC eluent solvents have higher values (water, IP = 12.6 eV; methanol, IP = 10.8 eV; acetonitrile, IP = 12.2 eV). This provides ionization of many analytes with lower IPs without interference from the eluent solvent. An example of an APPI interface is shown in Figure 10.16 (Robb *et al.*, 2000). Further enhancement of the ionization efficiency can sometimes be achieved by adding a dopant, such as acetone, anisole, benzene or toluene, to the auxiliary or sheath gas to aid indirect ionization.

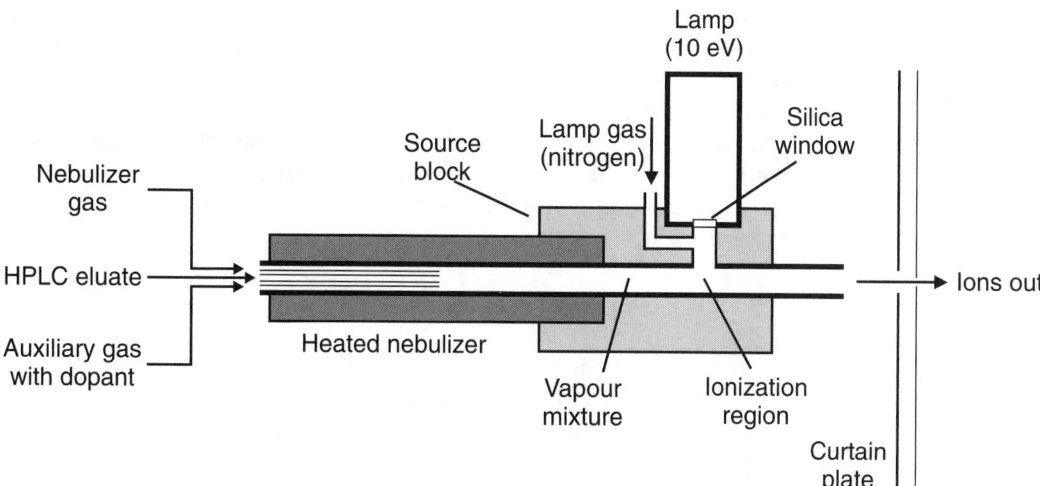

Figure 10.16 Schematic diagram of an APPI interface.

10.5.3 Electrospray or ionspray ionization

In these methods sprayed multicharged droplets are produced by high-voltage charging of a nebulizer. When the droplets are vaporized at atmospheric pressure, analyte ions are ejected. As

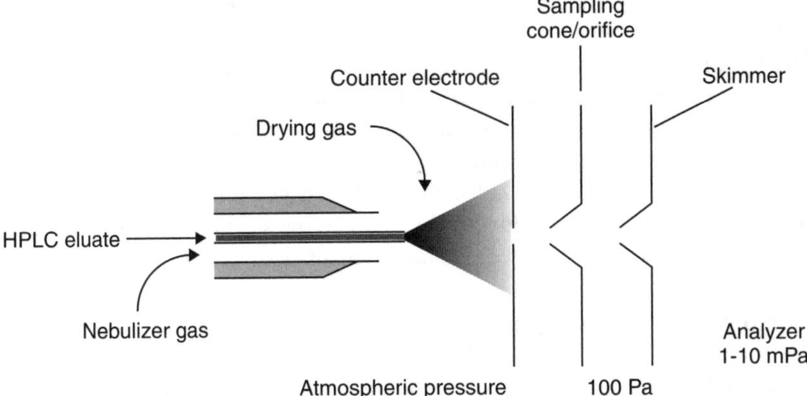

Figure 10.17 Schematic of an electrospray ionization (ESI) interface.

little energy is required to produce analyte ions, hardly any fragment ions are produced. In ISI gas pressure is normally used as the supplemental power for nebulization. The useful eluent flow rate is limited: electrospray can introduce up to 1 mL min^{-1}, whilst ionspray can only deal with a few μL min^{-1}. Therefore an effluent splitter is usually required with ISI.

In ESI the analyte solution flow passes through the electrospray needle maintained at a high potential with respect to a counter electrode (typically from 2.5 to 4 kV). Charged droplets are sprayed from the needle with a surface charge of the same polarity as the charge on the needle. Droplet formation is assisted by a flow of nebulizer gas, usually nitrogen. The droplets are repelled from the needle towards the source sampling cone on the counter electrode. As the droplets traverse the space between the needle tip and the cone, solvent evaporation occurs assisted by the flow of nebulizer gas (Figures 10.17 and 10.18).

As the solvent evaporates, the droplets shrink until they reach the point when surface tension can no longer sustain the charge (Rayleigh limit) at which point the droplets disintegrate, producing smaller droplets that can disintegrate further, as well as charged analyte molecules. This process is charge dependent and an electrolyte must be present in the eluent at a concentration of at least 0.05 mmol L^{-1} to promote ESI.

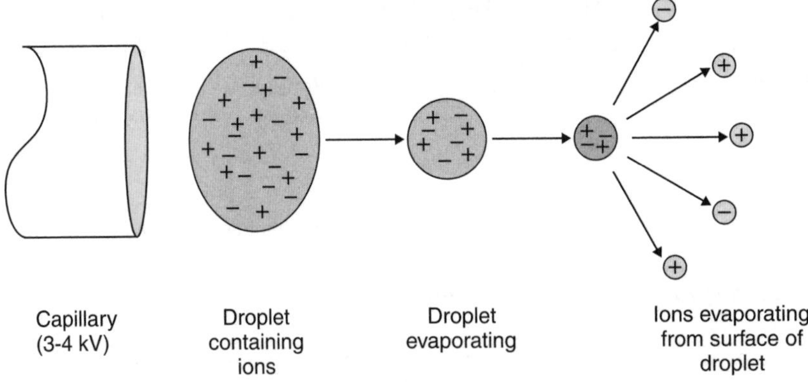

Figure 10.18 Electrospray ionization: schematic of the mechanism of ion formation.

This is a very soft ionization method as very little residual energy is retained by the analytes. The charged analyte molecules (they are not strictly ions) can be singly or multiply charged. The generation of multiply charged molecules enables high M_r compounds such as proteins to be analyzed as m/z is measured rather than mass *per se*. The major disadvantage of the technique is that there is very little (usually no) analyte fragmentation, although this may be overcome by using controlled fragmentation (MS-MS) techniques (Section 10.2.5).

10.5.4 Flow fast-atom bombardment ionization

Fast-atom bombardment (FAB) ionization is a very effective method for the analysis of compounds that are thermally labile, polar and have low volatility/high molecular mass. The solutes are generally dispersed in a matrix such as glycerine and are ionized by bombardment with high velocity argon or xenon gas. The interface consists of a metal frit at the tip of a fused-silica capillary. The solute–matrix dispersion penetrates onto the frit for ionization and the eluent solvents are rapidly vaporized. Eluent flow rates are limited to about 10 μL min^{-1}, and thus this interface is applicable to capillary HPLC using columns less than 0.5 mm i.d. and CE. It is also used for direct-insertion probe MS.

10.5.5 Particle-beam ionization

This interfacing method is also called monodispersive aerosol generation for introduction of liquid chromatographic effluent (MAGIC). It is similar to thermospray (TSP; Section 10.5.6) and uses a heater, nebulizer and skimmer resulting in an increase in analyte concentration. This technique differs from TSP in that the solutes are completely desolvated at the interface by a nozzle and skimmer, and is especially useful for interfacing to CI and EI sources.

10.5.6 Thermospray

The TSP interface consists of a nebulizer, an ion source and an exhaust system (Figure 10.19). The column effluent is heated as it passes through the nebulizer and is sprayed into the ion source as small droplets about 1 μm in diameter. The droplets are heated by the ion source block at approximately 300 °C and evaporated forming ions. If the eluent contains buffer solutions, complex reactions can be induced, and thus detection is influenced by the probe temperature, the vacuum applied to the nebulizing chamber and the eluent composition (Vekey *et al.*, 1989; Voyksner and Pack, 1989).

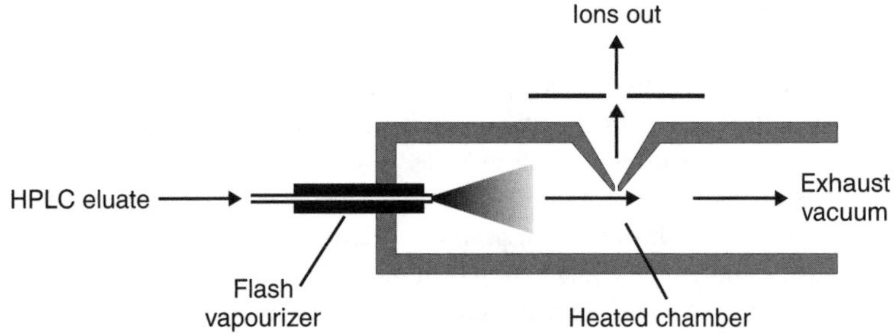

Figure 10.19 Schematic diagram of a thermospray interface.

TSP is suitable for reversed-phase HPLC separations, and mass spectra with little fragmentation can be obtained. When the proportion of an organic component in the eluent is high, as in normal-phase separations, the system is designed to be usable in conventional CI mode with a supplemental filament. If the eluent contains a high proportion of water, the system has a discharge electrode to produce a glow discharge for stable ionization.

10.5.7 Application in analytical toxicology

Although LC-MS is a very powerful technique (Figure 10.20), there are problems for the unwary (Maurer, 2004). There is potential for ion suppression or ion enhancement, interference from metabolites, and 'cross talk' (interference between SIM channels) when either analogues or stable isotope-labelled internal standards are used. Interference from metabolites may occur, for example, if metabolites with higher mass than the parent compound co-elute with and degrade to the parent compound in the MS interface. Interference from metabolites of one drug in the analysis of a second drug may also occur. This kind of problem is most likely to occur if chromatographic resolution is poor. Regulatory requirements include the need for the assessment and elimination of any matrix effect in methods used for supporting preclinical and clinical studies, although the experimental procedures necessary to assess the presence or absence of matrix effects may not be defined. In the pharmaceutical industry very short HPLC columns are the norm, but total reliance on the MS detection system is likely to lead to problems with clinical and forensic samples where an enormous range of drugs and metabolites may be present, especially if a nonselective sample preparation system has been used.

As an example of the problems that may occur, in an LC-MS-MS analysis developed to measure serum atropine in patients who had been supplied with cocaine containing, in one instance, 10 % w/w atropine, a very short (50 × 2.1 mm i.d. ODS, 5 μm aps) LC column was used. The atropine transition m/z 289 > 124 was employed to monitor atropine (Boermans *et al.*, 2006). However, benzoylecgonine (M_r 289.3) also gives a major peak at m/z 124 and was present at higher concentrations than atropine in the samples from patients who had ingested adulterated cocaine. Thus, extreme care was necessary to ensure chromatographic resolution of atopine from benzoylecgonine, especially since both compounds also give a secondary peak at m/z 82.

Matrix ion suppression (loss of signal for the ion being monitored) arises from competition between analyte and components in the matrix for ionization and is particularly prevalent with ESI. It can be an especial problem if protein precipitation is followed by a very short HPLC analysis time, for example cycle times of 1 min or less using 1 to 3 cm × 2 mm i.d. analytical columns. In other words, the problem of ion suppression is a consequence of attempting to achieve minimum sample clean-up with maximum sample throughput, using SIM to resolve the analyte from matrix components.

The matrix ions that cause ion suppression are often of low M_r. Temporary signal loss of 80–90 % has been observed as interfering component(s) elute from a column. Polar plasma components that cause matrix effects, such as salts, often elute quickly and can usually be separated from analytes of interest, but hydrophobic components such as triglycerides that may co-elute with an analyte are much more of a problem. Matrix effects can also be caused by exogenous materials, such as polymers contained in different brands of plastic tubes or lithium heparin, and may be a function of source design as well as the type of interface used. APCI may produce fewer deleterious effects than ESI, but this is not always the case (Mei *et al.*, 2003).

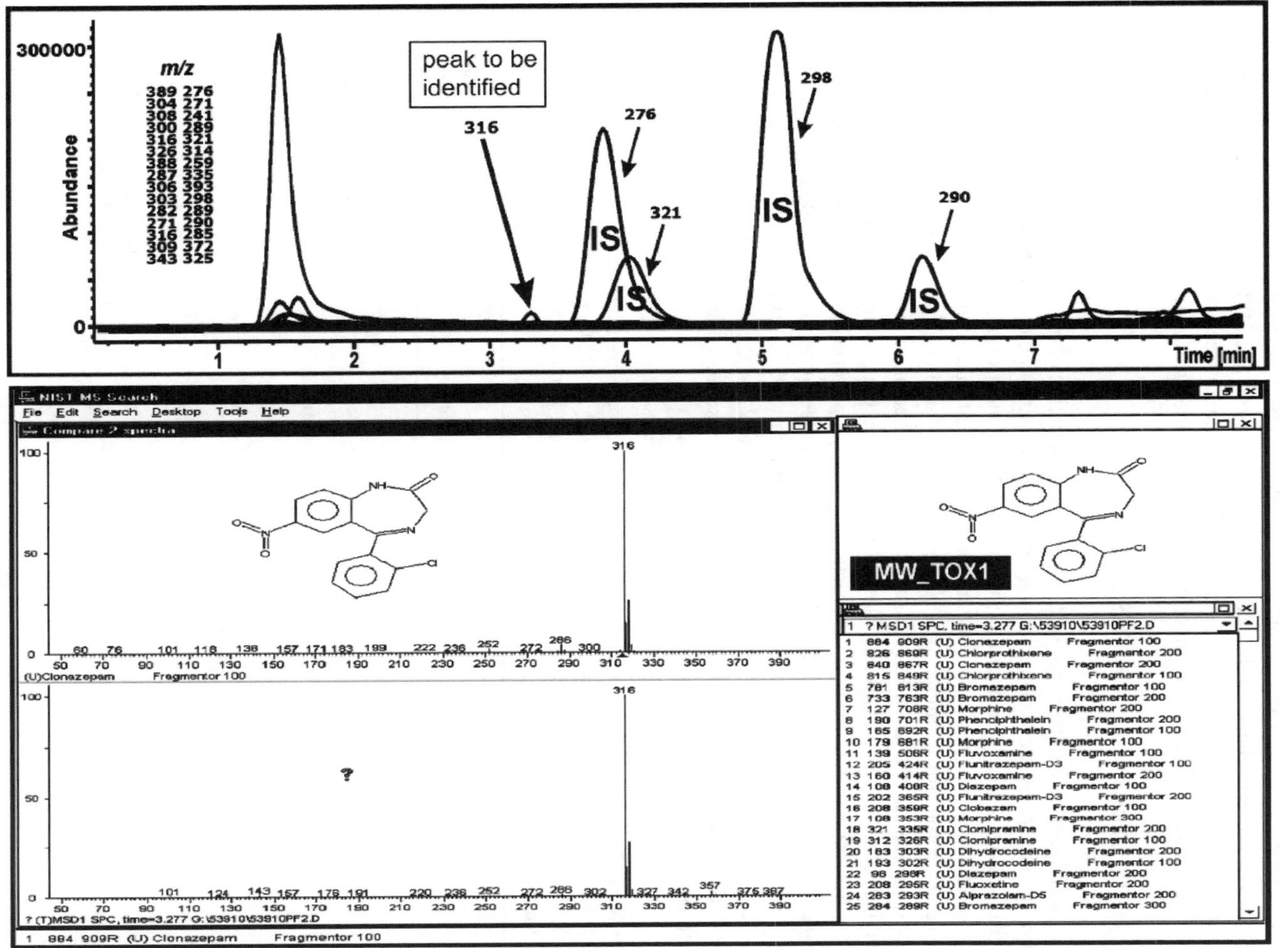

Figure 10.20 Gradient elution LC-MS-APCI screening for benzodiazepines using Merck LiChroCART column with Superspher 60 RP Select B stationary phase (reproduced from Kratzsch *et al.*, (2004), with permission from John Wiley & Sons Ltd). Smoothed and merged mass chromatograms (scan mode, 100 V) of an authentic plasma extract (upper part). Mass spectrum underlying the marked peak (lower spectrum), the reference spectrum (upper spectrum), the structure, and the hit list found by computer library search (lower part). Clonazepam concentration 0.02 mg L^{-1}.

Practical approaches for studying, identifying and eliminating the effect of the sample matrix on quantitative HPLC-MS-MS data have been described (Matuszewski *et al.*, 2003). For example, a direct comparison of the extent of any matrix effect may be made using a different interface (APCI, ESI or TSP) under otherwise the same sample preparation and HPLC conditions. Alternatively, standard addition of analyte and internal standard to analyte-free sample extracts and comparison with the results obtained from the reference solutions themselves may be performed.

10.6 Interpretation of mass spectra

The availability of databases notwithstanding, interpretation of mass spectra has to be undertaken with knowledge of the type of ionization used and other factors such as sample preparation, derivatization and LC eluent composition. The M_r of most organic analytes is an even number unless the molecule contains an odd number of (trivalent) nitrogen atoms. The presence of other elements, boron, for example, can also give an odd number M_r, but in analytical toxicology such elements are encountered infrequently. The presence of naturally occurring isotopes can aid identification (Table 10.1). With chlorine and bromine, the presence of ^{35}Cl and ^{37}Cl, and ^{79}Br and ^{81}Br, respectively, produces characteristic isotope patterns and fragments containing these elements can be identified easily. Frequently the average mass of a molecule in which isotopes contribute significantly is cited rather than an accurate mass. For example, hydrogen chloride has an average mass of 36.5 u, as it contains 75 % ^{35}Cl and 25 % ^{37}Cl. The presence of isotopes such as ^{13}C can lead to the occurrence of notable isotope peaks if an analyte contains a relatively large number of atoms of the appropriate element.

Table 10.1 Relative abundance of isotopes commonly encountered in analytical toxicology. Reproduced from Watson, D. (2004) Mass spectrometry, in *Clarke's Analysis of Drugs and Poisons* (eds A.C. Moffat, D. Osselton and B. Widdop), 3rd edn, Pharmaceutical Press, London, p. 380, with permission from RPS Publishing.

Isotope	M_r	Natural abundance (%)	Isotope	M_r	Natural abundance (%)
^{1}H	1.00782	99.985	^{29}Si	28.977	4.7
^{2}H	2.014	0.015	^{30}Si	29.974	3.1
^{12}C	12.0000000	98.9	^{31}P	30.974	100
^{13}C	13.003	1.1	^{32}S	31.972	95
^{14}N	14.0031	99.64	^{33}S	32.971	0.8
^{15}N	15.0001	0.36	^{34}S	33.968	4.2
^{16}O	15.9949	99.8	^{35}Cl	34.969	75.8
^{17}O	16.999	0.04	^{37}Cl	36.966	24.2
^{18}O	17.999	0.2	^{79}Br	78.918	50.5
^{19}F	18.998	100	^{81}Br	80.916	49.5
^{28}Si	27.977	92.2	^{127}I	126.904	100

Magnetic-sector instruments should have a mass accuracy of one part per million (ppm) or better, that is to say the mass of a fragment should be measurable to an accuracy of at least six significant figures. With this degree of accuracy it is possible to identify the atoms present in a fragment unequivocally provided the accurate masses of the isotopes are known. Accurate M_r values are arbitrarily standardized to ^{12}C (12.000000). Accurate mass measurement can be useful, for example, in identifying either the source of illicit drugs, or the origin of compounds used in the synthesis of such compounds.

Mass calibration of the instrument needs to be performed at the time of the analysis. Calibrants such as glycerol, perfluorokerosene (PFK), perfluorotributylamine (PFTBA), polyethylene glycol (PEG) and Ultramark 1621 (perfluoroalkylphosphazine) are used, depending on requirement. PFTBA is commonly employed in GC-MS work because it can be used in PCI and in NCI mode, and its fragmentation pattern covers a wide mass range. In addition, the M_r of fluorine is very close to its nominal mass of 19 (Table 10.1). Typically the ions at m/z 69, 219 and 502 (EI), 219, 414 and 652 (PCI), and 302, 452 and 633 (NCI) are used as calibrant ions. With quadrupole, QIT, or other low-resolution instruments (± 1 u), the nearest whole number is used, with exact masses to four decimal places being used in high resolution MS. In LC-MS instrument calibration is less well defined, and is usually developed for a particular instrument or application.

Poor availability of reference standards for designer drugs, metabolites and newly introduced substances is a continuing problem. Accurate mass measurement of sample components and comparison with databases of theoretical monoisotopic masses is one approach that has been adopted (Ojanperä *et al.*, 2005). Using LC-TOF-MS, a routine mass search window of 20–30 ppm was applied to urine samples. LC-FT-MS was capable of confirming the findings within a mass accuracy of 3 ppm. Thus, using a target database of 7640 compounds, the number of potential elemental formulae ranged from one to three with LC-TOF-MS, but it was always one with LC-FT-MS. In contrast to ordinary techniques requiring primary reference standards, the formula-based databases can be updated instantly with fresh numeric data from literature sources. Using current LC-TOF-MS instrumentation, a 10 ppm mass tolerance can be applied (Ojanperä *et al.*, 2006).

GC-MS fragmentation data are usually obtained under EI conditions at about 70 eV, although some such data may also be obtained by introduction of pure samples via a direct-insertion probe or by use of particle beam LC-MS. A fundamental requirement for producing EI mass spectra is that it must be possible to introduce the analyte into the MS in the vapour phase under reduced pressure. The resulting ions and ionized fragments, and their relative proportions, should be characteristic of the analyte and independent of the instrument used and mode of sample introduction.

General MS libraries, such as the Wiley Registry of Mass Spectral Data and that produced by the National Institute for Standards and Technology (NIST) in conjunction with the US Environmental Protection Agency (EPA) and National Institutes of Health (NIH) (McLafferty and US Department of Commerce, 2005), are available complete with database programmes that facilitate comparison of spectra from unknowns with those of known compounds. Various algorithms are available to help in compound identification. There are also libraries specifically created for use in analytical toxicology (Pfleger *et al.*, 2007; Kühnle, 2006; Rösner, 2006). If certain prerequisites are fulfilled, such libraries facilitate the identification of unknowns even in the absence of reference compounds (Maurer, 2006; Maurer and Peters, 2006).

When reference material is available, the ratios of several (usually three to five) selected ions from the unknown are compared with the ratios of those ions obtained from the standard, typically analyzed in the same batch (Figure 10.11). The presence of these ions combined with good agreement in the ion ratios with the ratios given by the reference standard, are generally accepted as sufficient to identify a compound positively, especially when considered together with chromatographic retention data. Use of selected ion ratios is less useful in identifying unknown compounds, and in such cases full-scan mass spectra are usually compared with library data, at least in the first instance. Several different search algorithms can be used to help identify compounds by comparison with the library spectra. If this approach fails, then study of the fragmentation pattern (Table 10.2) and consideration of other information such as any extraction and derivatization steps used

Table 10.2 Mass spectral interpretation: common fragment losses. Reproduced from Watson, D. (2004) Mass spectrometry, in *Clarke's Analysis of Drugs and Poisons* (eds A.C. Moffat, D. Osselton and B. Widdop), 3rd edn, Pharmaceutical Press, London, p. 386, with permission from RPS Publishing.

Mass loss (u)	Radical/neutral fragment	Likely origin
1	$H^•$	Amine or aldehyde
15	$CH_3^•$	Quaternary or TMS methyl
17	$OH^•$ or NH_3	
18	H_2O	Secondary or tertiary alcohol
19/20	$F^•/HF$	Fluoride
28	CO	Ketone/carboxylic acid
29	$C_2H_5^•$	Ethyl
30	CH_2O	Aromatic methyl ether
31	$CH_3O^•$	Methyl ester/methoxime
	CH_3NH_2	Secondary amine
32	CH_3OH	Methyl ester
33	$CH_3 + H_2O$	
35/36	$Cl^•/HCl$	Chloride
42	$CH_2{=}C{=}O$	Acetate
43	$C_3H_7^•$	(iso-)Propyl
	$CH_3CO^•$	Methyl ketone or acetate
	$CO + CH_3^•$	
44	CO_2	Ester
45	$CO_2H^•$	Carboxylic acid
46	$CO + H_2O$	
57	$C_4H_9^•$	Butyl
59/60	$CH_3COO^•/CH_3COOH$	Acetate
73	$(CH_3)_3Si^•$	TMS ether
90	$(CH_3)_3SiOH$	TMS ether

can often help in compound identification. A knowledge of likely metabolic pathways for a given analyte (phenolic hydroxylation, *N*-demethylation, etc.), may also be useful. Even so, fragment losses that are readily explained usually occur only within about 100 u of the molecular ion.

Problems with MS also arise in the case of volatile materials such as gases and many solvents. Some instruments have a low mass cut-off of 50 u, whilst the compounds themselves have relatively simple structures and thus produce very few diagnostic ions. Aliphatic hydrocarbons give very similar EI spectra, for example.

Spectra obtained using soft-ionization techniques such as CI can result in protonated and other adduct molecules that have a charge. Fragmentation of these ions is interpreted using the same general principles as used with EI spectra. Because these soft-ionization techniques are much gentler than EI, a molecular ion (or representation of it) is much more likely to be present in the spectrum, but fragmentation is usually much reduced and hence published libraries of CI spectra are generally less helpful than EI libraries in compound identification.

10.7 Quantitative mass spectrometry

Because of variations in the degree of ionization in MS, the use of internal standards is vital when performing quantitative measurements. Stable isotope-labelled analogues are used widely for this purpose (isotope dilution MS). Isotopic internal standards have virtually identical chemical and physical properties to the analyte and thus extraction, derivatization if needed, chromatography and fragmentation are often virtually identical. However, the site of isotopic labelling should be chosen such that the bonds linking the isotope are not broken during fragmentation as bonds involving heavier isotopes are more stable and the fragmentation pattern of the labelled compound could thus differ from that of the analyte. The internal standard may add to the degree of ion suppression in LC-MS, but generally both internal standard and analyte are affected equally by such phenomena (Sojo *et al.*, 2003). That labelled and unlabelled compounds may be partially resolved during the chromatographic analysis (e.g. deuterated analogues may elute slightly before the unlabelled analyte in GC-MS) must be borne in mind, not only with regard to choosing the correct integration parameters, but also because any ion suppression may differ between the internal standard and the analyte(s).

For MS analysis, the internal standard may be added at a higher concentration (e.g. 100-fold) than the analyte. This results in the assay being conducted over a narrower 'total' concentration range, thus minimizing the impact of possible concentration-related effects such as adsorption. Under these conditions the isotopic purity of the material will limit the amount that can be added for a particular lower limit of quantification (LLoQ). As the amount of nonlabelled material in the internal standard is constant, a calibration graph with a significant positive intercept is permissible providing the blank and lowest calibration standard can be distinguished (Section 14.2.4.6). Conversely, the presence of naturally occurring isotopes in the analyte can interfere with measurement of the internal standard ('cross talk'). This will be more prevalent at high analyte concentrations, when the masses of analyte and internal standard are only different by 1 or 2 u, and when the analyte contains elements with a high abundance of heavier isotopes (e.g. sulfur, Table 10.1).

Deuterium is the most commonly used stable isotope label. This is because of not only cost, but also the availability of high purity reagents. The presence of each deuterium atom in the molecule increases the M_r of the molecule by 1. For high-sensitivity analyses it is recommended that the mass of the internal standard is at least 3 u greater than the analyte to reduce interference from naturally occurring isotopes. It has been suggested that whenever possible ^{13}C-labelled materials should be used with MS detection (Chang *et al.*, 2002), but obviously this would add considerably to the cost of an assay and may make it more difficult to achieve the required mass difference between the analyte and internal standard. When stable isotope-labelled analogues are not available or the procedure can be used to quantify a large number of analytes simultaneously, internal standards that have chemical properties close enough to the analyte(s) to yield reliable quantitative data are used.

It was generally assumed that little could go wrong when isotopically labelled internal standards were used. Deuterated and tritiated internal standards are more likely to be affected by isotopic effects than, say, those containing ^{13}C, as such atoms are proportionally much heavier than hydrogen. The vibrational frequencies of carbon–deuterium bonds are less than those of the corresponding carbon–hydrogen bonds, so that deuterated compounds tend to be more stable than their unlabelled homologues. Partial or complete HPLC resolution has been reported for some ^{2}H internal standards (de Ridder and van Hal, 1976). Baseline resolution of 1-hydroxypyrene from [$^{2}H_9$]-1-hydroxypyrene (internal standard), for example, has allowed fluorescence rather than MS detection (Chetiyanukornkul *et al.*, 2002).

In MS-MS work, either the response from a single daughter ion from the analyte and from the internal standard may be monitored, or the ratio of response two daughter ions for each compound may be monitored (multiple reaction monitoring, MRM).

Resolution in MS is analogous to the separation of analytes in chromatography and is just as important if quantitative measurements are to be performed. Unit resolution implies that two mass peaks are completely separated (0 % peak:valley ratio). Magnetic-sector instruments can effect resolution of up to 20,000 at 50 % peak:valley, that is resolution between 20,000 and 20,001 u can be achieved with 50 % of the peak height between the peaks. ToF instruments can nowadays attain similar mass accuracy. However, simple quadrupole instruments can only provide resolution of about 500.

There is no specific test of sensitivity of a MS system other than those normally employed with a chromatographic system (LoD, LLoQ – see Chapter 14).

10.8 Summary

A thorough knowledge of GC- and LC-MS is fundamental to modern analytical toxicology. Capillary GC-EI-MS used in conjunction with appropriate sample preparation and temperature programming provides sensitivity at best on a par with GC-FID, but this is combined with unparalleled selectivity. There are, however, problems for the unwary. Although valuable in analyte identification, some analytes give poor EI spectra unless derivatized. However, many analytes are not amenable to derivatization. Daily mass calibration is essential. CI with SIM gives better sensitivity with targeted analytes, but all the restrictions surrounding GC remain. LC-MS and LC-MS^n are of increasing importance, in part because the need for temperature stability and/or derivatization is obviated, but here too thorough attention to detail is needed to avoid pitfalls such as signal suppression due to co-eluting sample components. Moreover, there is as yet no counterpart to EI in LC-MS.

References

Abian, J. (1999) The coupling of gas and liquid chromatography with mass spectrometry. *J Mass Spectrom*, **34**, 157–68.

Barwick, V., Langley, J., Mallet, T., Stein, B., and Webb, K. Best practice guide for generating mass spectra. LGC, 2007 (http://www.vam.org.uk/publications/publications_item.asp?)intPublicationID=1328, accessed 4 August 2007).

Boermans, P.A.M.M., Go, H.S., Wessels, A.M.A. and Uges, D.R.A. (2006) Quantification by HPLC-MS/MS of atropine in human serum and clinical presentation of six mild-to-moderate intoxicated atropine-adulterated-cocaine users. *Ther Drug Monit*, **28**, 295–8.

Chang, W.T., Smith J. and Liu, R.H. (2002) Isotopic analogs as internal standards for quantitative GC/MS analysis—molecular abundance and retention time difference as interference factors. *J Forensic Sci*, **47**, 873–81.

Chen, J., Korfmacher, W.A. and Hsieh, Y. (2005) Chiral liquid chromatography-tandem mass spectrometric methods for stereoisomeric pharmaceutical determinations. *J Chromatogr B*, **820**, 1–8.

Chetiyanukornkul, T., Toriba, A., Kizu, R., Makino, T., Nakazawa, H. and Hayakaw, K. (2002) Determination of 1-hydroxypyrene in human urine by high-performance liquid chromatography with fluorescence detection using a deuterated internal standard. *J Chromatogr*, **961**, 107–12.

Cody, J.T. (1992) Determination of methamphetamine enantiomer ratios in urine by gas chromatography-mass spectrometry. *J Chromatogr*, **580**, 77–95.

Downard, K. (2004) *Mass Spectrometry: A Foundation Course*, Royal Society of Chemistry, Cambridge.

Holt, D.W., Lee, T., Jones, K. and Johnston, A. (2000) Validation of an assay for routine monitoring of sirolimus using HPLC with mass spectrometric detection. *Clin Chem*, **46**, 1179–83.

Hopfgartner, G., Husser, C. and Zell, M. (2003) Rapid screening and characterization of drug metabolites using a new quadrupole-linear ion trap mass spectrometer. *J Mass Spectrom*, **38**, 138–50.

Kratzsch, C., Tenberken, O., Peters, F.T., Weber, A.A., Kraemer, T. and Maurer, H.H. (2004) Screening, library-assisted identification and validated quantification of 23 benzodiazepines, flumazenil, zaleplone, zolpidem and zopiclone in plasma by liquid chromatography/mass spectrometry with atmospheric pressure chemical ionization. *J Mass Spectrom*, **39**, 856–72.

Kühnle, R. (2006) *Mass Spectra of Pharmaceuticals and Agrochemicals*, Wiley-VCH, Weinheim.

Matuszewski, B.K., Constanzer, M.L. and Chavez-Eng, C.M. (2003) Strategies for the assessment of matrix effect in quantitative bioanalytical methods based on HPLC-MS/MS. *Anal Chem*, **75**, 3019–30.

Maurer, H.H. (2006) Hyphenated mass spectrometric techniques—indispensable tools in clinical and forensic toxicology and in doping control. *J Mass Spectrom*, **41**, 1399–413.

Maurer, H.H. (2005) Multi-analyte procedures for screening for and quantification of drugs in blood, plasma, or serum by liquid chromatography-single stage or tandem mass spectrometry (LC-MS or LC-MS/MS) relevant to clinical and forensic toxicology. *Clin Biochem*, **38**, 310–8.

Maurer, H.H. (2004) Position of chromatographic techniques in screening for detection of drugs or poisons in clinical and forensic toxicology and/or doping control. *Clin Chem Lab Med*, **42**, 1310–24.

Maurer, H.H. (2002) Role of gas chromatography-mass spectrometry with negative ion chemical ionization in clinical and forensic toxicology, doping control, and biomonitoring. *Ther Drug Monit*, **24**, 247–54.

Maurer, H.H. and Peters, F.T. (2006) Analyte identification using library searching in GC-MS and LC-MS, in *Encyclopedia of Mass Spectrometry* (eds M. Gross, R.M. Caprioli), Elsevier Science, Oxford, pp. 115–121.

McLafferty, F.W. and US Department of Commerce. (2005) *Wiley Registry of Mass Spectral Data* Edition 7 with *NIST/EPA/NIH 2005 Spectral Data*, Wiley, New York.

Mei, H., Hsieh, Y., Nardo, C., Xu, X., Wang, S., Ng, K. and Korfmacher, W.A. (2003) Investigation of matrix effects in bioanalytical high-performance liquid chromatography/tandem mass spectrometric assays: application to drug discovery. *Rapid Commun Mass Spectrom*, **17**, 97–103.

Ojanperä, I., Pelander, A., Laks, S., Gergov, M., Vuori, E. and Witt, M. (2005) Application of accurate mass measurement to urine drug screening. *J Anal Toxicol*, **29**, 34–40.

Ojanperä, S., Pelander, A., Pelzing, M., Krebs, I., Vuori, E. and Ojanperä, I. (2006) Isotopic pattern and accurate mass determination in urine drug screening by liquid chromatography/time-of-flight mass spectrometry. *Rapid Commun Mass Spectrom*, **20**, 1161–7.

Peters, F.T., Schaefer, S., Staack, R.F., Kraemer, T. and Maurer, H.H. (2003) Screening for and validated quantification of amphetamines and of amphetamine- and piperazine-derived designer drugs in human blood plasma by gas chromatography/mass spectrometry. *J Mass Spectrom*, **38**, 659–76.

Pfleger, K., Maurer, H.H. and Weber, A. (eds) (2007) *Mass Spectral and GC Data of Drugs, Poisons, Pesticides, Pollutants and Their Metabolites*, 3rd edn, Wiley-VCH, Weinheim.

de Ridder, J.J. and van Hal, H.J.M. (1976) Unexpected high-performance liquid chromatographic separation of Org CG 94 and [3,3,4,4–2H4]Org GC 94. *J Chromatogr*, **121**: 96–9.

Robb, D.B., Covey, T.R. and Bruins, A.P. (2000) Atmospheric pressure photoionization: an ionization method for liquid chromatography-mass spectrometry. *Anal Chem*, **72**, 3653–9.

Rösner, P. (2006) *Mass Spectra of Designer Drugs 2006*, Wiley-VCH, Weinheim.

Smith, R.M. (2004) *Understanding Mass Spectra—A Basic Approach*, 2nd edn, Wiley, New York.

Sojo, L.E., Lum, G. and Chee, P. (2003) Internal standard signal suppression by coeluting analyte in isotope dilution LC-ESI-MS. *Analyst*, **128**, 51–4.

Speranza, M., Satta, M., Piccirillo, S., Rondino, F., Paladini, A., Giardini, A., Filippi, A. and Catone, D. (2005) Chiral recognition by mass-resolved laser spectroscopy. *Mass Spectrom Rev*, **24**, 588–610.

Staack, R.F., Fritschi, G. and Maurer, H.H. (2003) New designer drug 1-(3-trifluoromethylphenyl)piperazine (TFMPP): gas chromatography/mass spectrometry and liquid chromatography/mass spectrometry studies on its phase I and II metabolism and on its toxicological detection in rat urine. *J Mass Spectrom*, **38**, 971–81.

Van Bocxlaer, J.F., Clauwaert, K.M., Lambert, W.E., Deforce, D.L., Van den Eeckhout, E.G. and De Leenheer, A.P. (2000) Liquid chromatography-mass spectrometry in forensic toxicology. *Mass Spectrom Rev*, **19**, 165–214.

Vekey, K., Edwards, D. and Zerilli, L.F. (1989) Liquid chromatography-mass spectrometry coupling and its application in pharmaceutical research. *J Chromatogr*, **488**, 73–85.

Voyksner, R.D. and Pack, T.W. (1989) Semi-quantitative method for high molecular weight neuropeptides by high-performance liquid chromatography/thermospray mass spectrometry. *Biomed Environ Mass Spectrom*, **18**, 897–903.

Wallemacq, P.E., Vanbinst, R., Asta, S. and Cooper, D.P. (2003) High-throughput liquid chromatography-tandem mass spectrometric analysis of sirolimus in whole blood. *Clin Chem Lab Med*, **41**: 921–5.

Watson, J.T. (ed) (1997) *Introduction to Mass Spectrometry*, 3rd edn, Lippincott-Raven, New York.

Webb, K.S. (1991) The identification of drugs by mass spectrometry, in *The Analysis of Drugs of Abuse* (ed T.A. Gough), Wiley, Chichester, pp. 175–224.

11 Trace Elements and Toxic Metals

11.1 Introduction

Spectroscopy is the study of interactions between matter and electromagnetic radiation. When applied to quantitative analysis, the term spectrometry is used. Different types of spectroscopy are concerned with different regions of the electromagnetic spectrum (X-ray, UV, IR, etc.), the properties of the matter with which the interactions occur, for example molecular vibration, electron transitions and so on, and the physical interactions involved, that is the scattering, absorption or emission of radiation (Haswell, 1991; Lobinski and Marczenko, 1996). An annual review of recent developments is published in the Journal of Analytical Atomic Spectrometry (Taylor *et al.*, 2007).

With modern analytical techniques it is possible to measure individual elements at concentrations of parts per thousand million ($\mu g\ L^{-1}$ or $\mu g\ kg^{-1}$) or less using microlitres or milligrams of sample. For many years, atomic absorption spectrometry (AAS) has been the technique of choice as it can give accurate results quickly and without the need for very expensive equipment. It can be used to measure more than 60 elements, but can only be used to measure one element at any one time.

The field of trace element analysis changed radically with the development of inductively coupled plasma-MS (ICP-MS) (Gray and Date, 1983). The detection limits and rate of sample analysis are equal to or better than those attainable with AAS, but, in addition, isotopes of most elements within the periodic table may be measured during the course of a single analysis. However, the costs associated with the purchase and operation of ICP-MS and, more importantly, the interferences that were evident in the analysis of biological samples, prevented the widespread application of the technique to clinical, toxicological and forensic situations until recently.

11.1.1 Historical development

As described in Chapter 1, the earliest chemical tests developed for toxicological analysis were those used to detect the presence of arsenic and other toxic metal oxides or salts. Arsenic was the principal homicidal poison in common use from medieval times up to the mid 1850s. Prior to 1836 only relatively insensitive and poorly specific colour tests were available (Male, 1816). Thus, a solution of silver nitrate in ammonium hydroxide, when added to a neutral solution of white arsenic (arsenious oxide, As^{III} oxide, As_4O_6) produced a yellow precipitate of what chemists of the period termed silver arsenite; in acidic solutions, the same reagent produced a red-brown precipitate of silver arsenate. In modern chemical terms, these were compounds of arsenic in two different oxidation states, As^{III} (arsenite) and As^{V} (arsenate). If hydrogen sulfide was bubbled through a solution of white arsenic, yellow arsenic sulfide (As_2S_3) was formed. In addition, when a basic solution of copper sulfate was mixed with arsenic it produced a green precipitate of copper arsenite (Scheele's green, $CuHAsO_3$). However, the introduction of the Marsh (1836) apparatus, in which metallic arsenic was separated from the sample matrix by sublimation, made these and similar tests largely redundant, except when used as confirmatory tests on material isolated from a biological sample.

The pioneering work of Henry Bence Jones, who used spectroscopy to study the absorption and distribution of lithium (1865), was mentioned in Chapter 4. Qualitative colorimetric assays were gradually refined to give quantitative or semiquantitative results, although sensitivity was limited to confirmation of gross exposure. Nevertheless, some important work was carried out, for example in the measurement of blood lead resulting from occupational exposure to this element (Kehoe, 1961). Introduction of AAS (Walsh, 1955), initially with the air–acetylene flame and then with the electrically heated graphite tube (Massman, 1968), started a new era characterized by lower LoDs and an increase in the range of elements that could be measured (Taylor, 2006a). However, to perform multi-element analyses it was necessary to use more complex and less sensitive methods such as direct current arc atomic emission spectroscopy (AES *et al.*, 1981). Other multi-element techniques (neutron activation analysis, NAA; X-ray fluorescence, XRF) have been available for some years, but these either lack the sensitivity of AAS, or require very specialized instrumentation that is not readily accessible.

The picture has changed yet again with the use of collision and dynamic reaction cells with ICP-MS. This technology eliminates many sources of interference and has been accompanied by a considerable reduction in instrument cost. Other important advances include the use of techniques that separate different molecular conformations of an element (speciation) and *in vivo* analysis, initially involving NAA, but now using XRF.

11.2 Sample collection and storage

By definition, trace elements are present at low concentrations in biological samples and many of the analytes that need to be measured occur naturally not only in such specimens, but also in the environment. Given that the analytical methods used are, in the main, inherently selective, the most important consideration in sample collection and storage is that of sample contamination (Taylor, 2006b). As a general rule, plastic containers are preferable to glass (Table 11.1). A 'blank' analysis guards against contamination from the sample tube going unrecognized. In the case of exhumations, samples of the surrounding soil should also be collected for analysis.

Anticoagulants and preservatives may be a source of contamination (Frank *et al.*, 2001). EDTA is the anticoagulant of choice for the collection of whole blood, as samples do not become viscous after a few days, unlike heparinized blood. However, if blood is allowed to clot and is then sent to the laboratory unseparated it will not be suitable for iron and zinc assays due to leakage of these analytes from erythrocytes.

As regards sample containers, possible sources of contamination include rubber (found in some stoppers) and the O-rings inside certain tube closures. Some bottle lids contain a liner, which may have a metal component. Vacuum blood collection systems represent another major source of contamination. Those containing clot activator and gel cell separation materials are likely to be unusable. There are sample collection systems marketed specifically for trace element analysis, but it is wise to check batches of such tubes for contamination before use with clinical samples. Needles may also be sources of contamination (e.g. with chromium), and it may be necessary to collect blood through a plastic cannula after discarding the first few millilitres. Elements most

Table 11.1 Sample requirements for measurements of metals/trace elements

Element	Sample requirements	Comments
Aluminium	5 mL whole blood in plastic tube (no anticoagulant/separating beads[a]) 5 mL dialysate/supply water in plastic bottle rinsed several times with portions of the intended sample[a]	Use plastic and not glass tubes. Blood should not be separated
Antimony	5 mL EDTA whole blood 20 mL urine	Urine preferred
Arsenic	5 mL EDTA whole blood 20 mL urine	To diagnose chronic poisoning exclude seafood (shell fish, etc.) from diet for five days before sample collection if total arsenic is to be measured
Bismuth	5 mL EDTA whole blood	
Bromide	5 mL EDTA whole blood	Interpretation of the result depends on whether an organobromine or inorganic bromide is the source of exposure
Cadmium	2 mL EDTA whole blood[a] 10 mL urine[a]	Blood preferred
Chromium	2 mL EDTA whole blood[a] 20 mL urine (hard plastic bottle)[a]	Use of a plastic cannula to collect blood is advisable
Copper	2 mL EDTA or clotted whole blood, or 1 mL plasma/serum	If Wilson's disease suspected, also send 10 mL urine
Fluoride	3 mL EDTA or clotted whole blood, or 1.5 mL plasma/serum, or 20 mL urine	
Iron	2 mL serum (no haemolysis)	To monitor use of a chelating agent send 20 mL urine
Lead	2 mL EDTA whole blood	Blood should not be separated and free of clots
Lithium	5 mL clotted blood or 2 mL serum	Do not use lithium heparin anticoagulant
Manganese	1 mL EDTA whole blood or 0.5 mL plasma[a]	Use of a plastic cannula to collect blood is advisable
Mercury	5 mL EDTA whole blood 20 mL urine (hard plastic bottle)	Send samples promptly to avoid loss of mercury on storage
Nickel	20 mL urine	
Selenium	2 mL EDTA whole blood or 1 mL plasma/serum	

Table 11.1 *(continued)*

Element	Sample requirements	Comments
Silver	2 mL EDTA whole blood or 1 mL plasma	
Strontium	2 mL EDTA whole blood or 1 mL plasma	
Thallium	5 mL EDTA whole blood 20 mL urine	
Zinc	1 mL plasma/serum (no haemolysis)	Do not use EDTA anticoagulant

[a]Send unused sample container from the same batch as used for sample collection to check for contamination

susceptible to contamination from an exogenous source are aluminium, iron, manganese and zinc, but nothing can be assumed to be unimportant.

Analyte concentrations are unlikely to alter on storage except in the case of mercury, but to avoid deterioration of the biological matrix, samples may be kept at 4 °C for short periods, and stored at −20 to −80 °C for longer-term preservation. In dilute solution, trace elements may be adsorbed onto the walls of a container. To prevent this from happening in urine, which normally contains little organic material to bind trace elements, it is advisable to acidify specimens as soon as possible after collection, for example by adding concentrated nitric acid to give a final concentration of 1 % (v/v).

Analysis of tissues other than blood is best performed using freshly collected material, although it is possible to examine specimens fixed in formalin and even embedded in wax. However, there are potential problems of both contamination and loss associated with fixing and embedding, hence careful monitoring for artefacts arising from pre-analytical treatment of the sample is essential.

11.3 Sample preparation

The aims of sample preparation are: (i) to transform the specimen into a suitable form (usually a liquid) for introduction to the analytical instrument, with the analytes at concentrations that will produce a measurable response and (ii) to reduce or eliminate possible interferences.

When instrument sampling systems include narrow capillary tubing it is important that the transfer rates of samples and calibration solutions are equal. The high concentrations of cells, proteins and other high M_r materials in samples such as serum and blood give rise to viscosity that is absent in aqueous calibration solutions. Therefore, assays may be subject to matrix interferences unless the viscosities of sample and calibration solution are matched. When the concentrations of analyte(s) are high, dilution with water may be all that is needed. Alternatively, protein precipitation can be performed by addition of, for example, 10 % (w/v) trichloroacetic acid followed by centrifugation (Section 3.2.2).

A particularly useful approach is addition of a chelating agent followed by solvent extraction of the resulting complex. Interfering material is retained in the aqueous phase and, if the volume of solvent is less than that of the sample, analyte enrichment is also achieved. Various chelating agents and solvents have been employed, but a common general-purpose combination is ammonium pyrrolidine dithiocarbamate (APDC) and 4-methyl-2-pentanone (methyl isobutyl ketone, MIBK) (Lopez-Artiguez *et al.*, 1993).

As with sample collection, the importance of avoiding contamination throughout the sample preparation procedure cannot be overemphasized. Equipment must be tested for cleanliness and reagents should be of the highest quality available. All reusable materials should be cleaned before use. Glassware should be first soaked in aqueous acid, for example 2–10 % (v/v) hydrochloric acid or nitric acid, and then rinsed with copious volumes of purified water. Pure water has an electrical resistance of more than 18 MΩ and ideally should be used at all times. Blank samples should be taken through the entire procedure and IQC samples analyzed routinely. In addition to guarding against false high results, it is necessary to check for losses of analyte due to volatilization, adsorption onto container surfaces, or the precipitation of insoluble complexes.

11.3.1 Analysis of tissues

Solid specimens have to be digested to give an aqueous solution prior to analysis. Acid digestion is used most commonly. Methods with addition of nitric acid, perchloric acid and hydrogen peroxide in varying combinations, sometimes with the inclusion of sulfuric acid, are widely employed and a number of possible procedures are available (Bazzi *et al.*, 2005). Care must be exercised when using the potentially explosive perchloric acid.

Heating may be performed in conical flasks or beakers, or in tubes that fit an aluminium heating block. Usually, the heater temperature is gradually increased to around 250 °C, although higher temperatures may be required for some applications. Care is necessary to avoid loss of more volatile elements; mercury will almost certainly evaporate using this approach. Digestion is normally complete within 2–6 h. Microwave heating in sealed digestion vessels is used widely (Grinberg *et al.*, 2005). Ovens have been specially designed for this purpose and include safety features to prevent damage from acid fumes and excessive pressure within digestion vessels. As many as 40 or 50 samples can be digested at the same time, often in less than 30 min, and because each vessel is sealed there should be no loss of volatile elements.

In a different approach, sample and acid can be placed in a PTFE vessel (pressure bomb), which fits within a stainless steel housing. The vessel is heated in either a microwave or conventional oven and the sample is effectively destroyed after an hour or so, although the unit must be allowed to cool before it is opened. Because pressure bombs are very expensive most laboratories have a limited number, which may restrict sample throughput. Again, there should be no loss of volatile elements.

A second technique involves dissolution of the sample in a concentrated alkaline solution similar to those used to prepare tissue samples for liquid scintillation counting. Tetramethylammonium hydroxide (TMAH) is widely used and samples are simply mixed with the reagent and heated for a few hours. By taking 50 mg of dried tissue and 2 mL 50 g L^{-1} TMAH into a screw-capped vial, complete solubilization is achieved by heating at 90 °C (2 h) with occasional shaking. In yet another approach, sample destruction may be achieved by heating the sample in a muffle furnace, typically to 400–450 °C, followed by cooling and dissolution of the ash in 1 % (v/v) nitric acid. Loss of volatile elements must be considered and this technique cannot be used if mercury is to be measured. These techniques for preparation of solid samples may also be used with liquid specimens, both to eliminate matrix interferences and to effect analyte enrichment.

11.3.2 Analyte enrichment

In the quest for lower LoDs, analyte enrichment has become important. Some examples have been mentioned above, but many methods have been developed for the specific purpose of concentrating the analyte. These involve trapping from a relatively large sample and subsequent elution into a smaller volume. The methods used include: (i) adsorption onto materials such as charcoal or silica, (ii) use of ion-exchange resins either with, or without functionalized groups to trap specific elements and (iii) size-exclusion chromatography (SEC; Loreti and Bettmer, 2004). Enrichment in this way can be off-line, but many methods incorporate on-line sample processing.

Experience with ion-exchange chromatography and SEC has led to a further important development in sample preparation, speciation. In some circumstances, measurement of the total concentration of an element can be misleading, hence it is necessary to measure particular species, for example Cr^{III} and Cr^{VI}, and As^{III}, monomethylarsonic acid and dimethylarsinic acid (Mandal *et al.*, 2004). Methods for speciation involving chromatographic separation or differential solvent extraction are used routinely (Taylor *et al.*, 2007).

11.4 Atomic spectrometry

Quantitative analytical atomic spectrometric techniques include AAS, AES, atomic fluorescence (AFS), inorganic MS and XRF. AAS, AES and AFS exploit interactions between UV/visible light and the outer-shell electrons of free, gaseous, uncharged atoms. In XRF, high-energy particles collide with inner-shell electrons of atoms to initiate transitions that culminate in the emission of X-ray photons. For inorganic MS, a magnetic field separates ionized analyte atoms according to their mass to charge (m/z) ratio.

11.4.1 General principles of AES, AAS and AFS

Uncharged atoms may exist at the most stable or *ground state* (E^0), having the lowest energy, or at any one of a series of *excited states* depending on how many electrons have been moved to higher energy levels, although it is usual to consider just the first transition (E'). This may be visualized in an energy level diagram (Figure 11.1). The energy levels, and the differences (ΔE) between them, are unique to each element.

The ΔE for movements of *outer-shell electrons* in most elements corresponds to the energy equivalent to UV/visible radiation. The energy of a photon is characterized by:

$$E = h\nu \tag{11.1}$$

where h is the Planck constant and ν is the frequency of the waveform corresponding to that photon. Furthermore, frequency and wavelength are related as:

$$\nu = \frac{c}{\lambda} \tag{11.2}$$

where c is the velocity of light and λ is the wavelength. Therefore,

$$E = \frac{hc}{\lambda} \tag{11.3}$$

and so a specific transition, ΔE, is associated with a unique wavelength.

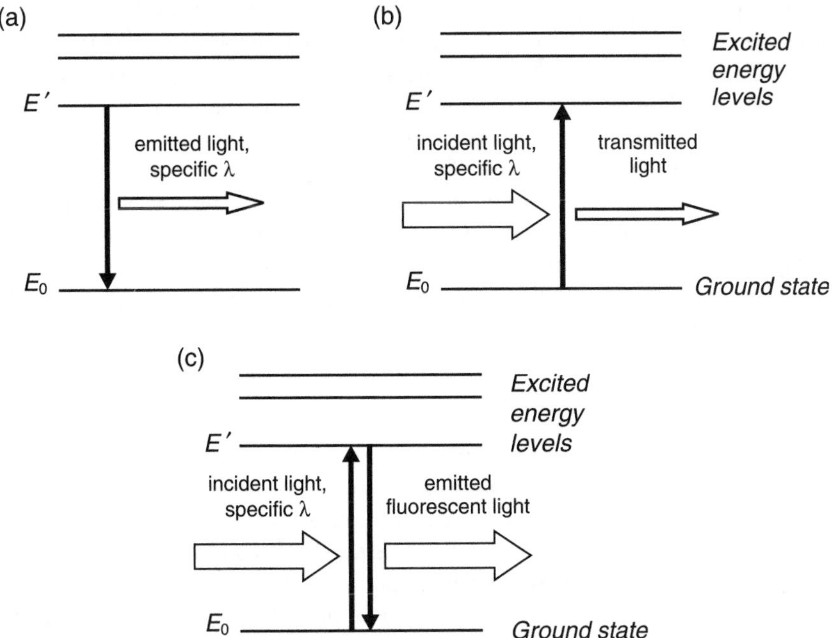

Figure 11.1 Energy level diagrams to show transitions associated with (a) AES, (b) AAS and (c) AFS. The vertical arrows indicate absorption or emission of radiant energy (light).

Under appropriate conditions, outer-shell electrons of vaporized, ground-state atoms within the analytical system may be excited by heating. As these atoms return to the more stable ground state, energy is lost. Some of this energy is emitted as light that can be measured. The intensity of the emitted light is proportional to the number of atoms present and the process is atomic emission spectrometry (AES), Figure 11.1(a).

When light of a specific wavelength enters an analytical system, outer-shell electrons of corresponding vaporized ground-state atoms within the light path will be excited as energy is absorbed. Consequently the amount of light transmitted through the system to the detector will be attenuated. The loss of light is proportional to the number of atoms present and the method is known as atomic absorption spectrometry (AAS), Figure 11.1(b).

Some of the radiant energy absorbed by ground-state atoms can be emitted as light as the atom returns to the ground state. This emission is known as resonance fluorescence and is again proportional to the number of atoms in the light path. The technique is known as atomic fluorescence spectrometry (AFS) Figure 11.1(c).

It follows from Equations (11.1)–(11.3) that the wavelengths of the absorbed and emitted light are the same, and are unique to any given element. This gives AAS, AES and AFS great specificity, so that one element can be measured accurately even in the presence of an enormous excess of a chemically similar element.

11.4.2 Atomic absorption spectrometry

AAS can be used to measure more than 60 elements with instrumentation that is comparatively inexpensive and simple to operate. The method has sufficient sensitivity to measure many of these elements at the concentrations present in clinical specimens (Haswell, 1991). The spectrometer consists of a light source, atomizer, monochromator, detector and readout/display. The essential feature of a good light source for AAS is to provide a high intensity, monochromatic output, which is achieved with hollow cathode or electrodeless discharge lamps. The monochromator, detector and display are similar to those of other spectrometers.

The atomizer is any device that will generate ground-state atoms as a vapour within the light path of the instrument. In the case of serum calcium, for example, the element is present bound to protein, complexed with phosphate and as free Ca^{2+}. Atomization requires: (i) removal of solvent (drying), (ii) separation from anionic or other components of the matrix to give Ca^{2+} and (iii) reduction ($Ca^{2+} + 2e^- \rightarrow Ca^0$). The necessary energy is supplied as heat from either a flame, or an electrically heated furnace.

11.4.2.1 Flame atomization

The typical arrangement involves a pneumatic nebulizer, mixing chamber and an air–acetylene laminar flame with a 10 cm path length (Figure 11.2). Acetylene burns in air at about 2000 °C. A nitrous oxide-acetylene flame, which reaches approximately 3000 °C, is used for elements such as aluminium and chromium that form refractory oxides and have no effective atomization in an air–acetylene flame. A high-speed auxiliary air flow causes the sample solution to be drawn continuously through the capillary due to the Venturi effect. The sample uptake rate through the nebulizer is usually about 5 mL min^{-1} and aspiration for several seconds is necessary to achieve a steady signal. The sample emerges from the nebulizer as an aerosol with a wide range of droplet sizes, is mixed with the combustion gases and transported to the flame for atomization. However, only droplets less than 10 μm diameter actually enter the flame, larger droplets falling to the sides of the chamber and running to waste. Thus, no more than about 15 % of the sample enters the flame. Hence, with the pneumatic nebulizer, the original sample undergoes dilution with the flame gases, a portion is lost in the mixing chamber and there is considerable thermal expansion (i.e. further dilution) within the flame. In addition to dispersion of sample through the flame, there are losses of atoms due to the formation of oxides or other species at the flame margins.

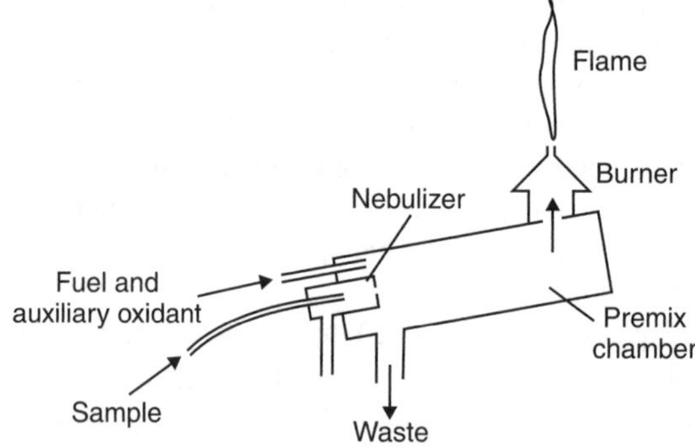

Figure 11.2 Pneumatic nebulizer for flame AAS.

The advantages and disadvantages of the pneumatic nebulizer-flame atomization system are shown in Box 11.1. Because of its simplicity, speed and freedom from interferences, this approach

Box 11.1 Advantages and disadvantages of pneumatic nebulization

Advantages

- Rapid
- Reproducible
- Few interferences
- Steady-state signal

Disadvantages

- Only about 15 % of the sample enters the flame
- Wide range of droplet sizes
- Low atomic density of sample in the flame
- Burner conditions impose limitations on nebulizer

should be used whenever possible. The lowest analyte concentrations that can be measured are typically about 1 mg L^{-1}.

Improved sensitivity is obtained with devices that overcome the limitations of pneumatic nebulizers. Such devices include those that: (i) trap atoms and give a greater density within the light path, (ii) by-pass the nebulizer so that all of the sample is atomized and (iii) introduce the sample as a single, rapid pulse rather than a continuous flow. Some employ a combination of these features. Devices used in a flame, for example the slotted quartz tube and Delves' cup, are most effective with more volatile elements such as zinc, cadmium and lead. These three approaches to improved sensitivity also feature in other atomizers used in AAS, AES, AFS and ICP-MS.

11.4.2.2 Electrothermal atomization

Most systems use an electrically heated graphite tube to vaporize the analyte and so this technique is often called graphite furnace atomization, although different materials are sometimes employed. Electrical contact is made to the ends of the tube and a voltage is applied. Resistance to the flow of current causes the temperature of the furnace to increase. A temperature program is used so that a solution placed inside the furnace is carefully dried, organic material is destroyed during an ashing step and the analyte ions dissociated from anions. With a rapid increase in temperature, analyte ions are vaporized and reduced to ground state atoms prior to spectrometric detection. A further temperature increase ensures that the graphite tube is clean for the next analysis.

The atomization temperatures achieved by this technique can be up to 3000 °C, allowing refractory elements such as aluminium and chromium to be measured. Typically only 10–50 μL of sample are needed for injection into the furnace and, because the entire sample is atomized within a small volume, a dense atom population is produced. The technique is, therefore, very sensitive and analyte concentrations of a few μg L^{-1} can be measured. However, electrothermal atomization AAS (ETAAS) is subject to greater potential interference than flame AAS (FAAS) and procedures to either eliminate, or compensate for such interference are necessary. Different forms of graphite (electrographite, pyrolitically coated graphite and total pyrolytic graphite) are used and the design of the furnace and its mode of heating are optimized to promote atomization and reduce interferences.

11.4.2.3 Sources of error

Devices that involve flow of solutions, such as nebulizers or flow injection systems, will give inaccurate results if the samples and calibrators have different viscosities, as this causes rates of flow into the analyzer to differ. If this occurs, internal standardization, standard addition or the addition of reagents to equalize analyte and calibrant flow rates can be used to improve accuracy. Sample viscosity can be reduced by dilution with water, but this approach may give very low absorbance readings. Different strategies to overcome this problem in the measurement of serum copper and zinc are outlined in Box 11.2. Note that when the method includes several steps the possibility for contamination to occur, especially with zinc, is increased. Whichever sample preparation method is used, it is important to measure the aspiration rates of samples and calibration solutions through the nebulizer premix chamber to ensure that they are equivalent.

Box 11.2 Actions to match sample and calibrator viscosity in the measurement of serum copper and zinc by FAAS

- Prepare the calibration solutions in 2 % (v/v) glycerol
- Add 1 part 10 % (w/v) TCA to 1 part serum or calibration solution. Mix well, centrifuge and remove the supernatant
- Dilute the samples and calibrators with 5 % (v/v) aqueous butanol
- Dilute the mixture 5- to 10-fold in purified water

Chemical interactions may influence the rate of atomization. During FAAS, calcium bound to phosphate in serum is not entirely separated at 2000 °C and gives a lower result than an equivalent concentration in an aqueous calibrant. Addition of a releasing agent such as lanthanum (La^{3+}), which binds preferentially to phosphate, avoids this interference.

Problems may also occur within the graphite furnace. Components in the sample matrix can cause the analyte to become volatile and be lost during the ashing step, for example lead can be lost as lead chloride, $PbCl_2$. However, the most difficult problems are those that develop in the vapour phase as the samples are heated. Ground-state atoms are vaporized into a rapidly changing thermal environment where the gas-phase temperature is lower than that of the tube wall and the ends of the tube are at a lower temperature than the centre. Consequently, vaporized atoms may condense on the cooler parts of the tube and then revaporize as the gas-phase temperature increases to the analyte appearance temperature, giving double peaks. This is because the molecular species formed give nonatomic absorption. Compensation for nonatomic absorption is provided by using chemical modifiers to stabilize the atoms and/or to promote destruction of the matrix at an earlier stage of heating (Table 11.2), by devices to establish isothermal atomization conditions and by background correction (BC) techniques (Table 11.3).

A typical method for the measurement of blood lead (Box 11.3) exploits each of these features. Matrix-matched calibration solutions are prepared in a sample of blood with a low concentration of lead. Blood is diluted 15-fold with a phosphate/Triton solution and introduced into the graphite tube furnace of the AAS. The instrument is temperature programmed to dry, ash and atomize samples at controlled rates. Oxygen is passed into the furnace during the ash phase to facilitate the destruction of organic material. Background correction is essential to compensate for the nonatomic absorption.

Table 11.2 Chemical modifiers used in ETAAS

Modifier	Function
Nickel; ammonium phosphate; palladium; ruthenium	Form a thermostable complex with the analyte ions to allow a higher ash temperature to be used and so remove interfering species
Gaseous oxygen	Assists the ashing of the organic matrix. Mediates formation of atoms via activation of the graphite surface
Nitric acid; ammonium nitrate	Promotes low-temperature volatilization of halides to prevent analyte–chloride vapour phase interferences
Magnesium nitrate	Delays atomization so that isothermal conditions can be established inside the furnace

Table 11.3 Approaches to eliminate nonatomic absorption in ETAAS

Technique	Rationale	Examples
Chemical modifiers	Promote the destruction of sample matrix. Delay atomization of analyte	Oxygen, magnesium nitrate, diammonium hydrogen orthophosphate
Isothermal atomization	Reduce vapour phase interactions by delaying atomization until furnace reaches constant temperature	L'Vov platform, graphite probe, novel furnaces
Background correction	Separately measure total and nonatomic absorption; difference = atomic absorption	Deuterium BC, Zeeman-effect BC, Smith-Heijfte BC

Box 11.3 Measurement of lead in blood by ETAAS

- Add whole blood (30 μL) to aqueous mixture [0.05 % (v/v) Triton X-100 + 0.5 % (w/v) diammonium hydrogen orthophosphate] (420 μL)
- Mix well and introduce to furnace
- Heating programme (instrument: Thermo 939)

Phase	Temp (°C)	Hold time (s)	Ramp ($°C\ s^{-1}$)	Gas	Gas setting	Remarks
1	80	2	100	Ar	2	Drying
2	130	35	2	Ar	2	Drying
3	500	10	0	O_2	2	O_2 ashing
4	600	15	10	Ar	3	O_2 desorbed from furnace
5	1400	2	max	Ar	0	'Temp control' and 'Read'
6	2800	3	max	Ar	3	Cool

11.4.3 Atomic emission and atomic fluorescence spectrometry

11.4.3.1 Atomic emission spectrometry

Flame AES (flame photometry) is convenient for the alkali metals at high concentrations, but AES is most useful with high-temperature energy sources when multi-element analysis can be undertaken. The heat source for atomization and excitation to a higher energy level can be a flame. Historical alternatives include arcs and sparks, but modern instruments use a plasma (argon or some other gas in an ionized state). The plasma is initiated by seeding from a high voltage spark to ionize the atoms:

$$Ar + e^- \rightleftharpoons Ar^+ + 2e^-$$

and is sustained with energy from an induction coil connected to a radiofrequency generator. This is known as an inductively coupled plasma (ICP).

Plasmas exist at temperatures of up to 10,000 °C and in the instrument have the appearance of a torch (Figure 11.3). Samples can be introduced via a nebulizer or, as for AAS, by hydride generation or cold vapour generation (Section 11.4.5), by electrothermal vaporization from a graphite atomizer or by laser ablation of solid specimens. Optical systems direct the emitted light either via a monochromator to a single detector, or to an array of monochromators and detectors positioned around the plasma. With the first arrangement a sequential series of readings can be made with the monochromator driven to give each of the wavelengths of interest in turn. Simultaneous readings can be made with the second arrangement as each of the monochromators transmit light of different required wavelengths. A sequential reading instrument is less expensive than a simultaneous reading instrument, but more sample is required to take a series of readings.

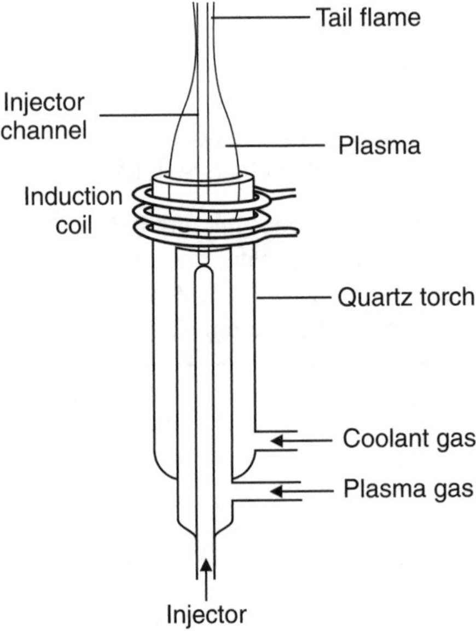

Figure 11.3 Inductively coupled plasma-mass spectrometry: ICP torch.

For most elements the analytical sensitivity for ICP-AES is similar to that obtained with FAAS.

At the high operating temperatures of the ICP many energy transitions take place, giving rise to the potential for spectral interferences when emission of light from different elements occurs at wavelengths that are too close to be separated by the monochromator. Most of these spectral interferences are known, so that when interference is suspected an alternative resonance line may be used for the measurement.

11.4.3.2 Atomic fluorescence spectrometry

Few commercial instruments for AFS are available and these are confined to the measurement of hydride-forming elements and mercury (Section 11.4.5). The components of the instrument are similar to those for AAS. Effective atomic fluorescence requires intense, stable light sources and these are difficult to construct reliably. Most success has been with electrodeless discharge lamps. The optical path of the emitted light is directed at 90° to that of the incident light so that only the emitted, fluorescent light reaches the detector. Very low detection limits can be achieved for the metalloid elements of Groups 4 to 6 of the periodic table.

11.4.4 Inductively coupled plasma-mass spectrometry

As samples are taken to high temperatures any organic component is destroyed and some or all of the inorganic elements are ionized. When these ions are directed into a MS they may be separated by a magnetic field established by a quadrupole or some other mass filter. The ions, separated according to mass-to-charge (m/z) ratio are detected and counted using an electron multiplier. This process is generally described as atomic or inorganic MS (Hill, 1999; Nelms, 2005; Bacon *et al.*, 2006).

Various ion sources have been employed, but for clinical analysis most recent work uses ICP-MS. Many elements can be measured simultaneously and detection limits are in the range of a few $\mu g\ L^{-1}$ or below. Elements that have hitherto been impossible to quantify can be measured reliably. Sensitivity is especially good for heavy elements including those of the actinide series, such as uranium, traditionally determined by α-spectrometry or related techniques, and ICP-MS is often the method of choice for measuring these elements. The other major feature of MS is the ability to measure isotopes of the same element (Al Saleh *et al.*, 1993). The technique is now well established for analysis of biological specimens (Heitland and Köster, 2004, 2006).

The modules required for inorganic MS include sample introduction and ion generation, ion focusing, ion separation, ion detection, and data collection and display (Figure 11.4).

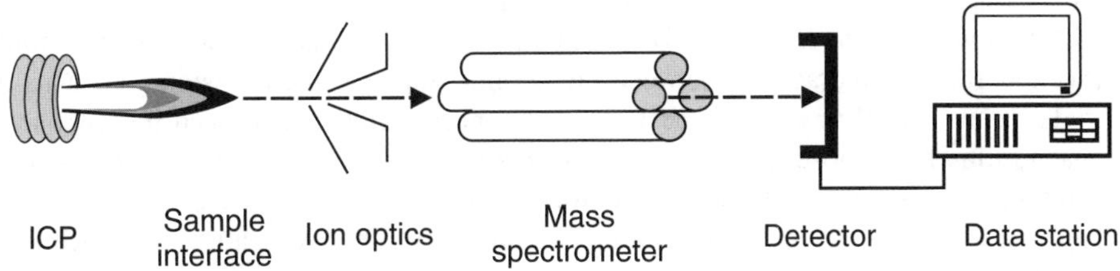

Figure 11.4 Components of an ICP-mass spectrometer.

11.4.4.1 *Ion sources*

As mentioned above there are a number of ion generation devices, but the most widely used is ICP. Samples are usually introduced to the plasma via a nebulizer or by chemical vaporization (Section 11.4.5). Electrothermal vaporization and vaporization by laser ablation of a solid sample may also be used.

11.4.4.2 *Mass analyzers*

The ICP torch is interfaced to the mass analyzer via two metallic cones (skimmer and sampler) through which ions are extracted into the ion-focusing unit. Here, a system of ion lenses direct ions to the analyzer. Several analyzer configurations are available commercially. Those with a quadrupole system have limited resolution allowing the separation of species on the basis of m/z (Figure 11.5). However, to exploit the full potential of the technique, a high-resolution sector field mass analyzer is required. Such instruments can achieve much greater resolution and higher sensitivity than a quadrupole ICP-MS. ToF-MS is particularly useful for the analysis of rapid, transient signals, such as those generated by electrothermal vaporization.

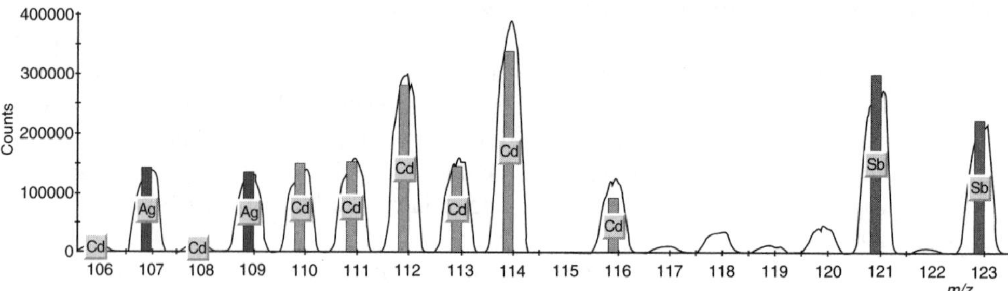

Figure 11.5 Example of ICP-MS mass spectrum. Reproduced by permission of Thermo Scientific.

11.4.4.3 *Interferences*

ICP-MS is subject to spectral interference caused by the presence of isotopes of either different elements, or ions formed from matrix components and the plasma gas. Examples important for clinical analyses include $^{40}Ar^+$ on ^{40}Ca, $^{31}P^{16}O_2^+$ on ^{63}Cu, $^{40}Ar^{35}Cl$ on ^{75}As and $^{40}Ar_2^+$ on ^{80}Se. Sector field ICP-MS is not subject to most of these interferences, although there may be some loss of sensitivity at high resolution, which may lead to problems with some applications such as the measurement of blood selenium (Klaue and Blum, 1999).

A relatively new development for use in quadrupole MS instruments is that of collision cells (Section 10.2.4). Interactions between polyatomic ions and the collision gas cause the former to dissociate, greatly reducing spectral interferences. Another approach involves separation of analyte ions from those involved in the formation of polyatomic species. Separation may be achieved by vaporization of the sample, for example by hydride generation or by electrothermal vaporization, or by prior use of a chromatographic or other separation step such as LLE. The addition of nitrogen, helium or methane to the carrier gas or an organic solvent to the diluent reduces some of the argon-based interferences (Branch *et al.*, 1994).

Non-spectral interferences, associated with sample introduction and fluctuations in the inductively coupled plasma, are effectively eliminated by using an internal standard. This should be an element not present in the original sample, not subject to spectral interferences and with a mass and ionization energy close to those of the analyte(s). Internal standards often used with biological specimens are scandium, indium and iridium for masses <80, 80–150 and >150 u, respectively.

11.4.5 Vapour generation approaches

A family of chemical vapour generation techniques has been described (Tsalev, 1999), but most practical applications involve either volatile hydride generation or the formation of mercury vapour.

11.4.5.1 Hydride generation

Elements such as arsenic, selenium, antimony and bismuth form gaseous hydrides, for example arsine (AsH_3), and this can be exploited in an assay. Using simple instrumentation, a reducing agent such as sodium borohydride is added to a reaction flask containing an acidified sample. Hydrogen is formed, which reacts with the analyte, gaseous hydride is evolved and transferred by a flow of inert gas to an ICP for AES or MS, or to a heated silica tube positioned in the light path for AAS or AFS (Heilier *et al.*, 2005). The tube is heated by either an air–acetylene flame or an electric current to a temperature that is sufficient to cause dissociation of the hydride and atomization of the analyte (Figure 11.6). Of relevance to AAS, there is no loss of specimen in the nebulizer, all the atoms enter the light path within a few seconds and they are trapped within the silica tube, which retards their dispersion. Instrument development to give a continuous flow has simplified automation.

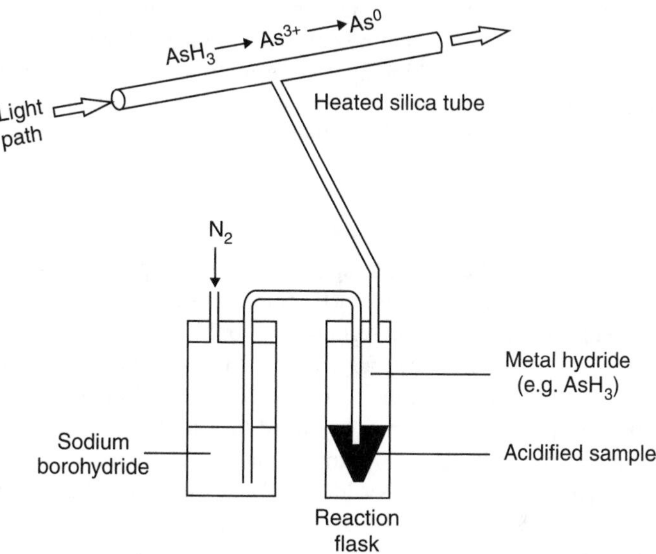

Figure 11.6 Hydride generation AAS.

Typically, the hydride-forming element will be present in a sample as a range of species (e.g. As^{III}, As^{V}, monomethylarsonic acid, dimethylarsinic acid, arsenobetaine). The rate of hydride formation is dependent on the species, so that to measure the total concentration of the element it is necessary to convert all of the different species present to the form that is the most reactive. This is usually achieved by heating with concentrated acids and then adding a reducing agent to give the reactive species (As^{III}, Se^{IV}, Sb^{III}, etc.) (Box 11.4).

Box 11.4 Sample preparation for the measurement of total selenium (Welz *et al.*, 1987)

- Place blood, urine or tissue (1 g) in a digestion tube. Add 5 mL nitric acid and heat (150 °C, 30 min)
- Add 0.2 mL perchloric acid and 0.5 mL sulfuric acid. Heat sequentially (15 min each temperature) at 150 °C, at 200 °C and at 250 °C
- Heat (320 °C, 20 min). Allow to cool to room temperature
- To reduce Se^{VI} to Se^{IV}, add 5 mL 50 % (v/v) hydrochloric acid. Heat (90 °C, 20 min)
- Cool, and add water (5 mL)

Seafoods are a rich source of nontoxic organoarsenic species and urinary total arsenic concentrations will be high for several days following consumption of fish. It is possible to differentiate between dietary and nondietary arsenic compounds by omitting the aggressive heating step and using a mild reducing agent such as an iodide salt or L-cysteine to convert any As^{III} metabolites to the hydride-forming species. The analysis will then reflect the exposure to toxic arsenic species (Cava-Montesinos *et al.*, 2005; Leermakers *et al.*, 2006).

Certain interferences are common to hydride generation whichever detector is used. If one hydride-forming element is present in the sample in large amounts it will consume the reducing agent so that other elements may not be detected (Welz and Stauss, 1993). High concentrations of transition element ions also inhibit hydride formation. If this is a problem, the use of masking agents, coprecipitation or chelation of the analyte followed by LLE of the chelate, can be used prior to the analysis (Welz and Melcher, 1984).

11.4.5.2 Mercury vapour generation

Mercury forms a vapour at ambient temperatures and this property is the basis for cold vapour generation. A reducing agent such as tin(II) chloride is added to the sample solution to convert Hg^{2+} to elemental mercury (Hg^0). Agitation or bubbling of gas through the solution causes rapid vaporization of the atomic mercury, which is transferred to a flow-through cell placed in a light path (Figure 11.7). To break the carbon–mercury bond in any organomercury compounds present, potassium permanganate may be added to acidified urine and the sample incubated at room temperature overnight. Excess permanganate is destroyed by addition of hydroxylamine hydrochloride. The reducing agent is added to the sample immediately before connection to the vapour generation accessory (Box 11.5). As with hydride generation, the LoD is a few nanograms and instruments to perform both procedures have been developed.

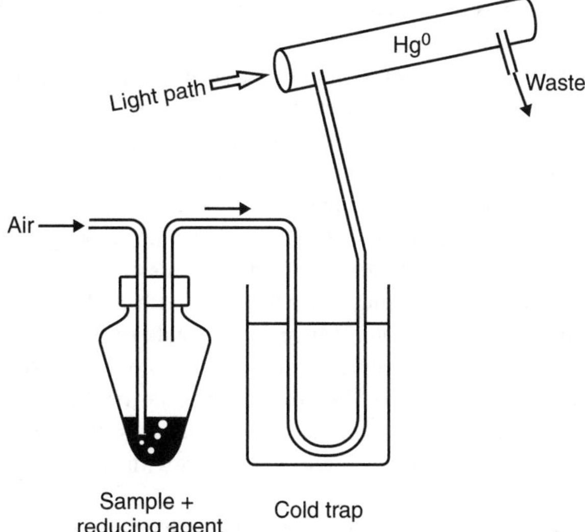

Figure 11.7 Cold vapour generation.

Box 11.5 Measurement of mercury in urine by cold vapour generation AAS

- Place urine (1 mL) in a tube (4 °C)
- Add 0.2 mL concentrated sulfuric acid and 1.5 mL 6 % (w/v) potassium permanganate
- Stand overnight and then add 0.3 mL 20 % (w/v) hydroxylamine hydrochloride
- Add 1 mL 20 % (w/v) tin(II) chloride in 50 % (v/v) hydrochloric acid to reduce Hg^{2+} to Hg^0
- Connect to accessory and purge with air or inert gas to volatilize the analyte

11.4.6 X-ray fluorescence

When high-energy photons, electrons or protons strike a solid sample, an electron from the inner shells (K, L or M) of a constituent atom may be displaced. The resulting orbital vacancy is filled by an outer-shell electron and an X-ray photon is emitted. The energy (wavelength) of the emitted photon is equal to the difference in the energy levels involved in the electron transition. This phenomenon is known as X-ray fluorescence (XRF). The wavelength is characteristic of the element from which it originated, whilst the intensity of the emission is related to the concentration of the atoms in the sample (Janssens and Adams, 1989; Potts *et al.*, 2006). According to the type of spectrometer used to measure the emission, XRF is characterized as wavelength dispersive (WDXRF) or energy dispersive (EDXRF). Total reflection XRF (TXRF) is usually described as a separate technique, although it may be considered as a variation of EDXRF (Klockenkämper and von Bohlen, 1992).

To perform an analysis, specimens are irradiated by high-energy photons, usually the polychromatic primary beam from an X-ray tube. However, the use of radioactive isotopes such as ^{244}Cm,

^{241}Am, ^{55}Fe and ^{109}Cd as sources has clinical application in semiportable instruments developed for *in vivo* XRF (Chettle, 2006).

The sample matrix can make a considerable contribution to signal intensity making calibration difficult, and requiring the use of reference materials, matrix-matched standardization and/or internal standardization. Fewer problems are encountered with samples prepared as very thin films, and in TXRF. In addition to the effect of the matrix, sensitivity is also influenced by wavelength and lower atomic mass elements are more difficult to measure accurately.

High intensity X-rays are employed in WDXRF. The fluorescence energy is dispersed into individual spectral lines by reflection using an analyzer crystal. The diffracted beams are collimated and directed onto a photomultiplier tube. As with ICP-AES, spectrometers may operate sequentially, with a number of interchangeable crystals to permit the measurement of the full range of elements, or in a multichannel (simultaneous) mode usually preset for specific analytes. Detection limits for light elements (silicon and below) are 10–100 times lower than with EDXRF. Resolution is good, although less so at shorter wavelengths. Sequential instruments require long analysis times to measure several elements compared with simultaneous instruments or EDXRF technology.

For EDXRF, X-rays emitted from the sample are directed into a crystal detector. Pulses of current are generated with intensities that are proportional to the energy of the X-ray photons. The different energies associated with the various elements in the sample are sorted electronically. Compared with WDXRF, lower-energy sources such as a low-power X-ray tube or a radioisotope (^{244}Cm, ^{241}Am or ^{109}Cd) can be used. The detector has to be maintained in a vacuum at the temperature of liquid nitrogen (77 K). Analysis times are 10–30 times longer than with WDXRF, but as EDXRF is a truly multi-element technique, the total analysis time is not necessarily increased.

When a collimated beam of X-rays is directed against an optically flat surface at a shallow angle (approximately 5 minutes), total reflection will occur. This is the principle of TXRF in which the sample is exposed to both primary and total reflected beams, and is excited to fluoresce. Emitted radiation is detected and resolved as an energy dispersive spectrum. As there is effectively no absorption by the matrix, measurement and calibration are much simpler and sensitivities are greater than with other X-ray techniques (Klockenkämper and von Bohlen, 1992). Useful applications of all types of XRF have been reviewed (Potts *et al.*, 2006).

11.5 Colorimetry and fluorimetry

A number of compounds react with metal ions to give coloured products. In some such cases (Table 11.4), a quantitative assay may be developed taking advantage of the Beer–Lambert Law. Applications to the analysis of clinical specimens are limited as sensitivity and selectivity are usually poor. Colorimetric measurement of calcium and magnesium in serum and urine, analytes that are present at concentrations of the order of mmol L^{-1} in these fluids, is routine. Measurement of zinc and copper in serum is also possible and commercial kits are available that can be used on modern clinical chemistry analyzers (Makino, 1999). However, there are interferences from the serum matrix and in EQA schemes the performance of these methods is poor (Taylor, 2006c).

Colorimetric methods for the measurement of iron(II) are available, but should be used with caution if chelating agents such as deferoxamine (desferrioxamine, DFO) have been given before the specimen was obtained, as the DFO-Fe complex will not react with the colour reagent. The LoD is 0.5 mg L^{-1}.

Some of the complexes formed between metals and dye compounds fluoresce. In addition, other compounds have been specially developed to exploit fluorimetric potential (Table 11.5).

Fluorimetric assays generally have greater sensitivity and selectivity than colorimetric methods. Nevertheless, in practice, only the measurement of selenium using 2.3-diaminonaphthylamine is of interest. This is as sensitive as the generally preferred AAS assays, but a much more complex sample preparation procedure involving acid digestion is needed. Hence, it is generally used only when there is no alternative.

Table 11.4 Some compounds that react with metals to give coloured products

Reagent	Metal
Dithizone	Pb, Hg
2,2′-Bipyridyl	Fe
o-Cresolphthalein complexone	Ca, Mg
Methylthymol blue	Ca
Sodium diethyldithiocarbamate	Cu
2-Carboxy-2′-hydroxy-5′-sulfoformazylbenzene (Zincon)	Zn
1-(2-Pyridylazo)-2-naphthol (PAN)	Zn
3,3′-Diaminobenzidine	Se
Catechol violet	Al
Eriochrome cyanide R	Al

Table 11.5 Some compounds that react with metal ions to produce fluorescent products

Reagent	Metal
2,3-Diaminonaphthylamine	Se
3,3′-Diaminobenzidine	Se
2′,3,4′,5,7-Pentahydroxyflavone (Morin)	Be
8-Hydroxyquinoline	Mg
3-(2,4-Dihydroxyphenylazo)-2-hydroxy-5-chlorobenzenesulfonic acid (Lumogallion)	Al

11.6 Electrochemical methods

11.6.1 Anodic stripping voltammetry

Anodic stripping voltammetry (ASV) is ideally suited to the analysis of dilute solutions where the sample volume is not a limiting factor. A reference electrode and a thin-film mercury graphite electrode are placed in the sample and a negative potential is applied to the mercury electrode, typically for periods of 2–30 s. This causes cations in the sample to concentrate ('plate out') on the surface of the mercury electrode (the anode). The direction of the potential is then reversed to give an increasingly larger positive potential over 2–30 min. As the voltage reaches the half-wave potential of an element, lead (Pb^{2+}), for example, all such ions are discharged (stripped) from the anode thereby producing a current that can be measured (Figure 11.8).

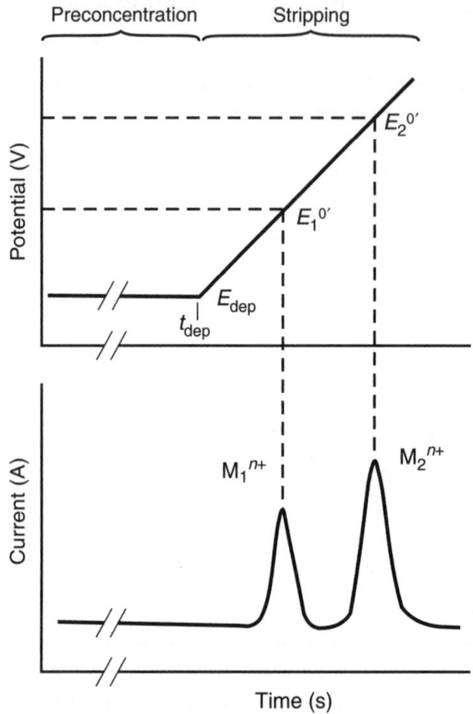

Figure 11.8 Schematic diagram of anodic stripping voltammetry. Metal ions plate out on the anode (potential $= E_{dep}$) during the preconcentration step (t_{dep}). As the potential is increased the ions ($M_1{}^{n+}$, $M_2{}^{n+}$) are discharged at characteristic half-wave potentials ($E_1^{0'}$, $E_2^{0'}$).

The current produced is proportional to the number of ions appearing at that voltage and is compared with those given by calibration solutions. While ASV is not widely used in the analysis of biological specimens, a niche application has developed, especially in the US, for the measurement of lead in blood. A chelating agent is added to the sample to mobilize lead bound to red cells and protein, and the measurement is performed using an instrument specifically designed for this application (Figure 11.9). The equipment is compact and can be set up in, for example, a medical examination room. Nowadays, a handheld instrument is available in which the electrode and sample container are provided as disposable units that are precalibrated by the manufacturer for each batch. Performance of this methodology as assessed by the NY State Blood Lead Proficiency Testing Scheme is generally acceptable (Parsons, 2006). A problem associated with the precalibration process was identified, but this has since been resolved (Stanton *et al.*, 2006).

11.6.2 Ion-selective electrodes

Ion-selective electrodes (ISEs), of which the pH meter is just one example, are widely used to measure major cations (Na^+, K^+) in biological specimens (Burnett *et al.*, 2000). Electrodes for other metal ions and also for anions are available. Lithium (Li^+) is commonly measured in this way,

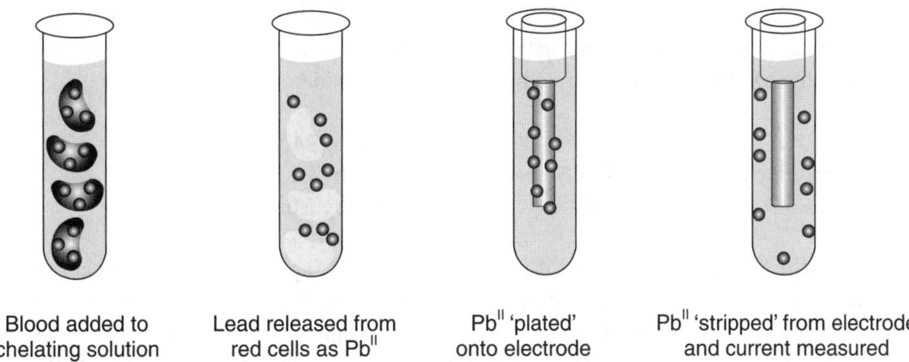
Blood added to chelating solution
Lead released from red cells as Pb^II
Pb^II 'plated' onto electrode
Pb^II 'stripped' from electrode and current measured

Figure 11.9 Measurement of lead in blood by anodic stripping voltammetry.

for example, in plasma or serum and in whole blood (Greil and Steller, 1992; Greffe and Gouget, 1996). Use of lithium-free blood collection tubes is, of course, vital (Section 2.2.2.2). Fluoride (F^-) may be measured in biological samples, not only to assess possible exposure to this poison or compounds giving rise to F^- by metabolism, but also in medicolegal work to establish that enough fluoride has been added to ensure inhibition of microbial growth (Kissa, 1987; Shajani, 1985).

ISEs are examples of membrane electrodes in which the membrane has physical or chemical properties that allow movement of only one kind of ion between the internal filling solution and any test solutions. If the activities of the ions in the two solutions are not equal there is a tendency for ions to cross the membrane towards the lower activity solution, and the electric charge thus generated will oppose the migration of ions until equilibrium is established. The actual number of ions involved is small and does not involve a gradual interdiffusion between the two compartments. The potential represented by the movement of ions may be measured and is logarithmically related to the activity in the test solution – it is important to remember that it is *activity* and not concentration that is measured by an ISE.

The preparation of selectively permeable membranes has led to many applications. The membranes may be solid with a fixed ionic structure, for example glass that responds to Na^+, or a water-immiscible liquid containing dissolved material that will actively exchange the selected ions in solution. An effective calcium exchanger is a calcium salt of an alkyl phosphate dissolved in dioctlyphenylphosphonate (Figure 11.10). The exchanger is prepared on a thin PVC layer to form the membrane.

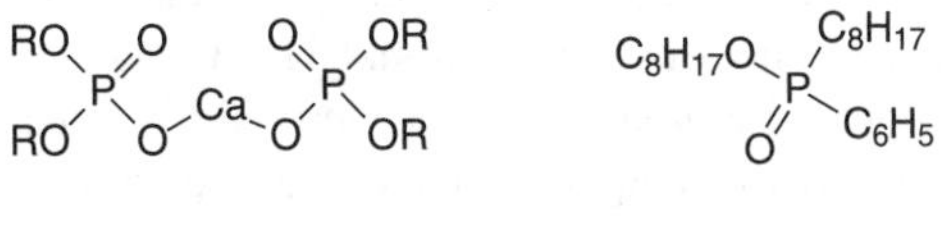

Alkyl phosphate Dioctylphenylphosphonate

Figure 11.10 An example of an effective calcium exchanger.

11.7 Catalytic methods

The concentration of an element can be measured from its catalytic role in a reaction, the rate of which is monitored using colorimetry or fluorimetry (Nakano *et al.*, 1983). The decolouration of ceric sulfate by arsenious acid, which requires iodine as catalyst, was once used to measure iodine in serum when investigating thyroid function. Most other examples of this type of reaction involve the oxidation of a substrate by hydrogen peroxide:

$H_2O_2 + KI \leftrightarrow I^{3-}$	catalyzed by Mo
o-Dianisidine oxidation by H_2O_2	catalyzed by Cr^{VI}
Acid blue 45 oxidation by H_2O_2	catalyzed by Mn

Chemiluminescence for the measurement of metals is a special example of a catalytic technique. Oxidation of luminol by hydrogen peroxide at alkaline pH is accompanied by the emission of light (Section 4.4.5). Chromium, cobalt, copper, iron and manganese have all been assayed in this way (Klopf and Nieman, 1983). Such methods are very sensitive and can detect a few $\mu g\ L^{-1}$ of analyte if interference from other metals can be excluded. However, these reactions have seldom been exploited in the analysis of biological samples.

11.8 Neutron activation analysis

NAA is sensitive and can be used to measure several elements simultaneously. However, because access to an atomic reactor is required, the technique is used only for special projects. In summary, the sample is bombarded with neutrons, which generate a range of radioisotopes. As these isotopes decay the emitted radiation can be measured using conventional β- or γ-counters. Although characteristic emissions from individual radioisotopes impart some selectivity, there may be interferences, especially from short-lived isotopes. Therefore, postactivation ion-exchange chromatography is often employed to give acceptable results.

NAA is expensive, as special equipment and highly trained operators are required, and in order to attain good detection limits irradiation times of several days may be needed. It is used, for example, in the characterization of reference materials when comparison with atomic spectrometric methods is to be performed. It may also be valuable in epidemiological and occupational toxicology studies (Porru *et al.*, 2001; Spallholz *et al.*, 2005), and in forensic work. An especial advantage in this latter area is that the sample is not consumed.

11.9 Chromatographic methods

11.9.1 Chromatography

In order to use chromatography to measure metal ions the problem that had to be addressed initially was that of detection using conventional HPLC and GC systems. It was found that chelates with, for example, trifluoroacetylacetonate could be measured by GC-ECD or HPLC-UV. However, sensitivity was inadequate for most applications. Nevertheless, work continues on the development of new chelating agents that afford improved detection limits (Hu *et al.*, 2005). A similar strategy has been used for ion chromatography (Pohl *et al.*, 1997). Although this latter technique is more usually associated with the measurement of anions, it is possible to measure metal ions using anion exchange columns, derivatization with a reagent such as 4-(2-pyridylazo)resorcinol and colorimetric detection. Detection limits of 0.02–0.5 ng using a 50 μL aqueous sample have been reported. Other detection systems may also be used including AES and ICP-MS (Heitkemper *et al.*, 2001).

At a time when ICP-MS was both expensive and subject to interferences, a novel approach was adopted in some laboratories in which a chelate of the element of interest, for example molybdenum with sodium bis(trifluoroethyl)dithiocarbamate, was prepared and analyzed by GC-MS. With this procedure any ions that might give rise to interferences in the MS were separated on the column, while the detector provided the necessary sensitivity. Subsequent developments with chromatography and ICP-MS also feature the use of a column to remove interferences, such as separation of Ar from Cl, which otherwise forms $^{40}Ar^{35}Cl$ and distorts the reading for ^{75}As (Sheppard *et al.*, 1990).

11.9.2 Speciation

Measuring the total concentration of an element in a specimen usually provides the information that is required for toxicological investigations, but there are situations where this is inadequate or misleading. The presence of nontoxic arsenic compounds in blood and urine, derived from seafood was mentioned above (Section 8.4.5.1), and measurement of arsenic species to differentiate between dietary and other sources of exposure may be needed. Another example is that of methylmercury and inorganic mercury where selective measurement of the mercury species present can give information as to the source of exposure. Speciation analysis is also a feature of more fundamental work to investigate the metabolism of an element (Devos *et al.*, 2002; Kremer *et al.*, 2005). Thus, speciation is an important topic for the analyst and features prominently in recent literature, much of it involving use of chromatography to effect the separations prior to measurement (Taylor *et al.*, 2007). Detection may involve any analytical technique, but it is AAS, AES, AFS and ICP-MS that are employed most frequently. Typically the chromatographic and detection systems are linked in series.

Analyses involve any or all of the following steps: extraction (e.g. from tissue samples), formation of a volatile derivative (for GC), chromatographic analysis and measurement. The extraction procedure should extract all the element in the sample and no species transformation should occur. Reviews of arsenic and of selenium speciation are available (Francesconi and Kuehnelt, 2004; Francesconi and Pannier, 2004; Polatajko *et al.*, 2006).

11.10 Quality assurance

Unlike many organic compounds, inorganic analytes are generally stable. In addition, calibration solutions can be prepared reliably and with high purity, and specialist reference laboratories are able to make measurements using, for example, isotope dilution analysis, that are traceable to international metrological standards. In turn, this means that certified reference materials are readily available for accuracy control and for method validation.

Apart from features such as inherent sensitivity, speed of operation and cost of the analytical method used, contamination is the factor that has the greatest impact on the quality of results. Contamination can occur during: (i) the collection and storage of specimens, (ii) sample preparation and (iii) the spectrometric measurement itself. Scrupulous attention to cleanliness and to methodological detail is essential to obtain accurate and reproducible results. Data from EQA schemes indicate that whilst some routine laboratories can obtain good results, it is specialist trace element centres that tend to maintain the highest standards of performance. Such centres also accumulate experience with unusual clinical and forensic cases and are well placed to provide advice and interpretation (Taylor and Walker, 1992).

11.11 Summary

The analytical toxicologist has a range of techniques available for the quantitative measurement of metal ions and other species in biological and related samples. Depending on the questions being asked, the analyst can accurately measure a specific element at concentrations down to ng g^{-1} (ppb) or even less, or can produce a multi-element profile to indicate whether a metal is unexpectedly present or absent. For many years AAS has been the most important technique and while this continues to be widely used, ICP-MS is increasingly the method of choice. In addition, techniques are being developed and validated that work together with AAS and ICP-MS to separate and identify molecules in the sample with which the metal is associated. Because metals are ubiquitous in the environment it is vital to exercise extraordinary caution to ensure that there is no contamination at any stage from collection of the specimen to the generation of results.

References

Al Saleh, I.A., Fellows, C., Delves, H.T. and Taylor, A. (1993) Identification of sources of lead exposure among children in Arar, Saudi Arabia. *Ann Clin Biochem*, **30**, 142–5.

Bacon, J.R., Linge, K.L., Parrish, R.R. and Van Vaeck, L. (2006) Atomic spectrometry update. Atomic mass spectrometry. *J Anal At Spectrom*, **21**, 785–818 (updated annually).

Bazzi, A., Nriagu, J.O., Inhorn, M.C. and Linder, A.M. (2005) Determination of antimony in human blood with inductively coupled plasma-mass spectrometry. *J Environ Monit*, **7**, 1251–4. Erratum 2005; **7**: 1388.

Blacklock, E.C. and Sadler, P.A. (1981) A rapid screening method for heavy metals in biological materials by emission spectroscopy. *Clin Chim Acta*, **113**, 87–94.

Branch, S., Ebdon, L. and O'Neill, P. (1994) Determination of arsenic species in fish by directly coupled high-performance liquid chromatography-inductively coupled plasma mass spectrometry. *J Anal At Spectrom*, **9**, 33–7.

Burnett, R.W., Covington, A.K., Fogh-Andersen, N., Kulpmann, W.R., Lewenstam, A., Maas, A.H., Muller-Plathe, O., VanKessel, A.L. and Zijlstra, W.G. (2000) Use of ion-selective electrodes for blood-electrolyte analysis. Recommendations for nomenclature, definitions and conventions. *Clin Chem Lab Med*, **38**, 363–70.

Cava-Montesinos, P., Nilles, K., Cervera, M.L. and de la Guardia, M. (2005) Non-chromatographic speciation of toxic arsenic in fish. *Talanta*, **66**, 895–901.

Chettle, D.R. (2006) Occupational nuclear medicine: Trace element analysis of living human subjects. *J Radioanal Nucl Chem*, **268**, 653–61.

Devos, C., Sandra, K. and Sandra, P. (2002) Capillary gas chromatography inductively coupled plasma mass spectrometry (CGC-ICPMS) for the enantiomeric analysis of D,L-selenomethionine in food supplements and urine. *J Pharm Biomed Anal*, **27**, 507–14.

Francesconi, K.A. and Kuehnelt, D. (2004) Determination of arsenic species: A critical review of methods and applications, 2000–2003. *Analyst*, **129**, 373–95.

Francesconi, K.A. and Pannier, F. (2004) Selenium metabolites in urine: A critical overview of past work and current status. *Clin Chem*, **50**, 2240–53.

Frank, E.L., Hughes, M.P., Bankson, D.D. and Roberts, W.L. (2001) Effects of anticoagulants and contemporary blood collection containers on aluminum, copper, and zinc results. *Clin Chem*, **47**, 1109–11.

Gray, A.L. and Date, A.R. (1983) Inductively coupled plasma source mass spectrometry using continuum flow ion extraction. *Analyst*, **108**, 1033–50.

Greffe, J. and Gouget, B. (1996) Red cell effects on lithium measurements by ion-selective electrode. *Scand J Clin Lab Invest*, **224** (Suppl), 187–91.

Greil, W. and Steller, B. (1992) Lithium determination in outpatient clinics by an ion-selective electrode in venous and capillary whole blood. *Psychiatr Res*, **44**, 71–7.

Grinberg, P., Goncalves, R.A. and de Campos, R.C. (2005) The determination of total Se in urine and serum by graphite furnace atomic absorption spectrometry using Ir as permanent modifier and in situ oxidation for complete trimethylselenonium recovery. *Anal Bioanal Chem*, **383**, 1044–51.

Haswell, S.J. (ed.) (1991) *Atomic Absorption Spectrometry. Theory, Design and Applications*, Elsevier, Amsterdam.

Heilier, J.F., Buchet, J.P., Haufroid, V. and Lison, D. (2005) Comparison of atomic absorption and fluorescence spectroscopic methods for the routine determination of urinary arsenic. *Int Arch Occup Environ Health*, **78**, 51–9.

Heitland, P. and Köster, H.D. (2006) Biomonitoring of 30 trace elements in urine of children and adults by ICP-MS. *Clin Chim Acta*, **365**, 310–8.

Heitland, P. and Köster, H.D. (2004) Fast, simple and reliable routine determination of 23 elements in urine by ICP-MS. *J Anal At Spectrom*, **19**, 1552–8.

Heitkemper, D.T., Vela, N.P., Stewart, K.R. and Westphal, C.S. (2001) Determination of total and speciated arsenic in rice by ion chromatography and inductively coupled plasma mass spectrometry. *J Anal At Spectrom*, **16**, 299–306.

Hill, S.J. (1999) *Inductively Coupled Plasma Spectrometry and Its Applications*, Sheffield Academic Press, Sheffield.

Hu, Q., Yang, X., Huang, Z., Chen, J. and Yang, G. (2005) Simultaneous determination of palladium, platinum, rhodium and gold by on-line solid phase extraction and high performance liquid chromatography with 5-(2-hydroxy-5-nitrophenylazo)thiorhodanine as pre-column derivatization regents. *J Chromatogr A*, **1094**, 77–82.

Janssens, K.H. and Adams, F. (1989) New trends in elemental analysis using X-ray fluorescence spectrometry. *J Anal At Spectrom*, **4**, 123–35.

Kehoe, R.A. (1961) The Harben Lectures. The metabolism of lead in health and disease. *J Royal Inst Pub Hlth Hyg*, **24**, 81–97.

Klockenkämper, R. and von Bohlen, A. (1992) Total reflection X-ray fluorescence—an efficient method for micro-, trace and surface layer analysis. *J Anal At Spectrom*, **7**, 273–9.

Klopf, L.L. and Nieman, T.A. (1983) Effect of iron(II), cobalt(II), copper(II) and manganese(II) on the chemiluminescence of luminal in the absence of hydrogen peroxide. *Anal Chem*, **55**, 1080–3.

Kissa, E. (1987) Determination of inorganic fluoride in blood with a fluoride ion-selective electrode. *Clin Chem*, **33**, 253–5.

Klaue, B. and Blum, J.D. (1999) Trace analysis of arsenic in drinking water by inductively coupled plasma mass spectrometry: high resolution versus hydride generation. *Anal Chem*, **71**, 1408–14.

Kremer, D., Ilgen, G. and Feldmann, J. (2005) GC-ICP-MS determination of dimethylselenide in human breath after ingestion of (77)Se-enriched selenite: monitoring of in-vivo methylation of selenium. *Anal Bioanal Chem*, **383**, 509–15.

Leermakers, M., Baeyens, W., De Gieter, M., Smedts, B., Meert, C., De Bisschop, H.C., Morabito, R. and Quevauviller, Ph. (2006) Toxic arsenic compounds in environmental samples: Speciation and validation. *Tr Anal Chem*, **25**, 1–10.

Lobinski, R. and Marczenko, Z. (1996) *Spectrochemical Trace Analysis for Metals and Metalloids*, Elsevier, Amsterdam.

Lopez-Artiguez, M., Camean, A. and Repetto, M. (1993) Preconcentration of heavy metals in urine and quantification by inductively coupled plasma atomic emission spectrometry. *J Anal Toxicol*, **17**, 18–22.

Loreti, V. and Bettmer, J. (2004) Determination of the MRI contrast agent Gd-DTPA by SEC-ICP-MS. *Anal Bioanal Chem*, **379**, 1050–4.

Makino, T. (1999) A simple and sensitive colorimetric assay of zinc in serum using cationic porphyrin. *Clin Chim Acta*, **282**, 65–76.

Male, G.E. (1816) *An Epitome of Judicial or Forensic Medicine for the Use of Medical Men, Coroners and Barristers*, E. Cox, London.

Mandal, B.K., Ogra, Y., Anzai, K. and Suzuki, K.T. (2004) Speciation of arsenic in biological samples. *Toxicol Appl Pharmacol*, **198**, 307–18.

Marsh, J. (1836) Account of a method of separating small quantities of arsenic from substances with which it may be mixed. *Edin N Philosoph J*, **21**: 229–36.

Massman, H. (1968) Vergleich von atomabsorption und atomfluoreszenz in der graphitküvette. *Spectrochim Acta B*, **23**, 215–26.

Nakano, S., Tanaka, M., Fushihara, M. and Kawashima, T. (1983) Catalytic determination of nanogram amounts of Cu(II) by the oxidative coupling reaction of *N*-phenyl-*p*-phenyldiamine with *N*,*N*-dimethylaniline. *Mikrochim Acta*, **1**, 457–65.

Nelms, S.M. (ed.) (2005) *Inductively Coupled Plasma Mass Spectrometry Handbook*, Blackwell, Oxford.

Parsons, P. (2006) Reports from the NY State Blood Lead Proficiency Testing Scheme (http://www.wadsworth.org/testing/lead/ptresults.htm, accessed 15 July 2007).

Pohl, C.A., John, R., Stillian, J.R. and Jackson, P.E. (1997) Factors controlling ion-exchange selectivity in suppressed ion chromatography. *J Chromatogr A*, **789**, 29–41.

Polatajko, A., Jakubowski, N. and Szpunar, J. (2006) State of the art report of selenium speciation in biological samples. *J Anal At Spectrom*, **21**, 639–54.

Porru, S., Placidi, D., Quarta, C., Sabbioni, E., Pietra, R. and Fortaner, S. (2001) The potential role of rare earths in the pathogenesis of interstitial lung disease: a case report of movie projectionist as investigated by neutron activation analysis. *J Trace Elem Med Biol*, **14**, 232–6.

Potts, P.J., Ellis, A.T., Kregsamer, P., Streli, C., Vanhoof, C., West, M. and Wobrauschek, P. (2006) Atomic spectrometry update. X-ray fluorescence spectrometry. *J Anal At Spectrom*, **21**, 1076–1107.

Shajani, N.K. (1985) Determination of fluoride in blood samples for analysis of ethanol. *Can Soc Forensic Sci J*, **18**, 49–52.

Sheppard, B.S., Shen, W.L., Caruso, J.A., Heitkemper, D.T. and Fricke, F.L. (1990) Elimination of the argon chloride interference on arsenic speciation in inductively coupled plasma mass spectrometry using ion chromatography. *J Anal At Spectrom*, **5**, 431–5.

Spallholz, J.E., Boylan, L.M., Palace, V., Chen, J., Smith, L., Rahman, M.M. and Robertson, J.D. (2005) Arsenic and selenium in human hair: a comparison of five countries with and without arsenicosis. *Biol Trace Elem Res*, **106**, 133–44.

Stanton, N.V., Fritsch, T., Geraghty, C., Verostek, M.F., Weiner, B. and Parsons, P.J. (2006) The role of proficiency testing in the detection and resolution of calibration bias in the LeadCare®blood lead analyzer; limitations of peer-group assessment. *Accred Qual Assur*, **11**, 590–2.

Taylor, A. (2006a) Atomic spectrometry and the clinical chemistry of trace elements. *J Anal At Spectrom*, **21**, 381–3.

Taylor, A. (ed.) (2006b) *SAS Trace Element Laboratories Clinical and Analytical Handbook*, 4th edn, Royal Surrey County Hospital, Guildford.

Taylor, A. (2006c) Reports from the UK NEQAS for Trace Elements. Guildford: School of Biomedical and Molecular Sciences, University of Surrey (www.surrey.ac.uk/sbms/eqas/pdf/TQreport.pdf, accessed 15 July 2007; updated regularly).

Taylor, A. and Walker, A.W. (1992) Measurement of aluminium in clinical samples. *Ann Clin Biochem*, **29**, 377–89.

Taylor, A., Branch, S., Day, M., Patriarca, M. and White, M. (2007) Atomic spectrometry update. Clinical and biological materials, foods and beverages. *J Anal At Spectrom*, **22**, 415–56 (updated annually).

Tsalev, D.L. (1999) Hyphenated vapour generation atomic absorption spectrometric techniques. *J Anal At Spectrom*, **14**, 147–62.

Walsh, A. (1955) The application of atomic absorption spectra to chemical analysis. *Spectrochim Acta*, **7**, 108–17.

Welz, B. and Melcher, M. (1984) Mechanisms of transition metal interferences in hydride generation atomic-absorption spectrometry. Part 3. Releasing effect of iron (III) on nickel interference on arsenic and selenium. *Analyst*, **109**, 577–9.

Welz, B. and Stauss, P. (1993) Interferences from hydride-forming elements on selenium in hydride-generation atomic absorption spectrometry with a heated quartz tube atomiser. *Spectrochim Acta B*, **48**, 951–76.

Welz, B., Wolynetz, M. and Verlinden, M. (1987) Determination of selenium in lyophilised human serum, blood and urine using hydride generation atomic absorption spectrometry. *Pure Appl Chem*, **59**, 927–36.

12 Immunoassays and Enzyme-Based Assays

12.1 Introduction

Immunoassay depends on the principle of competition between antigen and labelled antigen for binding sites on an antibody raised against the antigen. The proportionality of the resulting signal reflects the concentration of the antigen (analyte) in the sample. Immunoassays may be classified on the basis of the need to separate the bound and free antigen prior to measuring the signal. Heterogeneous assays require separation of the antibody–antigen complex from the unbound antigen prior to measurement, whereas homogeneous assays do not. Homogenous assays, although technically simpler to perform, may suffer deleterious effects from the presence of other components of the sample (matrix effects). Nevertheless, many homogeneous immunoassays have been successfully automated and are in use in a number of different high and low throughput clinical chemistry systems. Some practical aspects of the use of immunoassays in analytical toxicology are summarized in Box 12.1.

Box 12.1 Immunoassays: practicalities

General

- Have to use confirmatory techniques if medicolegal implications
- Need commercial assay

Therapeutic drugs

- Many drugs (e.g. cardioactives, psychoactives) have active metabolites: need separate antibody for separate assay or carefully crafted antibodies to measure both
- Cannot raise antibodies to some drugs (lithium) or selective antibodies to others (e.g. amiodarone, cross reacts with thyroxine)

Drugs of abuse (urine testing)

- Have to do separate test for each drug/drug group
- Opiate and amfetamine assays poorly selective – need confirmatory techniques

Clinical toxicology

- Cannot raise antibodies to some inorganic or very small organic poisons (carbon monoxide, cyanide ion, ethanol, ethylene glycol)
- Condition of sample may affect result due to nonspecific binding (matrix effects) or the presence of heterophilic antibodies, but can minimize interferences by solvent extraction

12.1.1 Historical development

The first system to exploit the competition between antigen and antibody *in vitro* was developed by Rosalyn Yalow (1921–) and Solomon Berson (1918–1972) in New York (1959) to measure insulin using radioimmunoassay (RIA). A competitive serum protein-binding assay ('saturation analysis') for thyroxine (T4) was developed by Roger Ekins in London (1960). Yalow was awarded the 1977 Nobel Prize in Physiology or Medicine for her part in these discoveries. The first RIA for a low M_r exogenous agent in biological samples was that developed to measure digitoxin (Oliver *et al.*, 1968). By 1971 a group led by Edwin Ullman had devised a technique called spin immunoassay that was based on electron spin resonance detection of stable nitroxide radicals. This process was the basis for the first commercial homogeneous immunoassay, launched under the name free radical association technique (FRAT), which was used for the detection of drugs of abuse in urine, primarily for US armed forces stationed in Vietnam. Enzyme multiplied immunoassay technique (EMIT), also developed by Ullman's group, was the first viable nonisotopic drug immunoassay and revolutionized detection of drugs of abuse in urine (Rubenstein *et al.*, 1972). Other homogeneous technologies followed,including fluorescence polarization immunoassay (FPIA), substrate labelled fluorescence, fluorescence excitation transfer and cloned enzyme donor immunoassay (CEDIA) (Price, 1998).

12.2 Basic principles of competitive binding assays

Immunoassays such as RIA are competitive binding assays, in that the analyte and a second molecule, the label, compete for binding sites on an antibody that has been raised against the analyte. The label is a molecule that can be measured by counting radioactivity, by a change in spectral properties, or via an enzyme that can be reacted with a substrate and the rate of product formation monitored. Because the number of binding sites is finite, competitive binding assays produce non-linear calibration curves and so have defined working ranges. Antibodies, rather than some other macromolecule to which analyte and label bind, are used impart specificity (Section 12.2.3). RIAs are heterogeneous assays in which the analyte and label have to be separated. This step is avoided in homogenous assays such as EMIT, although separation steps such as solvent extraction may be used prior to the immunoassay to enhance selectivity or reliability (Section 12.4.1).

12.2.1 Antibody formation

Most drugs are relatively small molecules and therefore are not inherently immunogenic. Hence immunogenicity has to be induced by conjugating the analyte, or a derivative of the analyte, to a protein such as bovine serum albumin (Figure 12.1), before anti-analyte antibodies can be produced by injecting the modified protein into an animal. Usually rabbit, goat or sheep, rather than rat or mouse, are used, as relatively large amounts of blood need to be taken repeatedly. Antibodies raised against the analyte–protein complex are immunoglobulins (IgG). The heavy chain component, the F_{ab} portion, is the antigen-binding portion. The degree of antibody–antigen binding is known as the antibody titre.

The antibody recognizes the antigen by its three-dimensional structure and its peptide sequence. The degree of antigen binding, the affinity, reflects the 'goodness-of-fit' of the antigen and antibody. The selectivity of an immunoassay is dependent on the portion of the analyte molecule exposed on the surface of the protein following hapten synthesis and therefore available to promote antibody formation. Generally, this exposed portion of the analyte should be chosen to maximize selectivity (i.e. to minimize cross reactivity with structurally similar drugs and metabolites). Careful selection of the orientation of the analyte during immunogen formation to

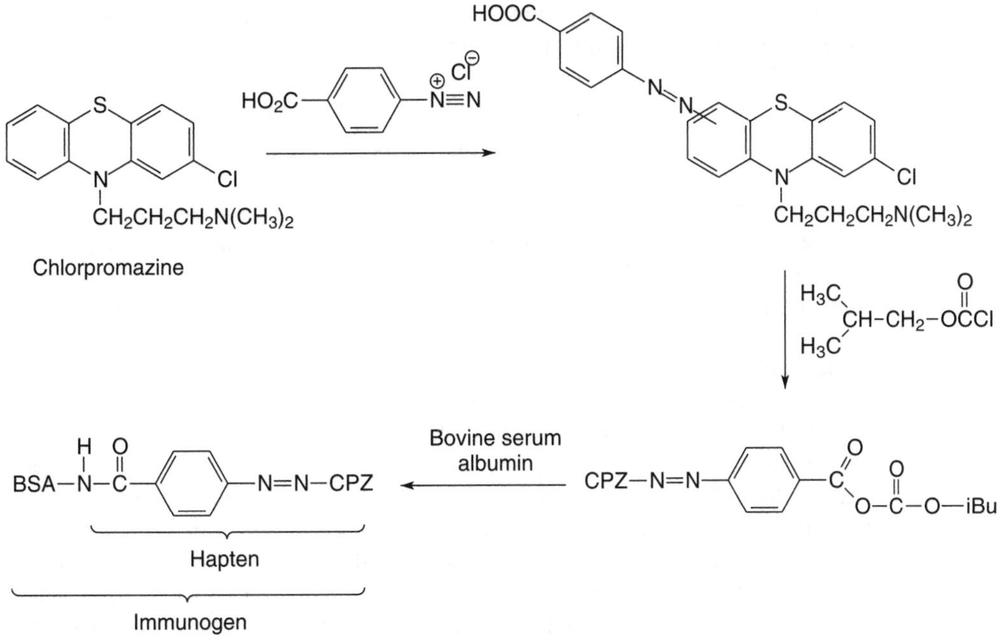

Figure 12.1 Preparation of immunogen for chlorpromazine radioimmunoassay (after Kawashima *et al.*, 1975).

expose unique portions of the analyte molecule, if possible, improves selectivity. Selectivity may also be improved if the antibody is raised against an analyte that has several 'spacer bonds' added to its linkage with the protein.

Antibodies vary in their affinity for the antigen, and the mix of antibodies produced on inoculation varies between different animals and between species. This variability causes difficulties in maintaining a reproducible supply of such polyclonal antisera over time. The development of monoclonal antibodies proved an important advance. In the hybridoma technique, murine myeloma cells and lymphocytes from immunized mice are fused and propagated. Selection of a single cell for propagation provides reproducibility, continuity and most importantly good selectivity, although sensitivity may be an issue. The affinity of the antibody for the analyte is greater for polyclonal antibodies than monoclonal antibodies.

12.2.2 Specificity

An important issue when evaluating an immunoassay is specificity. For assays such as RIA, this is usually defined by the cross reactivity, that is the extent to which other molecules will displace the label and be quantified (in error) as the analyte. The percentage cross-reactivity (*CR*) can be defined in terms of the concentration of substance that displaces 50 % of the antibody-bound label:

$$CR\ (\%) = \frac{\text{Apparent concentration of analyte}}{\text{Concentration of displacing drug}} \times 100 \qquad (12.1)$$

For example, if (−)-amfetamine (2 mg L^{-1}) cross reacts with an antibody raised to dexamfetamine to give an apparent (+)-amfetamine concentration of 1 mg L^{-1}, then:

$$CR\ (\%) = \frac{\text{Apparent concentration of (+) -amfetamine (1 mg L}^{-1}\text{)}}{\text{Concentration of (−) -amfetamine (2 mg L}^{-1}\text{)}} \times 100 = 50\ \%$$

It is usual to test potentially interfering compounds at several concentrations, over several orders of magnitude, and to plot log (concentration)-displacement curves to assess the magnitude of any problem. It should be remembered that the potencies, and hence the concentrations of related drugs encountered in clinical samples, may vary widely (compare the potencies of fentanyl and codeine) and even low cross-reactivity *in vitro* does not guarantee that interference from other drugs or metabolites, for example, will not occur with real samples.

The way in which the immunogen is formed can affect cross-reactivity (Table 12.1). Cross reactivity can be harnessed to detect a drug class, a feature that is particularly useful when screening for drugs of abuse. By linking a barbiturate via a 5-substituent, for example, antibodies can be raised that will recognize most barbiturates, whereas for a more specific assay, the 5-substituent must be available to the antibody (Figure 12.2). However in TDM, interference by other drugs, drug metabolites or endogenous compounds can be problematic, as in the case of digoxin (Section 12.7.1). In addition, non-specific binding can occur in poorly designed systems limiting the sensitivity of the assay. This can often be counteracted to an extent by adding BSA or analyte-free neonatal calf serum to the incubation mixture to bind unwanted matrix components,

Table 12.1 Drugs of abuse in urine 'opiates screen': some compounds detected

Codeine	Hydrocodone	Morphine + metabolites
Dextromethorphan	Hydromorphone	Pholcodine
Dihydrocodeine	6-Monoacetylmorphine (6-MAM)	'Poppy seed' metabolites (includes morphine)

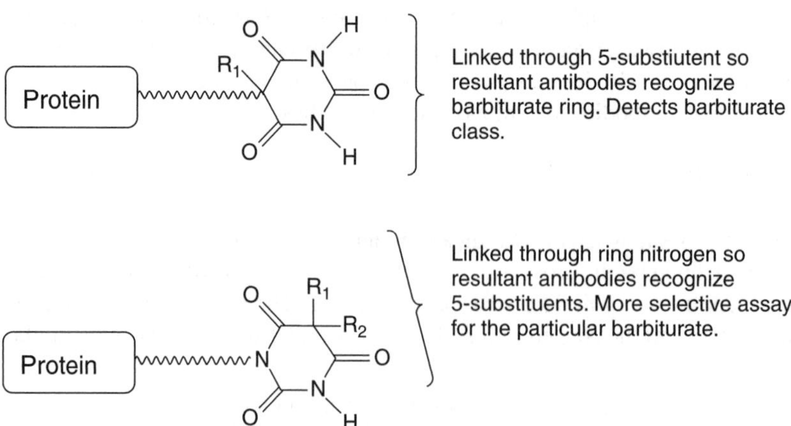

Figure 12.2 Directing antibody formation for a class-selective and an analyte-selective assay of a chosen barbiturate.

thus limiting interference in an assay. Serial dilution of the sample will usually give differing results if the binding is nonspecific.

12.2.3 Performing the assay

The principles of competitive binding assays can be understood most easily by considering RIA, which is a heterogeneous method, that is to say separation of antibody-bound and nonbound label is required. Homogeneous assays do not require a separation step, but the principle of competitive binding is the same.

12.2.3.1 Classical radioimmunoassay

In classical RIA, radiolabelled analyte was added to a solution of the antiserum, that is serum containing the antibody to the analyte, that would bind the radiolabelled analyte. Adding unlabelled analyte (in a sample) displaced some of the radioactive analyte bound to the antibody, the degree of displacement being a function of the analyte concentration in the sample. Separation of bound and free analyte, for example by adsorption onto activated charcoal (Figure 12.3), and counting

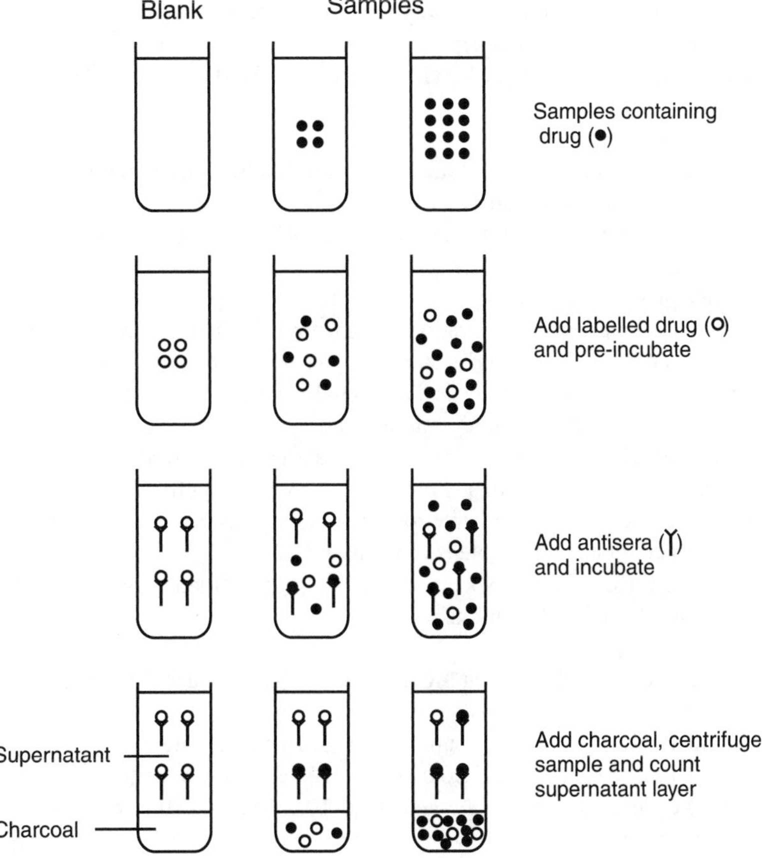

Figure 12.3 Principle of competitive binding.

the radioactivity, usually in the supernatant layer, enabled calculation of the analyte concentration in the sample using an appropriate calibration graph.

Initally, ^{3}H- or ^{14}C-labelled analytes were used and because the chemical structure of the labelled and unlabelled analytes are identical, such compounds would be expected to be ideal for competitive assays such as RIA. However, these 'soft' β-emitters have to be 'counted' (quantified) by liquid scintillation spectrometry (Box 12.2), which requires large and relatively expensive apparatus. Moreover, the disposal of the scintillation cocktail can be a problem. Furthermore,

Box 12.2 Radioactive counting

Scintillation counting

- Substances that emit light when exposed to ionizing radiations are called scintillators (e.g. sodium iodide irradiated with γ-rays gives out flashes of light)
- Soft (low-energy) β-particles from ^{3}H are too weak to penetrate container walls – must be counted in solution
- Energy from β-particles transferred via solvent molecules to organic scintillator dissolved in solvent
- Emitted light measured by pair of photomultiplier tubes–coincidence counting reduces background noise as counts are only recorded when both PM tubes activated
- Number of photons emitted is proportional to energy of β-irradiation–gives a spectrum characteristic of the nuclide. Maximum energies ^{3}H: 0.018 MeV; ^{14}C: 0.155 MeV; ^{32}P: 1.71 MeV

Background counts

- Electronic noise in counting system and natural radiation, for example ^{40}K in glass, give counts not arising from sample
- Background counts should be measured and subtracted from sample counts prior to any further calculation

Counting efficiency – quenching

- Not every radioactive disintegration is measured due to detector geometry, electronic circuitry and inefficient transfer of energy to the scintillant. Quenching is due to colour or the presence of certain chemicals such as halogenated hydrocarbons
- Counts min^{-1} less than disintegrations min^{-1}
- Counting efficiency (%) = Counts min^{-1}/disintegrations $min^{-1} \times 100$ %
- As quenching can vary from sample to sample counting efficiency should be measured and counts min^{-1} converted to disintegrations min^{-1}. This is most important with tritium–the maximum efficiency attainable is typically 65 %
- Modern liquid scintillation spectrometers are programmed to measure efficiency and produce result as disintegrations min^{-1}

Counting error

- Higher precision is obtained at higher count rates. Standard deviation of N counts = $\sqrt{N}$
- Example of sample with 100 counts min^{-1}
  - Count 1 min ⇒ 100 ± 10 counts = 100 ± 10 counts min^{-1}
  - Count 10 min ⇒ 1000 ± 31.6 counts = 100 ± 3.16 counts min^{-1}
- Better to collect set number of counts (e.g. 10,000) than count for set time

for maximum sensitivity the amount of added tracer should be as low as possible requiring radioactive nuclides of high specific activity, a property that is inversely related to the half-life of the radioisotope. Tritum ($t_{0.5} = 12.3$ years) can be used, but the half-life of ^{14}C (5568 years) is too long for this label to be of practical use and so such labels were soon replaced by ^{131}I ($t_{0.5} = 8.08$ d) and ^{125}I ($t_{0.5} = 60.14$ d). These γ- and X-ray emitting isotopes do not require the use of liquid scintillation counting, but disadvantages are that stock solutions require lead shielding and accidental ingestion is dangerous, as radioactive iodine is sequestered in the thyroid gland thus giving a high dose of radiation to a small area. Shelf-life too may be a problem with rarely requested assays. Generally, ^{125}I is used because its longer half-life makes it more convenient and it gives faster and more efficient counting than tritium. The pecision of radioactive counting is related to the total number of counts measured (Box 12.2).

12.2.3.2 Modern radioimmunoassay (RIA)

Clearly, separating nonbound radioactive label by charcoal adsorption and centrifugation was not only inconvenient, but also the nature of the binding had to be controlled to prevent charcoal adsorption of the antibody-bound label. During method development it was important to test various modified charcoals including dextran-coated charcoal (Bartlett *et al.*, 1980). With the introduction of antibody-coated tubes, the analysis mixture could be decanted, thus easily separating the bound and free fractions. Using ^{125}I tracers, the tubes can be placed directly into a γ-counter (Box 12.3).

Box 12.3 Radioimmunoassay

- Tubes precoated with antibody to the analyte
- Standard, blank or sample, and radiolabelled (^{125}I) analyte added
- Analyte in the sample competes with the radiolabelled analyte for binding sites
- Fluid poured from tubes and residual radioactivity measured (γ-counter) – the more analyte added, the lower the residual radioactivity
- Problems: cost, short shelf-life, radioactive handling and disposal

Use of a second antibody precipitation step can also accomplish separation of bound and free fractions. Centrifugation gives a pellet of material including bound analyte that can be counted after decanting or aspirating the free fraction. Magnetic beads and other particles coated with a second antibody have also been used to help separate bound and free fractions. Assay kits to measure buprenorphine, fentanyl, insulin, LSD and free (unconjugated) morphine in plasma or urine have been available commercially.

12.2.4 Non-isotopic immunoassay

Immunoradiometric assays (IRMA) where the antibody itself was labelled resulted in significant improvements in sensitivity. However, difficulties associated with the safe handling and storage of radioactivity, the disposal of radioactive waste, the short shelf-life of some radioactive isotopes used as labels due to continuing radioactive decay, lengthy assay times and the difficulty of automating the assay have resulted in RIA being replaced largely by non-isotopic immunoasssays (Law, 1996). Such systems also offer reduced costs, speed and simplicity of operation (Box 12.4).

Box 12.4 Non-isotopic immunoassays in analytical toxicology

- Fast, simple to use, automated equipment
- Various manufacturers of assay kits, but can be expensive
- Ideal for preliminary screening ('presumptive screen')
- Important role, especially in screening for illicit drug use and in TDM of certain drugs
- Qualitative and quantitative
- Sensitivity adequate for purpose (reasonable 'cut-off limits')

12.2.5 Assay sensitivity and selectivity

Immunoassays for small molecules such as most drugs are prone to cross-reaction with structurally related molecules and may also suffer from heterophilic antibody interference resulting in inappropriately high results (Selby, 1999; Armstrong and Oellerich, 2006). The same issues apply in analytical toxicology (Sasse, 1997; Wu *et al.*, 2003). Thus, all immunoassay procedures used for detecting drugs of abuse or other poisons require confirmation of positive results using a second method, ideally a chromatographic method, if the findings are to withstand scrutiny. This lack of selectivity can be turned to advantage when screening for a number of related compounds such as opiates/opiate analogues (Table 12.1), but can cause problems for the unwary. A further factor is that not all compounds that cross-react do so to the same extent at the same concentration. There can also be significant batch-to-batch variation in cross-reactivities for kits produced at different times.

In some cases cross-reactivity is not confined to compounds of the same structural class. Some urinary amfetamine immunoassays, for example, also give positive results with tranylcypromine, proguanil, isoxuprine, labetalol and phenylethylamine, amongst other compounds (Table 12.2). This poor selectivity is because amfetamine (1-methylphenylethylamine) is a relatively simple molecule. Small changes to the side chain, while markedly affecting pharmacological activity,

Table 12.2 Urinary amfetamine immunoassays: potential cross reactants

L-Amfetamine	D,L-Methylenedioxyethylamfetamine (MDEA)
Benzathine phenoxymethylpenicillin	D,L-Methylenedioxymetamfetamine (MDMA)
Chlorpromazine metabolites	Phenmetrazine
Diethylpropion	Phentermine
Ephedrine	Phenylethylamine
Ethylaniline	Phenylpropanolamine
Isometheptene	Proguanil
Isoxuprine	Pseudoephedrine
Labetalol (+ metabolites)	Ranitidine
D,L-Metamfetamine	Tranylcypromine
D,L-Methylenedioxyamfetamine (MDA)	

do not facilitate the production of antibodies of greater selectivity. Even with larger molecules problems still occur. The Syva (now Dade-Behring) EMIT antidepressant assay, for example, was reported to cross react with phenothiazines, which are also tricyclic compounds, after overdosage (Schroeder *et al.*, 1986) and more recently quetiapine has been shown to give positive results in not only the EMIT tricyclic assay, but also with other immunoassays (Hendrickson and Morocco 2003; Caravati *et al.*, 2005). Fluoroquinolone antibiotics interfere in a number of opiate immunoassays (Zacher and Givone, 2004; Straley *et al.*, 2006). Oxaprozin unexpectedly interfered in most urine benzodiazepine immunoassays (Camara *et al.*, 1995). On the other hand antibodies to the cocaine metabolite benzoylecgonine are relatively selective (Cone *et al.*, 1990).

12.2.6 Immunoassay development

Immunoassay development requires a degree of expertise no longer readily found in routine clinical or forensic toxicology laboratories where, typically, developments focus on chromatographic methods. Consequently, commercial assays, either homogenous or immunochromatographic methods, abound despite sometimes having poor selectivity. Except perhaps for in-house assays developed by, for example, pharmaceutical companies, availability is restricted to assays that are available as kits. Abbott, Dade-Behring, Microgenics and Roche (Table 12.3) are amongst the principal manufacturers, although there are others, particularly of immunochromatographic devices utilized in point-of-care testing (POCT) and related areas (Chapter 13).

Table 12.3 Immunoassay kits: web sites

Syva EMIT	http://www.dadebehring.com/
Abbott TD_X	http://www.abbottdiagnostics.com/
Roche Abuscreen	http://www.roche-diagnostics.com/
Microgenics CEDIA	http://www.microgenics.com/
Siemens Immulite	http://www.smed.com/

Immunoassay kits do have advantages in that factors such as selectivity, sensitivity and precision (reproducibility) will have been investigated beforehand to an extent, but may appear expensive and may not be readily applicable to specimens other than those for which they were developed, usually plasma or urine. Some drugs-of-abuse assays are listed in Table 12.4. The issues surrounding use of immunoassays in POCT drug/poison detection have been reviewed (George and Braithwaite, 2002; George, 2004; Schütz *et al.*, 2004) and are discussed further in Chapter 13 and in the National Academy of Clinical Biochemistry Laboratory

Table 12.4 Some immunoassays for drugs of abuse in urine

Amfetamines	Cannabinoids	Opiates
Barbiturates[a]	Methadone	Phencyclidine (PCP)[a]
Benzodiazepines	Methaqualone[a]	(Dextro)propoxyphene
Cocaine (as benzoylecgonine)		

[a]Rarely encountered in the United Kingdom

Management Practice Guideline – Evidence Based Practice for Point of Care Testing (http://www.nacb.org/lmpg/poct/POCT_LMPG_final_rev41706.pdf, accessed 22 July 2007).

12.2.7 Radioreceptor assays

Rather than raising antibodies to an analyte, competitive binding assays can be performed using purified receptors. Clearly, such an approach will detect drugs and metabolites by pharmacological class and is likely to be of limited use in STA. A comparison of GC and radioreceptor assays for diazepam showed reasonable correlation between the GC method and the *sum* of the diazepam and nordazepam concentrations (Tuomisto *et al.*, 1984). In the absence of a suitably sensitive alternative, Metcalfe (1981) developed a radioreceptor assay for atropine using the muscarinic antagonist, ^{3}H-quinuclidinyl benzilate, which was available with high specific activity, as the label. Radioreceptor assays are important in drug screening programmes such as that of the NIMH Psychoactive Drug Screening Program (http://pdsp.med.unc.edu/indexR.html, accessed 22 July 2007) – this web site can be a useful source of information if the development of a radioreceptor assay is contemplated.

12.3 Heterogeneous immunoassays

An enzyme immunoassay (EIA) is a non-isotopic assay that uses an enzyme and a suitable substrate as the end step to quantify the amount of analyte that was present in the original sample. EMIT is a homogeneous assay (Section 12.4.1), whereas an enzyme-linked immunosorbent assay (ELISA) is a heterogenous assay that requires a separation step that can be simplied by the use of 96-well format plastic (polystyrene or PVC) microplates. Various formats of ELISA have been described including direct, indirect, sandwich and competitive. Unfortunately these terms may be misleading as the same method may be described by more than one term. An advantage of using heterogenous EIAs is that some interfering substances, including pigments from biological samples, are washed from the plate before colour development.

12.3.1 Tetramethylbenzidine reporter system

A frequently used system for quantifing enzyme activity in ELISA assays is 3,3′,5,5′-tetramethylbenzidine (TMB) with hydrogen peroxide (Figure 12.4). When reacted with horseradish peroxidase (EC 1.12.1.7) or phosphatase (EC 3.1.3.1), the rate of oxidation of TMB can be monitored at 370 or 630–650 nm. Alternatively, the reaction can be stopped after 20–30 min by adding 1 mol L^{-1} sulfuric acid, which gives a shift in absorbance to 450 nm. The yellow colour is stable for at least an hour and can be measured using a standard microplate reader. Horseradish

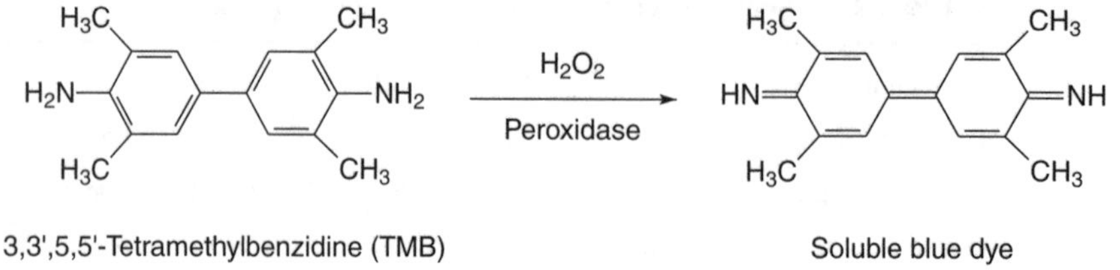

H_3C CH_3 H_2N NH_2 H_3C CH_3 —H_2O_2 / Peroxidase→ H_3C CH_3 HN =NH H_3C CH_3

3,3',5,5'-Tetramethylbenzidine (TMB)   Soluble blue dye

Figure 12.4 Reaction of tetramethylbenzidine with hydrogen peroxide in the presence of peroxidase.

peroxidase may be prefered to alkaline phosphatase because it has a lower M_r. However, it should not come into contact with solutions stablized with sodium azide, as it is very sensitive to this compound. Alkaline phosphatase, on the other hand, can be used with solutions containing sodium azide, but as zinc and magnesium ions are required as cofactors, concomitant use of solutions containing high concentrations of chelating agents such as EDTA must be avoided. TMB is also used with enzyme systems that produce hydrogen peroxide. It forms the basis of saliva alcohol test strips, when the intensity of the colour is used to quantify ethanol (Section 13.3.1.2).

12.3.2 *Antigen-labelled competitive ELISA*

This term is used to describe ELISA-type assays in which the antibody is bound to the microplate and the analyte (antigen) is labelled with enzyme. Preparation of the wells is relatively simple. A precise amount of antibody solution (in bicarbonate buffer, pH 9.0) is added to each well and incubated for an appropriate time, typically 4 h at ambient temperature or overnight at 4 °C. Once the antibodies are bound, the plate is washed and dried. In order to increase the stability of the bound antibody and to reduce nonspecific binding of assay components, a second nonspecifc protein coating may be added. Commercial EIA kits are provided as dry microplates precoated with antibody.

Sample (10–50 μL) is added to an antibody-coated microplate well, followed by a buffered solution (e.g. 100 μL) containing enzyme-labelled analyte. The plate is incubated at ambient temperature to allow the enzyme-labelled analyte, analyte and antibody to equilbrate [Figure 12.5(a)]. After the incubation period, the plate is washed with buffer (the separation step) to remove unbound enzyme. Bound enzyme conjugate and bound drug are left on the microplate. TMB/hydrogen peroxide solution is added and the plate incubated to develop the colour as described above. The higher the concentration of analyte in the sample, the less labelled enzyme will be bound on the plate and so the calibration curve has a negative slope, there being no colour at very high analyte concentrations.

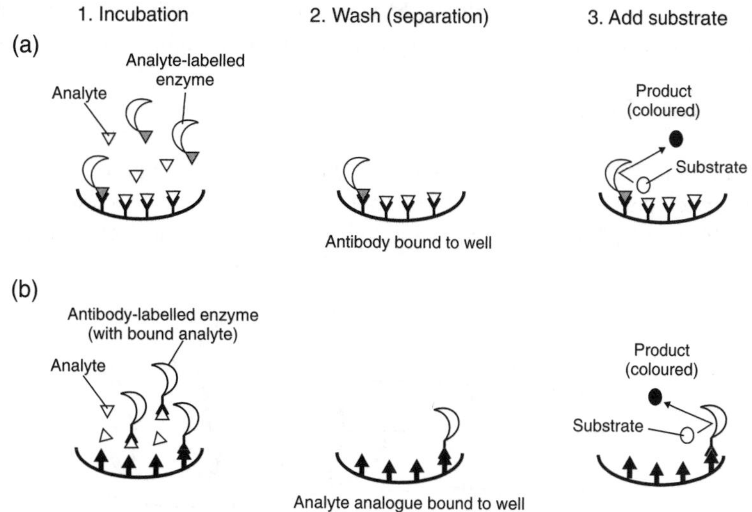

Figure 12.5 Comparison of (a) antigen-labelled and (b) antibody-labelled ELISA.

12.3.3 Antibody-labelled competitive ELISA

In this ELISA the antibody is directly coupled to the enzyme, the antigen being bound to the microplate. To do this the analyte is conjugated to a protein using a process similar to that used to prepare an immunogen, but the protein must be different from the one used to make the immunogen as otherwise the antibody may 'recognize' the protein and bind to it. The protein–analyte conjugate can be coated to the microplate well in the same was as antibodies are coated (Section 12.3.2). During the assay, competition occurs between the analyte in the sample and the immobilized analyte for binding to the enzyme-labelled antibody [Figure 12.5(b)]. Following incubation, a washing step separates bound and free fractions and leaves enzyme-labelled antibody bound to the analyte derivative in the well (Box 12.5). TMB/hydrogen peroxide can be used as the reporter system as decribed above, and because the amount of labelled-antibody that is bound is inversely related to the analyte concentration in the sample, the calibration curve has a negative slope.

Box 12.5 Enzyme-Linked Immunosorbent Assay (ELISA)

- Wells of ELISA plate precoated with immobilized analyte
- Sample and enzyme conjugate (analyte labelled-horseradish peroxidase) incubated in the well (30 min)
- Wells washed, then substrate (3,3′,5,5′-tetramethylbenzidine) added
- After further 30 min, reaction stopped by adding 1 mol L^{-1} sulfuric acid
- Absorbance of each well measured (460 nm)

ELISA kits are available for a range of drugs including: amfetamines, barbiturates, benzodiazepines, benzoylecgonine, cannabinoids, fentanyl, flunitrazepam, phencyclidine (PCP), opiates, methadone and tricyclic antidepressants (TCAs). ELISA is sensitive and has a good dynamic range. As an ELISA is a heterogeneous assay there is less opportunity for interference than with many homogenous assays, and although easily automated, such assays are performed typically in 96 well plates and thus are not compatible with high-throughput clinical chemistry analyzers.

12.3.4 Sandwich ELISA

Sandwich (or indirect) ELISA uses two or more antibodies. This requires the analyte to have more than one binding site and so, in general, is not useful for small analytes such as most drugs. In the example (Figure 12.6) excess primary antibody has been bound to the well and incubated with the

Figure 12.6 Representation of sandwich ELISA microplate well after its second washing. Note that the amount of enzyme on the plate is proportional to the amount of analyte present.

sample. After washing a second enzyme-labelled antibody was added and allowed to equilibrate. After a second wash to remove excess label, reporter solution was added as described above. Advantages of this approach are that the amount of enzyme bound to the plate is proportional to the amount of analyte in the sample, hence: (i) the calibration curve has a positive slope and (ii) the assays are potentially much more sensitive than competitive ELISAs. By using excess antibody, equilibration times are short, leading to rapid assays. An obvious disadvantage is that generally more antibody is required. Sandwich ELISAs are used frequently to test for hormones, for example human chorionic gonadotrophin (HCG) in pregnancy testing kits.

12.3.5 Lateral flow competitive ELISA

Lateral flow systems are used in many POCT kits as the technique lends itself to miniaturization and is discussed more fully in Chapter 13. Briefly, drug is immobilized as a line across a test strip, which has a reservoir of labelled antibody towards one end. When sample (e.g. urine or saliva) is added, the antibody is carried along the strip by capillary action. In the absence of drug in the sample all the labelled antibody is free to react with the immobilized drug to produce a visible line. Any drug in the sample reduces the amount of free antibody and so reduces the intensity of the line. In such a system, a strong positive result would appear as the absence of a visible line.

12.3.6 Chemiluminescent immunoassays (CLIA)

These assays may be extensions of the ELISA assays described above, but use chemiluminescence as the endpoint, for example the luminol/hydrogen peroxide system (Section 4.4.5). Advantages of chemiluminescence include (i) greater sensitivity because less sample may be required or diluted samples may be used (which may have the benefit of diluting out potential interferences), (ii) lower background signals and (iii) shorter analysis times. Furthermore, the number of chemiluminescent counts (relative chemiluminescent units, RLU) are usually greater than the counts obtained using radioactive labels and so competitive CLIAs have a larger working range, giving accurate concentration values up to 95 % analyte binging. A second mode of CLIA that uses superparamagnetic beads instead of an enzyme is described in Section 12.5.2.

12.4 Homogenous immunoassays

12.4.1 Enzyme-multiplied immunoassay technique (EMIT)

EMIT assays were introduced by Syva, now part of Dade-Behring. By attaching the analyte to an enzyme it is possible to inhibit the enzyme activity by introducing an antibody raised to that analyte. This is the principle of EMIT. Lysozyme was the enzyme used originally (Rubenstein *et al.*, 1972), but it has a low K_m, resulting in an insensitive turbidimetric assay that was subject to interference from urinary lysozyme. The commercial assay is now based on hydrolysis of glucose-6-phosphate (G-6-P) in the presence of bacterial G-6-P dehydrogenase (G-6-PDH) to which analyte has been conjugated. The coenzyme nicotine adenine dinucleotide (NAD) is reduced to NADH and the resulting increase in absorbance is monitored at 340 nm (Figure 12.7).

Antibody bound to the analyte–enzyme conjugate prevents substrate binding and reduces the rate of formation of NADH. Analyte in the sample competes with the analyte-labelled enzyme for binding to the antibody, which increases the fraction of unbound enzyme and thereby increases the rate of change of absorbance (Box 12.6). Use of bacterial G-6-PDH (NAD coenzyme), avoids interference from endogenous G-6-PDH, which requires nicotine adenine dinucleotide phosphate (NADP) as cofactor.

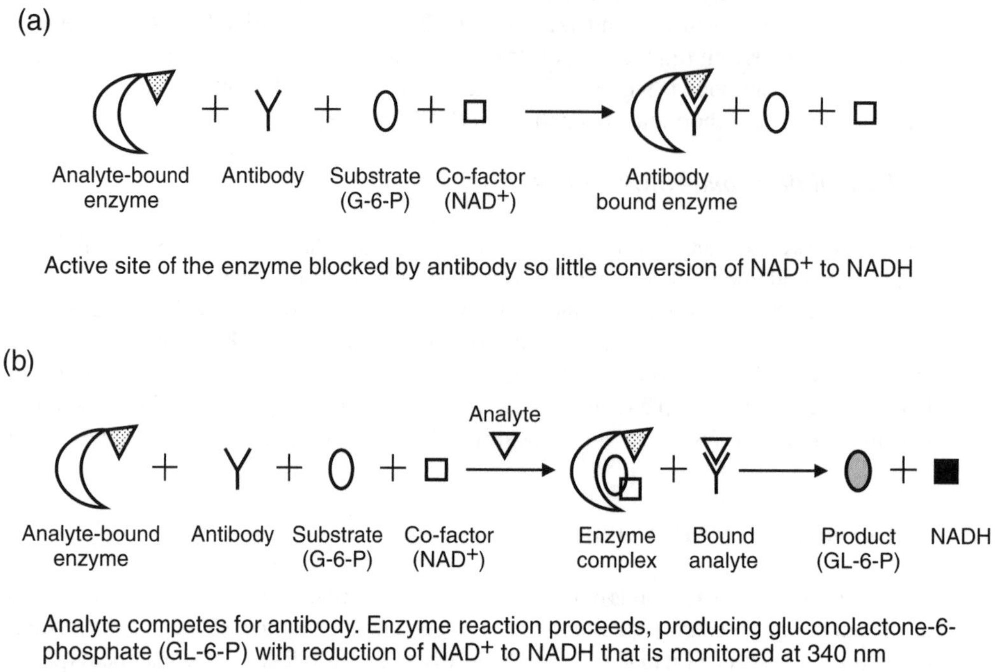

(a)

Analyte-bound enzyme + Antibody + Substrate (G-6-P) + Co-factor (NAD^+) ⟶ Antibody bound enzyme + Substrate + Co-factor

Active site of the enzyme blocked by antibody so little conversion of NAD^+ to NADH

(b)

Analyte-bound enzyme + Antibody + Substrate (G-6-P) + Co-factor (NAD^+) —Analyte⟶ Enzyme complex + Bound analyte ⟶ Product (GL-6-P) + NADH

Analyte competes for antibody. Enzyme reaction proceeds, producing gluconolactone-6-phosphate (GL-6-P) with reduction of NAD^+ to NADH that is monitored at 340 nm

Figure 12.7 Principle of enzyme-multiplied immunoassay technique (EMIT): (a) without additional analyte and (b) in the presence of analyte.

Box 12.6 Enzyme-Multiplied Immunoassay Technique

- Homogenous assay – little if any sample preparation
- Mainly used for urine or plasma (but can use for other samples after e.g. solvent extraction)
- System: antibody to analyte, analyte bonded to glucose-6-phosphate dehydrogenase (G-6-PDH), glucose-6-phosphate and NAD
- Antibody–enzyme complex inactive – added analyte in sample displaces antibody from enzyme giving enhanced enzyme activity
- Measure enzyme activity by monitoring NAD to NADH conversion (340 nm)

There is now extensive experience of EMIT, particularly as regards interference in drugs-of-abuse assays. It is simple, has adequate sensitivity for compounds with M_r < 200 present in biological fluids at moderate concentrations and, most importantly, avoids the use of radioactive labels. Although the initial rate of the reaction is proportional to the concentration of enzyme, the amount of enzyme is not directly proportional to the analyte concentration so the calibration curve is not linear, but has a positive slope. Rate monitoring can be performed using high-throughput clinical chemistry analyzers, which makes this technique and other homogenous assays attractive technically and commercially. Plasma assays are available for a number of TDM analytes (Table 12.5). Whilst not particularly reliable for analytes in the μg L^{-1} range, there are also EMIT II Plus assays for a range of drugs of abuse in urine including: amfetamines, barbiturates, benzodiazepines, benzoylecgonine, cannabinoids, opiates, methadone, (dextro)propoxyphene and TCAs. In general, these assays may be used with serum if assay calibrators of lower concentration are used, a stratagem made possible because of the lower background results obtained as compared with urine. Other modifications of the basic assay system have been suggested to facilitate use of EMIT assays with whole blood, for example (Asselin and Leslie, 1992).

Table 12.5 Dade-Behring EMIT assays for therapeutic drug monitoring

Drug class	Examples
Antiasthmatic	Theophylline, caffeine
Anticonvulsant	Phenytoin, phenobarbital, primidone, ethosuximide, carbamazepine, valproate
Antidepressant (group specific)	Amitriptyline, nortriptyline, imipramine, desipramine
Antimicrobial	Amikacin, gentamicin, tobramycin, chloramphenicol
Antineoplastic	Methotrexate
Cardioactive	Digoxin, lidocaine, procainamide/NAPA, quinidine, disopyramide
Immunosuppressive	Ciclosporin, mycophenolic acid, tacrolimus

12.4.2 Fluorescence polarization immunoassay (FPIA)

When fluorescent molecules are irradiated with polarized light of the approprate wavelength, freely rotating molcules emit light in different planes. However, slowly rotating antibody-bound fluorophores emit more light in a similar plane to the incident light and this can be measured via use of a polarizing filter. This is the basis of homogeneous competitive FPIA (Box 12.7; Figure 12.8). Fluorescein (excitation wavelength 485 nm, emission 525–550 nm) was chosen as the fluorophore (Dandliker *et al*., 1973). The background fluorescence present in biological samples means it is usual to take a reading of the sample and reagents before the addition of the fluorescent tracer.

Box 12.7 Fluorescence Polarization Immunoassay (FPIA)

- Use with plasma or urine – advantages/disadvantages similar to EMIT except that not compatible with clinical chemistry analyzers
- Fluorescein-labelled analyte rotates rapidly in solution (Brownian motion) – if irradiated with polarized light, emitted light not polarized
- When antibody added to analyte, bound analyte rotates more slowly – emitted light retains polarization
- Added analyte in sample competes with labelled drug for antibody sites, increasing depolarization of emitted light – measure on polarimeter
- Amount of depolarization related to the concentration of analyte in the sample

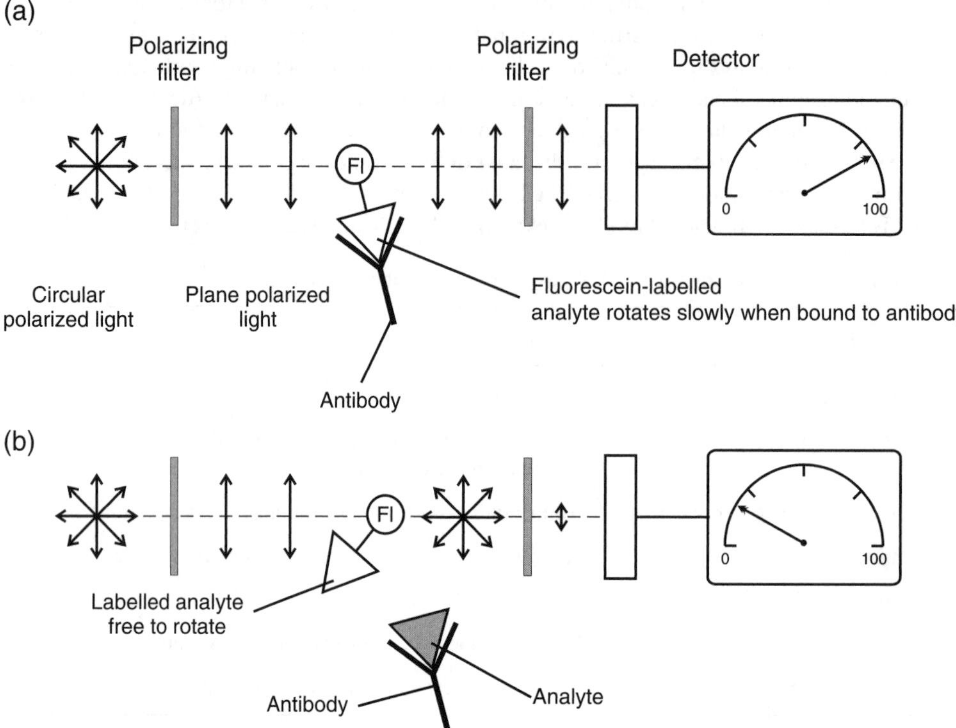

Figure 12.8 Principle of fluorescence polarization immunoassay: (a) without added analyte (b) with analyte.

Abbott introduced FPIA commercially as the basis of the AD_x analyser for urine drugs of abuse work and the TD_x analyzer for TDM. Urine assays included amfetamines, barbiturates, benzodiazepines, benzoylecgonine, cannabinoids, opiates and (dextro)propoxyphene, and serum assays carbamazepine, digoxin, paracetamol (for overdose work), phenobarbital, phenytoin, primidone, theophylline and valproate. FPIA has been used as the basis of assay systems in addition to those produced by Abbott (Colbert *et al.*, 1985), but such systems are no longer available commercially. The improved precision and reagent stability offered by FPIA are advantages over EMIT, but FPIA is not compatible with the clinical chemistry analyzers for which EMIT is readily suited. Moreover, although fluorescence methods are inherently sensitive, in FPIA sensitivity is limited by protein concentration hence the digoxin assay incorporates a protein precipitation step (Porter *et al.*, 1984).

12.4.3 Cloned enzyme donor immunoassay (CEDIA)

As with EMIT, CEDIA exploits antigen–antibody binding to influence spectrophotometrically measured enzyme activity (Henderson *et al.*, 1986). β-Galactosidase from *Escherichia coli* is supplied as inactive fragments. The large fragment (some 95 % of the enzyme) is termed the enzyme acceptor (EA), and the smaller fragment is termed the enzyme donor (ED). By conjugating hapten (analyte) to the ED fragment, antibodies to the hapten can prevent the formation of intact, active enzyme. Any analyte present in the sample competes for binding sites on the antibody, hence an increase in analyte concentration will decrease binding of antibody to the ED fragment and increase enzyme activity, which can be monitored by production of chlorophenol red (CPR) from CPR-β-galactoside (Box 12.8; Figure 12.9).

Box 12.8 Cloned Enzyme Donor Immunoassay (CEDIA)

- β-Galactosidase is split into two inactive fragments: larger fragment enzyme acceptor (EA), small fragment enzyme donor (ED)
- When EA and ED mixed they combine to form the active enzyme
- Sample and Reagent 1 (EA/antibody) first placed in well
- Reagent 2 (analyte labelled with ED and substrate) then added
- Analyte in sample binds to the antibody preventing ED-drug conjugate from binding
- The higher the analyte concentration in the sample, the lower the number of ED-drug conjugates bound to antibody
- More ED-drug conjugate available to combine with an EA fragment – this combination forms the active enzyme which in turn hydrolyzes the substrate
- Hydrolyzed product detected spectrophotometrically; absorbance proportional to analyte concentration

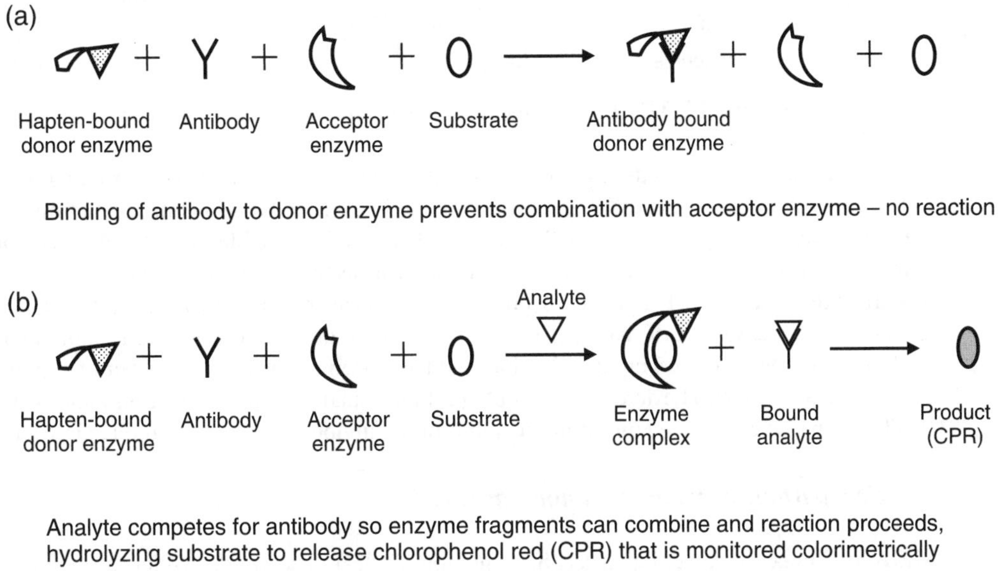

Figure 12.9 Principle of cloned enzyme donor immunoassay (CEDIA): (a) no added analyte and (b) with added analyte.

As with EMIT, CEDIAs are rate assays and may be run on high-throughput clinical chemistry analyzers. The technique has a wide dynamic range. The repertoire includes: amfetamines, barbiturates, benzodiazepines, cannabinoids, cocaine, LSD, methadone, opiates, paracetamol, PCP, (dextro) propoxyphene, salicylates, TCAs and some therapeutic drugs. CEDIA assays for haloperidol and bromperidol have also been described (Yasui-Furukori *et al.*, 2004). Weak points of CEDIA include the fact that the ED and EA fragments are not as stable as naturally occurring proteins, and the need to assemble the complex means that it is susceptible to physico-chemical disruption.

12.5 Microparticulate and turbidimetric immunoassays

The aggregation of microparticles as a diagnostic test is by no means new, and microsphere or latex agglutination tests (LAT) were described some 50 years ago (Singer and Plotz, 1956). LATs are portable, robust, efficient and work under primitive conditions making them one of the first POCT techniques (http://www.bangslabs.com/technotes/301.pdf, accessed 5 May 2006). Submicron sized polystyrene microspheres, often referred to as 'latex particles', are used for the solid support, to which usually antigen (analogue of the analyte) is bound. When antibody is introduced, several microspheres bind to each antibody producing aggregates that scatter incident light. Front- or back-scattered light can be measured with a nephelometer or by scanning laser microscopy, respectively, or the transmitted light can be measured with a spectrophotometer (Figure 12.10).

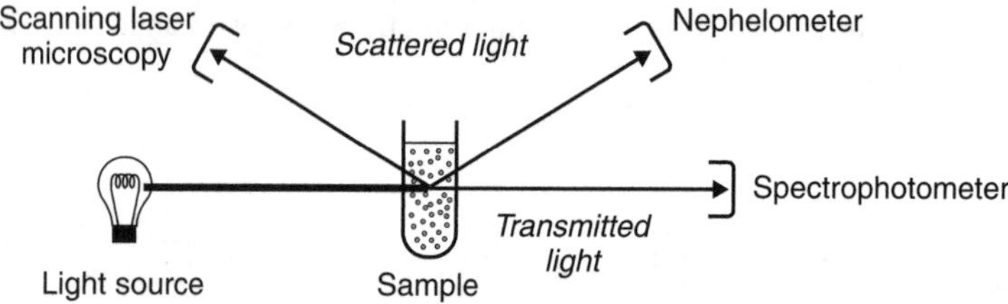

Figure 12.10 Different modes of measurement for tubidimetric assays.

The greatest degree of scattering is obtained with particles with diameters approximately equal to the wavelength of the incident light. So for visible light ($\lambda = 390$–760 nm) the optimal size of the aggregates should be in the range 0.4–0.8 μm. It would be possible to start with microspheres of this size and to observe the reduction in light scattering as they aggregate, but it is more common to start with smaller (<0.1 μm) microspheres and to measure the increase in light scattering – an apparent increase in absorbance if using a spectrophotometer. Turbidimetric immunoassays (TIA) provide sensitive and rapid end points and have been adopted by a number of suppliers of commercial assay kits. The acronyms PETIA (particle enhanced turbidimetric immunoassay) and PETINIA (particle enhanced turbidimetric inhibition immunoassay) were introduced by Du Pont.

12.5.1 Microparticle enzyme immunoassay (MEIA)

MEIA is similar to a sandwich ELISA (Section 12.3.4). The beads, which are coated with antibodies against the analyte of interest, capture the analyte. The separation step is to add the reaction mixture to a fibreglass mat. An anti-analyte antibody labelled with an enzyme such as alkaline phosphatase is incubated with the beads and a suitable substrate, 4-methylumbelliferone phosphate, for example, is added. The fluorescence of the product, 4-methylumbelliferone, is measured. As it is not a competitive assay, the signal is proportional to analyte concentration in the sample.

12.5.2 Chemiluminescent magnetic immunoassay (CMIA)

This CLIA avoids the use of an enzyme by labelling the antibody with a chemiluminescent precursor. The antigen is bound to superparamagnetic particles that are not magnets, but can be attracted to a magnet, making them very easy to separate and wash. The beads retain no residual magnetism when the magnet is removed. For example, in a competitive heterogeneous digoxin assay, a digoxin analogue was bound to superparamagnetic particles and a murine monoclonal antibody labelled with an acridinium ester added. After incubation with the sample, the beads were collected on a magnet, washed and a 'trigger solution' (hydrogen peroxide and a promoter) added to start the chemiluminescent reaction. Using acridinium labels, reaction times can be as short as 2 s, as opposed to the 30 min typical for colorimetric assays.

Alternative approaches include sandwich-type assays in which superparamagnetic beads with bound primary antibodies against the analyte are incubated with sample and, after washing, a second antibody labelled with a luminescent compound is added. Various acridinium derivates may be encountered as different manufacturers develop and patent their own labels. Superparamagnetic particles can be used for ELISAs rather than using microplates as described above.

12.6 Assay calibration, quality control and quality assurance

Quantitative immunoassays require calibration solutions (calibrators) that are traceable to an identifiable primary standard. This is rarely the case with proprietary material. Assay calibrators need to be prepared in the same matrix as the sample to be assayed. Indeed, with proprietary assays this will be defined by the manufacturer and certified by CE marking (EU) and by the FDA (US). IQC should be independent of the materials supplied by the manufacturer and carefully monitored. Participation in EQA schemes is good practice and is required by most laboratory accreditation schemes. EQA schemes challenge assays with different concentrations of: (i) analytes and (ii) potentially interfering substances, exposing the strengths and weaknesses of analytical systems, particularly drug immunoassays.

12.6.1 Immunoassay calibration

Calibration graphs for immunoassays tend to be complex, particularly if the calibration range is large. For EIAs the *rate* of the enzyme reaction should be directly proportional to the enzyme concentration. However, in competitive assays the *amount* of active enzyme present will not be directly proportional to the analyte concentration in the sample. Binding of analyte to antibody is analogous to drug–receptor binding and can be treated by the Law of Mass Action in the same way. The amount bound, B is:

$$B = \frac{B_{\max} C}{K + C} \tag{12.2}$$

where C is the concentration of analyte, K is the binding constant and $B_{\max}$ the maximum amount that can be bound. This is the equation of a rectangular hyperbola.

As can be seen from Figure 12.11(a), calibration plots of bound label versus analyte concentration are not easy to read and the data may be transformed, for example by plotting log C. This gives an approximately linear region between 20 and 80 % binding [Figure 12.11(b)]. A linear

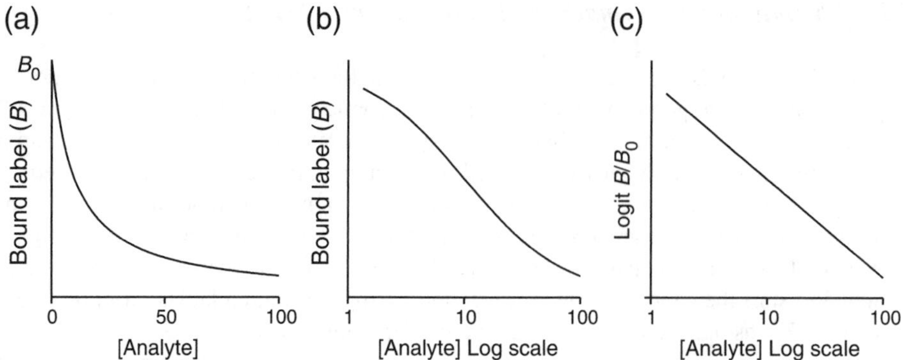

Figure 12.11 RIA calibration graphs showing original data (a) transformed to (b) log C versus B and (c) logit B/B_0 *vs* log C. B_0 = bound label in absence of added analyte.

plot is obtained using the log–logit plot where:

$$\ln\left[\frac{B/B_0}{1 - B/B_0}\right]$$

is plotted against log C [Figure 12.11(c)]. Such approaches were useful when calibration graphs were drawn manually and the B versus log C plot was helpful in defining the working range of the assay. However, the errors associated with the log–logit plot are complex and production of a linear calibration line should not lead to complacency; the results should not be extrapolated above or below the higher and lower LoQs. With the availability of curve-fitting programs there is no reason for data transformation (Figure 12.12).

A problem that can occur with calibration curves is the 'hook' effect. This may occur when samples having extraordinarily high concentrations of analyte, far exceeding the highest calibration standard, appear to have much lower concentrations, that is an extended calibration

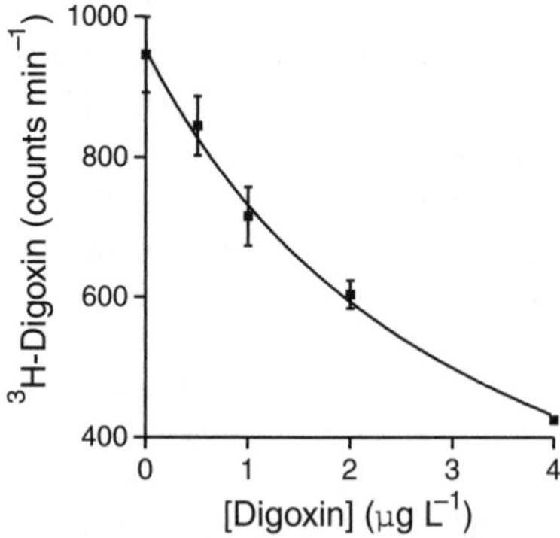

Figure 12.12 Calibration curve for digoxin RIA fitted to a hyperbola (solid-line, $B_0 = 953$ counts min^{-1}, $K = 3.31$ µg L^{-1}). Each point mean of duplicates. Reproduced from Curry, S.H. and Whelpton, R. (1983) *Manual of Laboratory Pharmacokinetics*, Wiley, Chichester, with permission from John Wiley & Sons Ltd.

curve would be hook-shaped (Selby, 1999). This is similar to the situation with fluorescence assays when two concentrations can produce the same signal (Section 4.4.2). Several explanations have been suggested for the phenomenon, including, in sandwich-type assays, binding of analyte to the secondary antibody preventing binding to the analyte that is bound to the primary antibody.

12.6.2 Drug screening

Immunoassays are convenient and rapid if screening for the presence of drugs and other substances. The requirement for only small volumes of sample has made immunoassays popular for POCT and workplace testing (Chapter 13). A full calibration line is normally unnecessary. The aim is to indicate if a substance is present, usually with the intention of confirming its identity with a more selective quantitative technique. 'Cut-off' limits have been defined, chiefly for drugs of abuse, below which the drug is deemed not to be present. Thus, the complication of calibration curves is avoided as only a calibrator at the cut-off concentration need be used. However, assays using cut-offs have to have good reproducibility at the nominated limit and have good within-batch precision (of the order of 12 %).

The US Substance Abuse and Mental Health Services Administration (SAMHSA, 2004) have defined cut-off values for five drugs of abuse, the 'SAMHSA five' (Table 12.6), in urine, oral fluid and sweat. These do not necessarily represent the LLoQ of an assay, but take account of other considerations. For example, the limit for morphine in urine was set at 300 $\mu g\ L^{-1}$ for a number of years, but was changed to 2 $mg\ L^{-1}$ to avoid confusing heroin abuse with the ingestion of poppy seed cake.

Table 12.6 Proposed SAMHSA 'cut-off' values for drugs of abuse screening

	Intial 'cut-off' concentration		
Analyte	Urine ($\mu g\ L^{-1}$)	Oral Fluid ($\mu g\ L^{-1}$)	Sweat (ng $patch^{-1}$)
Amfetamine[a]	500	50	25
Cannabis metabolites	50	4	4
Cocaine metabolites	150	20	25
Opiate metabolites	2000[b]	40[c]	25
Phencyclidine	25	10	20

[a]Target analyte must be (+)-metamfetamine and cross react (50–150 %) with MDMA, MDA and MDEA [b]Initial screen may be 6-MAM (cut-off 10 $\mu g\ L^{-1}$) [c]Initial screen may be 6-MAM (cut-off 4 $\mu g\ L^{-1}$)

12.7 Interferences and assay failures

Different systems are prone to different forms of interference (Selby, 1999). This is a particular issue in drugs of abuse work, and in digoxin and insulin assays. EMIT is particularly prone to disruption by the presence of non-physiological concentrations of salts, acid, bleach and other oxidizing agents, but other immunoassay systems are also susceptible to sample adulteration to a greater or lesser extent.

Metabolites and other structurally related compounds often cross react in immunoassays. This can be helpful in qualitative work such as drug abuse screening, but is obviously undesirable in quantitative work unless exploited, as in the cross-reaction in digoxin assays with other digitalis glycosides (Section 12.7.1.2). Many opiate immunoassays, for example, detect codeine as well as morphine and metabolites, necessitating further analysis to prove the presence of morphine. Some morphine immunoassays detect inactive metabolites such as morphine-3-glucuronide ('total morphine') whilst some just detect morphine ('free morphine'). Clearly the interpretation of results varies dramatically depending on the assay system used. Cross-reactivity with structurally similar compounds (congeners) is a further common problem – ephedrine, for example, cross-reacts in many amfetamine immunoassays (Table 12.2). Interference in a number of CEDIA measurements by a range of drugs has been reported (Table 12.7).

Table 12.7 Drug interference reported in CEDIA (Sonntag and Scholer, 2001)

Analyte	Interfering compounds
Carbamazepine	Doxycycline, levodopa, methyldopa, metronidazole
Digitoxin	Rifampicin
Phenytoin	Doxycycline, ibuprofen, metronidazole, theophylline
Theophylline	Cefoxitin, doxycycline, levodopa, paracetamol, phenylbutazone, rifampicin
Tobramycin	Cefoxitin, doxycycline, levodopa, rifampicin, phenylbutazone
Valproate	Phenylbutazone

12.7.1 Digoxin

12.7.1.1 Digoxin-like immunoreactive substances (DLIS)

DLIS were first reported in volume-expanded dogs. Elevated DLIS concentrations are encountered in patients with a variety of volume-expanded conditions, such as diabetes, uraemia, essential hypertension, liver disease and pre-eclampsia (Tzou *et al.*, 1997). DLIS, which is more than one substance, cross react with many anti-digoxin antibodies and may falsely elevate serum digoxin concentrations in immunoassays. The association of DLIS with volume expansion led to speculation that they could be natriuretic hormones. Other structures that have been proposed include non-esterified fatty acids, phospholipids, lysophospholipids, bile acids, bile salts and steroids.

Most reported endogenous DLIS are highly protein bound, whilst only 20–30 % of digoxin is bound. It has thus been suggested that measurement of digoxin in plasma ultrafiltrates can be used to assess possible interference from endogenous DLIS (Dasgupta, 2002). More recently, an association with DLIS measured using FPIA (digoxin and digitoxin assays) with endogenous ouabain has been suggested (Berendes *et al.*, 2003). Steroid hormones and bilirubin are other possible candidates for DLIS as measured by FPIA (Ijiri *et al.*, 2003, 2004); such interference did not occur using a digoxin MEIA.

12.7.1.2 Other digoxin-like immunoreactive substances

Spironolactone, canrenone and potassium canrenoate cross reacted in earlier Abbott immunoassays (Steimer *et al.*, 2002), as do various plant and other naturally occurring materials. For example, anti-digoxin F_{ab} antibody fragments have been used to reverse toxicity from cardiac glycosides present in plants such as *Apocynum cannabinum* (Indian hemp), *Digitalis purpurea* (Purple foxglove), *Nerium oleander* (Common or Pink Oleander) and *Thevetia peruviana* (Yellow Oleander). All of these substances may cross react in digoxin immunossays (Flanagan and Jones, 2004). Anti-digoxin F_{ab} antibody fragments have also been used to treat poisoning with toad venom, the most toxic components of which are cardioactive sterols (bufadienolides, notably bufalin, cinobufotalin and cinobufagin). However, monoclonal digoxin immunoassays may fail to cross react with the cardioactive sterols, and thus should not be relied upon to confirm exposure (Brubacher *et al.*, 1996; Panesar *et al.*, 2005).

12.7.1.3 Measurement of plasma digoxin after F_{ab} antibody fragment administration

Conventional serum immunoassays of glycoside concentration are no longer useful when the patient has been treated with F_{ab} fragments because the digoxin is already bound and not available for competition in an assay system. Equilibrium dialysis or ultrafiltration (Section 3.3) is required to measure free, pharmacologically active digoxin. Digestion of the F_{ab} antibody fragment–digoxin complex using a proteolytic enzyme is also required before measurement of 'total' digoxin as the affinity of the F_{ab} fragment for digoxin may well be similar to or greater than the affinity of the antibody used in the immunoassay. Plasma digoxin measurements using conventional methodology may not be reliable for up to two weeks post-treatment especially in patients with impaired renal function (Miller *et al.*, 1996; Flanagan and Jones, 2004). The use of a physical method such as HPLC-MS to measure the free fraction would be appropriate provided that sensitivity was adequate.

12.7.2 Insulin and C-peptide

Insulin is initially synthesized as proinsulin, a form of insulin in which the α- and β-chains of insulin are linked to a third polypeptide chain (connecting peptide, C-peptide). Hydrolysis of proinsulin in the pancreas leads to secretion of insulin and C-peptide in equimolar amounts, and therefore in theory plasma insulin and C-peptide can be measured and the ratio of the two analytes used as an indicator of endogenous insulin production when exogenous insulin, which lacks C-peptide, may have been given (Faber and Binder, 1986; Koskinen *et al.*, 1999; Iwase *et al.*, 2001).

But it may not be that straightforward (Gama *et al.*, 2003). Firstly, secretion of insulin and C-peptide is accompanied by the release of small amounts of proinsulins (proinsulin and other peptides related to proinsulin, cleaved in a variety of positions). The first insulin assays, also called immunoreactive insulin (IRI) assays, were RIAs with polyclonal antisera (Section 12.1.1), which cross-reacted with proinsulins. With the development of monoclonal antibodies, two-site insulin assays were introduced, that is immuno-radiometric (IRMA), -enzymometric (IEMA) and -fluorimetric (IFMA) methods (Sobey *et al.*, 1989). 'Insulin' values given by most IRMA/IEMA methods are around 20–40 % lower than corresponding IRI results (Chevenne *et al.*, 1999). Secondly, insulin has a shorter plasma half-life than C-peptide, although the half-life of C-peptide is prolonged in renal impairment.

As regards exogenous insulin, most IRI assays recognize most animal insulins and human insulin analogs such as lispro (from its structure: [LYS(B28), PRO(B29)], Humalog). Immunometric assays using monoclonal antibodies generally recognize porcine insulin, but not rat insulin. Cross reactivity with other animal insulins and lispro varies according to the assay (Chevenne *et al.*, 1999). It has been reported that most commercially available insulin immunoassays do not detect recently introduced recombinant insulins (Heald *et al.*, 2006), but the reason for the results obtained is unclear (Morovat, 2007).

The insulin-degrading enzyme (IDE: EC 3.4.24.56), specific for insulin, is widely distributed in tissues, including erythrocytes. However, haemolysis increases plasma insulin hence analysis of haemolyzed samples is unwise on two counts. Use of EDTA anticoagulant is also associated with an artefactual rise in plasma insulin. The presence of hepatic or renal insufficiency, or of anti-insulin or anti-proinsulin antibodies (analogous to the situation with digoxin assay if anti-digoxin F_{ab} antibody fragments have been given) may be further possible sources of error. The nature of the interference by anti-insulin antibodies in RIA depends on their affinity for insulin and the method used to separate bound from free radioligand, thus leading to either falsely high or falsely low results. In two-site immunoassays, the presence of anti-insulin antibodies can give rise to overestimation of plasma free insulin. The degree of overestimation depends on the comparative affinity of the autoantibodies and of the antibodies used in the assay. If the affinity of the assay antibodies exceeds that of the autoantibodies, displacement of the autoantibody–insulin complex may occur, leading to overestimation of the insulin concentration (Sapin, 1997).

Measurement of free insulin requires the removal of anti-insulin antibodies and antibody-bound insulin. Total insulin (free + bound) can be assayed by dissociating bound insulin by acidification before the antibodies are separated from the reaction mixture. Most methods use polyethylene glycol to precipitate the antibodies. In two-site assays, other types of antibody may lead to falsely elevated results. These include anti-mouse antibodies (if mouse monoclonal antibodies are used) and autoantibodies such as rheumatoid factor. These types of interference are not specific to insulin assays, but apply to all two-site assays (Marks, 2002).

Plasma C-peptide is stable for about 2–3 weeks at −20 °C and for up to 6 months at −80 °C, whereas plasma insulin is more stable (about 5 h at room temperature, one week at 4 °C, several months at −20 °C). On the other hand, C-peptide is not degraded by IDE. Anti-insulin antibodies bind proinsulin via its insulin moiety and greatly retard its clearance from the circulation. Because of cross reaction with proinsulin in some C-peptide immunoassays, proinsulin bound to anti(pro)insulin antibodies can interfere (Chevenne *et al.*, 1999).

12.8 Enzyme-based assays

Ease of automation and rapid throughput make homogeneous enzyme assays appealing, but specificity must be addressed. Two analytes for which enzyme-based assays are used quantitatively are paracetamol and ethanol; the procedures can also be adapted for qualitative use.

12.8.1 Paracetamol

An enzyme assay for paracetamol was described some 25 years ago (Hammond *et al.*, 1981; Price *et al.*, 1983). Bacterial aryl acylamide amidohydrolase (EC 3.5.1.13) was used to hydrolyze paracetamol to 4-aminophenol, which was measured colorimetrically after reaction with *o*-cresol and ammoniacal copper sulfate (Box 12.9). The method was easily automated using commercially

Box 12.9 Enzymatic assay of paracetamol (Price *et al.*, 1983)

Reagents

- Enzyme/buffer solution: acyl acylamide amidohydrolase diluted in glycerol:tris buffer (100 mol L^{-1}, pH 8.6) 1 + 1; 70,000–80,000 U L^{-1}
- Colour reagent: *o*-cresol (3.4 mmol L^{-1}), copper(II) sulfate (0.46 mmol L^{-1}) and ammonium hydroxide (9.9 mmol L^{-1}) in deionized water

Method

- Mix serum (0.1 mL) with enzyme solution (0.1 mL)
- Stand for 5 min at room temperature
- Add colour reagent (2.5 mL) and stand (3 min)
- Read absorbance (615 nm) against reagent blank

available systems (Higgins, 1987) and later modified to use 8-hydroxyquinoline in the presence of manganese ions as the chromogenic reagent forming a blue dye in the presence of 4-aminophenol (Morris *et al.*, 1990). These assays have good specificity, but have suffered from interference (false lowering of plasma paracetamol concentration) from *N*-acetylcysteine (NAC) at the concentrations attained during the early stages of a NAC infusion in the treatment of paracetamol poisoning (Tyhach *et al.*, 1999). However, a recent (2005) UK NEQAS circulation showed no such difficulty with current assays. Paracetamol immunoassays do not suffer from interference from NAC.

12.8.2 Ethanol

A number of kits are available commercially for measuring ethanol enzymatically. Most are based on the oxidation of ethanol to acetaldehyde catalyzed by yeast alcohol dehydrogenase (ADH, EC 1.1.1.1) with subsequent reduction of the cofactor NAD to NADH and monitoring of the reaction at 340 nm (Figure 12.13). Generally the reaction mixture is incubated (21–37 °C, 10–30 min) to allow the reaction to go to completion. A 'trapping agent', such as semicarbazide or hydrazine, which reacts with acetaldehyde, may be included (Poklis and Mackell, 1982).

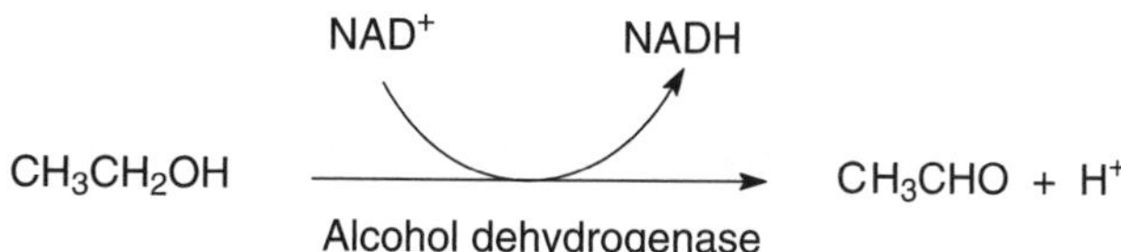

Figure 12.13 Alcohol dehydrogenase catalyzed oxidation of ethanol.

Plasma or serum can be assayed directly, but whole blood has to be treated before analysis with perchloric acid, for example. Calibration standards (0.5–4.0 g L^{-1}) should be prepared in heparinized whole blood containing sodium fluoride (10 g L^{-1}). Although ADH is insensitive to the presence of fluoride at this concentration, solutions of the enzyme should not be shaken, for example during reconstitution, to avoid foaming and denaturation. Methanol and acetone do not interfere, but higher alcohols and ethylene glycol may cross react to some extent. As these latter compounds react more slowly than ethanol, interference will be more apparent if the samples are incubated for longer than normal.

A colorimetric enzyme assay, which forms the basis of some POCT protocols for ethanol, uses alcohol oxidase (EC 1.1.3.13) to produce hydrogen peroxide (Figure 12.14), which is detected by its reaction with horseradish peroxidase in the presence of a suitable chromogen such as TMB (Figure 12.4) or 2,2′-azino-bis(3-ethylbenzthiazoline-6-sulfonic acid) (ABTS, Boehringer Mannheim). However, methanol shows cross reactivity in this type of assay.

$$CH_3CH_2OH + O_2 \xrightarrow{\text{Alcohol oxidase}} CH_3CHO + H_2O_2$$

Figure 12.14 Oxidation of ethanol catalyzed by alcohol oxidase.

12.8.3 Anticholinesterases

A qualitative test for organophosphorus and other anticholinesterases is based on the inhibition of plasma cholinesterase. Plasma cholinesterase activity can be assessed using acetyl- or butyryl-thiocholine as substrate and 5,5′-dithiobis(2-nitrobenzoic acid), Ellman's reagent. This latter compound reacts with thiols to produce 5-thio-2-nitrobenzoic acid, which is yellow in alkaline solution (Figure 12.15). The presence of anticholinesterases inhibits the hydrolysis of the thiocholine esters and so reduces the intensity of the yellow colour formed on incubation. Oximes such as pralidoxime can be added to a further patient sample – such a sample should give the same reaction as samples from unexposed individuals provided that irreversible inactivation ('ageing') of the enzyme has not occurred.

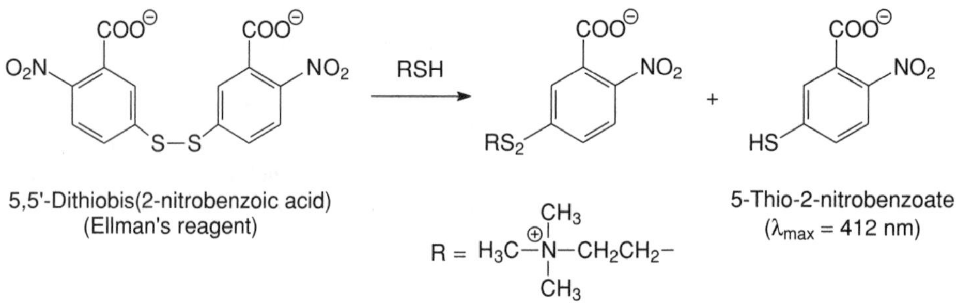

Figure 12.15 Reaction of Ellman's reagent with thiols.

12.9 Summary

All immunoassays need to be validated for use. If an assay is to be used with patient samples, then EQA samples from patients should also be analyzed. The analyst must be aware of possible matrix effects, and of the likelihood of interference from metabolites, co-prescribed or illicit drugs, or dietary or other sources. Manufacturers sometimes change their product characteristics without warning and this may alter the likelihood of interference in drugs of abuse assays. Also be aware of possible unexpected effects if in-house modifications are made to assays – dilution of reagents, for example, may affect selectivity, invalidate the product warranty and is in breach of CE marking, that is to say any results produced are at the analyst's risk.

References

Armstrong, V.W. and Oellerich, M. (2006) Critical evaluation of methods for therapeutic drug monitoring, in *Applied Pharmacokinetics and Pharmacodynamics: Principles of Therapeutic Drug Monitoring* (eds M.E. Burton, L.M. Shaw, J.J. Schentag and W.E. Evans), Baltimore, Lippincott Williams & Wilkins, pp. 30–39.

Asselin, W.M. and Leslie, J.M. (1992) Modification of EMIT assay reagents for improved sensitivity and cost effectiveness in the analysis of hemolyzed whole blood. *J Anal Toxicol*, **16**, 381–8.

Bartlett, A.J., Lloyd-Jones, J.G., Rance, M.J., Flockhart, I.R., Dockray, G.J., Bennett, M.R. and Moore, R.A. (1980) The radioimmunoassay of buprenorphine. *Eur J Clin Pharmacol*, **18**, 339–45.

Berendes, E., Cullen, P., Van Aken, H., Zidek, W., Erren, M., Hubschen, M., Weber, T., Wirtz, S., Tepel, M. and Walter, M. (2003) Endogenous glycosides in critically ill patients. *Crit Care Med*, **31**, 1331–7.

Brubacher, J.R., Hoffman, R.S. and Kile, T. (1996) Toad venom poisoning: failure of a monoclonal digoxin immunoassay to cross-react with the cardioactive steroids. *J Toxicol Clin Toxicol*, **34**, 529–30.

Camara, P.D., Audette, L., Velletri, K., Breitenbecher, P., Rosner, M. and Griffiths, W.C. (1995) False-positive immunoassay results for urine benzodiazepine in patients receiving oxaprozin (Daypro). *Clin Chem*, **41**, 115–6.

Caravati, E.M., Juenke, J.M., Crouch, B.I. and Anderson, K.T. (2005) Quetiapine cross-reactivity with plasma tricyclic antidepressant immunoassays. *Ann Pharmacother*, **39**, 1446–9.

Chevenne, D., Trivin, F. and Porquet, D. (1999) Insulin assays and reference values. *Diabetes Metab*, **25**, 459–76.

Colbert, D.L., Gallacher, G. and Mainwaring-Burton, R.W. (1985) Single-reagent polarization fluoroimmunoassay for amfetamine in urine. *Clin Chem*, **31**, 1193–5.

Cone, E.J., Yousefnejad, D. and Dickerson, S.L. (1990) Validity testing of commercial urine cocaine metabolite assays: IV. Evaluation of the EMIT d.a.u. cocaine metabolite assay in a quantitative mode for detection of cocaine metabolite. *J Forensic Sci*, **35**, 786–91.

Dandliker, W.B., Kelly, R.J., Dandliker, J., Farquahar, J. and Levin, J. (1973) Fluorescence polarisation immunoassay. Theory and experimental method. *Immunochemistry*, **10**, 219–27.

Dasgupta, A. (2002) Endogenous and exogenous digoxin-like immunoreactive substances: impact on therapeutic drug monitoring of digoxin. *Am J Clin Pathol*, **118**, 132–40.

Faber, O.K. and Binder, C. (1986) C-peptide: an index of insulin secretion. *Diabetes Metab Rev*, **2**, 331–45.

Flanagan, R.J. and Jones, A.L. (2004) Fab antibody fragments: some applications in clinical toxicology. *Drug Saf*, **27**, 1115–33.

Gama, R., Teale, J.D. and Marks, V. (2003) Best practice No 173: clinical and laboratory investigation of adult spontaneous hypoglycaemia. *J Clin Pathol*, **56**, 641–6.

George, S. (2004) Position of immunological techniques in screening in clinical toxicology. *Clin Chem Lab Med*, **42**, 1288–309.

George, S. and Braithwaite, R.A. (2002) Use of on-site testing for drugs of abuse. *Clin Chem*, **48**, 1639–46.

Hammond, P.M., Scawen, M.D. and Price, C.P. (1981) Enzyme based paracetamol estimation. *Lancet*, **i**, 391–2.

Heald, A.H., Bhattacharya, B., Cooper, H., Ullah, A., McCulloch, A., Smellie, S. and Wark, G. (2006) Most commercial insulin assays fail to detect recombinant insulin analogues. *Ann Clin Biochem*, **43**, 306–8.

Henderson, D.R., Friedman, S.B., Harris, J.D., Manning, W.B. and Zoccoli, M.A. (1986) CEDIA, a new homogeneous immunoassay system. *Clin Chem*, **32**, 1637–41.

Hendrickson, R.G. and Morocco, A.P. (2003) Quetiapine cross-reactivity among three tricyclic antidepressant immunoassays. *J Toxicol Clin Toxicol*, **41**, 105–8.

Higgins, T. (1987) Enzymic method for acetaminophen adapted to an Abbott ABA-200 analyzer. *Clin Chem*, **33**, 612.

Ijiri, Y., Hayahi, T., Ogihara, T., Ohi, K., Suzuki, K., Tamai, H., Kitaura, Y., Takenaka, H. and Tanaka, K. (2004) Increased digitalis-like immunoreactive substances in neonatal plasma measured using fluorescence polarization immunoassay. *J Clin Pharm Ther*, **29**, 565–71.

Ijiri, Y., Hayashi, T., Kamegai, H., Ohi, K., Suzuki, K., Kitaura, Y. and Takenaka, H. (2003) Digitalis-like immunoreactive substances in maternal and umbilical cord plasma: a comparative sensitivity study of fluorescence polarization immunoassay and microparticle enzyme immunoassay. *Ther Drug Monit*, **25**, 234–9.

Iwase, H., Kobayashi, M., Nakajima, M. and Takatori, T. (2001) The ratio of insulin to C-peptide can be used to make a forensic diagnosis of exogenous insulin overdosage. *Forensic Sci Int*, **115**, 123–7.

Kawashima, K., Dixon, R. and Spector, S. (1975) Development of radioimmunoassay for chlorpromazine. *Eur J Pharmacol*, **32**, 195–202.

Koskinen, P.J., Nuutinen, H.M., Laaksonen, H., Klossner, J.A., Irjala, K.M., Kalimo, H. and Viikari, J.S. (1999) Importance of storing emergency serum samples for uncovering murder with insulin. *Forensic Sci Int*, **105**, 61–6.

Law, B. (1996) *Immunoassay: a Practical Guide*, Taylor & Francis, London.

Marks, V. (2002) False-positive immunoassay results: a multicenter survey of erroneous immunoassay results from assays of 74 analytes in 10 donors from 66 laboratories in seven countries. *Clin Chem*, **48**, 2008–16.

Metcalfe, R.F. (1981) A sensitive radioreceptor assay for atropine in plasma. *Biochem Pharmacol*, **30**, 209–12.

Miller, J.J., Straub, R.W. and Valdes, R. (1996) Analytical performance of a monoclonal digoxin assay with increased specificity on the ACS:180. *Ther Drug Monit*, **18**, 65–72.

Morovat, A. (2007) Detection of Actropid by insulin assays. *Ann Clin Biochem*, **44**, 315–6.

Morris, H.C., Overton, P.D., Ramsay, J.R., Campbell, R.S., Hammond, P.M., Atkinson, T. and Price, C.P. (1990) Development and validation of an automated enzyme assay for paracetamol (acetaminophen). *Clin Chim Acta*, **187**, 95–104.

Oliver, G.C., Parker, B.M., Brasfield, D.L. and Parker, C.W. (1968) The measurement of digitoxin in human serum by radioimmunoassay. *J Clin Invest*, **47**, 1035–42.

Panesar, N.S., Chan, K.W. and Law, L.K. (2005) Changing characteristics of the TDx digoxin II assay in detecting bufadienolides in a traditional Chinese medicine: for better or worse? *Ther Drug Monit*, **27**, 677–9.

Poklis, A. and Mackell, M.A. (1982) Evaluation of a modified alcohol dehydrogenase assay for the determination of ethanol in blood. *Clin Chem*, **28**, 2125–7.

Porter, W.H., Hanver, V.M. and Bush, B.A. (1984) Effects of protein concentrations on the determination of digoxin in serum by fluorescence polarisation immunoassay. *Clin Chem*, **39**, 1826–9.

Price, C.P. (1998) AACC 50th anniversary retrospective. The evolution of immunoassay as seen through the journal Clinical Chemistry. *Clin Chem*, **44**, 2071–4.

Price, C.P., Hammond, P.M. and Scawen, M.D. (1983) Evaluation of an enzymic procedure for the measurement of acetaminophen. *Clin Chem*, **29**, 358–61.

Rubenstein, K.E., Schneider, R.S. and Ullman, E.F. (1972) "Homogeneous" enzyme immunoassay. A new immunochemical technique. *Biochem Biophys Res Commun*, **47**, 846–51.

SAMHSA (Substance Abuse and Mental Health Services Administration). (2004) *Mandatory Guidelines for Federal Workplace Drug Testing Programs*. SAMHSA, 2004 (http://www.workplace.samhsa.gov/FedPages/Pgms/Mand_Guid_04.aspx, accessed 20 July 2007).

Sapin, R. (1997) Anti-insulin antibodies in insulin immunometric assays: a still possible pitfall. *Eur J Clin Chem Clin Biochem*, **35**, 365–7.

Sasse, E.A. (1997) Immunoassays and immunoassay analyzers for analytical toxicology, in *Handbook of Analytical Therapeutic Drug Monitoring and Toxicology*, (eds S.H.Y. Wong and I. Sunshine), Boca Raton, CRC Press, pp. 223–235.

Schroeder, T.J., Tasset, J.J., Otten, E.J. and Hedges, J.R. (1986) Evaluation of Syva EMIT toxicological serum tricyclic antidepressant assay. *J Anal Toxicol*, **10**, 221–4.

Schütz, H., Verhoff, M.A., Risse, M., Auch, J. and Weiler, G. (2004) [Risks of drug screening using immunoassays]. *Dtsch Med Wochenschr*, **129**, 1931–4.

Selby, C. (1999) Interference in immunoassay. *Ann Clin Biochem*, **36**, 704–21.

Singer, J.M. and Plotz, C.M. (1956) The latex fixation test. I. Application to the serologic diagnosis of rheumatoid arthritis. *Am J Med*, **21**, 888–95.

Sobey, W.J., Beer, S.F., Carrington, C.A., Clark, P.M., Frank, B.H., Gray, I.P. and Luzio, S.D., Owens, D.R., Schneider, A.E., Siddle, K. (1989) Sensitive and specific two-site immunoradiometric assays for human insulin, proinsulin, 65–66 split and 32–33 split proinsulins. *Biochem J*, **260**, 535–41.

Sonntag, O. and Scholer, A. (2001) Drug interference in clinical chemistry: recommendation of drugs and their concentrations to be used in drug interference studies. *Ann Clin Biochem*, **38**, 376–85. Erratum: 2001; **38**: 731.

Steimer, W., Muller, C. and Eber, B. (2002) Digoxin assays: frequent, substantial, and potentially dangerous interference by spironolactone, canrenone, and other steroids. *Clin Chem*, **48**, 507–16.

Straley, C.M., Cecil, E.J. and Herriman, M.P. (2006) Gatifloxacin interference with opiate urine drug screen. *Pharmacotherapy*, **26**, 435–9.

Tuomisto, J., Tuomainen, P. and Saano, V. (1984) Comparison of gas chromatography and receptor bioassay in the determination of diazepam in plasma after conventional tablets and controlled release capsules. *Acta Pharmacol Toxicol (Copenh)*, **55**, 50–7.

Tyhach, R.J., Mayer, M. and Salpeter, L. (1999) More on interference of *N*-acetylcysteine in measurement of acetaminophen. *Clin Chem*, **45**, 584–5.

Tzou, M.C., Reuning, R.H. and Sams, R.A. (1997) Quantitation of interference in digoxin immunoassay in renal, hepatic, and diabetic disease. *Clin Pharmacol Ther*, **61**, 429–41.

Wu, A.H., McKay, C., Broussard, L.A., Hoffman, R.S., Kwong, T.C., Moyer, T.P., Otten, E.M., Welch, S.L. and Wax, P. (2003) National academy of clinical biochemistry laboratory medicine practice guidelines: recommendations for the use of laboratory tests to support poisoned patients who present to the emergency department. *Clin Chem*, **49**, 357–79.

Yasui-Furukori, N., Saito, M., Furukori, H., Inoue, Y., Someya, T., Kaneko, S. and Tateishi, T. (2004) Establishment of new cloned enzyme donor immunoassays (CEDIA) for haloperidol and bromperidol. *Ther Drug Monit*, **26**, 336–41.

Zacher, J.L. and Givone, D.M. (2004) False-positive urine opiate screening associated with fluoroquinolone use. *Ann Pharmacother*, **38**, 1525–8.

13 Toxicology Testing at the Point of Care

13.1 Introduction

Point-of-care testing (POCT), as its name implies, refers to testing carried out in close proximity to the patient or subject, usually with the aim of providing an almost immediate result. A National Academy of Clinical Biochemistry (NACB) Laboratory Medicine Practice Guideline (LMPG) defines POCT as 'clinical laboratory testing conducted close to the site of patient care, typically by clinical personnel whose primary training is not in clinical laboratory sciences or by patients (self-testing)' (http://www.nacb.org/lmpg/poct/POCT_LMPG_final_rev41706.pdf, accessed 7 July 2006). Thus, POCT may be considered as any testing performed outside of the traditional, central laboratory and may also be referred to as point-of-care laboratory testing (POCLT), bedside testing, near-patient testing (NPT), home testing and patient self-management (Table 13.1).

Table 13.1 Examples of point-of-care testing

Type of monitoring	Analyte	Test sample
Emergency department, trauma clinic	Paracetamol, salicylates, drugs of abuse, blood gases, ethanol, carbon monoxide	Urine, blood, saliva, breath
Therapeutic monitoring	Lithium, theophylline	Blood, plasma
Roadside testing	Ethanol, cannabis, cocaine	Breath, saliva, sweat
Workplace testing		
Pre-employment	Drugs of abuse	Urine, saliva
On site	Ethanol, drugs of abuse	Breath, saliva
Drug rehabilitation clinics	Ethanol, drugs of abuse	Urine, saliva, sweat
Self-management	Warfarin	Blood

With the use of POCT systems for testing for driving under the influence (DUI) of drugs (including ethanol) or screening for impairment due to drugs at work, roadside and workplace (field) testing can be added to the list. The NACB LMPG helped draw attention to important, but sometimes neglected, considerations such as staff training, QA, and issues surrounding the interpretation of results (Meier and Jones, 2005), and the corporate responsibilities inherent in a decision to introduce POCT.

13.1.1 Historical development

The first breath alcohol tester, the Drunkometer was introduced by Rolla Harger (1890–1983), but the first portable device was the Breathalyzer invented by Robert Borkenstein (1912–2002), a captain with the Indiana State Police, in 1954. To ensure that the sample was deep alveolar air, the subject had first to blow up a balloon, the air from which was then sampled by passing it over potassium dichromate crystals in a glass tube. As the ethanol was oxidized the dichromate was reduced, turning green; the intensity of the colour indicating the breath alcohol concentration (BrAC).

POCT for drugs was probably first widely used in emergency department (ED) side-room tests for salicylate using acidified iron(III) nitrate; a purple colour confirmed exposure to salicylates (Section 4.2). Use of this test with a coloured glass comparator to give a semi-quantitative value is still a viable option (King *et al.*, 1995). Early devices for DoA testing of urine gave significant numbers of false positives and false negatives. A paper chromatography system, the KDI Quik Test, was found to be unsafe (Bogema *et al.*, 1988), and the Technology Resources Inc (TRI) 'dipstick' test was no better (Jukofsky *et al.*, 1981).

With the development of non-isotopic immunoassay methods, meaningful POCT for drugs of abuse became feasible. EMIT, introduced in the early 1970s, was used to detect target analytes present above a defined 'cut-off' value in urine samples. Initially this was a laboratory method (Section 12.4.1), but the demand for field testing resulted in the development of the EMIT-ETS pack, which had the apparent benefit of utilizing the same methodological principle as the EMIT assays used in laboratories. The pack consisted of a basic photometer and dry reagents. All that was needed was the addition of the sample and buffer. Although the results required laboratory confirmation, the product was successful, albeit expensive, and set the scene for the subsequent development of POCT for drugs of abuse into the major international business that it is today (Price, 1998).

13.2 Use of POCT

The use of a POCT system should be given careful consideration before a POCT programme is initiated (Box 13.1). POCT may be performed to aid diagnosis of disease, guide therapy, or detect poisons. However, in the context of analytical toxicology, the term is more generally applied to screening for illicit or otherwise proscribed substances. The nature of the tests performed will also depend on where they are being carried out. An ED will have access to blood gas analyzers, and these may be fitted with co-oximeters thus facilitating COHb assay (Section 4.3.3.2). On the other hand, roadside or point-of-entry testing kits for use by trained police or customs officials need to be portable, robust and reliable, whilst POCT kits for workplace testing need to be simple to use and give unambiguous results within the limitations of the device.

The immediacy of the result is the major difference between POCT and laboratory-based testing and this in turn can affect the doctor/patient or tester/subject relationship, particularly if the results may lead to punitive action (George, 2004). Under these circumstances, it is important that samples be taken for confirmatory analysis by GC-MS or HPLC-MS of positive POCT drugs-of-abuse (DoA) results. In the case of testing for impairment due to alcohol or other drugs, it is preferable to arrest (roadside testing) or suspend from duty (workplace testing) until the confirmatory results are known.

Box 13.1 Questions when deciding to implement point-of-care testing

- **What are the reasons for testing?**
  - Therapeutic monitoring/emergency department testing
  - Law enforcement
  - Workplace testing/pre-employment screening
- **Why are immediate results required?**
  - Initiate early treatment
  - Effect an arrest
  - Suspend from work/do not employ
- **What is the window of detection to be?**
  - Will influence choice of fluid to be sampled
  - Currently under the influence
  - Detect a regular substance abuser
- **Which drugs should be detected?**
  - Lithium/theophylline
  - Ethanol
  - Drugs of abuse
- **Where are the tests to be done?**
  - Clinic
  - Roadside
  - Workplace
- **What tests are available?**
  - Are the kits suitable for the venue?
  - Will the operators be able to use the system correctly?
  - Is storage/shelf life a consideration?
- **Will it be cost effective?**
  - Avoid need to transport samples to laboratory
  - Avoid need for employee to visit test laboratory
  - Reduce the number of samples to be assayed by laboratory – only test putative positives
  - Cost of kits/numbers of samples

13.2.1 Samples and sample collection

POCT uses samples that require minimal manipulation prior to the analysis, common matrices being urine, blood, oral fluid or sweat. Additionally, breath is used to detect and quantify ethanol and sometimes carbon monoxide. The issues surrounding the use of such fluids are the same as, or similar to, those when traditional analyses are to be conducted (Chapter 2).

DoA testing is usually performed with urine, using POCT systems based on immunoassays similar to those described in Chapter 12. Oral fluid (saliva) has been suggested as an alternative as supervised saliva collection is more acceptable to the donor and there is less possibility of tampering and adulteration (Box 13.2). However, saliva samples are more likely to be infectious than urine and the volume of sample may be limited. DoA analysis in oral fluids and sweat has been reviewed (Kidwell *et al.*, 1998), although not in the context of POCT.

Box 13.2 Considerations in using oral fluid to detect drugs of abuse

Advantages
- Less invasive
- Less opportunity for adulteration
- Relatively easy to collect
- Detect recent drug use
- Potential for correlation between drug concentration and impairment

Disadvantages
- Flow- and pH-dependent effects on drug partition
- Narrower window of detection than urine
- Greater potential for passive exposure to smoke, for example cocaine/cannabis
- Oral residues
- Some drugs reduce saliva flow
- May be more difficult to assay (viscous sample)
- Potentially infective

Saliva has an average pH of 6.5, but stimulation of salivary flow can increase this to pH 8 which could have marked effects on the plasma:saliva ratios of weak electrolytes. Lipophilic compounds more readily diffuse from plasma to saliva and so parent drug:metabolite ratios may be different in saliva compared with plasma and urine. Cocaine:benzoylecgonine ratios, for example, are higher in saliva than in plasma and urine. Thus, POCT kits designed for use with oral fluid samples may need to target different analytes than those used when testing urine and certainly the target concentrations will be lower. SAMHSA has proposed new cut-off limits for workplace testing using urine, oral fluid or sweat samples (Table 12.7). For DUI roadside testing, an advantage of oral fluid is that the result relates more directly to concentrations of drug in blood, and hence impairment, than does the presence of drug in urine.

As with saliva, sweat production is not uniform in either amount, or composition. 'Insensible sweat' is moisture that is lost from the body through the skin and does not form droplets. 'Sensible sweat', that is sweat that can be seen as liquid, arises from two types of gland, apocrine and eccrine. The former, which tend to be located in the axillae, pubic and mammary areas, are larger and secrete a thicker substance. Also the surface of the skin is covered with sebaceous secretions, chiefly lipids, higher concentrations being found on the scalp and forehead. Thus, the fluid collected for analysis is generally a mixture of secretions. As with hair, surface contamination by exposure to drug use by others (e.g. smoking cocaine or cannabis) is a potential problem. There are two methods of collecting and testing sweat. One is the Drugwipe, which may also be used with saliva, and has found use for roadside testing. The other is a tamper-proof 'sticking-plaster' that may be used to collect sweat over several days and is generally used in detoxification clinics and US prisons. Problems with skin contamination have been investigated (Kidwell and Smith, 2001). There was good consistency between results for methadone, morphine and opiates, but not benzoylecgonine, in sweat collected at different sites (Taylor *et al*., 1998). The patches are not amenable to POCT testing.

Blood, used for POCT glucose measurements by diabetics, is not used for DoA analysis, although there is at least one 'in-office' test for lithium that uses whole blood (Section 13.3.4). Other matrices, such as hair and nail, are not particularly amenable to current POCT technology.

13.3 Analytes

13.3.1 Ethanol

To establish impairment whilst driving or operating machinery it is necessary to measure or derive the concentration of ethanol in blood. GC-FID is the most reliable method for measuring blood ethanol (Section 7.6.2.2), but this is clearly unsuitable for roadside or workplace testing. However, at equilibrium, breath, urine and oral fluid alcohol concentrations are correlated with blood concentrations and these matrices may be used as alternatives to blood. Correlating blood alcohol concentrations (BAC) with those in sweat is less reliable and so this sample is best used for qualitative purposes only.

13.3.1.1 Breath ethanol

The rationale for breath ethanol analysis is Henry's Law, which states that in a closed container at a given temperature and pressure, a solute in solution will be in equilibrium with air in the space above. The distribution of ethanol between blood and alveolar air at 34 °C is 2100:1 and this is the factor that is used to convert BrAC to BAC. Thus, 2100 litres of alveolar air will contain the same amount of ethanol as 1 litre of blood.

Modern instruments use fuel cells in which oxidation of ethanol produces a current (e.g. the Intoxilyzer) or semiconductor oxide devices in which an ethanol-specific sensor is used. Semiconductor oxide sensors are claimed to offer many benefits, including low cost, low power consumption and small size, although they need calibrating more frequently than the fuel cell devices. Laboratory-based (evidential) instruments use infrared absorption at two wavelengths (3.37 and 3.44 μm) to identify and measure ethanol. Using the ratio of wavelengths reduces the risk of interferences.

13.3.1.2 Saliva ethanol

The ethanol saliva:blood ratio equilibrates about 30 min after cessation of drinking and remains at 1.077 for at least 6 h (Jones, 1979). Thus, oral fluid provides an alternative to blood- and breath-ethanol measurement and reagent sticks for semi-quantitative measurement of ethanol have been produced. With the Alco-Screen, ethanol is oxidized by alcohol oxidase and the resulting hydrogen peroxide reacted with tetramethylbenzidine in the presence of peroxidase. Methanol, ethanol and allyl alcohol give positive results; strong oxidizing agents will give false positives and reducing agents, including ascorbic acid and L-DOPA, will reduce the signal.

The OraSure QED (quantitative ethanol detector) is based on ADH oxidation and production of a coloured end-point. Basically, it consists of a capillary tube, with a saliva reservoir at one end and a 'control spot' at the other. The spot turns purple when wetted with saliva, thereby ensuring that sufficient saliva has been applied. As the sample is drawn up the capillary the ethanol is oxidized turning a marker dye purple until all the ethanol has been consumed. The length of colour in the capillary is proportional to the original concentration of ethanol and the concentration in the saliva can be read off a scale 'like reading a thermometer'. Acetone, methanol, butanone and ethylene glycol in aqueous solution (1 g L^{-1}) showed no interference, but propanol and 2-propanol (both 1 g L^{-1}) gave apparent ethanol results of 0.6 and 0.2 g L^{-1}, respectively (Jones, 1995).

The QED is available in two sizes, 0–1.45 g L^{-1} (A150) and 0–3.45 g L^{-1} (A350) for use in roadside testing and the emergency room, respectively. Several studies have been conducted comparing the QED with alternative ethanol assays. Jones (1995), for example, reported good agreement between HS-GC and the QED for aqueous standards and for venous blood and end-expired breath samples. However, in a clinical setting the same group had difficulty obtaining sufficient saliva from some intoxicated subjects (Bendtsen *et al.*, 1999). Engelhart and Jenkins (2001) evaluated the QED for rapid postmortem saliva ethanol measurements, but not surprisingly, had difficulty in obtaining sufficient volumes of uncontaminated saliva. However, they did conclude the QED could be used for rapid measurement of ethanol in vitreous humour.

13.3.2 Drugs of abuse

More recent devices for DoA screening are based on lateral flow immunoassays (Figure 13.1). The strips usually have a wick for dipping into the sample or a reservoir (well) into which sample is pipetted (Figure 13.2). As the sample migrates along the strip, antibodies labelled with colloidal gold, latex beads or some other suitable visualizing label, are carried in the stream of liquid. Immobilized antigen, usually bound as a line on the strip at a suitable distance from the origin, will capture any nondrug-bound antibody to produce a visible line of labelled antibody. The intensity of the line will be maximal when there is no drug in the sample. When there is enough analyte in the sample to bind all the antibody, no line will be visible. Thus, by selecting the appropriate amount of antibody, devices can be manufactured to provide the required cut-off values.

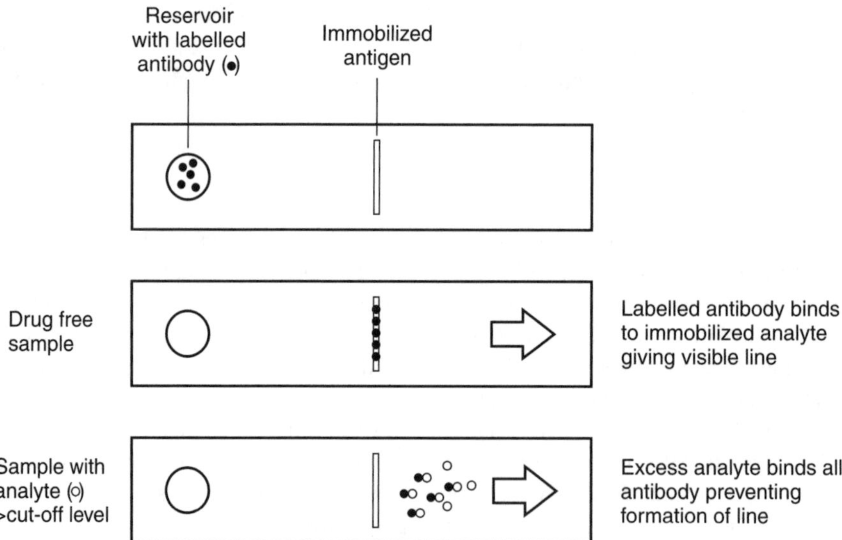

Figure 13.1 Principle of lateral flow immunoassay.

A control strip or spot is frequently included beyond the bound antigen(s). A 'positive' reaction indicates that the sample has reached the zone (sufficient sample) and the reactants are functioning properly. Several antigens can be placed on one strip, giving an array of tests. One system uses labelled drug derivatives, the immobilized antibodies being bound onto the membrane, so that a positive result is indicated by the appearance of a line. With either type of device faint lines should

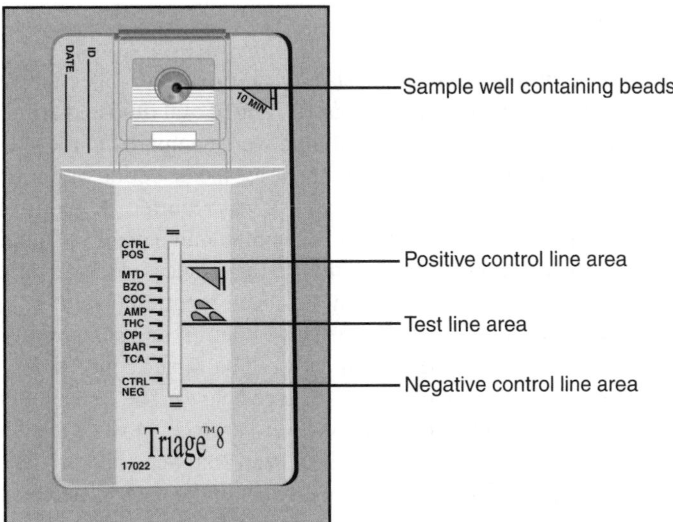

Figure 13.2 Example of a cassette-type device designed to test for eight analytes in urine (from Burtonwood *et al.*, 2003).

be read as negative. Finally, as with other immunoassay methods (Chapter 12) the specificities of the antibodies vary and many of the devices react to classes of drug, points that need to be remembered when interpreting and reporting results.

13.3.2.1 Urine testing

On behalf of the UK Medicines and Healthcare Products Regulatory Agency (MHRA) 16 devices for the detection of drugs of abuse in urine were compared against a laboratory automated immunoassay analysis. Instructions for use, manufacturer support and training, safety, ease of use, batch-to-batch variation, result interpretation and interferences were also compared (Burtonwood *et al.*, 2003). There was wide variation in performance, quality of support and cross-reactivities between the different devices. Most of the interfering substances were tested at relatively high concentrations spiked into urine, although urine from volunteers who had ingested either poppy seed bars, or codeine linctus were also used (Table 13.2). Note that the cut-off values for these devices are different from the SAMHSA values (Table 12.7).

Table 13.2 Possible interferences for DoA test kits (composite results from Burtonwood *et al.*, 2003)

Analyte	Cut off (μg L^{-1})	Interference/Cross-reactant
Amfetamine	1000	Metamfetamine, MDMA, fenfluramine, phentermine
Benzodiazepines	300	Ecgonine methyl ester, flunitrazepam
Cannabinoids	50	11-Hydroxy-Δ^9-tetrahydrocannabinol
Cocaine	300	Ecgonine
Methadone	300	Dipipanone, levacetylmethadol, diphenhydramine
Opiates	300	Poppy seed bar, morphine-3-glucuronide, codeine linctus, pholcodine

13.3.2.2 Oral fluid testing

The ORALscreen system (Avitar) consists of an oral fluid collection device and a test device based on lateral flow membrane immunoassay. Paired oral fluid and urine samples were collected from drug users and the results from oral fluid were compared to those from laboratory analysis of urine (Barrett *et al.*, 2001). Good agreement for the detection of cocaine and opiates for up to 2.5 days, and for THC for up to 1 day, post-use was reported. A good correlation between urine and oral fluid results was also reported for metamfetamine positive samples.

More recently, evaluation of six oral fluid testing devices, including the Drugwipe, which can be used with oral fluids as well as sweat, has been reported (Walsh *et al.*, 2003; Crouch *et al.*, 2005). The devices were tested with a negative control, and samples at 0.5, 2 and 10 times the target concentration spiked into human saliva. The target concentrations were the SAMHSA proposed cut-offs for saliva, apart from THC (see below). The test solutions were assayed by MS techniques to verify that the concentrations were correct. Most of the devices gave a visual end point, although the OraSure Uplink and the Cozart RapiScan had digital displays – the latter can be connected to a printer for record keeping. None of the devices detected THC at the SAMHSA cut-off of 4 $\mu g\ L^{-1}$, a cut-off set to identify recent marijuana use.

13.3.2.3 Sweat testing

Drugwipe (Figure 13.3) is a pen-size detector that can be used to detect drugs on surfaces, in sweat or in saliva. It was developed for customs use in identifying illicit drugs. Second generation Drugwipes, with increased sensitivities, were developed specifically for road traffic police testing for DUI. Forehead sweat is usually sampled. Several devices are available: Drugwipe II detects single analytes: opiates, cocaine, amfetamines (metamfetamine/MDMA), cannabis, or benzodiazepines. Drugwipe II Twin detects pairs of drug types: opiates/cocaine or cannabis/amfetamines. Drugwipe 5 simultaneously detects five analytes. Quoted sensitivities are 20–300 $\mu g\ L^{-1}$ for sweat or saliva and 2–50 $ng\ cm^{-2}$ for surfaces.

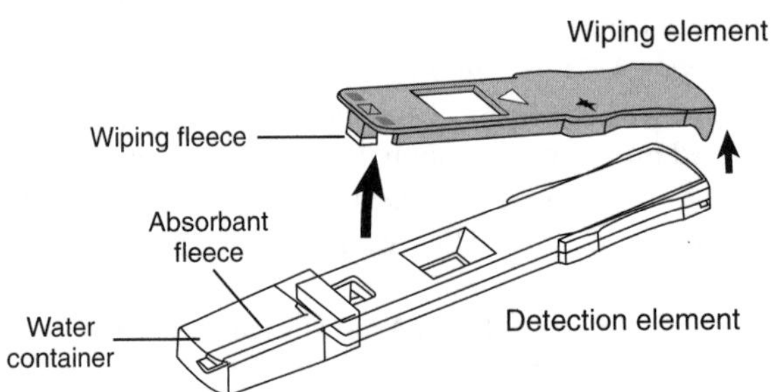

Figure 13.3 Diagram of Drugwipe.

13.3.3 Paracetamol and salicylates

Colorimetric measurement of urine salicylate in the emergency department was mentioned above (Section 13.1.1). A quantitative paracetamol assay based on reflectance photometry with a test specific dry reagent card (AcetaSite, http://www.surrey.ac.uk/MHRA/

Pages/GuildfordEvaluations/PreviousReports/Reports/StatSite.html, accessed 7 July 2006), has been evaluated (Egleston *et al.*, 1997). There was poor agreement with a standard laboratory assay (Quantase, Shield Diagnostics). Jones *et al.* (1998) found that results from AcetaSite were consistently high when compared with the Cobas (Roche) paracetamol assay or HPLC, and suggested that there may have been a calibration problem with the AcetaSite. Reasonable accuracy was claimed in another study if results <20 and >250 mg L^{-1}, were excluded (FitzPatrick *et al.*, 1998). However, it was concluded that standard laboratory tests were cheaper.

More recently, a lateral flow immunoassay POCT kit has been introduced for screening paracetamol (limit of accurate measurement 25 mg L^{-1}) and salicylate (100 mg L^{-1}) in blood (Song and Dou, 2003). Dale *et al.* (2005) observed two false negatives, which laboratory testing reported as 28 and 50 mg L^{-1}, respectively, for paracetamol, but it was thought that the test could prove useful to rule out paracetamol overdosage provided the cut off was reduced to 12.5 mg L^{-1}.

13.3.4 Snake envenomation

Identification of the snake responsible for an envenomation can guide the use of antiserum and help assess prognosis. Only rarely does the offending animal accompany the victim to hospital to facilitate visual identification, hence there is often a need for simple methods for snake venom detection and identification in biological samples. Several approaches to this problem have been reported. The Commonwealth Serum Laboratories (CSL, Australia) released a snake venom detection kit (SVDK) in 1990 aimed at detecting toxic venom in wound exudate. This is based on a two-step EIA in which the wells of an ELISA plate are coated with antibodies against various types of snake venom. Using a swab, some venom is taken from the bite wound (patient or pet) and identified. Blood and urine can also be used, but are less reliable. A positive SVDK result *per se* is no indication for antivenom, that is to say the results must always be interpreted in the clinical setting. There may be some cross reactivity with wound exudate from mildly or non-venomous snakes (Jelinek *et al.*, 2004), for example.

A kit for detecting Thai cobra (*Naja kaouthia*) venom has been developed using latex particles sensitized with venom specific immunoglobulin (Khow *et al.*, 1999). Sensitivity was 25–50 $\mu g\ L^{-1}$ of venom in plasma. There was no cross reactivity with venom from *Vipera russelli, Calloselasma rhodostoma, Trimeresurus albolabris, Naja siamensis, Ophiophagus hannah* and *Bungarus fasciatus.*

Another approach uses optical immunoassay (OIA). The underlying principle is that of detection of physical changes in the thickness of a molecular thin film as a result of specific binding events on an optical silicon chip (SILAS-I, ThermoBioStar). A prototype kit for the simultaneous identification of species and semi-quantitative measurement of venom from four venomous snakes occurring in South Vietnam (*Trimeresurus albolabris, Calloselasma rhodostoma, Naja kaouthia* and *Ophiophagus hannah*) has been developed (Dong *et al.*, 2004). The kit can detect venom analytes in blood, plasma, urine, wound exudates, blister fluid and tissue homogenates. The efficacy of the test kit in snakebite diagnosis has been demonstrated in experimental envenomations and in clinical samples from snakebite victims.

13.3.5 Therapeutic drug monitoring

TDM to ensure that plasma concentrations are in an optimum range requires quantitative assay hence the simple cut-off approach used for DoA screening is not appropriate. A number of colorimetric or immunochromatographic methods have thus been developed. However, POCT for TDM has not found widespread use in the UK, although devices are available for the American market. Use of POCT in TDM has been reviewed (Oles, 1990), although not recently.

13.3.5.1 Lithium

Capillary (finger-prick) blood has been used to measure lithium after membrane separation of the erythrocytes (Lithium System, Akers Biosciences). The residual fluid is reacted with porphyrin and the increase in absorbance measured (505 nm). The results are said to be comparable with those obtained by alternative assays (Glazer *et al.*, 2004).

13.3.5.2 Theophylline

'Immunochromatography' is the term used to describe an antibody-based test that typically uses capillary flow through an absorbent membrane to mix and subsequently separate the various components of the test mixture. This concept was marketed as AccuLevel (Syva) for TDM in the 1980s and was successful, if expensive, initially (Chandler *et al.*, 1989). However, enthusiasm was not sustained, and manufacture ceased. Asmus *et al.* (1998) concluded that the AccuMeter (ChemTrak) demonstrated good precision and minimal bias in comparison to TDx and the AccuLevel. AccuTech now sell the AccuMeter for whole blood theophylline. It is marketed in Japan (Nikken Chemicals) and has been evaluated for peri-operative use in asthma patients (Iwasaki *et al.*, 2005).

13.3.5.3 Anticonvulsants

A clinical chemistry analyzer that also measured drugs, the Ames Seralyzer, used in an epilepsy clinic to measure phenobarbital and phenytoin, significantly improved service efficiency and was convenient for patients. However reliable results required analysts trained to laboratory standards; the device is no longer available (Elliot *et al.*, 1990). Similarly, the Syva AccuLevel kit for phenobarbital, phenytoin and carbamazepine is no longer manufactured, despite the fact that, provided due care was taken, the results showed good agreement with those from the Abbot TDx (Nielsen *et al.*, 1992).

13.4 Interferences and adulterants

The problems of interferences and adulterants are not, of course, confined to POCT, but the use of POCT in workplace testing in particular, makes consideration of these issues important. Immunoassay-based kits would be expected to suffer from the same, or similar, interferences to those seen when immunoassay is used in the laboratory. Some sources of interference are not obvious. For example, ecgonine methyl ester interacts with some benzodiazepine assays, and diphenhydramine with methadone screens. Known sources of interference should be quoted in the manufacturer's literature.

The three means used by substance abusers in attempting to avoid giving a positive urine test are dilution, adulteration, and substitution. Because urine screens are based on *concentration* cut-offs rather than the *amount* of drug present in the sample, the simple expedient of drinking large amounts of water can dilute the urine so that the cut-off is not exceeded. Counter measures are to measure the relative density (RD) of the sample and the creatinine concentration. Urine samples with creatinine concentration <1.8 mmol L^{-1} (200 mg L^{-1}) should be reported as 'too dilute' and the test repeated using a more concentrated specimen. Dyes to colour the water in toilet bowls are available, and access to running water during a test can be limited to remove obvious sources of sample diluent.

Chemicals may be added to urine in an attempt to mask a test result, or to destroy the analyte. The chief objective of adulteration is to produce a negative result, but even an invalid test may lead to a second collection, thereby 'buying time' for drug and/or metabolites to be eliminated before the second test. There is a large 'underground' sale of adulterants. Commonly used compounds include household products, such as bleach, vinegar and ammonia, and chemicals that are commercially available via the Internet (Table 13.3). Soap and bleach disrupt the screening process and are easily detectable. Glutaraldehyde is also detectable by the effects observed on the screening process. Nitrite is used to conceal THC metabolites, primarily in GC-MS confirmatory tests, although at low pH values it may also reduce the concentration to below cut-off values in screening tests. Another oxidizing agent, pyridinium chlorochromate, which can be detected by the presence of chromium or pyridine, has been shown to interfere with EMIT assays and in some cases Roche Abuscreen assays (Wu *et al.*, 1999; Paul *et al.*, 2000). 'Stealth' (Table 13.3), which has been described as a mixture of peroxidase and peroxide, has been shown to reduce the concentrations of morphine and codeine in some samples (Cody *et al.*, 2001). The presence of these oxidizing agents was detected when the amounts of analyte and ^{2}H-labelled internal standards present in GC-MS confirmatory tests were reduced markedly.

Table 13.3 Examples of adulterants used to mask urine drug screens

Adulterant	Examples
Potassium nitrite	Klear, Whizzes
Glutaraldehyde	UrinAid, Instant Clean ADD-IT-ive
Pyridinium chlorochromate	Urine Luck
Peroxidase/peroxide	Stealth

It is good practice to check routinely for evasive measures, and test strips are available that will measure creatinine, nitrite, glutaraldehyde, pH, RD and two oxidants, including bleach and pyridinium chlorochromate, simultaneously. Three of these test kits have been evaluated (Peace and Tarnai, 2002). Some POCT kits for DoA include tests for adulterants.

As a consequence of testing for adulterants, urine substitution has become more prevalent. Urine substitutes are commercially available – such as a product described as 'ultra pure premixed synthetic unisex (sic) laboratory urine' that can be supplied with a heating pad to avoid detection by collection cups with built-in temperature strips. Greater supervision (i.e. invasion of privacy) during sample collection may overcome some of these measures, but it is difficult to see how one can defeat those prepared to use a catheter to inject drug-free urine into their bladders prior to testing.

13.5 Quality assurance

The fact that POCT devices are designed to be simple to use by non-laboratory staff does not mean that that such devices are foolproof and that rigorous QA measures are not needed. Some of the issues surrounding POCT are addressed in the ISO standard (ISO, 2006). A major QA issue is staff training. Unlike automated immunoassay-based laboratory methods, most of the steps in POCT require operator intervention, including sample application, timing of the reaction, reading/interpreting a visual endpoint, and recording and documenting the result. Usually, trained laboratory staff will have evaluated POCT kits, and no study to date has shown 100 % concordance with validated laboratory tests. Suppliers of POCT kits generally provide staff training, but, in general, staff with no laboratory experience have more difficulty in performing the tests and reporting results than laboratory staff. Manufacturers have tried to address the issue of incubation time by providing devices that include indicators to show when the test is ready to be read. POCT devices should be designed to facilitate the required regulatory agency documentation and retrieval of data, including QA data.

It is important that users understand the limitations of POCT kits and the interferences that may occur. For example, because a test kit is labelled 'morphine' it does not necessarily mean that a positive result unequivocally indicates the presence of morphine.

Control samples should be analyzed when a new batch of test kits arrives and at specified intervals thereafter, for example 30 days or as otherwise directed by the manufacturer. The kits should be stored under appropriate conditions. All sites where POCT is used should be included in EQA schemes. The most effective approach is through circulation of samples of known composition with known amounts of substance or interfering compounds, and assessment of the results against: (i) those of all participants and (ii) the known sample composition. The UK National Quality Assessment Scheme for Drugs runs such a programme that is tailored to the needs of those performing initial screening as well as those offering a confirmation service.

13.6 Summary

POCT applied to drugs and other poisons has made important contributions to road safety, detection of drug trafficking and emergency clinical care. However, it finds its principal application in testing for substance abuse, and is widely used in employment and pre-employment screening, in prisons, in the enforcement of Drug Treatment and Testing Orders (DTTOs) and in drug treatment and rehabilitation centres. The management of the use of POCT devices is, however, of vital importance in order to ensure that reliable results are obtained within the limits of the devices employed. Training in the use of the devices, QA and both confirmation and interpretation of results may fall within the responsibility of the analytical toxicology laboratory. Hence it is important that laboratory staff are up-to-date with developments in both the technology used in POCT devices, and their use in the area the laboratory serves.

References

Asmus, M.J., Milavetz, G., Teresi, M.E. and Weinberger, M.M. (1998) Evaluation of a noninstrumented disposable method for quantifying serum theophylline concentrations. *Pharmacotherapy*, **18**, 30–4.

Barrett, C., Good, C. and Moore, C. (2001) Comparison of point-of-collection screening of drugs of abuse in oral fluid with a laboratory-based urine screen. *Forensic Sci Int*, **122**, 163–6.

Bendtsen, P., Hultberg, J., Carlsson, M. and Jones, A.W. (1999) Monitoring ethanol exposure in a clinical setting by analysis of blood, breath, saliva, and urine. *Alcohol Clin Exp Res*, **23**, 1446–51.

Bogema, S., Schwartz, R. and Godwin I. (1988) Evaluation of the Keystone Diagnostics Quik Test using previously screened urine specimens. *J Anal Toxicol*, **12**, 272–3.

Burtonwood, C.A., Marsh, A., Halloran, S.P. and Smith, B.L. (2003) *Sixteen devices for the detection of drugs of abuse in urine*. Medicines and Healthcare Products Regulatory Agency, 2003 (see: http://aic.mhra.gov.uk/

mda/mdawebsitev2.nsf/7b31481c2988df5800256a7600410663/c71fbf70dcddf0b680256e1c0041c13b/$FILE/MHRA %2003078.pdf, accessed 7 July 2006).

Chandler, M.H., Hughes, J.A. and Clifton, G.D. (1989) Utility of rapid theophylline assay in clinical research. *Ther Drug Monit*, **11**, 365–7.

Cody, J.T., Valtier, S. and Kuhlman, J. (2001) Analysis of morphine and codeine in samples adulterated with Stealth. *J Anal Toxicol*, **25**, 572–5.

Crouch, D.J., Walsh, J.M., Flegel, R., Cangianelli, L., Baudys, J. and Atkins, R. (2005) An evaluation of selected oral fluid point-of-collection drug-testing devices. *J Anal Toxicol*, **29**, 244–8.

Dale, C., Aulaqi, A.A., Baker, J., Hobbs, R.C., Tan, M.E., Tovey, C., Walker, I.A. and Henry, J.A. (2005) Assessment of a point-of-care test for paracetamol and salicylate in blood. *Q J Med*, **98**, 113–8.

Dong, le V., Eng, K.H., Quyen, le K. and Gopalakrishnakone, P. (2004) Optical immunoassay for snake venom detection. *Biosens Bioelectron*, **19**, 1285–94.

Egleston, C.V., Browning, C., Hamdi, I., Campbell-Hewson, G. and Robinson, S.M. (1997) Comparison of two assays for measuring plasma concentrations of paracetamol. *Br Med J*, **315**, 991–2.

Elliot, K., Watson, I.D., Tsintis, P., Gray, J., Stewart, M.J., Kadr, H. and Lawson, D.H. (1990) The impact of near-patient testing on the organisation and costs of an anticonvulsant clinic. *Ther Drug Monit*, **12**, 434–7.

Engelhart, D.A. and Jenkins, A.J. (2001) Evaluation of an onsite alcohol testing device for use in postmortem forensic toxicology. *J Anal Toxicol*, **25**, 612–5.

FitzPatrick, R., Hassan, T., Ward, V. and Bodiwala, G. (1998) Comparison of assays for measuring plasma paracetamol. Training and education in use of assay are important. *Br Med J*, **316**, 475–6.

George, S. (2004) Position of immunological techniques in screening in clinical toxicology. *Clin Chem Lab Med*, **42**, 1288–309.

Glazer, W.M., Sonnenberg, J.G., Reinstein, M.J. and Akers, R.F. (2004) A novel, point-of-care test for lithium levels: description and reliability. *J Clin Psychiatr*, **65**, 652–5.

ISO (International Organization for Standardization). ISO 22870:2006. Point-of-care testing (POCT) – Requirements for quality and competence. ISO: Geneva, 2006.

Iwasaki, S., Yamakage, M., Satoh, J. and Namiki, A. (2005) Perioperative evaluation of a kit for simplified measurement of blood theophylline concentration (Accumeter Theophylline). *Masui*, **54**, 1385–91 [Japanese, English abstract].

Jelinek, G.A., Tweed, C., Lynch, D., Celenza, T., Bush, B. and Michalopoulos, N. (2004) Cross reactivity between venomous, mildly venomous, and nonvenomous snake venoms with the Commonwealth Serum Laboratories Venom Detection Kit. *Emerg Med Australas*, **16**, 459–64.

Jones, A.W. (1979) Inter- and intra-individual variations in the saliva/blood alcohol ratio during ethanol metabolism in man. *Clin Chem*, **25**, 1394–8.

Jones, A.W. (1995) Measuring ethanol in saliva with the QED enzymatic test device: comparison of results with blood- and breath-alcohol concentrations. *J Anal Toxicol*, **19**, 169–74.

Jones, A.L., Jarvie, D.R., Simpson, D. and Prescott, L.F. (1998) Comparison of assays for measuring plasma paracetamol. Possibility of calibration error needs evaluation. *Br Med J*, **316**, 475.

Jukofsky, D., Kramer, A. and Mule, S.J. (1981) Evaluation of the TRI 'dipstick' test for the detection of drugs of abuse in urine. *J Anal Toxicol*, **5**, 14–9.

Khow, O., Wongtongkam, N., Pakmanee, N., Omori-Satoh, T. and Sitprija, V. (1999) Development of reversed passive latex agglutination for detection of Thai cobra (Naja kaouthia) venom. *J Nat Toxins*, **8**, 213–20.

Kidwell, D.A., Holland, J.C. and Athanaselis, S. (1998) Testing for drugs of abuse in saliva and sweat. *J Chromatogr B Biomed Sci Appl*, **713**, 111–35.

Kidwell, D.A. and Smith, F.P. (2001) Susceptibility of PharmChek drugs of abuse patch to environmental contamination. *Forensic Sci Int*, **116**, 89–106.

King, J.A., Storrow, A.B. and Finkelstein, J.A. (1995) Urine Trinder spot test: a rapid salicylate screen for the emergency department. *Ann Emerg Med*, **26**, 330–3.

Meier, F.A. and Jones, B.A. (2005) Point-of-care testing error: sources and amplifiers, taxonomy, prevention strategies, and detection monitors. *Arch Pathol Lab Med*, **129**, 1262–7.

Nielsen, I.M., Gram, L. and Dam, M. (1992) Comparison of AccuLevel and TDx: evaluation of on-site monitoring of antiepileptic drugs. *Epilepsia*, **33**, 558–63.

Oles, K.S. (1990) Therapeutic drug monitoring analysis systems for the physician office laboratory: a review of the literature. *Drug Intell Clin Parm*, **24**, 1070–7.

Paul, B.D., Martin, K.K., Maguilo, J. and Smith, M.L. (2000) Effects of pyridinium chlorochromate adulterant (urine luck) on testing for drugs of abuse and a method for quantitative detection of chromium (VI) in urine. *J Anal Toxicol*, **24**, 233–7.

Peace, M.R. and Tarnai, L.D. (2002) Performance evaluation of three on-site adulterant detection devices for urine specimens. *J Anal Toxicol*, **26**, 464–70.

Price, C.P. (1998) Progress in immunoassay technology. *Clin Chem Lab Med*, **36**, 341–7.

Song, W. and Dou, C. (2003) One-step immunoassay for acetaminophen and salicylate in serum, plasma, and whole blood. *J Anal Toxicol*, **27**, 366–71.

Taylor, J.R., Watson, I.D., Tames, F.J. and Lowe, D. (1998) Detection of drug use in a methadone maintenance clinic: sweat patches versus urine testing. *Addiction*, **93**, 847–53

Walsh, J.M., Flegel, R., Crouch, D.J., Cangianelli, L. and Baudys, J. (2003) An evaluation of rapid point-of-collection oral fluid drug-testing devices. *J Anal Toxicol*, **27**, 429–39.

Wu, A.H., Bristol, B., Sexton, K., Cassella-McLane, G., Holtman, V. and Hill, D.W. (1999) Adulteration of urine by "Urine Luck". *Clin Chem*, **45**, 1051–7.

14 Basic Laboratory Operations

14.1 Introduction

Quality management including laboratory accreditation, that is the inspection and independent certification of laboratories to ensure as far as possible the quality and reliability of the work produced, is becoming increasingly important (Burnett, 2002). This chapter discusses issues surrounding accreditation and quality, especially as regards quantitative analysis.

Written procedures, usually known as standard operating procedures (SOPs), or laboratory procedures (LPs), should describe all aspects of the laboratory operation, including laboratory management. The International Laboratory Accreditation Cooperation web site (http://www.ilac.org/, accessed 27 June 2006) gives details of laboratory accreditation procedures. The International Organization for Standardization (ISO) gives details of quality management systems (ISO 9000, ISO 14,000) which can be used to describe all types of operations, including laboratory operation (http://www.iso.ch/iso/en/iso9000-14000/index.html, accessed 27 June 2006). ISO Standard 15189 defines standards for the operation of a medical laboratory, and is consistent with ISO 9000. The Society of Forensic Toxicologists (SOFT) and American Academy of Forensic Sciences (AAFS) have published detailed guidelines for the operation of forensic toxicology laboratories, much of which is applicable to clinical toxicology laboratories (SOFT/AAFS, 2006).

As discussed in Chapter 1, laboratory operations can be divided into pre-analytical, analytical and post-analytical phases (Box 14.1). The actual analytical methods used will depend on local circumstances. It is not essential that uniform methodology is employed, only that the method used gives accurate, reliable, reproducible results when used for its designated purpose. Assay

Box 14.1 Stages in analytical toxicology laboratory operation

- *Pre-analytical*
  Procedures must be in place to advise on appropriate sample collection (including sample tubes) and to ensure the safe transport, receipt and storage of biological samples once in the laboratory, and for arranging the priority for the analysis
- *Analytical*
  Validated (i.e. tried and tested) procedures must be used to perform the requested or appropriate analyses to the required degree of accuracy and reliability in an appropriate timescale
- *Post-analytical*
  A mechanism for reporting results and maintaining confidentiality by telephone, fax, or other electronic means and in writing must be in place. Proper interpretation of results, especially for less common analytes, must be provided. Full records of the analysis must be kept for at least 5 years (10 or more years if the case has medicolegal implications). Residues of samples must be stored appropriately until disposed of safely in an agreed time-frame

validation should conform as far as possible with the US Food and Drug Administration (FDA) Center for Drug Evaluation and Research (CDER) guidance for bioanalytical method validation (FDA/CDER, 2001). Data for within-day (repeatability), between-day and total precision should be calculated according to the protocol proposed by the Clinical and Laboratory Standards Institute (2004).

14.1.1 Reagents and standard solutions

Chemicals obtained from a reputable supplier are normally graded as to purity [analytical reagent grade (AR or AnalaR), general purpose reagent (GPR) or laboratory reagent grade and so on]. In general, material of the highest available purity should be used in analytical work. Often the maximum limits of well-known or important impurities will be stated on the label together with recommended storage conditions. Some chemicals readily absorb atmospheric water vapour and either remain solid (hygroscopic, e.g. the sodium salt of phenytoin) or enter solution (deliquescent, e.g. trichloroacetic acid), and thus should be stored in a desiccator. Others (e.g. sodium hydroxide) readily absorb atmospheric carbon dioxide either when solid or in solution. Alkaline solutions readily absorb carbon dioxide with a subsequent decrease in pH. Bacteria, often visible as a cloudy precipitate, will grow in most aqueous solutions, especially phosphate buffer solutions. Such growth will not only change the composition of the solution, but also may be a source of contamination, particularly if biological molecules (amino acids, prostaglandins) are to be assayed.

The concentrations of buffer solutions are frequently prepared as mol L^{-1} (older literature may use the symbol, M, for 'molar', i.e. mol L^{-1}). The relative molecular mass of the salt (M_r, sometimes referred to as formula weight, FWt) is usually to be found on the bottle label or in the supplier's catalogue. To prepare a 1 mol L^{-1} solution of disodium hydrogen orthophosphate, for example, the appropriate M_r (in grams) of the salt available should be weighed and dissolved in less than one litre of purified water and made up to volume with water. If the protocol requires a specified weight of salt to be taken, then it is important to use the specified salt or make the appropriate correction should that not be available. Silly mistakes can occur, such as taking disodium hydrogen orthophosphate dihydrate (Soresen's salt) to be a misprint for the more commonly encounted dodecahydrate.

Preparing standard solutions of drugs and poisons can be even more complex. For drugs that are weak electrolytes it is usual to report the result of an assay in terms of the concentration of 'free' acid or base. However, many compounds are supplied as salts, and there may be several alternatives available. One supplier offers codeine as either the base (CAS 76-57-3; M_r 299.36) or as the sulfate (CAS 1420-53-7; M_r 696.81). Morphine may be available as the anhydrous sulfate, the hydrochloride, the sulfate pentahydrate or the anhydrous tartrate (CAS 302-31-8). Although tartaric acid is tribasic, the salt is actually bimorphine hydrogen tartrate. The other crucial information that is required is the number of molecules of water of hydration per molecule of salt so that the appropriate weights can be calculated (Box 14.2).

Note that unless specifically defined otherwise, the use of the terms 'free' and 'total' drug in reports should be confined to nonprotein-bound drug measurements (Section 3.3) and to measurements performed after hydrolysis of conjugated metabolites (Section 3.4), respectively.

14.1.2 Reference compounds

The supplier, batch or lot number, purity, expiry date and any other relevant information supplied with a compound should be recorded. Compounds should be stored in the dark under conditions

Box 14.2 Weights of morphine required to make 1 litre of 100 μmol L^{-1} solution

Compound	CAS number	M_r	Weight (mg) containing 10 mg free base	Weight (mg) for 100 μmol L^{-1}
Morphine	57-27-2	285.34	10.00	28.5
Morphine hydrochloride	52-26-6	321.80	11.28	32.1
Morphine sulfate	64-31-3	668.76	11.72	33.4
Morphine sulfate pentahydrate	6211-15-0	758.83	14.32	37.9
Morphine hydrogen tartrate trihydrate	6032-59-3	774.82	14.58	38.7

Explanations:
Morphine hydrochloride Every 321.80 g of morphine hydrochloride contains 285.34 g morphine base, therefore the correction is:

$$321.80 \div 285.34 = 1.128$$

Morphine sulfate Every 668.76 g of morphine sulfate contains (285.34 × 2) g morphine base, therefore the correction is:

$$668.76 \div (285.34 \times 2) = 1.172$$

Morphine sulfate pentahydrate Every 758.83 g of morphine sulfate contains (285.34 × 2) g morphine base, therefore the correction is:

$$758.83 \div (285.34 \times 2) = 1.332$$

Morphine hydrogen tartrate trihydrate Every 774.82 g of morphine tartrate contains (285.34 × 2) g morphine base, therefore the correction is:

$$774.82 \div (285.34 \times 2) = 1.358$$

recommended by the supplier. Every effort should be made to obtain a certificate of analysis or other appropriate documentation when a new supply of a compound is obtained, but supply of such documentation is at the discretion of the manufacturer. Reference compounds withdrawn from use should be disposed of safely according to local protocols unless return to the supplier is requested.

Reference materials, and indeed reagents, can be divided into portions when first received and stored in air-tight containers with minimal headspace. This may be appropriate for deliquescent and hydroscopic materials and those that are easily oxidized because the reference material is not so frequently exposed to the atmosphere. However, this may not be appropriate if only very small quantities of material are available, or storage space is limited. Reference compounds, and so on, that have been stored in a refrigerator or freezer should be allowed to come to ambient temperature before the containers are opened to minimize the risk of water condensing on the material. To ensure good accuracy, the highest quality (purest) material available should be used to prepare calibration standards. The material must have been stored appropriately and dried before use if necessary.

Obtaining reference samples of rarely encountered compounds and of metabolites can prove very difficult. In addition to chemical synthesis using a precursor or analogue as starting material, a further approach that has been suggested for metabolite synthesis is incubation of parent drug with fission yeast (*Schizosaccharomyces pombe*) genetically modified to express human CYP2D6 followed by metabolite isolation (Peters *et al.*, 2005).

In any event, if primary standards, such as drugs, have been obtained from ill-defined sources, it is important to have some idea of the purity of each sample. Visual inspection may show inconsistencies in colour, which may indicate decomposition. Useful information can often be obtained by carrying out a qualitative chromatographic analysis (TLC, HPLC) or spectroscopic investigation (UV, IR, NMR). Drying a sample to constant weight will indicate if it contains moisture. A UV spectrum may be valuable. It is also possible to measure the absorbance of a solution of the drug and compare the result with literature values for specific absorbance, A_1^1 (Section 4.3.1). It should be noted that wherever possible A_1^1 values are best measured at the λ_{max} so that errors in the wavelength calibration of the instrument are minimized. Also, the literature values may be in error, or not sufficiently accurate for the purpose, but a large discrepancy between the quoted and observed value probably indicates a gross error. It is good practice to store the UV spectra for future reference.

14.1.3 Preparation and storage of calibration solutions

Assay calibration is normally by analysis of standard solutions containing each analyte over an appropriate range of concentrations prepared in analyte-free plasma, urine or other appropriate fluid. The chosen medium should be analyzed prior to the addition of the analyte(s) to ensure the absence of interferences

Balances for weighing reagents or calibration standards, and automatic and semi-automatic pipettes, must be kept clean and checked for accuracy on a regular basis. Weighing of balance check weights should be recorded. Pipette accuracy can be documented by dispensing purified water and recording the weight dispensed. Weighing of reference compounds should normally be performed by one analyst and witnessed by a second analyst, as should other steps in preparing calibration standards and IQC solutions such as calculating and performing dilutions.

Unless the reference material is in short supply, then an 'appropriate' amount should be weighed. It is difficult to give an exact figure because this will depend on the accuracy of the balance used. If a well-maintained four-place balance, accurate to ±0.1 mg, is used then 100 mg can be weighed to ±0.1 %. This may be sufficiently accurate, particularly if the accuracies of subsequent volumetric steps are not better than this. *Ideally*, a portion of standard reference compound should be removed (tipped) into a clean container and samples of this taken for weighing. Any excess should be disposed of, not returned to the stock container. This approach also prevents dirty spatulas contaminating the stock reference material.

To minimize operator bias, an approximate amount should be weighed accurately rather than attempting to weigh to a pre-determined value. For example, ideally a weight of 98.4 mg should be accepted and recorded rather than attempting to weigh exactly 100.0 mg. Although this can lead to complex calculation of calibration standard concentrations, the ready availability of calculators and spreadsheets means that it is not the problem it used to be when calibration graphs were drawn by hand. To avoid errors caused by transfer losses, samples can be weighed directly into volumetric flasks. This is particularly useful when only a small amount of material is available, which is often the case in analytical toxicology. It is, of course, important to consider the solubility of the material in the chosen solvent when choosing the appropriate volumetric flask.

The volumetric part of preparing calibration curves is usually less accurate than the gravimetric (weighing) part. Good quality volumetric glassware and pipettes are required and must be used and cleaned correctly – volumetric glassware must *not* be dried by heating in an oven. Glass (bulb) pipettes may be appropriate, but biological fluids (plasma and blood) are more viscous than water and take longer to drain. Semi-automatic pipettes are normally calibrated to measure aqueous fluids (RD approximately 1), and thus should not be used for organic solvents or other solutions with RDs or viscosities greatly different from those of water. Positive displacement pipettes should be used for very viscous fluids such as whole blood. Using the same pipettor, with clean, properly fitted tips will ensure that any systematic error in the volume delivered will be cancelled out.

The most accurate dilutions are obtained by diluting 1 + 1. However, this is often impractical, especially if the final dilution is into plasma or whole blood, when a minimal amount of water should be added else the blood will be markedly diluted. To avoid diluting the matrix, primary standards can be prepared in a suitable volatile solvent and the required volume of solution pipetted into a volumetric flask. The solvent is evaporated under a stream of compressed air or nitrogen and the dry residue taken up in the matrix. This can be very successful with blood and plasma when protein binding helps to dissolve the analyte. However, different compounds can take different amounts of time to dissolve so it is important to test that dissolution is complete. The solvent should not interact with the analyte. Acetone would react with primary amines and use of methanol, although usually suitable, may result in transesterification of esters.

The range of the calibration curve should cover the range of concentrations expected in the samples. The calibration curve should not normally be extrapolated beyond the lowest or highest standard solution. If the concentrations of some of the samples being analyzed are below the lowest calibration standard, then the assay may be repeated with the inclusion of lower concentration standards, if that is possible, or the result reported as less than the lowest standard. High concentration samples may be diluted with an appropriate 'blank' matrix to fit within the calibration range, provided the validity of so doing has been demonstrated during method development and if sufficient sample is available. Dilutions should be made with the same (e.g. blank plasma from the subject), or very similar matrix (e.g. the matrix used for the calibration standards) as the sample.

Neonatal calf serum (Sigma, Poole, UK or other appropriate supplier) is often used in the preparation of calibration solutions for plasma or serum assays. Newly prepared calibration solutions should be validated before use by comparison either with existing calibration solutions, or with IQC solutions prepared in human plasma or serum, and the results recorded. If for any reason neonatal calf serum proves unsuitable, human plasma or serum from an appropriate source should be used.

Blood-bank whole blood (transfusion blood) is sometimes used to prepare calibration standards. Such blood, and plasma derived from it, will: (i) usually be diluted with citrate solution, which has a high buffering capacity, (ii) contain lidocaine and sometimes lidocaine metabolites and (iii) may well contain plasticizers and other contaminants that may interfere in chromatographic and possibly other assays, and may alter the distribution of drugs between red cells and plasma by altering protein binding. Commercially available equine or bovine blood may suffer from some of these same problems.

If newly prepared calibration standards and the IQCs meet the criteria for acceptance, they should be transferred to labelled 3 mL plastic tubes and stored (−20 °C) until used. Cleaning records and lists of contents should be posted on all refrigerators and freezers to meet health and safety requirements. These instruments should also be fitted with failure alarms and temperature monitoring devices and temperature records kept (Box 2.5).

14.2 Aspects of quantitative analysis

14.2.1 Analytical error

Replicate analyses of a sample usually do not produce identical values, due to errors that are associated with the analysis. These errors may be either random or systematic. Systematic errors (e.g. an incorrectly set pipetting device, variable recoveries during LLE, etc.) should be investigated, and either eliminated, or an appropriate correction made. Some systematic errors can be reduced by the way the work is performed such as use of the same pipette for calibration solutions and for samples. Gross errors such as dropping a sample or adding the wrong reagent may be obvious to the eye or be revealed by a very poor recovery of internal standard, for example.

Any measurement is an estimate of the 'true' value. Thus, there is an error associated with that measurement – the difference between the true and the measured value. Consequently, when repeated measurements are made the results cover a range of values. The result could be reported as the average value of the measurements ± the range of values obtained (largest to smallest value), however, the values at the extremes will not give a good representation of the distribution of the results about the mean value. If a sufficiently large number of replicates are used then the distribution can be seen from a frequency diagram. A set of all possible measurements (an infinite number) is known as the *population* and the histogram of results (Figure 14.1) represents a *sample* from that population.

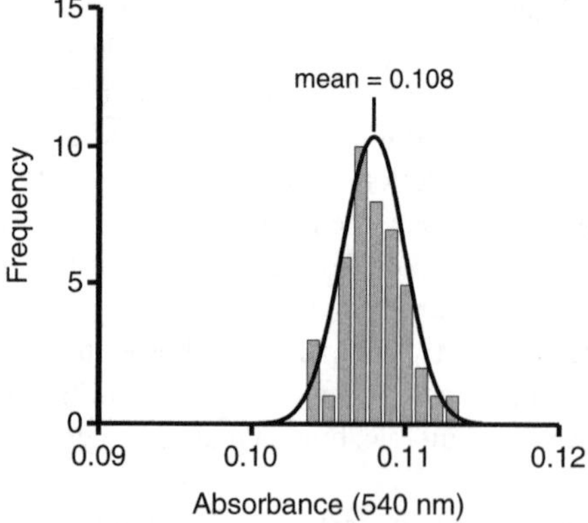

Figure 14.1 Histogram of replicate absorbance measurements ($n = 44$) for plasma sulfadimidine. The solid line is a Gaussian curve fitted to the data; mean = 0.108, SD = 0.002.

Population distributions for analytical measurements are frequently a normal or Gaussian curve and, in the absence of any systematic errors, the population mean (μ) is the true value. Statistics based on such distributions are referred to as *parametric*. The spread of the curve can be described in terms of the population SD (σ): the greater the SD, the wider the curve will be. Furthermore, for a normal distribution, approximately 68 % of the results lie within $\pm 1\ \sigma$ of the mean, ~95 % are within $\pm 2\ \sigma$ and ~99.7 % are within $\pm 3\ \sigma$ of the mean. If values of μ and σ are known, then it is possible to calculate the exact proportion of values that lie within any interval. Firstly the standardized normal variable is obtained:

$$z = \frac{(x - \mu)}{\sigma} \tag{14.1}$$

and then tables of z-values can be used to ascertain the proportion of values lying outside the range, or the NORMSDIST(z) function in Microsoft Excel can be used, which calculates:

$$f(z) = \frac{1}{\sqrt{2\pi}} \exp\left(-\frac{z^2}{2}\right) \tag{14.2}$$

Using values of -1.96 and 1.96, returns 0.025 and 0.975, respectively. Thus 95 % of the results lie within $\pm 1.96\ \sigma$. Similarly, the formula can be used to show that 99 % of results lie within $\pm 2.58\ \sigma$ of the mean and 99.9 % within $\pm 3.29\ \sigma$.

Clearly, μ and σ cannot be measured directly, but have to be estimated from sample values. The sample mean is:

$$\bar{x} = \frac{\sum x}{n} \tag{14.3}$$

and the sample SD, s is:

$$s = \sqrt{\frac{\sum (\bar{x} - x)^2}{n - 1}} \tag{14.4}$$

and the sample variance is s^2. In this case $(n - 1)$ is the number of degrees of freedom and so variance is the sum of squares divided by the degrees of freedom. Note that population values take Greek characters, whereas sample values are distinguished by Roman characters.

Unless purely by chance, the calculated sample mean will not be exactly the same value as the population mean, but the more measurements that are included in its calculation the closer the two will become. The error in the sample mean is known as the standard error of the mean:

$$s.e.m. = \frac{s}{\sqrt{n}} \tag{14.5}$$

and this gives a measure of the variability of $\bar{x}$. It is often more informative to describe the SD as a proportion or percentage of the mean. This is the relative standard deviation, RSD, also known as the coefficient of variation, CV:

$$\text{RSD} = \frac{s}{\bar{x}}\ x\ 100\,\% \tag{14.6}$$

14.2.1.1 *Confidence intervals*

When n is large (>200), s is considered to be a reliable estimate of σ, and the general formula for the confidence interval is given by:

$$\bar{x} \pm \frac{zs}{\sqrt{n}} \tag{14.7}$$

The appropriate value of z is selected for the level of confidence that is required: 1.96 for 95 %, 2.326 for 99.0 % and so on. For smaller sampler sizes, the normal distribution does not hold true and s is a less reliable estimate of σ and so an alternative distribution, the t-distribution or Student's distribution is used. The confidence intervals are:

$$\bar{x} \pm \frac{t_\nu s}{\sqrt{n}} \tag{14.8}$$

where $\nu = n - 1$ and is known as the degrees of freedom. Values of t_ν are obtained from statistical tables for the required level of confidence, examples of which are shown in Table 14.1.

Table 14.1 Examples of values of t for three confidence levels

		Confidence level		
Number of replicates (n)	Degrees of freedom ($n-1$)	90 %	95 %	99 %
2	1	6.31	12.71	63.66
3	2	2.92	4.30	9.92
4	3	2.35	3.18	5.84
5	4	2.13	2.78	4.6
6	5	2.02	2.57	4.03
10	9	1.83	2.26	3.25
20	19	1.73	2.09	2.86

A much-debated topic is the need for replicate analyses (n) in clinical and forensic analysis. In statistical terms, by taking the mean of n replicates, the variability of the analysis (RSD) is reduced by the square root of n (Table 14.2). In some RIA, for example, the variability was such that triplicates or even quadruplicates were needed. In practice, with chromatographic methods especially and 'one-off' clinical/forensic samples, duplicates usually give a reasonable compromise between

Table 14.2 Effect of replicate analyses (n) on RSD (arbitrary assay RSD of 10 %)

n	Square root	RSD (%)	n	Square root	RSD (%)
1	1.00	10.00	6	2.45	4.08
2	1.41	7.07	7	2.65	3.78
3	1.73	5.77	8	2.83	3.54
4	2.00	5.00	9	3.00	3.33
5	2.24	4.47	10	3.16	3.16

reagent and other costs, time, sample volume availability and the need for accuracy. Also, should one of a pair of tubes be broken during the analysis, one at least remains. A marked difference between duplicates indicates gross error and the need for the test to be repeated, if possible. On the other hand, it is often argued that in pharmacokinetic profiling a single analysis is all that is needed, as gross error will be revealed by comparison with the results from adjacent samples. A number of duplicates may be assayed for QA purposes, but these data should not be included in the profile as the errors differ from those of the singleton analyses. One of the results from each duplicate could be included, but which one of the pair should be decided before the assay.

If insufficient sample to permit a duplicate analysis is available, then use of a smaller sample volume in one or both replicates may be acceptable, provided that the volume is made up using 'blank' plasma/serum and such dilutions have been validated for the assay.

14.2.2 *Minimizing random errors*

Each time a measurement is made there will be an associated error and the total error will be a function of the total number of measurements and how they relate to each other. When measured values are added or subtracted, the variance (s^2) of the answer is the sum of the variances for each measurement. For example, when weighing a sample, if the mass of the weighing boat is $M_1 \pm s_1$ and the mass with added sample is $M_2 \pm s_2$ then:

$$\text{Variance}(M_2 - M_1) = \text{Variance}(M_1) + \text{Variance}(M_2)$$

So the combined SD, s_{diff} is:

$$s_{\text{diff}} = \sqrt{s_2^2 + s_1^2} \tag{14.9}$$

The same relationship is used for the combined SD when values are added.

The relationship for multiplication and division is more complex, but the approximate values can be obtained using the RSDs. Thus, for a ratio of measurements:

$$\text{RSD}_{\text{ratio}} \approx \sqrt{\text{RSD}_1^2 + \text{RSD}_2^2} \tag{14.10}$$

14.2.2.1 Preparation of a solution of known concentration

In this example, 10 mg of analyte is weighed into a 10 mL flask and made up to volume. Assuming that the weight of the flask is 12 g and that the weighings have a SD of 0.2 mg, then:

$$s_w = \sqrt{0.2^2 + 0.2^2} = 0.283 \text{ mg} \tag{14.11}$$

The estimate of the weight of drug is 10 ± 0.283 mg, so the RSD_w is $\pm 0.283/10 \times 100 = 2.83$ %. If the volume added to the flask is 10 ± 0.1 mL (RSD = 1 %), then the RSD for the concentration is:

$$RSD_{concn} = \sqrt{2.83^2 + 1^2} = \sqrt{9.01} = 3\,\% \tag{14.12}$$

and the concentration is 1 ± 0.03 mg mL^{-1} (g L^{-1}).

If a RSD of 3 % is deemed to be too high, then inspection of Equation (14.12) and the data of Table 14.3 show that the larger RSD is associated with the weight of analyte. Weighing 25 mg reduces the RSD to 1.5 %, but increasing the weight to 50 mg gives only a marginal improvement. Weighing 100 mg would be a waste of compound because for this weight most of the error in the concentration is due to the error in measuring the volume. This illustrates the important point that if the accuracy of an assay is to be improved, then the steps with the largest errors need to be identified and appropriate measures taken.

Table 14.3 Effect of amount of drug weighed on overall assay error

Weight of drug (mg)	RSD_w (%)	$(RSD_w)^2$	RSD_{concn} (%)
5	4.76	22.7	4.9
10	2.83	8.0	3.0
25	1.13	1.28	1.5
50	0.57	0.32	1.15
100	0.28	0.078	1.04

14.2.3 Accuracy and Precision

When a sample is analyzed, the closer the measured concentration is to the actual concentration then the more *accurate* is the result. If repeat analyses are conducted, then the less variation there is between the observations, that is to say the less scatter there is about the mean value, then the greater is the *precision*. Thus, precision defines random errors, *bias* defines systematic errors, and accuracy defines how close a measured value (or mean of values) is to the 'true' value, and contains both types of error. The 'true' value can never be ascertained as there must always be some error in measuring it. A working definition is that accuracy is the closeness of agreement between a test result and an accepted reference value.

Repeatability is used to describe within-batch precision (i.e. when the same standard solutions and apparatus are used by a single analyst) whereas *reproducibility* refers to between-batch precision, when, usually, different standard solutions, apparatus, operators and possibly laboratories are being compared.

14.2.3.1 *Assessing precision and accuracy*

To assess assay precision, replicate assays (e.g. $n = 8$) are performed, usually at several concentrations, and the means and RSDs calculated. It is usual to quote the results as RSDs because the errors are generally proportional to the concentrations being measured. Indeed, most assays are designed and performed to ensure that the RSD does not exceed 15 % (or some other suitably chosen value) irrespective of the concentration. Intra-batch (within-day) precisions are determined from a single analytical run and inter-batch (between-day) values from the comparison of values from several analytical runs. It is generally assumed that inter-batch RSDs will be greater than corresponding intra-batch values, but this does not always prove to be the case.

Accuracy is more difficult to assess, as a sample of known concentration is needed with which to compare results. EQA schemes provide one method of assessing accuracy (Section 17.7.2). However, such schemes are not available for all analytes.

14.2.3.2 *Detecting systematic error (fixed bias)*

If n replicate measurements have a mean concentration of $\bar{x}$ and a standard deviation, s, then the result can be tested to see if it is significantly different from the target concentration, μ, using a t-test:

$$t_{(n-1)} = \frac{(\bar{x} - \mu)s}{\sqrt{n}} \tag{14.13}$$

If $t_{(n-1)}$ is greater than the critical value (Table 14.3) then $\bar{x}$ is significantly different from μ. Thus, if the target value is for a QA sample, then the estimate contains a systematic error, also referred to as bias. In a similar way, the results from two analysts can be compared to assess if the differences in the mean values were due to random errors, in which case they should not be significantly different. If the samples have equal size variances, then the appropriate t-test is:

$$t_\nu = \frac{\bar{x}_1 - \bar{x}_2}{s\sqrt{\frac{1}{n_1} + \frac{1}{n_2}}} \tag{14.14}$$

where $\nu = n_1 + n_2 - 2$ and s is given by:

$$s = \sqrt{\frac{(n_1 - 1)s_1^2 + (n_2 - 1)s_2^2}{\nu}} \tag{14.15}$$

The F-test is used to ascertain whether or not the variances are significantly different:

$$F = \frac{s_1^2}{s_2^2} \tag{14.16}$$

Note that the variance ratio is always written such that $s_1 > s_2$ to ensure $F \geq 1$. Critical values of F are tabulated with ν_1 [= $(n_1 - 1)$] along one axis and ν_2 [= $(n_2 - 1)$] along the other. A two-tailed test should be used. When the sample variances are not equal the equation is more complex, but the principle of the test is the same.

14.2.3.3 Identifying sources of variation: analysis of variance

There are situations when more than two means need to be compared, for example when comparing results from several laboratories. There will be within-laboratory variation, but there may also be additional variation due to between-laboratory variation. This can be tested for using analysis of variance (ANOVA). As an example, data from an experiment testing the effects of storage conditions (Table 14.4) give rise to mean values for replicate analyses of the freshly prepared and stored samples. The question is 'Do all the means come from the same population?' With each treatment having n replicates:

Total sum of squares (TSS) $= \Sigma$(Measured concentration $-$ Overall mean)^2
Within-treatment sum of squares (WSS) $= \Sigma$(Measured concentration $-$ Treatment mean)^2
Between-treatment sum of squares (BSS) $= \Sigma n$ (Treatment mean $-$ Overall mean)^2

Table 14.4 Data from a typical stability experiment

Treatment	Measured concentration					Mean	s	s^2
Freshly prepared	104	100	98	95	101	99.6	3.36	11.3
Stored −20 °C (18 h)	100	97	95	103	98	98.6	3.05	9.3
Stored 4 °C (18 h)	98	94	89	100	95	95.2	4.21	17.7
Room temperature (4 h)	99	89	93	85	88	90.8	5.04	29.2
Overall mean						96.05	4.01	16.88

In the example, the within-treatment variance, σ_w^2, has 16 degrees of freedom as each replicate of n samples has $(n - 1)$ degrees of freedom and there are four treatments, and so:

$$\sigma_w^2 = MS_w = \frac{WSS}{m(n-1)} \tag{14.17}$$

where m is the number of treatments. It is also referred to as the within-treatment *mean square* (MS) and it can be shown that it is the mean within-treatment variance (Table 14.4). The between-treatment variance has (number of treatments − 1) degrees of freedom, represented by the between-treatment mean square MS_b, hence:

$$MS_b = \frac{BSS}{m-1} \tag{14.18}$$

If there was no additional variation due to different storage conditions, then MS_b should equal MS_w and

$$F = \frac{MS_b}{MS_w} \tag{14.19}$$

would not be significantly different from 1.

In the example, $MS_b = 78.98$ and $MS_w = 16.88$, so $F = 4.68$. The critical value ($P = 0.05$) of F for 3 and 16 degrees of freedom obtained from statistical tables is 3.239. Note that one-tailed tables are always used for one-way ANOVA. As 4.48 > 3.239, the storage conditions do have a significant effect on the mean concentrations measured.

ANOVA is included in the statistics available with Microsoft Excel. The output includes *BSS*, *WSS*, *TSS*, the degrees of freedom, MS_b, MS_w, F and the level of significance, P (Figure 14.2).

Anova: Single Factor

SUMMARY

Groups	Count	Sum	Average	Variance
Row 1	5	498	99.6	11.3
Row 2	5	493	98.6	9.3
Row 3	5	476	95.2	17.7
Row 4	5	454	90.8	29.2

ANOVA

Source of Variation	SS	df	MS	F	P-value	F crit
Between Groups	236.95	3	78.98333	4.680494	0.015666	3.238867
Within Groups	270	16	16.875			
Total	506.95	19				

Figure 14.2 Excel table of results for one-way ANOVA of the data of Table 14.4.

Having shown that there is a statistically significant effect between storage conditions, further testing can be performed to test for differences between the treatments. Statistical programs usually include this option, so that the tests are automatically applied if F is significant. For example, Prism (GraphPad Software) will perform Bonferroni's t-test for multiple comparisons or Dunnett's test, where the treatment means are compared to a control mean (the freshly prepared solution in this example). Both tests showed that there was a significant difference between the mean value given by the freshly prepared solution and that given by a sample stored at room temperature. There is also the option to test for a linear trend, which is useful when the effect of time, for example, on a parameter is being examined.

14.2.4 Calibration graphs

Normally a calibration graph of analyte response versus concentration in the calibration standards is constructed. In chromatographic assays the response may be peak height or area, or peak height or area ratios to an internal standard (Section 14.4). The relationship may be linear (a straight line) or curved. Before ready access to curve-fitting programs, calibration curves were drawn on graph paper and the concentrations interpolated manually. Having a straight line made the task considerably easier and analysts tried to work within a 'linear' region of the curve even for techniques (fluorescence, ECD) that are not inherently linear. Least-squares fitting of calibration curves can now be performed with hand-held calculators and this has the advantage that statistical fitting of the data allows an assessment of the quality of the results in terms of a confidence interval or SD.

14.2.4.1 Linear regression

Least-squares linear regression is a statistical method to obtain the best fit of a straight line ($y = a + bx$) to experimental data. The formula minimizes the sum of the squares of the residuals, $(\hat{y}_i - y_i)$ where $\hat{y}_i$ is the y-value calculated from the regression line at x_i (Figure 14.3).

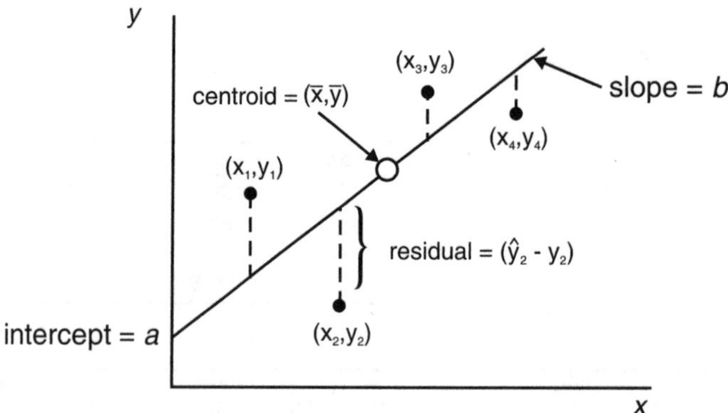

Figure 14.3 The principle of least-squares regression.

The sum of squares of the residuals is used because the sum of the residuals will be zero. The slope of the regression line, b, is:

$$b = \frac{\sum_i (x_i - \bar{x})(y_i - \bar{y})}{\sum_i (x_i - \bar{x})^2} \tag{14.20}$$

The intercept, a, is:

$$a = \bar{y} - b\bar{x} \tag{14.21}$$

Note that the regression line goes through the mean value of x and y, a point which can exploited when plotting a graph manually.

Estimates of a and b can be assessed from Equations (14.20) and (14.21) or from a hand-drawn graph and the unknown values of concentrations can be interpolated from:

$$x = \frac{(y - a)}{b} \tag{14.22}$$

The correlation coefficient, r, can be calculated from:

$$r = \frac{\sum_i (x_i - \bar{x})(y_i - \bar{y})}{\sqrt{\sum_i (x_i - \bar{x})^2 (y_i - \bar{y})^2}} \tag{14.23}$$

Values of r can range from 1 to -1, which represent perfect positive and negative correlations, respectively. The correlation coefficient gives an indication of how y varies as a function of x and r^2 represents the explained variance. For example, if $r = 0.95$, $r^2 = 0.903$ and so ~10 % of the change in y cannot be ascribed to a change in x, or the *percentage* fit is 90.3 %. Unless the data are a perfect fit to the regression line, then clearly there must be errors associated with the estimates of a and b. The errors in x (the concentrations of the standard solutions) are assumed to be negligible compared with those in y, and the sum of the squares of the residuals is used to calculate:

$$s_{x/y} = \sqrt{\frac{\sum_i (y_i - \hat{y})^2}{n-2}} \tag{14.24}$$

from which the standard error of a is:

$$s_a = s_{x/y}\sqrt{\frac{\sum_i x_i^2}{n\sum_i (x_i - \bar{x})^2}} \tag{14.25}$$

and the standard error of the slope, b, is:

$$s_b = \frac{s_{x/y}}{\sqrt{\sum_i (x_i - \bar{x})^2}} \tag{14.26}$$

Confidence intervals for the intercept and slope, respectively, can be calculated from:

$$a \pm t_{(n-2)}s_a \quad \text{and} \quad b \pm t_{(n-2)}s_b$$

Note that for linear regression the number of degrees of freedom is $n-2$, because a straight line will pass through any two points and so a fit cannot be derived statistically from only two points. Having obtained values for a and b, concentrations in the test samples can be calculated from:

$$\text{concentration} = \frac{\text{response} - a}{b} \tag{14.27}$$

The equation to calculate the confidence intervals for the interpolated concentrations is complex because the error will arise from errors in the estimate of the slope and the intercept and, of course, the response will have random errors. Provided that b and the sum of squares of x are relatively large (which they are in a well-constructed calibration curve) then an approximation may be used to calculate the confidence interval (Miller, 1991):

$$\text{CI} = \pm\frac{t_{(n-2)}s_{x/y}}{b}\sqrt{\frac{1}{m} + \frac{1}{n} + \frac{(\text{response} - \bar{y})^2}{b^2\sum_i (x_i - \bar{x})^2}} \tag{14.28}$$

where m is the number of replicates used to obtain a value for the response of the unknown sample. Features of Equation (14.28) are: (i) that CI is smallest for signal responses that equal the mean of the responses in the calibration curve, (ii) replicate measurements can have a marked effect on reducing the CI and (iii) increasing the number of calibration points reduces CI. Curve plotting and fitting programs usually produce the required statistics and will plot confidence intervals on the calibration graph (Figure 14.4). Note how the confidence intervals are symmetrically curved about the regression line and the effect that the number of calibration points has on the confidence intervals. Not surprisingly, the curve with only three points has very large confidence intervals; there is only one degree of freedom ($n-2$) and t_ν is 12.71. Moreover, the correlation coefficient tested using:

$$t = \frac{r\sqrt{n-2}}{\sqrt{1-r^2}} \tag{14.29}$$

was not significantly different from zero. With two additional standards [Figure 14.4(b)], $\nu = 3$ and $t_\nu = 3.18$.

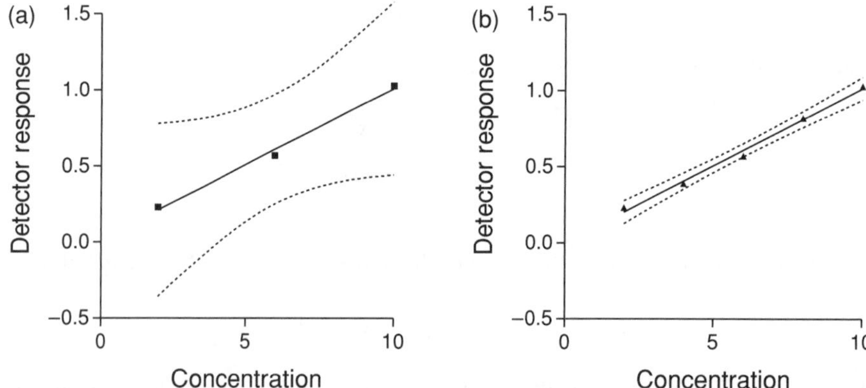

Figure 14.4 Demonstration of the effect of the number of data points on the 95 % confidence interval: (a) three points, (b) five points.

14.2.4.2 Testing for linearity

Fitting calibration data to a straight line as described above makes several assumptions about the data, including the obvious one that the response increases linearly with increasing concentration. But how does one know that the relationship is indeed linear? The first approach to answering the question is the use of 'pattern recognition' and the best pattern recognition is to look critically at the data. Calibration data should always be plotted or displayed on a computer screen and scrutinized.

At first sight the data of Figure 14.5(a) look as though they should be fitted to a straight line (least squares linear regression gave a value of $r = 0.998$). However, on closer inspection the graph is obviously curved, emphasising that the commonly held belief that a high value of r is proof that the calibration curve is linear is not true. It is the case that if the calibration curve were

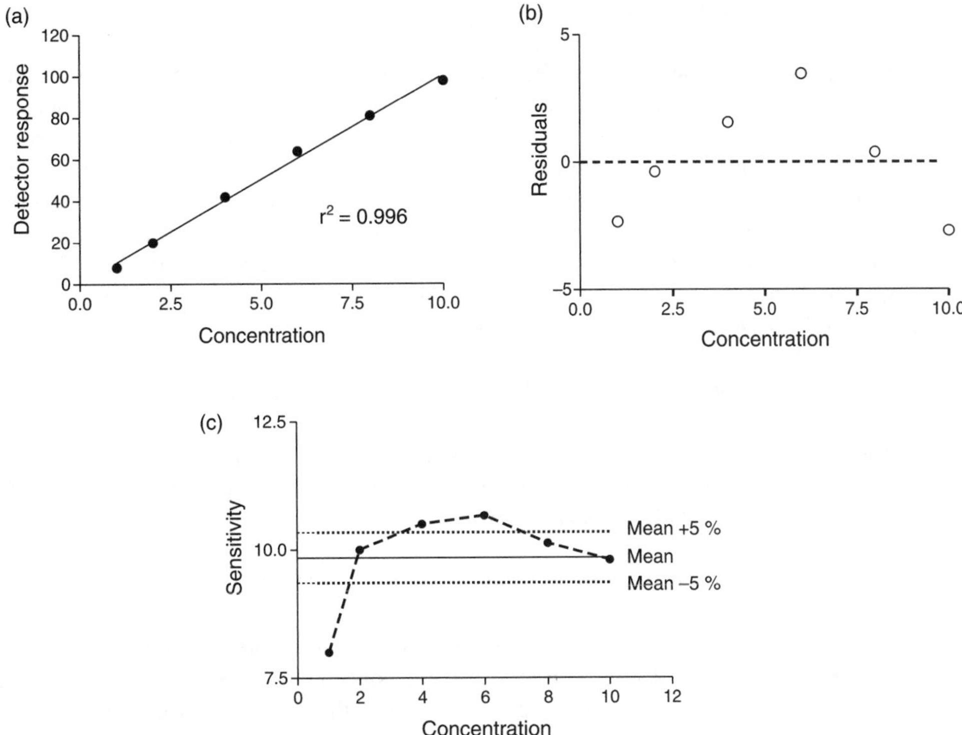

Figure 14.5 Testing calibration curves for linearity: (a) data fitted to a straight line, (b) residual plot, (c) plot of sensitivity (detector response/concentration) versus concentration.

a perfect straight line r would be unity, but a correlation coefficient approaching unity does not of itself prove that the calibration line is linear (Analytical Methods Committee, 1988). Inspection of the points about the line shows two points below the line, then a 'run' of points above and a point below. This pattern is seen more clearly if the y-residuals are plotted against concentration [Figure 14.5(b)]. If the calibration data fitted a straight line then the residuals, whose sum must equal zero, would be normally distributed about zero. The U-shape distribution in Figure 14.5(b) is typical of data that should be fitted to a curve.

A further way to prove that a curve is nonlinear is to plot sensitivity against concentration (or log concentration in the case of a calibration curve that spans a wide range of concentrations). Sensitivity can be defined as detector response divided by concentration (http://www.iupac.org/goldbook/S05606.pdf, accessed 27 June 2006). If the calibration is perfectly linear then sensitivity will be the same at every concentration. However, if sensitivity varies in a systematic manner then fitting a straight line would be incorrect. For a calibration graph to be considered linear the sensitivity should not vary by more than ±5 % of the mean value [Figure 14.5(c)]. Clearly the data from the example has not been derived from linear data. An advantage of examining sensitivity as a function of concentration is that such plots can be used to define a working linear range. This is particularly useful for curvilinear plots often seen with fluorescence and MS detection and has been applied to defining the linear working ranges of detectors (Dorschel *et al.*, 1989).

14.2.4.3 Weighted linear regression

Simple linear regression described above assumes equal variance at all concentrations, that is to say the data are homoscedastic. However, for calibration data the variance generally increases with concentration – indeed by ensuring that the RSD values cover a limited range and never exceed a predefined limit, the errors are going to be approximately proportional to the concentration. Thus, the errors associated with high concentrations are higher and these may unduly affect the slope of the calibration line. However, unweighted regression treats all points equally, which in effect means that lower concentration points are more likely not to be fitted to the line. This can be a major problem if the calibration curve extends over several orders of magnitude. The answer is to weight the data and to minimize the sum of weighted squares:

$$\text{Weighted sum of residual squares} = \Sigma[(\text{weight})(\text{residual})^2] \tag{14.30}$$

The size of the random errors in the measured response is given by the variance (s^2) and it is usual to weight the data by $1/s^2$, or some value that is directly proportional to the reciprocal of the variance. Equation (14.30) can thus be written:

$$\text{Weighted sum of residual squares} = \Sigma[1/s^2(\text{residual})^2] \tag{14.31}$$

In other words the points with the lowest errors assume more importance than those with the largest errors. If the RSD is constant over the concentration range used for the calibration curve then the size of s is proportional to the size of the signal and, for a linear relationship proportional to the concentration of the sample. Under these conditions one could weight the data by $1/(\text{response})^2$ or by $1/(\text{concentration})^2$. Many commercially available statistical programs allow the option of weighting data by$1/y^2$ or by $1/x^2$.

The weighted calibration line passes through the weighted centroid (cf. Figure 14.3):

$$\frac{\sum_i w_i x_i}{n}, \frac{\sum_i w_i y_i}{n}$$

The confidence intervals are smallest at this point, but unlike the homoscedastic case are not symmetrical about the line, but increase disproportionately as the concentration increases, as would be expected from the model.

Weighting data, whether calibration data or pharmacokinetic data (Chapter 16), is not some method of manipulating the result to make it appear more acceptable, *it is the correct statistical treatment for heteroscedastic data.* However, this being said, when the errors are small and the concentration range limited, then simple linear regression may be shown to be fit for purpose.

14.2.4.4 Nonlinear calibration curves

There is no reason why a calibration line should be a straight line over the entire calibration range. Indeed some detectors, such as electron-capture and fluorescence detectors, are known to have a limited linear range, and even the Beer–Lambert law breaks down at high analyte concentration. MS and some immunoassays, such as EMIT, give a curvilinear response with changes in concentration. If the concentration–response relationship has been shown to be nonlinear then a decision as to which type of curve should be fitted will have to be made.

The data of Figure 14.5(a), which are clearly not linear, can be fitted to a quadratic equation (Figure 14.6). Not only does the quadratic look like a better fit to the data, but also the correlation coefficient is greater ($r = 0.9995$) confirming that r-values approaching 1.0 cannot be taken as being indicative of linearity. Inspection of the residuals, which are close to, and randomly distributed about, zero, confirm that the quadratic model is a better choice than the linear one.

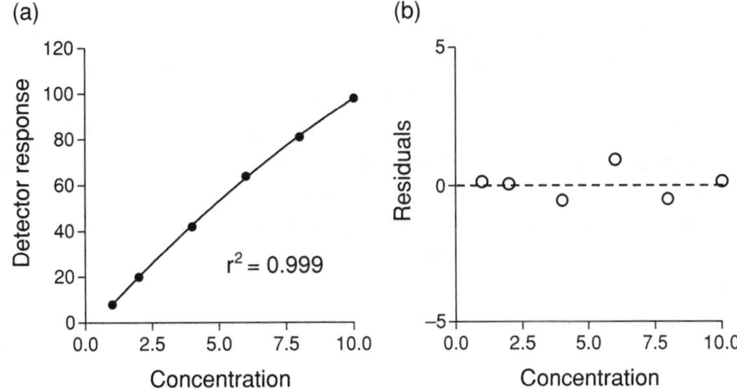

Figure 14.6 The data of Figure 14.5a fitted (a) to a quadratic function and (b) the resulting residuals.

Sometimes a model for the relationship between detector response and concentration can be proposed. Calibration data for medazepam measured by GC-ECD in DC mode are shown in Figure 14.7. As the electrons are captured the standing current falls and so the curve might be expected to be defined by a hyperbola [Figure 14.7(a)]. The hyperbola passes though the points and the origin and asymptotes to a maximum value (the point where there is no longer any standing current). However, the quadratic fit [Figure 14.7(b)] is clearly a suitable alternative over the working range of the calibration data. With non-linear curves it is unusual to have a model for the relationship

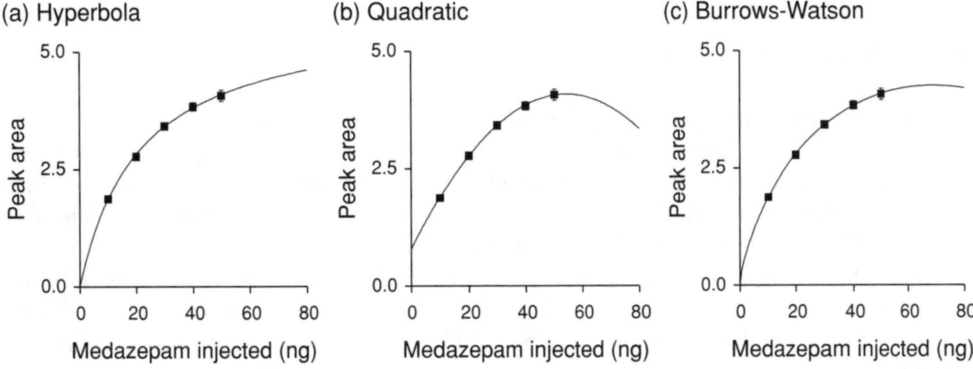

Figure 14.7 Calibration data for GC-ECD of medazepam fitted to (a) a hyperbola, (b) a quadratic function and (c) Burrows–Watson equation.

between the concentration and response and there may be several components leading to non-linearity. Thus, the choice of an equation is empirical, and a quadratic or cubic expression may be suitable. For curvilinear plots of the type frequently seen in MS assays, Burrows and Watson (1994) recommend [Figure 14.7(c)]:

$$y = a + bx + cx \log x \tag{14.32}$$

14.2.4.5 Residuals and standardized residuals

Inspection and plotting the residuals can be very useful in demonstrating that the most suitable equation and method of weighting have been applied to the calibration data, and may indicate the presence of outliers. Ideally, the residuals should be randomly distributed about zero [Figure 14.6(b)]. Curves such as that of [Figure 14.5(b)] indicate that an incorrect equation has been chosen. If the residual versus concentration plot shows an approximately constant value of residual over the concentration range then no weighting is required [Figure 14.8(a)]. However, the sizes of the residuals are often proportional to analyte concentration [Figure 14.8(c)]. Indeed, this is almost invariably the case as methods are developed to have more or less constant RSDs over the calibration range.

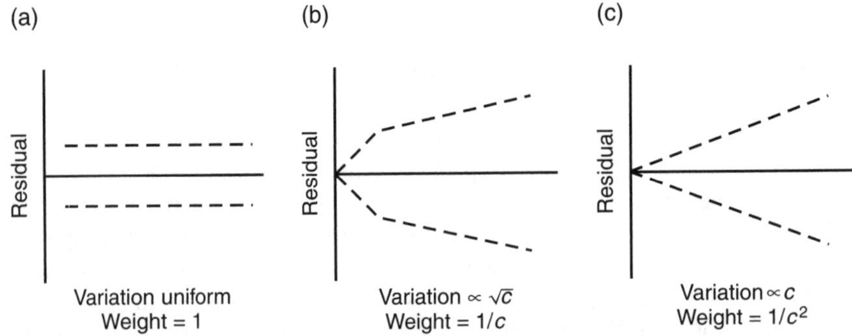

Figure 14.8 Use of residual plots to ascertain the most appropriate method to weight calibration data.

14.2.4.6 Blank samples and the intercept

Blank samples, usually the drug-free matrix, should be prepared and analyzed along with the calibrators and unknown samples. The blank is an important part of QC as it should be free of interfering substances. In a chromatographic assay, for example, there should be no interfering peaks in the region of the analyte(s) or internal standard(s). The response of the blank should not be so high as to limit the working range of the assay. The blank signal should not be subtracted from the responses to the calibrators as this alters the error model (Section 14.5.2). Similarly, quantitative photometric assays should not be run against blank samples placed in the reference cell of a double-beam spectrometer. Note, this is different from the qualitative investigations described in Chapter 4. The blank is not a zero concentration calibrator, and must not be included in the calibration data. Sample analyte values below the lowest calibrator should not be quantified. Replicate analyses of 'blank' signal are necessary to define the limit of detection (Section 14.6.3.7) and so it is important when using electronic data capture to ensure the calibration curve is not being forced though zero. Because of memory limitations, some older integrators had one-point calibration and assumed a 0,0 origin.

In practice, calibration curves rarely pass through the origin (0,0), although the origin may be within the 95 % confidence intervals of the intercept. A positive intercept is most likely due to a contribution from background signals and would be expected to be close to the value given by the blank sample (this is a useful QC check). Thus, the intercept is the average contribution that the blank signal makes to calibrator signals. Negative intercepts are often indicative of adsorptive losses, which are proportionally greater at lower analyte concentrations. Although statistical methods can be used to 'force' a calibration though the origin, this is not necessary as it is simple to calculate the concentrations in the unknown samples, for example from Equation (14.27).

14.2.4.7 Method of standard additions

In the preparation of calibration curves described above, it has been assumed that the analyte-free plasma, urine and so on are representative of the matrix to be analyzed. However, there may be situations when there are wide variations in the matrix or when it is impossible to obtain suitable analyte-free matrix, for example when measuring endogenous compounds. In the method of standard additions the calibration curve is prepared using the sample to be analyzed. The calibration curve is drawn or calculated as usual and the unknown concentration obtained from the intercept on the concentration (x) axis (Figure 14.9). This has two obvious drawbacks: (i) the volume of the unknown sample must be large enough to prepare the required number of calibrators and (ii) it is time consuming because each unknown sample requires its own calibration curve. The unknown sample could be divided and a known amount of analyte added to one portion, but this single point calibration approach assumes a perfectly linear response between concentration and response.

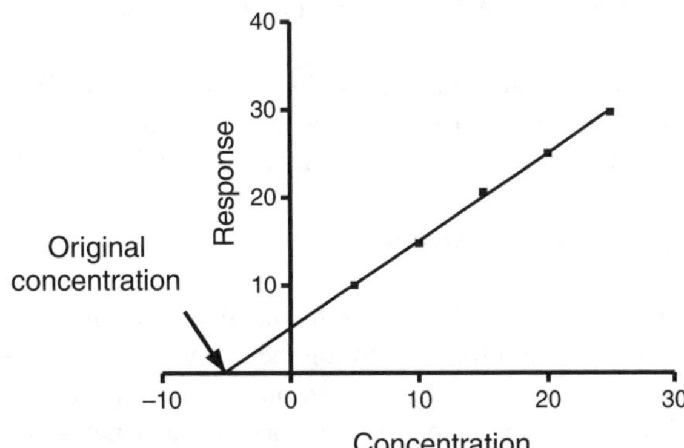

Figure 14.9 Calibration curve for the method of standard additions.

14.2.4.8 Limits of detection and quantitation

An important criterion for an analytical method is the limit of detection (LoD), that is the minimum concentration (amount) of analyte that can be detected reliably and differentiated from any background signals measured in analyte-free samples. Defining the LoD is not as simple as it might at first seem, as there are random errors associated with both the blank sample and samples at or near the LoD. For chromatographic assays the LoD might be quoted as some arbitrary multiple of the signal-to-noise ratio (S/N) such as 3, 5, or 10 times. This, of course, presupposes that the S/N can be measured and with biological samples the limiting factor is rarely instrument noise, but rather signals due (usually) to endogenous interfering species.

Although there is no universally agreed definition of LoD, the IUPAC formula is frequently used:

$$\text{LoD} = \bar{y}_B + k\sigma_B \tag{14.33}$$

where $\bar{y}_B$ is the mean signal of the blank and σ_B is the SD of the blank signal. If the mean signal for the blank is close to zero, then some responses will be negative and it is important that these are measured. Using a value of 3 in Equation (14.33) (sometimes referred to as the 3σ convention) there is approximately a 7 % probability that the analyte will be deemed to be present in the sample when in fact it is not. To reduce the error to the 5 % level requires the factor to be increased to 3.28 (Miller and Miller, 2000). Having derived an estimate for LoD, results below this can be reported as < LoD, but the best practice in analytical toxicology is to report them as 'not detected' because this does not imply that the analyte is present when in fact it may not be.

For quantitative analyses it is usual to quote the lowest concentration or amount that can be measured with defined values of precision and accuracy; this is the lowest limit of quantitation (LLoQ). Again the criteria are arbitrary, and the RSDs for precision and accuracy typically range from 10 to 15 %, depending on the requirement of the assay. For trace analyses and pharmacokinetic experiments it may be necessary to accept values as high as 20 %, although the RSDs for concentrations above the LLoQ, would normally be considerably lower. The concentration of the lowest calibrator may be set to the LLoQ, although for some applications it may not be necessary to measure such low concentrations. In any event, concentrations below the lowest standard must not be reported. An alternative approach, using a value of $k = 10$ in Equation (14.33), is not widely used.

Reports do not always quote a higher limit of quantitation (HLoQ), particularly when the assay has a wide linear range. However for some techniques, such as immunoassay and fluorescence detection in HPLC, the working range of the assay should be defined by both LLoQ and HLoQ.

14.2.4.9 Curve fitting and choice of equation

How the calibration data are used to derive the analyte concentrations will depend on a number of factors including the reason for performing the assay, the type of assay and whether the assay is a routine one involving large numbers of samples, or a single quantification of an analyte that is required rarely. It will also depend on local factors such as what is available. In some instances it may not be necessary to 'draw' a calibration curve, for example in the case of using cut-off values in immunoassays (Chapter 12). The data station or integrator may have the calibration routine built in and the results expressed as concentrations (or amounts) directly. When there are only a few samples, drawing a calibration on graph paper and interpolating the unknown concentrations by eye may be adequate, but for larger numbers this is tedious, time consuming and prone to gross errors. Furthermore it is not amenable to statistical analysis. Many inexpensive hand-held calculators will perform linear regression, and once the slope (b) and intercept (a) of the calibration line are known it is simple to calculate the concentrations in the test samples using Equation (14.27).

For more complex calibration relationships computer programs may be needed. These can include commercially available statistical and curve-drawing packages. Linear and quadratic equations have discrete solutions and calibration curves can be generated using a spreadsheet such as Microsoft Excel. Some relationships may not have a mathematical solution and have to be solved iteratively using software that can be found in many statistical packages. A useful source code for iterative curve fitting was that written in BASIC (Neilsen-Kudsk, 1983). Not only could this be modified to run under different forms of BASIC, but also it was relatively easy to modify the code so that it could be used to fit a wide variety of equations including calibration curves, most compartmental pharmacokinetic relationships (Chapter 16), and pH extraction curves (Whelpton, 1989; 2004).

The choice of calibration equation (model) should be dictated by the data. Statistical fitting allows the various equations, for example quadratic and cubic, to be compared. Simply choosing the equation which gives the lowest residual sum of squares (*SS*) is unhelpful because this will be the equation with the largest number of parameters. Consequently, most statistical packages compute 'goodness of fit' parameters, which take into account the number of parameters in the equation. Some programs will fit two equations simultaneously and compare them using the F-test. Neilsen-Kudsk (1983) used the Akaike information criterion (AIC):

$$\mathrm{AIC} = N \ln(SS) + 2M \tag{14.34}$$

where N is the number of data points and M is the number of parameters. The equation with the lowest AIC is statistically the most appropriate.

14.2.4.10 Single point calibration

Multiple point calibration curves as described above are suitable for batch analyses. However, there are occasions when single point calibration is appropriate, such as emergency requests for assays that are not routine in the laboratory. Single point calibration methods should be validated and the results compared with those from multipoint calibration (Peters *et al.*, 2005). It is important to show that 'blank' samples from a variety of sources are free from interferences, to define the linear range of the assay, which should encompass the lowest and highest expected therapeutic concentrations, and assess the size of the intercept. The concentration of the single point calibrator should be chosen so that the 99 % confidence intervals for mean values are within 50 % of the target value.

With commercially produced immunoassays, manufacturers often develop an equation apparently enabling straight-line calibration to be used. This can be 'checked' routinely with single point calibration and use of a zero intercept. However, it is important to use multiple calibrators to confirm continued performance if, for example, a different batch of reagent is used, as assay calibrators may be 'adjusted' to be reagent-lot specific. Failure to perform multi-level calibration may result in calibration drift.

14.2.5 Batch analyses

Whatever the actual instrumentation employed, quantitative batch assay calibration should normally be by analysis of standard solutions of each analyte (6–8 concentrations across the calibration range) prepared in, for example, analyte-free neonatal calf or human serum. IQC procedures should be instituted. This involves the analysis of independently prepared standard solutions of known composition that are not used in assay calibration; normally low, medium and high concentrations of each analyte are prepared in analyte-free human serum. Calibration standards are normally analyzed in duplicate, once at the beginning and once at the end of the batch. IQC samples are analyzed at the beginning and end of the batch and also after every 5–10 patient sample, as appropriate. EQA samples are analyzed as appropriate to conform to the requirements of particular EQA schemes.

The performance of batch analyses (analysis of a number of samples in the same analytical sequence) and analysis acceptance criteria should be as set out in the method validation guidance (FDA/CDER, 2001). Typical assay acceptance criteria are: (i) chromatography (reproducible peak shape and retention time, stable baseline), (ii) calibration graph ($r = 0.98$ or greater, intercept not markedly different from zero) and mean IQC results within acceptable limits (generally within 10 % of nominal value). Acceptance criteria for patient samples are: (i) 'clean' chromatogram (i.e. absence, as far as can be ascertained, of interferences) and (ii) duplicate values (peak height ratio to the internal standard) within 10 % except when approaching the limit of sensitivity of the assay when duplicates within 20 % may be acceptable and (iii) results within the calibration range of the assay. Clinical sample analyses falling outside acceptance limits may be repeated if sufficient sample is available.

While immunoassays may be performed as batch analyses, they may, depending on the instrument used, be subjected to random access analysis. IQC in such circumstances is best performed at timed intervals at low, medium and high concentration. Some immunoassays are used as screens for the presence of drugs or metabolites, and 'cut-off' concentration values are used to define positives. It is particularly important to ensure that the analytical performance at this cut-off point is properly investigated (±25 % of cut-off value) to minimize the risk of false positives and negatives arising from poor assay performance.

If a result above the calibration range is obtained, then, if possible, a portion of the sample should be diluted with 'blank' plasma/serum and reanalyzed. If a sample is from a suspected overdose patient, then sample dilutions ($1 + 1$, $1 + 3$, $1 + 9$) should be made using 'blank' serum at the time of the initial analysis (if the available sample volume permits) and analyzed at the same time as the normal sample analysis. Postmortem whole blood samples from suspected overdose cases should be diluted $1 + 3$ and $1 + 9$ with 'blank' human serum prior to analysis using standard methods. If the information available suggests a massive overdose, further dilutions may be made as appropriate prior to the analysis.

14.3 Use of internal standards

An internal standard is a compound added in known amount at some stage in an assay with the aim of demonstrating method integrity and correcting for systematic errors (Section 1.3). Obviously internal standards are mainly used in chromatographic, electrophoretic or MS assays (Chapters 6–10) where the resolving power of the analytical system permits independent measurement of the added compound. Internal standards can be used with other kind of assays, for example a radioactive internal standard could be used in almost any kind of assay, but the apparatus required and the poor precision of radioactive counting means that this approach was uncommon and, with concerns over radiation safety and disposal, is no longer used.

A constant amount of an internal standard must be added with a high degree of precision to each sample, with thorough mixing. Absolute accuracy in the amount added is generally less important and, within limits, an internal standard does not need to be as pure as the analyte. Depending on the reasons for its inclusion, the internal standard may be added at the beginning or at a later stage of a procedure. Requirements for an internal standard for chromatographic assays have been summarized (Box 14.3).

Box 14.3 Requirements for an internal standard (Haefelfinger, 1981)

An internal standard must:

- Be completely resolved from the known and unknown substances in the chromatogram
- Elute near to (preferably just after the last) peak(s) of interest
- Have a similar detector response (peak height or area) to the analyte(s)
- Have similar chemical and physical properties to the analyte(s)
- Undergo any derivatization reaction in the same way as the analyte(s)
- Be chemically and physically stable on storage in solution and during the analysis
- Be easily available with adequate purity

In general, it is good practice to use an internal standard in chromatographic assays. However, as use of an internal standard may increase the analysis time it is worth ensuring that a suitable compound is employed. It should be shown that the use of internal standardization is beneficial, or at least is not deleterious, to assay performance.

If a suitable internal standard cannot be found, it will be necessary to use external calibration. This does not mean that the method will inevitably have poor precision and accuracy. The increased use of reproducible injection systems, stable detectors and extraction methods that gave good, reproducible recoveries can give acceptable results (Table 14.5). The points highlighted in Box 14.4 are particularly relevant when working without an internal standard. The 'solvent-flush' injection technique is when a small volume of solvent is drawn into a syringe and air is then drawn in until the meniscus can be seen. The sample is drawn in next using the movement of the meniscus to measure the volume, and finally more air is drawn in until the bolus of sample can be seen. The volume of sample, seen between the menisci, can be checked before injection. This approach also minimizes analyte carry over. Automated extraction systems are more amenable to

Table 14.5 Comparison of intra-assay accuracy and precision of an assay with and without an internal standard (data are means of 6 replicate samples) (Wilson, 1990)

	Found concentration (mg L^{-1})		RSD (%)	
Nominal concentration (mg L^{-1})	Without IS	With IS	Without IS	With IS
1	1.3	1.3	15.4	15.4
3	3.6	3.5	11.1	11.4
10	10.6	10.3	5.7	4.9
30	35	33.9	4.3	5.3
100	107	104	2.4	2.8
300	319	316	2.5	3
1000	1051	1037	2.8	0.4

Box 14.4 Points to be considered when using external calibration

- Minimize the number of extract transfer and evaporation steps; consider injecting the sample or sample extract directly or after dilution, increasing the detector sensitivity/selectivity if necessary
- If analyte concentration steps are required, consider SPE as this often gives high recovery with good reproducibility
- When LLE is used, extract transfer volumes should be measured
- Avoid volatile extraction solvents as it is difficult to assess the volume change during handling
- If feasible, concentration of the analyte in sample extracts by back-extraction into a small volume of liquid, is preferable to solvent evaporation
- Completely (over) fill the sample loop when injecting in HPLC
- Inject as large a volume as possible in GC to reduce the errors associated with measuring small volumes, and/or use the 'solvent-flush' technique

external calibration, obviating the need for an internal standard, provided the extraction system has been properly validated.

Whenever possible the internal standard should be added to the biological sample in (preferably aqueous) solution. If there are extraction step(s), the internal standard should show similar behaviour to the analyte(s), that is the partition coefficients should be similar. It is self-evident that the internal standard should show a similar detector response to, and be resolved from, the analyte(s) and any other compounds likely to be present on the chromatogram or mass spectrum. In particular, the internal standard should be neither a drug that is likely to be present in the samples to be analyzed, nor a known or possible metabolite or decomposition product of the analyte(s).

14.3.1 Advantages of Internal Standardization

14.3.1.1 Reproducibility of injection volume

Manual injection of small volumes of an extract in GC especially is not very precise. Provided that the detector shows a linear response to the internal standard, the signal given by the internal standard should be directly proportional to the volume of sample injected. Thus, if all the samples contain the same concentration of internal standard a calibration curve of:

$$\frac{\text{analyte response}}{\text{internal standard response}}$$

versus analyte concentration can be constructed. If the sample analyses are treated in the same way, response ratios can be used to calculate the analyte concentration in the sample, due allowance for differences in injection volume having been made. The same reasoning applies to situations where a volatile injection solvent is used. In this case the drug and internal standard concentrations will rise as the solvent evaporates, the response given by the internal standard being inversely proportional to the volume of solution remaining. Use of internal standardization is important in headspace analysis (Section 3.2.4).

14.3.1.2 Instability of the detection system

Some detection systems are less stable than others. In GC, ECD and NPD systems show reduced response if contaminated. Electrochemical detectors in HPLC behave similarly. A suitably chosen internal standard may compensate for such changes. However, the analyte:internal standard signal ratio *must* remain constant for a given mixture. This is illustrated by the use of thioridazine to compensate for loss of NPD response to repeated injections of fluphenazine trimethylsilyl ether. Although there was marked loss of signal, the peak-height ratio of fluphenazine to thioridazine in test solutions varied by <2 %, even when the fluphenazine response (peak height) had declined by as much as 80 %. The detector sensitivity was restored by periodically increasing the temperature of the rubidium silicate bead for a few minutes. The validity of the approach was confirmed by investigating the effect of changing the bead temperature (Table 14.6).

Table 14.6 Effect of increasing the GC-NPD rubidium bead temperature on the response to repeated injections of a solution containing fluphenazine TMS ether and thioridazine

	Peak height (mm)[a]		
Arbitrary bead temperature setting	Fluphenazine	Thioridazine	Peak-height ratio
620	61.5	36.5	1.68
640	75.5	45.5	1.66
650	105	62.5	1.68
660	109.5	66	1.67
Mean	87.9	52.6	1.67
Range (% of Mean)	54.6 %	56.1 %	1.20 %

[a]Mean of two injections

When the hydrogen flow rate was similarly varied the peak-height ratios were in the range 1.66–1.68. If detector contamination results in the responses to different compounds being affected differently, then internal standardization is of no value.

14.3.1.3 Pipetting errors and evaporation of extraction solvent

Once an internal standard has been added, losses on extract transfer between vessels or extract concentration due to evaporation of a volatile solvent during the procedure should be reflected in the recovery of the internal standard. It can be argued, therefore, that inclusion of an internal standard obviates the need for quantitative transfer, at least in part, and so cheaper glassware, for example Pasteur pipettes, or quicker methods, such as pouring, may be used when separating the phases. If the transfer volumes are measured, then good precision can usually be obtained with external standardization. Be this as it may, obtaining quantitative phase transfer and use of relatively nonvolatile extraction solvents is to be recommended as the internal standard response on the chromatogram should be relatively constant and any major reduction in response should alert the analyst to a potential problem.

14.3.1.4 Extraction efficiency

For an internal standard to compensate for extraction losses it must be as chemically similar to the analyte as possible. Thus in MS, stable isotope-labelled internal standards are most appropriate (Section 10.7), but if such compounds are unsuitable, as in non-MS methods, or unavailable, then the closest appropriate homologue should be selected. This is usually more successful than using an analogue or 'generic' internal standard (Figure 14.10, Table 14.7).

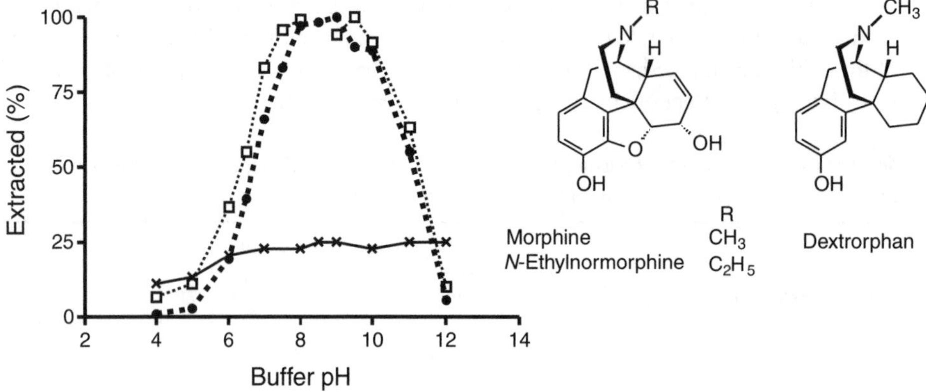

Figure 14.10 Extraction characteristics of morphine (•), *N*-ethylnormorphine (□) and dextrorphan (×) used as internal standard in a published morphine assay. Aqueous phase: Britton–Robinson buffers (0.1 mol L^{-1}). Extraction solvent: butanol:toluene (1 + 9). Solvent:buffer ratio 10:1. Reprinted from Whelpton, R. (1983) Morphine homologues as internal standards in liquid chromatographic assay, in Methods of Morphine Estimation in Biological Fluids and the Concept of Free Morphine (ed. J.F.B. Stuart), Academic Press & Royal Society of Medicine, London, pp. 15–20, with permission from Elsevier.

Table 14.7 Intra-assay RSD for morphine assay by HPLC, with and without *N*-ethylnormorphine as internal standard

Concentration ($\mu g\ L^{-1}$)	*n*	Mean peak height	RSD (%)	Mean peak ratio	RSD (%)
100	6	10	3.58	0.985	0.55
50	6	4.75	5.11	0.510	1.24
10	6	0.75	5.96	0.075	3.52

A problem arises when several compounds, for example a drug and its metabolites are to be quantified, as the extraction characteristics of each compound are likely to differ. Some studies may require the use of more than one internal standard, for example if two different detectors are used or if there is a long chromatographic run, such as in LC-MS-MS screening for abused drugs. If the extraction efficiency is high (95 % or so), then inclusion of an internal standard may give only marginal improvement in assay precision. The same is true of situations were the extraction efficiency is lower, but reproducible. The inclusion of internal standard(s) to correct for differences in protein binding, losses on protein precipitation and adsorptive losses on glassware is likely to be unhelpful unless an isotopically labelled internal standard can be used.

14.3.1.5 Derivatization and non-stoichiometric reactions

Other factors being equal, only isotopically labelled internal standards should be used in analyte derivatization, as chemically distinct entities are less likely to react identically. The causes of any imprecision should be investigated and the reaction conditions modified to make them suitably robust rather than rely on the inclusion of a second compound. There may even be problems with isotopically labelled internal standards, particularly if the bond(s) linking the isotope to the rest of the molecule are involved in the derivatization reaction. It may be necessary to use compounds with alternative sites of labelling.

14.3.2 Internal standard availability

Compounds suitable for use as internal standards may be available from chemical suppliers, pharmaceutical companies or elsewhere. Deuterated or ^{13}C analogues of a number of drugs and metabolites are also available commercially for MS work. If the desired compound is not immediately available, it may be possible to have it synthesized or to prepare it in the analytical laboratory. An internal standard need not be 100 % chemically pure and a relatively small quantity will often suffice to assess the value of using a given compound in this way, prior to synthesizing a larger quantity. Often the reaction(s) required to synthesize a particular compound are not dissimilar to those used in common derivatization reactions. An internal standard for paracetamol assay can be prepared easily from 4-aminophenol hydrochloride and propionic acid anhydride, for example.

The synthesis of an internal standard for physostigmine (eserine) assay necessitated hydrolysis of the drug under nitrogen and reaction of the resultant eseroline with *N*,*N*-dimethylcarbamoyl chloride (Figure 14.11). The internal standard is now available commercially (*N*-methylphysostigmine, Sigma).

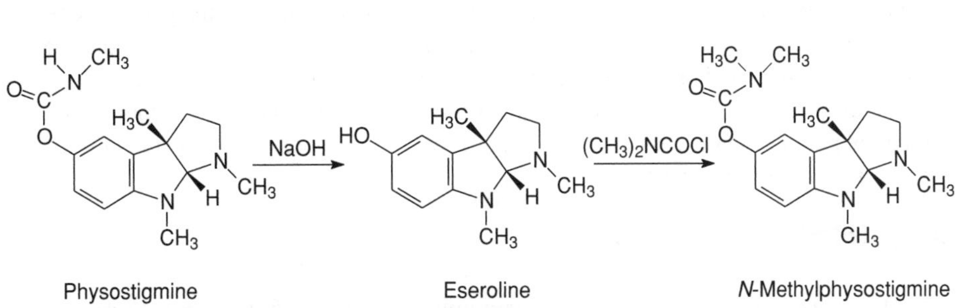

Figure 14.11 Synthesis of internal standard for HPLC assay of physostigmine.

Reactions can be monitored using the analytical tools that are to be employed in the proposed analysis and, for trace analysis, it is often possible to isolate sufficient product by TLC or HPLC (Whelpton, 2005).

14.3.3 Potential disadvantages of internal standardization

It should not be assumed that the inclusion of an internal standard will inevitably improve the precision and accuracy of a GC or HPLC assay. An internal standard cannot reduce the imprecision caused by random errors. Indeed, the random errors associated with the quantification of the internal standard are likely to *increase* the overall imprecision rather than reduce it (Equation 14.28). From replicate assays of analyte and internal standard, mean responses (e.g. peak height or area) and their associated RSDs (RSD_a and RSD_{is}) can be obtained. The relationship between the deviations in the measurement of the responses of the individual compounds and the ratio as used in a typical assay is given by:

$$(\mathrm{RSD}_r)^2 \approx (\mathrm{RSD}_a)^2 + (\mathrm{RSD}_{is})^2 - 2r.(\mathrm{RSD}_a.\mathrm{RSD}_{is}) \qquad (14.35)$$

where r is the correlation coefficient between the responses to the analyte and internal standard. If the precision of the assay has been improved by the introduction of the internal standard then:

$$\mathrm{RSD}_r < \mathrm{RSD}_a \qquad (14.36)$$

Substituting in Equation (14.35) and rearranging, gives:

$$\mathrm{RSD}_{is} < 2r.\mathrm{RSD}_a \qquad (14.37)$$

Thus, depending on the correlation between the responses to analyte and internal standard, and the imprecision associated with those responses, inclusion of an internal standard may reduce or increase the overall imprecision. Even when $r = 1$, the RSD for the internal standard must be less than half that of the analyte to give an overall improvement in precision. If the analyte can be measured with high precision in the absence of the internal standard, then the above equation confirms that addition of an internal standard will not improve the situation. A large RSD for the internal standard would indicate that the compound was a poor choice. The effect of an internal standard on assay accuracy and precision can be calculated using analyte response data. The example given in Table 14.7 indicates that, apart from a beneficial effect at the highest concentration, the use of an internal standard was unnecessary for this particular assay.

Crucial general assumptions are that a plot of internal standard concentration versus detector response is linear and passes through a (0,0) origin. The plot of concentration versus response for the analyte need not be linear, but if the internal standard concentration–response plot is non-linear then the result of using a calibration graph of analyte response/internal standard response versus concentration will be bizarre.

14.4 Method comparison

Methods may need to be compared for a number of reasons. It may be necessary to demonstrate that a new method is either equivalent, or superior, to an existing method, either in the same laboratory or in a different laboratory. Alternatively, the requirement may be to compare the same procedure carried out either by different analysts, or by different laboratories.

Linear (least squares) regression analysis assumes that all the errors are associated with the dependent variable, y, there being negligible error in the independent variable. However, it has been shown that the flaw inherent in this approach if used to compare methods is inconsequential if at least 10 data pairs, spread uniformly over the concentration range under study, are used (Miller, 1991). Results from the method with the greater precision should be plotted on the x-axis. If both methods gave identical results then the regression analysis will give slope of 1 and $r = +1$. Such a line, referred to as the line of identity may be drawn on the $x - y$ plot (Figure 14.12).

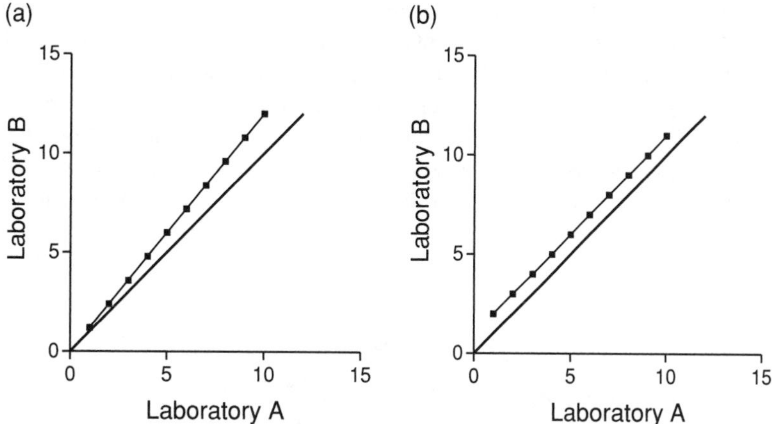

Figure 14.12 Linear regression to compare results from two laboratories showing systematic errors with proportionate bias (a) and constant bias (b). Line of identity (bold) is shown. See text for limitations of this approach.

If there is clearly a systematic error in the results from one of the laboratories [Figure 14.12(a)], this is clear from the $x-y$ plot, although there is no way of knowing which method is the more accurate from the data. Because there is such good correlation between the results from the two laboratories, but a clear relative bias, a possible explanation could include an error in the stock solution used to prepare the calibrators. For example, the stock solution used in one laboratory may have partially decomposed, leading to overestimation of the test concentrations. Alternatively one of the laboratories may have used a salt to prepare the standards and not taken this into consideration when reporting the data. In such situations, exchanging calibrators may shed light on the reasons for the discrepancy.

Linear regression should not be used without considering and stating the provisos above. For collaborative trials, that is when a method is tested in several laboratories, Youden plots can be used to detect bias (Youden and Steiner, 1975). For a rigorous estimation of reproducibility, analysis of variance should be used (Caulcutt and Boddy, 1983). Different measurement methods are usually compared using Bland–Altman plots as discussed below.

14.4.1 Bland–Altman plots

The Bland–Altman plot (Bland and Altman, 1986; 1999) is a method that is used to compare two measurement techniques. The differences between measurements performed using two methods are plotted against the averages of the methods. Alternatively the differences can be plotted as the % of the average – this is useful when there is an increase in variability of the differences as the magnitude of the measurement increases. Finally, the ratios of the measurements may be plotted instead of the differences – this avoids the need for log transformation of the data, and is also useful when there is an increase in variability of the differences as the magnitude of the measurement increases. The Bland–Altman plot is useful to reveal a relationship between the differences and the averages, to look for systematic bias and to identify possible outliers. If there is a consistent bias, it can be adjusted for by subtracting the mean difference from the new method.

Comparisons performed using linear regression analysis and a Bland–Altman plot of two methods of measuring plasma clozapine are shown in Figure 14.13. Linear regression suggests that there may be both fixed and proportionate bias; the 95 % CI for the intercept (0.04–0.12) and slope (0.79–0.94), do not encompass 0 and 1, respectively. Plotting the differences in concentrations measured by the two methods as a percentage [Bland–Altman plot, Figure 14.13(b)] confirms a bias of ~5 % for the LC-MS method. With the Bland–Altman method, there are no formal statistical tests, so one has to decide whether the differences within the mean ± 2 SD are sufficiently close (some 25 %, in this example) for the two methods to be considered equivalent.

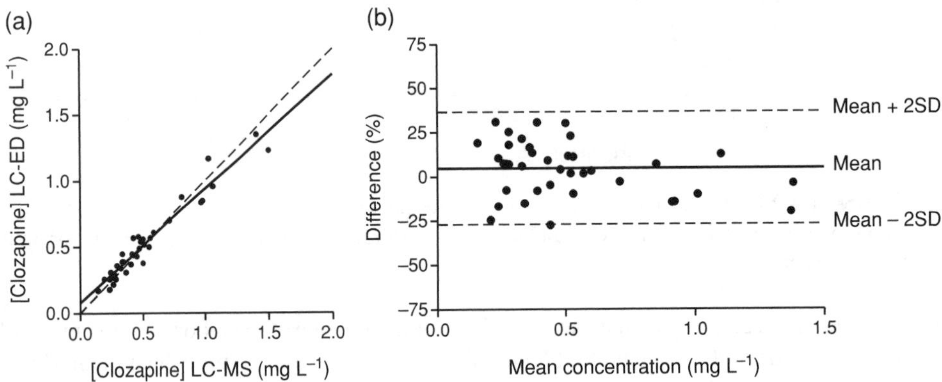

Figure 14.13 Comparison of linear regression (a) and Bland–Altman plot (b) to compare two analytical methods (LC-MS versus LC-ED) for measurement of clozapine in plasma. Broken line in (a) is the line of identity.

Bland–Altman plots may also be used to assess the repeatability of a method by comparing a number of measurements using one single method on a series of subjects. The graph can then also be used to check whether the variability or precision of a method is related to the size of the characteristic being measured. Because for the repeated measurements the same method is used, the mean difference should be zero. Therefore the Coefficient of Repeatability (CR) can be calculated as 1.96 times the SDs of the differences between the two measurements (d_2 and d_1):

$$CR = 1.96 \times \sqrt{\frac{\sum(d_2 - d_1)^2}{n - 1}} \tag{14.38}$$

14.5 Non-parametric statistics

Parametric statistics require the data to be from a population that is normally distributed about the mean. In many instances it is assumed that the data are normally distributed, although it is possible to increase the sample size and to test for normality, for example as in Figure 14.1. If skewed data can be transformed to a normal distribution, for example by taking logarithms, then parametric statistics can be applied to the transformed data. If it is suspected that the data are normally distributed, but are skewed because of an outlier, then Grubbs' test can be applied (Box 14.5). However, if it is not possible to prove normality then a range of non-parametric statistical tests are available that use ranked data and are thus less prone to the influence of outliers than parametric tests (Table 14.8).

Box 14.5 Grubbs' test for outliers

Compares the deviation of a suspect outlier from the sample mean with the standard deviation (s), which is calculated with the suspect value included:

$$G = \frac{|\text{suspected outlier} - \bar{x}|}{s}$$

If G > critical value (from statistical table) then the suspect point may be omitted and the remaining data analyzed using parametric tests

Table 14.8 Comparison of parametric and non-parametric tests to compare means/medians

Reasons for test	Parametric	Non-parametric
Comparison of mean/median with standard value	t-test	Sign test Wilcoxon signed rank test
Comparison of means/medians of two samples	t-test	Mann–Whitney U-test
Comparison of two sets of paired data	Paired t-test	Wilcoxon signed rank test
Comparison of means/medians > 2 samples	ANOVA	Kruskal–Wallis test
Outliers	Grubbs' test	

For skewed data the *median* rather than the mean, is a better indicator of 'central tendency'. The median is sometimes called the middle quartile, quartiles being the lines that divide a data set into four equal portions when the data are ranked in order. For an odd-numbered data set, the median is the $(n+1)/2$ value, for an even-numbered data set it is the average of the 'middle two'. The dispersion of the data can be described by the range and/or the upper and lower quartiles, which can be presented visually using box-and-whisker plots (Figure 14.14).

14.5.1 Sign Tests

Although the sign test is the crudest and most insensitive non-parametric test, it is also the most convincing and easiest to apply. The null hypothesis is that in a distribution, values larger (+) and smaller (−) than the median are equally likely. When matched pairs are used, the probability of observing (A > B) is equal to that of observing (B > A) and the value of A − B has median value 0. If the number of positives is $n+$ and the number of negatives is $n-$, then these are binomially distributed, with $p = q = {}^1\!/_2$, where p and q represent the probability of a positive or negative, respectively. An alternative notation is to express the probability $P(r)$ of obtaining r observations

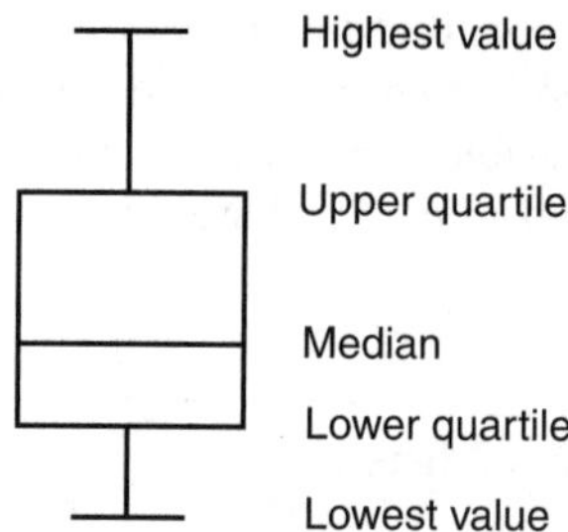

Figure 14.14 Box-and-whisker plot to illustrate the spread and symmetry of data.

(e.g. heads in coin tossing) from a total of n events:

$$P(r) = {}^{n}C_r p^r q^{(n-r)} \tag{14.39}$$

where:

$${}^{n}C_r = \frac{(n)!}{(r)!(n-r)!}$$

Alternatively, values of P can be obtained from statistical tables, derived using the binomial distribution with $p = q = ½$. With regular use it is possible to remember combinations that indicate significant differences – 6 out of 6, 8 out of 9, 10 out of 12 and so on (Table 14.9).

Table 14.9 Values of P for the sign test (two-tailed)

Total (n)	6	9	12	15	17	20
r	0	1	2	3	4	5
$n-r$	6	8	10	12	13	15
P	0.0313	0.039	0.0386	0.0352	0.0490	0.044

14.5.1.1 Wilcoxon signed rank test

The sign test described above takes no account of the size of the deviation – it is simply positive or negative. The Wilcoxon test assumes that the data are distributed symmetrically about the median – this is not to say that the data must be normally distributed, although of course for symmetrical distributions the mean and median will be the same. The test may be used to investigate whether a median/mean value has come from a given set of data (cf. t-test), or paired data can be used (matched-pairs rank test) (cf. paired t-test).

Differences either from the median, or between data pairs are arranged in order without regard to the sign, ignoring any zero differences, and ranked. Any observations with the same value are given the same rank – the mean of the ranks that would have been assigned if they had been different. The original signs are given to the rank numbers, and positive and negative ranks summed, ($W+$) and ($W-$). The lower absolute value is used as the test statistic for comparison with critical values in the appropriate statistical table.

14.5.2 Runs test

The sign test examines the number of positive and negative results, but sometimes it is important to examine whether the *sequence* of results is random. In Figure 14.5(a), for example, the sequence is: $- - + + +-$. The data contain three negative and three positive values, and there are three sequences or 'runs'. Statistical tables provide values for the numbers of runs that would be considered to be significantly different from the number arising by chance. Unfortunately, the numbers of points required for data with three runs to be significant is >11. This is typical of non-parametric statistics, which generally require larger numbers of observations than parametric tests to demonstrate significant differences. Had more data points (≥ 12) been included in the calibration data of Figure 14.5, then the residuals could have been tested to detect non-linearity. Some linear regression programs include the option of performing a runs test, although of course the caveat about the number of data points still applies. The runs test can also be used to test for non-random fluctuations about the line, for example $+ - + - + - + - + - +-$.

14.5.3 Mann–Whitney U-test

This test is used to compare two sets of data. To give an example, the data from Figure 14.15 were ranked according to the procedure outlined in Box 14.6 using Microsoft Excel, and assigned a rank. This process is simplified if the data from the two treatments are colour coded and the Excel sort function used. The ranks of the placebo ($n_1 = 7$) and the treated ($n_2 = 8$) groups were summed (Table 14.10) and U_1 and U_2 calculated, giving 50.5 and 5.5, respectively. Comparison of the lower value with critical values for U showed that the scores from the treated patients were significantly different from the scores from the group given placebo.

14.5.4 Spearman rank correlation

The Spearman rank correlation coefficient, r_s, is probably the best-known non-parametric test for correlation between pairs of observations. It is used commonly for assessing clinical treatments

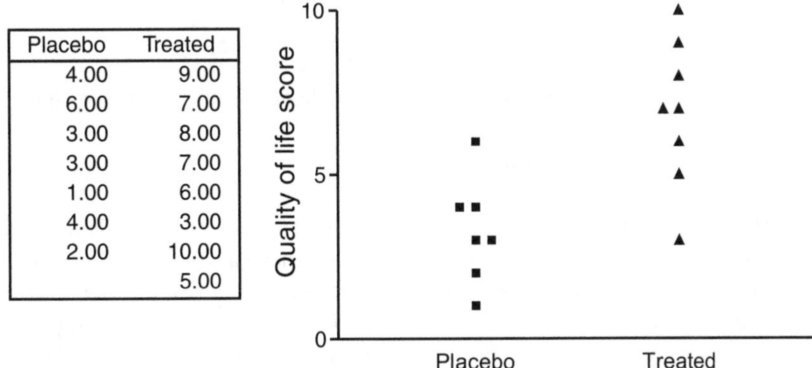

Placebo	Treated
4.00	9.00
6.00	7.00
3.00	8.00
3.00	7.00
1.00	6.00
4.00	3.00
2.00	10.00
	5.00

Figure 14.15 Quality of life scores from patients receiving placebo or drug treatment.

Box 14.6 Procedure for calculating U-values

- Define the smaller group as n_1 and the other as n_2
- Rank the combined scores of $n_1 + n_2$ from low to high, that is the lowest score gets a rank of 1. Where ranks are tied give each score the average rank of the tied scores
- Find the sum of the ranks of the smaller group (S_1)
- Find the sum of the ranks of the larger group (S_2)
- Calculate U_1 and U_2 from:

$$U_1 = n_1.n_2 + \frac{n_1(n_1-1)}{2} - S_1 \quad \text{and} \quad U_2 = n_1.n_2 + \frac{n_2(n_2-1)}{2} - S_2$$

Alternatively U_2 may be calculated from:

$$U_2 = n_1.n_2 - U_1$$

- Use the lower of the U-values as the test statistic

Table 14.10 Data of Figure 14.15 ranked according to procedure of Box 14.6

Treatment[a]	P	P	P	P	D	P	P	D	P	D	D	D	D	D	D	Sum
Quality score	1	2	3	3	3	4	4	5	6	6	7	7	8	9	10	
Rank (placebo)	1	2	4	4		6.5	6.5		9.5							33.5
Rank (treated)					4			8		9.5	11.5	11.5	13	14	15	86.5

[a]P = placebo, D = Drug treated.

and can be used when observations are expressed in rank order rather than quantitatively. Both sets of data are ranked and the difference in rank, d, obtained for each data pair. Average values should be used for tied ranks. The correlation coefficient is calculated from:

$$r_s = 1 - \frac{6\sum d^2}{n^3 - n} \tag{14.40}$$

where n is the number of data pairs. The significance of r_s is obtained from statistical tables.

14.5.5 Non-parametric regression

Although there are curve-fitting methods for non-parametric data, Theil's incomplete method is suitable for use with a spreadsheet. It is important to have an even number of calibration points (n) so that $n/2$ estimates of the slope can be calculated using pairs of points (Box 14.7, Figure 14.16). The median of the values for the slope is used to calculate n values of the intercept, using all the data points. The median values for the intercept and slope are used to calculate the unknown concentrations. Median values in a list can be located using the median function in Microsoft Excel, so templates can be created to perform the necessary steps. Advantages of the non-parametric approach to calibration include: (i) it does not assume that the errors are normally distributed, (ii) it does not assume all the errors are in the y-direction with none in the x-direction and (iii) it is less influenced by outliers compared to least squares regression.

Box 14.7 Steps for Theil's incomplete method for nonparametric calibration curves

- Rank data in order
  - Must have even number of points [Figure 14.16(a)]
  - Discard median if odd number of points
  - Calculate new median value
- Obtain median slope, by calculating slopes:
  - 1st point (x_1, y_1) to first point beyond median
  - 2nd point (x_2, y_2) to second point beyond median
    and so on...
  - First point before median $(x_{(n/2)}, y_{(n/2)})$ to last data point [Figure 14.16(b)]
  - Rank the slopes in order
  - Calculate the median slope (b)
- Obtain median intercept
  - Calculate values (a_i) for every data point (x_i, y_i) using $a_i = y - (b \cdot x_i)$
  - Rank a_i values in order and obtain the median value (a)
- Use median values of a and b to construct calibration line [Figure 14.16(c)]

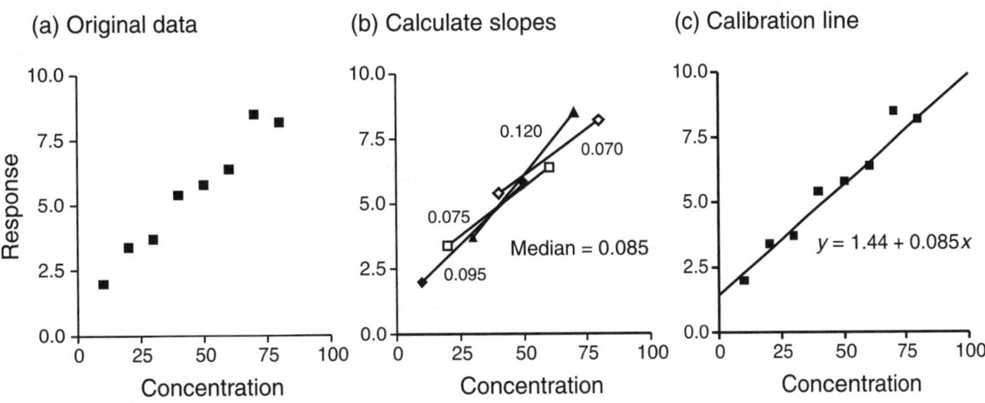

Figure 14.16 Example of Theil's incomplete method for non-parametric calibration.

14.6 Quality control and proficiency testing

Having developed or applied an analytical method and shown that it is 'fit for purpose' using appropriate validation procedures, it is important to demonstrate that the method continues to perform as required. Internal quality control (IQC) samples, ideally prepared independently by someone else using a different stock of reference material (often not feasible), different stock solutions and different volumetric apparatus, should be randomly distributed in a batch of samples to be analyzed. FDA guidelines require duplicate IQCs at three concentrations (high, medium and low) to ensure that the assay is performing satisfactorily across the calibration range. The assay batch is deemed acceptable provided four of the six controls (with at least one at each concentration) are within specification. The FDA guidelines are applicable to batch processing – different assay acceptance criteria may apply for emergency work (Section 14.3.1.10).

14.6.1 Quality control charts

QC charts are valuable in that they: (i) produce evidence of satisfactory assay performance and (ii) give visual warning if assay performance begins to deteriorate. As with any QC method it is important that reliable estimates of the parameters defining the control material are established. The mean value, μ, should be obtained from a minimum of 10 observations and the SD, σ, should be the between-assay value (inter-assay value). Obviously, these parameters must be measured when the analysis is performing satisfactorily.

14.6.1.1 Shewhart charts

On the Shewhart chart lines are drawn representing the target value, and the 95 and 99.7 % confidence limits (Figure 14.17). Values between the lower and upper warning lines (99.7 % lines) are considered acceptable, but if two consecutive values fall outside the same warning line or a single IQC falls outside an action line then the assay batch should be rejected and the reasons for the assay failure investigated. Such charts can also be used to detect trends in results, such as assays giving sequentially increasing (or decreasing) values. This can be built into the QC procedure, so that, for example, an assay is investigated if a defined number of IQC values produce increasing results.

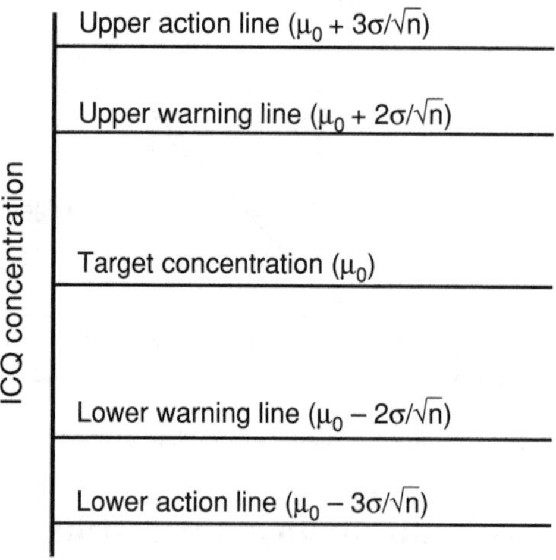

Figure 14.17 Example of a Shewhart Chart.

14.6.1.2 Cusum charts

A limitation of Shewhart plots is they are relatively insensitive to systematic drift of results. It may take as many as 50 analyses before an IQC value falls outside an action line. Cusum (cumulative sum) plots address this problem. Rather than plot the observed value, the difference between the control and target value is calculated for each run and the cumulative sum calculated. If the method is under control, the cumulative sum should fluctuate about zero, whereas a systematic trend would be evident as either a positive or negative divergence from zero. The ways of assessing when the process is out of control are either: (i) the V-mask or (ii) the tabular cusum, the latter being the preferred method (Montgomery, 2005).

14.6.1.3 J-chart

The J-chart, also known as a zone control chart, combines the properties of the Shewhart and cusum charts so that is it responsive to both abrupt changes and drift in results (Analytical Methods Committee, 2003). The mean and three equal-sized bands or zones (1 σ wide) are drawn on either side of the mean (Figure 14.18). Each observation is ascribed a weight, 0, 2, 4 or 8, depending on the band in which it falls. The weights of successive observations are cumulated and the value written on the chart at the appropriate point. When an observation falls on the opposite side of the centre line (mean) to the previous one the cumulative total is reset to zero and the process continues.

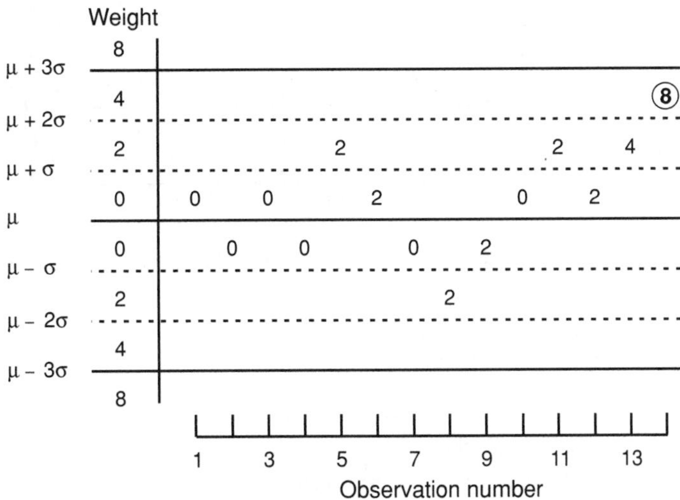

Figure 14.18 Example of a J-chart.

The process is considered out of control if the cumulative total is 8 or greater, so an isolated value greater than $\mu \pm 3\sigma$ would require the analysis be suspended and the reasons for the divergent observation investigated. The example of Figure 14.18 illustrates how a drift to high values triggers the alarm. Although observation 12 is $<1\sigma$ from the mean, the previous observation was between 1 and 2 SDs away and on the same side of the centre line, giving a cumulative weight of 2. Observation 13, is also between $\mu + 1\sigma$ and $\mu + 2\sigma$, so the new total is 4. Observation 14, is between $\mu + 2\sigma$ and $\mu + 3\sigma$, which attracts a weight of 4, giving a cumulative value of 8, the critical number for the method to deemed out of control. Had observation 14 remained between $\mu + 1\sigma$ and $\mu + 2\sigma$, the new cumulative total would have been 6, and a further value in this band would trigger the alarm, as the cumulative total for observation 15 would be 8.

14.6.1.4 Westgard rules

Clinical laboratories commonly use Westgard rules to establish that assay performance is adequate (Westgard *et al.*, 1981). These 'multirules', which are a development of the Shewhart chart, use five different control rules to access the acceptability of an analytical sequence. The rules have been given their own notation (Figure 14.19). Thus, for example, 1_{2s} indicates that 1 measurement is >2 SD from the control mean and 1_{3s} indicates that one measurement is >3 SD from the mean (Table 14.11). Note that 1_{2s} is a warning not a criterion for rejection of the analysis. The $10_{\bar{x}}$ rule may be replaced with $8_{\bar{x}}$ or $12_{\bar{x}}$, as these are more suitable to situations where one or two IQC samples are measured once or twice. Other rules may be applied, for example if there are three IQC samples. Further details and a multirule worksheet can be obtained from http://www.westgard.com/mltirule.htm (accessed 12th October 2006).

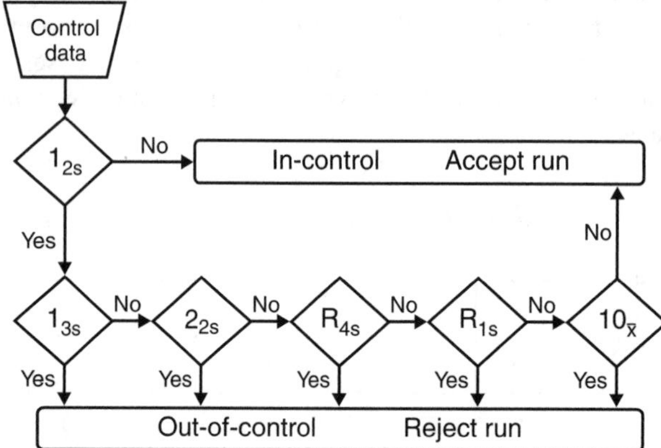

Figure 14.19 Decision tree for application of Westgard rules.

Table 14.11 Some Westgard rules

Rule	Explanation	Action
1_{2s}	A value falls outside the control mean $\pm$ 2*s*	Warning
1_{3s}	A value falls outside the control mean $\pm$ 3*s*	Reject
2_{2s}	Two consecutive control measurements exceed the control + 2*s* or the control mean minus −2*s* control limit	Reject
R_{4s}	One control measurement in a group exceeds the mean +2*s* and another exceeds the mean −2*s*	Reject
4_{1s}	When 4 consecutive control measurements exceed the mean +1*s* or the same mean −1*s* control limit	Reject
$10_{\bar{x}}$	When 10 consecutive control measurements fall on one side of the mean	Reject

14.6.2 External quality assurance

EQA schemes measure inter-laboratory performance and allow individual laboratories to detect and correct systematic errors. Portions of homogenous samples are distributed to participating laboratories. The samples should resemble clinical samples as closely as possible. The laboratories do not all have to use the same analytical method, as is usually the case with collaborative trials that are designed to test the reproducibility of a particular method. The results of EQA schemes are usually given as the z-score:

$$z = \frac{x - x_{\mathrm{a}}}{\sigma_{\mathrm{p}}} \tag{14.41}$$

where x is an individual result, x_{a} is the accepted, 'true' value and σ_{p} is known as the 'target value of SD' which should be decided on the basis of what is required of the test, and should be circulated in advance. If the result needs to be measured with high precision, then a low value of σ_{p} would be used. Thus, z is a measure of a laboratory's accuracy and the organizer's judgement as to what is 'fit for purpose'. If the results of an EQA scheme are normally distributed with a mean of x_{a} and variance of 1, then z-scores <2 would be deemed acceptable whereas those >3 would not.

In many countries it is a condition of laboratory accreditation that laboratories offering services to patients participate in all relevant EQA schemes. It is important that the results from EQA analyses are recorded graphically, discussed regularly with laboratory staff and action taken in the event of poor performance. Consistently poor performance below an agreed performance threshold, or a poor EQA result return rate, is likely to result in the issue of a warning notice to the laboratory and if there are continued failings the laboratory may be subject to an intervention. In some countries such continued failure results in withdrawal of the right of the laboratory to be recognized to perform and be paid for analyses.

14.7 Operational considerations

14.7.1 Staff training

Clinicians, especially accident and emergency staff, need guidance not only as to what toxicological assays are clinically useful in a given set of circumstances, but also on sample collection, transport and storage, local assay availability and turn-around time, and the interpretation of results.

Not only do laboratory staff in hospitals providing emergency analytical toxicology services need training in providing these services and in the interpretation of results, they will also need some training in dealing with requests for tests that will need to be referred to other analytical centres. Staff in regional analytical centres will require extensive training in the more complicated analytical methods they will be called upon to use. This training must encompass not only the techniques themselves (TLC, HPLC, etc.), but also their application in analytical toxicology and related areas. Knowledge of the role of local hospital laboratories and the local poisons centre is also important.

There are no internationally recognized training programmes in analytical toxicology. However, details of a training scheme for a graduate clinical scientist specializing in analytical toxicology are available (http://www.acb.org.uk/training/documents/training_tox.htm, accessed 27 June 2006). The training programme comprises four years of full-time study, followed by a period of higher specialist training, in some cases leading to the award of a research degree such as Doctor of Philosophy. The American Board of Clinical Chemistry hosts an examination in toxicological chemistry (http://www.aacc.org/abcc/tox_chem.stm, accessed 22 July 2007).

Participation in continuing education or continuing professional development (CPD) programmes is important when staff reach career grades (i.e. when initial and higher specialist training has been completed), and may be necessary for continued specialist registration in countries where such registration is mandatory. Compliance with CPD programmes necessitates maintenance and external audit of personal records listing educational activities such as scientific meetings attended, papers published, lectures given, and so on. Details of such a scheme maintained by the UK Royal College of Pathologists are available (http://www.rcpath.org/index.asp?PageID=620, accessed 27 June 2006). Many other countries run similar schemes.

Non-graduate scientific staff will be normally trained in house in specific aspects of laboratory operation, although training in the operation of newer specialized instruments may sometimes be provided by manufacturers. Proper recording of training is important.

14.7.2 Recording and reporting results

All results should be recorded in laboratory notebooks or on worksheets together with information such as the date, the name of the analyst, the name of the patient and other relevant information, the number and nature of the specimens received for analysis and the tests performed. All specimens received in laboratories are normally allocated a unique identifying number. This number is used when referring to the tests performed on the specimen.

UV spectra, chromatograms, calibration graphs and other documents generated during an analysis should always be kept for a period of time after the results have been reported. Recording the results of colour tests and TLC analyses is more difficult – digital photography is proving helpful here. When reporting the results of tests in which no compounds were detected in plasma/serum or in urine, the limit of sensitivity of the test (detection limit) should always be known, at least to the laboratory, and the scope of generic tests (benzodiazepines, opioids) should be defined.

The results of urgent (emergency) analyses must be communicated directly to the clinician without delay, and should be followed by a written report as soon as possible. Ideally, confirmation from a second, independent method, or failing this, an independent duplicate, should be obtained before reporting positive findings. However, this may not always be practicable, especially if only simple methods are available. In such cases it is vital that the appropriate positive and negative controls have been analyzed together with the specimen (Section 4.2).

When reporting quantitative results it is important to state clearly the units of measurement used. In addition, any information necessary to ensure that the clinical implications of the result are fully understood must be available and should also be noted on the written report.

Although it is often easy for the analyst to interpret the results of analyses in which no compounds are detected, such results are sometimes difficult to convey to others, especially in writing. This is because it is important to give information as to the poisons *excluded* by the tests performed with all the attendant complications of the scope, sensitivity and selectivity of the analyses and other factors such as sampling variations. Because of the potential medico-legal and other implications of any toxicological analysis, it is important not to use laboratory jargon such as 'negative' or sweeping statements such as 'absent' or 'not present'. The phrase '*not detected*' should convey precisely the laboratory result, especially when accompanied by a statement of the specimen analyzed and the LLoQ.

Whatever phraseology is adopted, it can still be difficult to convey the scope of analyses such as that of TLC for acidic and basic drugs as discussed above. Even with relatively simple tests such as Trinder's test, normally used to detect aspirin ingestion, a number of other salicylic acid derivatives also react if the test is performed on the appropriate sample. One way of giving at least some of this information in a written report is to create a numbered list of the compounds or groups of compounds normally detected by commonly used procedures. If these groups are listed on the back of the report then it is relatively simple to refer to the qualitative tests performed by number and thus to convey at least some of the information required.

14.7.3 Toxicology EQA schemes

Quantitative EQA schemes are available for a wide range of TDM analytes and some other poisons in many countries. The European Network of Forensic Science Institutes (ENFSI) web site (http://www.enfsi.org/standingcommittees/qcc/ptce, accessed 18 July 2007) and European Proficiency Testing Information System (http://www.eptis.bam.de/, accessed 18 July 2007) list EQA schemes for a range of analytes including drugs that are available in Europe, the Americas and in Australia. The United Kingdom National External Quality Assessment Scheme (UKNEQAS) for therapeutic drug assays and the Dutch KKGT [Association for Quality Assessment in Therapeutic Drug Monitoring (TDM) and Clinical Toxicology] schemes, for example, have been running for over 25 years.

In one quantitative UK scheme, results are usually reported as a bar chart and/or a graph so that participants can see where their result(s) lie on the distribution for a particular method (Figure 14.20, left hand panel). The target value is typically a trimmed 'consensus' mean, that is to say a mean calculated for all participants or group of assays after any outliers have been removed (Healy, 1979). The intended ('spiked') value may also be given. 'CCV' is the chosen RSD used in the calculation of the bias index score (BIS) (Bullock and Wilde, 1985). CCV varies with analyte concentration and is derived by interpolation at the consensus mean concentration into a mathematical function fitted to historic data of variation in SD (y-variable) versus consensus mean (x-variable). As the mathematical function is fixed, changes in laboratory performance with time can be identified. The BIS is calculated using the formula:

$$\text{BIS} = \frac{(\text{measurement} - \text{consensus mean})}{\text{concensus mean}} \times \frac{100}{\text{CCV}} \tag{14.42}$$

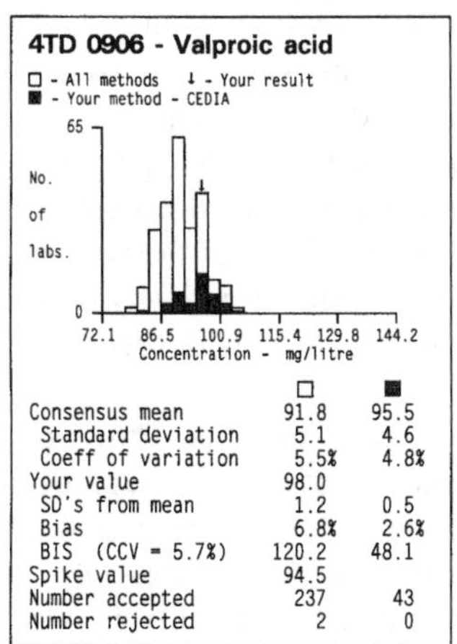

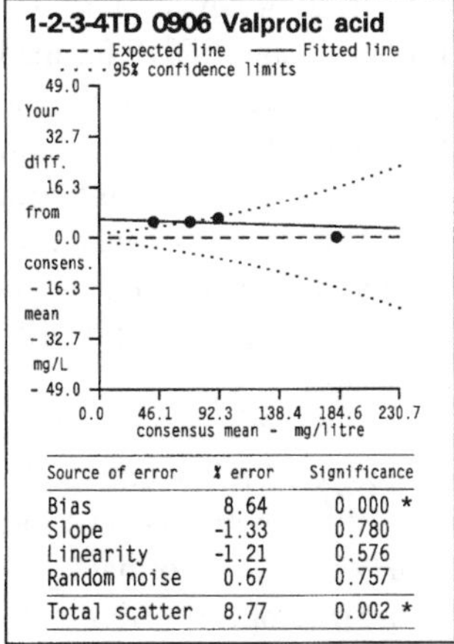

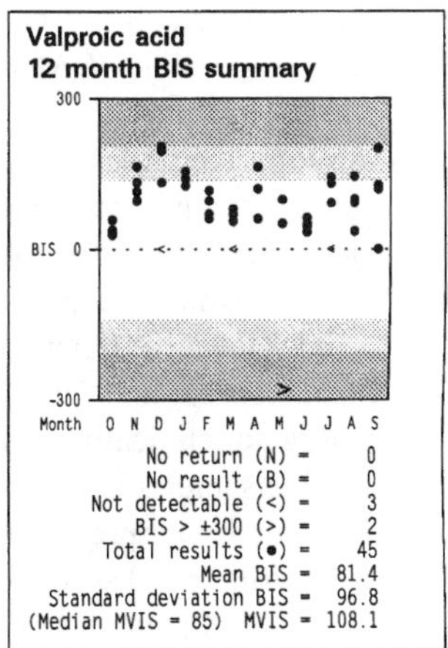

Figure 14.20 An example of a report from the United Kingdom National External Quality Assessment Scheme (UKNEQAS) for Therapeutic Drug Assays.

For some common analytes, typically four different samples are supplied at any one time (Figure 14.20, central panel). This approach not only gives an indication of relative accuracy across a range of values, but also can provide a measure of precision (Wilson *et al.*, 1984).

Cumulative assessment of a laboratory's data, for example over the last 12 months, provides information on bias. Figure 14.20 (right-hand panel) is a BIS plot for a 12-month period (four samples circulated monthly). The shaded regions indicate where laboratory measurements would be approximately >2 and >3 within-laboratory SDs from the consensus mean. Mean BIS and SD provide measures of accuracy and precision, and mean absolute BIS value (MVIS) an overall performance indicator, for which the median MVIS for all scheme participants is a useful reference.

There are, of course, sometimes concerns such as the possible effects of freeze-drying on analyte stability, and issues concerning the matrix used to prepare material for circulation (due to the cost of analyte-free human plasma or serum, neonatal calf serum may be used for some analytes). Nevertheless, from the datasets generated, scheme organizers can ascertain the methods that give the best performance (Tsanaclis and Wilson, 1997), investigate sources of interference (Wilson *et al.*, 2002) or bias (Neef *et al.*, 2006), and in extreme cases report poorly performing methods to regulatory authorities. In the 30 years since the inception of these schemes, poorly performing assays have been identified and participants advised accordingly. The mean values reported have moved nearer to the intended value and the spread of results about the mean has been reduced (Wilson, 2002).

EQA of qualitative work is also invaluable, particularly in STA as in drugs-of-abuse screening (Wilson and Smith, 1999). Here, poor selectivity due to immunoassay cross-reactivity or GC injection port artefacts, as well as manipulation of the sample by the patient, are real problems. Even with the ongoing educative role of EQA, errors (such as failure to detect morphine at 1 mg L^{-1} in urine using LC-MS-MS) still occur. Often human error is the cause of such mistakes.

Well-run qualitative EQA schemes not only assess routine performance, but also add in hard-to-detect compounds or 'spike' at concentrations around nationally agreed 'cut-off' values. Good screening and confirmatory analytical performance from laboratories is required to ensure that the correct result is reported. Scoring of performance for a set of commonly encountered drugs such as morphine, 6-MAM for heroin and EDDP for methadone, for example, helps laboratories address any problems in their procedures. Urine is the matrix used most commonly, but schemes to support analyses in oral fluid and indeed in other matrices such as hair are emerging. Scheme cut-offs are quite generous, but laboratories may also be scored against agreed workplace cut-offs (see: http://www.wdtforum.org.uk, accessed 18 July 2007). A breakdown of the performance of each recognized method type should be given to help laboratories understand why false positive or false negative reports have occurred in a given distribution. Meetings with users and scheme organizers are helpful to both groups.

Regular clinical/forensic toxicology case schemes have also been instituted. Provision of case specific samples and a brief history enable not only the assessment of analytical approaches, but also of the interpretation given with the results. The initial UK experience indicated that there was a trimodal distribution in laboratory performance (Watson *et al.*, 2002). Poorly performing laboratories have: (i) improved, (ii) stopped offering a clinical service or (iii) dropped out of the scheme. In a ground-breaking agreement, laboratories are held accountable for a report, even though it may be the work of one person – it is the service that is being examined, not an individual's performance. If the service is found wanting, appropriate corrective action is expected.

14.8 Summary

The importance of a thorough knowledge of laboratory operations is crucial if reliable results are to be obtained and, if necessary, defended in court. Even in laboratories offering a relatively simple repertoire of tests, adoption of the principles of quality management as outlined above and expanded upon in the references will provide a framework to support the growth of the laboratory. With the increasing use of POCT devices, the principles of quality management also have to be applied to such procedures in order to ensure the reliability of results. Although quality management is more difficult to apply to the interpretation of analytical results, it is vital that this aspect of the work of the laboratory is not neglected. The next three chapters aim to provide some of the information needed in this task.

References

Analytical Methods Committee (1994) Is my calibration linear? *Analyst*, **119**, 2363–6.

Analytical Methods Committee (2003) The J-chart: a simple plot that combines the capabilities of Shewhart and cusum charts, for use in analytical quality control. AMC Technical Brief 2003; 12 (http://www.rsc.org/images/brief12_tcm18-25954.pdf, accessed 27 June 2006).

Analytical Methods Committee (1988) Uses (proper and improper) of correlation coefficients. *Analyst*, **113**, 1469–71.

Bland, J.M. and Altman, D.G. (1986) Statistical methods for assessing agreement between two methods of clinical measurement. *Lancet*, **1**, 307–10.

Bland, J.M. and Altman, D.G. (1999) Measuring agreement in method comparison studies. *Statistical Meth Med Res*, **8**, 135–60.

Bullock, D.G. and Wilde, C.E. (1985) External quality assessment of urinary pregnancy oestrogen assay: further experience in the United Kingdom. *Ann Clin Biochem*, **22**, 273–82.

Burnett D. (2002) *A Practical Guide to Accreditation in Laboratory Medicine*, 2nd edn, ACB Venture Publications, London.

Burrows, J.L. and Watson, K.V. (1994) Development and application of a calibration regression routine in conjunction with linear and nonlinear chromatographic detector responses. *J Pharm Biomed Anal*, **12**, 523–31.

Caulcutt, R. and Boddy, R. (1983) *Statistics for Analytical Chemists*, Chapman & Hall, London.

Clinical and Laboratory Standards Institute (2004) *Evaluation of Precision Performance of Quantitative Measurement Methods: Approved Guideline*, 2nd edn, Document EP05-A 2, 2004 (http://webstore.ansi.org/ansidocstore/product.asp?sku=EP05 %2DA2, accessed 27 June 2006).

Dorschel, C.A., Ekmanis, J.L., Oberholtzer, J.E., Warren, F.V. and Bidlingmeyer, B.A. (1989) LC detectors: Evaluation and practical implications of linearity. *Anal Chem*, **61**, A951–68.

FDA/CDER (Food and Drug Administration/Center for Drug Evaluation and Research) (FDA) (2001) Guidance for industry. *Bioanalytical method validation* (http://www.fda.gov/cder/guidance/4252fnl.htm, accessed 28 June 2006).

Haefelfinger P. (1981) Limits of the internal standard technique in chromatography. *J Chromatogr*, **218**, 73–81.

Healy, M.J. (1979) Outliers in clinical chemistry quality-control schemes. *Clin Chem*, **25**, 675–7.

Miller, J.N. (1991) Basic statistical-methods for analytical chemistry Part 2. Calibration and regression methods. A review. *Analyst*, **116**, 3–14.

Miller, J.N. and Miller, J.C. (2000) *Statistics and Chemometrics for Analytical Chemists*, 4th edn, Pearson Education, Harlow.

Montgomery, D.C. (2005) *Introduction to Statistical Quality Control*, 5th edn, Wiley, New York.

Neef, C., Touw, D.J., Harteveld, A.R., Eerland, J.J., Uges, D.R. (2006) Pitfalls in TDM of antibiotic drugs: analytical and modelling issues. *Ther Drug Monit*, **28**, 686–9.

Neilsen-Kudsk F. (1983) A microcomputer program in BASIC for iterative, non-linear data-fitting to pharmacokinetic functions. *Int J Biomed Computing*, **14**, 95–107.

Peters, F.T., Jung, J., Kraemer, T. and Maurer, H.H. (2005) Fast, simple, and validated gas chromatographic-mass spectrometric assay for quantification of drugs relevant to diagnosis of brain death in human blood plasma samples. *Ther Drug Monit*, **27**, 334–44.

SOFT/AAFS (Society of Forensic Toxicologists/American Academy of Forensic Sciences) (2006) *Forensic toxicology laboratory guidelines*. (http://www.soft-tox.org/docs/Guidelines%202006%20Final.pdf, accessed 22 July 2007)

Tsanaclis, L.M. and Wilson, J.F. (1997) Comparison by external quality assessment of performance of analytical systems for measurement of therapeutic drugs in serum. *Ther Drug Monit*, **19**, 420–6.

Watson, I.D., Wilson, J.F., Toseland, P.A., Smith, B.L., Williams, J., Capps, N.E., Thomson, A.H., Sweeney, G. and Sandle, L.N. (2002) Scoring and performance of analysis and interpretation of toxicologic cases assessed by external quality assurance. *Ther Drug Monit*, **24**, 156–8.

Westgard, J.O., Barry, P.L., Hunt, M.R. and Groth T. (1981) A multi-rule Shewhart chart for quality control in clinical chemistry. *Clin Chem*, **27**, 493–501.

Whelpton, R. (1989) Iterative least squares fitting of pH-partition data. *Trends Pharmacol Sci*, **10**, 182–3.

Whelpton, R. (2004) Iterative least squares fitting of pH-extraction curves using an object-orientated language. *Chromatographia*, **59**, S193–6.

Whelpton, R. (2005) Internal standards, in *Encyclopedia of Analytical Science* (eds P. Worsfold, A. Townshend, C. Poole), Academic Press, London, pp. 432–441.

Wilson, J.F. (2002) External quality assessment schemes for toxicology. *Forensic Sci Int*, **128**, 98–103.

Wilson, J.F., Newcombe, R.G., Marshall, R.W., Williams, J. and Richens, A. (1984) Analysis of assay errors in drug measurements from the Heathcontrol interlaboratory quality assessment schemes. *Clin Chim Acta*, **143**, 203–16.

Wilson, J.F. and Smith, B.L. (1999) Evaluation of detection techniques and laboratory proficiency in testing for drugs of abuse in urine: an external quality assessment scheme using clinically realistic urine samples. *Ann Clin Biochem*, **36**, 592–600.

Wilson, J.F., Watson, I.D., Williams, J., Toseland, P.A., Thomson, A.H., Sweeney, G., Smith, B.L., Sandle, L.N., Ramsey, J.D. and Capps, N.E. (2002) Primary standardization of assays for anticonvulsant drugs: comparison of accuracy and precision. *Clin Chem*, **48**, 1963–9.

Wilson, I.D. (1990) Observations on the usefulness of internal standards in the analysis of drugs in biological fluids, in *Analysis for Drugs and Metabolites including Anti-infective Agents* (eds E. Reid, I.D. Wilson), Royal Society of Chemistry, Cambridge, pp. 79–82.

Youden, W.J. and Steiner, E.H. (1975) *Statistical Manual of the Association Official Analytical Chemists*, AOAC, Arlington.

15 Absorption, Distribution, Metabolism and Excretion of Xenobiotic Compounds

15.1 Introduction

Although some substances used as drugs are either naturally occurring compounds, such as noradrenaline (norepinephrine), or essential molecules, for example oxygen, most drugs and other poisons are foreign to the body hence the general term xenobiotic (foreign compound) is often used to describe them. Almost all substances can be poisonous, that is can cause deleterious effects on body function, depending not only on the dose, but also on a variety of other factors such as the manner of exposure, whether the subject has been exposed to the compound recently, or whether other poisons have been taken at the same time. Hence in order to understand the role of xenobiotic and other measurements in clinical or forensic samples in the diagnosis, treatment, prognosis and prevention of poisoning, it is important to understand the processes by which drugs and other poisons are absorbed, distributed, in some cases metabolized, and eventually excreted and removed from the body. Not all drugs pass through all of these stages.

Absorption, distribution, metabolism and excretion studies are frequently given the acronym ADME; sometimes liberation of the drug from the pharmaceutical preparation is also included, giving LADME. The term disposition may be used to refer to the various processes that occur after administration of, or exposure to, a given compound. The terms pharmacokinetics (for drugs) and toxicokinetics are used to describe the rate at which these various processes occur, although the term pharmacokinetics is to be preferred. The terms pharmacogenetics and toxicogenetics are applied to the study of genetic influences on these processes. Much of the material presented here is inter-related and has been derived from the study of drugs given in therapy, but in the main it is applicable to xenobiotics in general. The important parameters, plasma half-life ($t_{0.5}$), apparent volume of distribution (V), a measure of the extent of tissue distribution, and clearance (C), a measure of the ability of organs such as the liver and kidney to remove the xenobiotic, are discussed further in Chapter 16.

15.1.1 Historical development

Friedrich Wöhler (1800–1882) won a prize whilst a medical student at Marburg University for his studies on the passage of metabolic products into urine, many of the investigations being carried out on either himself or his dog. Alexander Ure (1810–1866) working in London reported the formation of hippurate from benzoate administered to man (1841), a reaction predicted by Wöhler. Systematic studies of xenobiotic metabolism were conducted in Germany from the 1840s (http://www.issx.org/pages/page04a.html, accessed 20 July 2007). Oxidation, reduction and several conjugation reactions had been reported before the end of the century. Gerhard Domagk

(1895–1964) was awarded the Nobel Prize for Physiology or Medicine (1939) for his work on the antistreptococcal drug prontosil. He knew this compound was effective *in vivo* although not *in vitro*, and thought that activity must reside in a metabolite. The *in vivo* reduction of prontosil was confirmed when sulfanilamide was isolated from patient urine (Colebrooke and Kenny, 1936).

Developments in drug metabolism and pharmacogenetics in Britain have been reviewed (Caldwell, 2006). Glutathione was discovered by Sir Frederick Gowland Hopkins (1861–1947) in 1921, although it was some time before Sybil James and colleagues elucidated the pathway of mercapturate formation (Barnes *et al.*, 1959). Dutton and Storey meanwhile demonstrated that uridine diphosphate (UDP) was the cofactor required for glucuronide conjugation (1953). However, probably the greatest influence was that of Richard Tecwyn Williams (1909–1979), whose acclaimed book *Detoxication Mechanisms: The Metabolism of Drugs, Toxic Substances and Other Organic Compounds* appeared in 1947 (edition 2, 1959). In the preface, it was acknowledged that the title was something of a misnomer as there are many examples of metabolism that result in increased toxicity.

The contribution of Brodie and co-workers in the US to the understanding of drug metabolism in the later 1940s and 1950s was noted in Chapter 4. At this time groups working mainly in the US and in Japan identified enzymes, including those present in liver microsomal preparations, which required molecular oxygen and NADPH as cofactors (Mason, 1957). These enzymes became known as mixed-function oxygenases or oxidases (MFO). Omura and Sato described the spectral properties of cytochrome P450 in 1964, whilst Estabrook and colleagues investigated the properties of these enzymes (1963). With the realization that the cytochromes were a family of enzymes a convention for naming the isoforms systematically (Section 15.5.1.1) was proposed (Nebert *et al.*, 1987).

15.2 Routes of administration

The route by which a drug is administered will depend to an extent on the indication for a particular treatment. For example, thiopental is used to induce anaesthesia and is given intravenously (i.v.). Drugs to treat glaucoma are applied to the cornea. Most drugs are given orally, but for some drugs the physiochemical properties of the drug are such that the oral route cannot be used and a suitable alternative mode of administration has to be found.

15.2.1 Oral dosage

The drug in tablet or capsule form is swallowed and, after release from the formulation, is absorbed across the mucosa of the GI tract unless the intention is to treat a disorder of, or in, the tract itself. This form of administration is sometimes referred to as enteral, which suggests that routes that do not involve the GI tract should be termed parenteral. However to some, parenteral is synonymous with ‘by injection’, and hence these terms are best avoided unless clearly explained. Giving drugs by mouth does not require the use of sterile preparations, special skills or special apparatus. If swallowing is difficult, for example in the young or the elderly, then drugs may be formulated as solutions or suspensions. However, not all drugs can be given orally (Figure 15.1).

Once swallowed, the drug first reaches the acidic environment of the stomach. Acid labile drugs may be destroyed, and proteins and peptides digested by proteolytic enzymes. Most absorption occurs in the small intestine. Highly ionized and polar molecules may not be absorbed unless

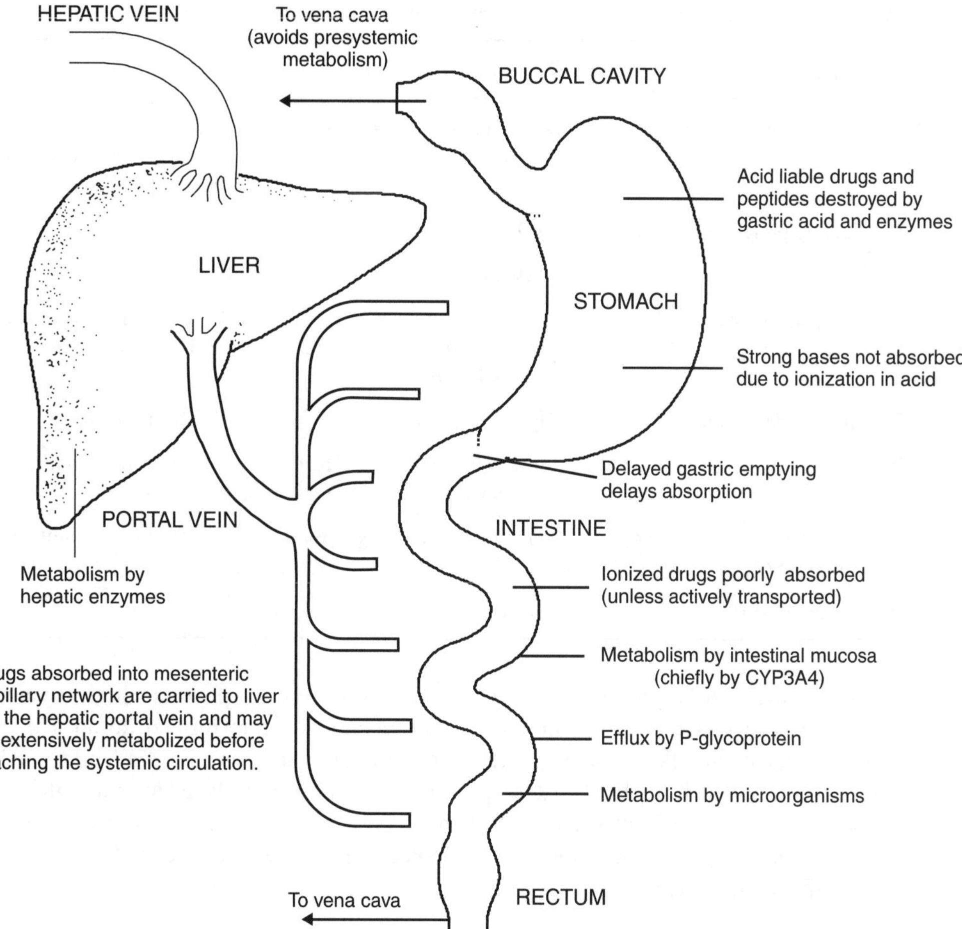

Figure 15.1 Diagram showing some of the factors that affect oral bioavailability.

they can utilize specific transport mechanisms (Table 15.1). There is no guarantee that substances absorbed across the mucosal membrane will reach the systemic circulation. Efflux pumps (Section 15.2.1.1) may return them to the gut lumen or those that are absorbed may be subjected to extensive first-pass hepatic metabolism (Section 15.2.1.2).

Drug interactions are another source of delayed or erratic absorption. Opioids and antimuscarinic drugs may cause delayed gastric emptying and reduced gut motility, and this may reduce the bioavailability of co-administered compounds. Drugs may also interact in other ways. Tetracycline antibiotics chelate divalent metal ions such as Ca^{2+} and Mg^{2+}, and interactions between antacid preparations and tetracyclines have been reported. If ferrous sulfate and a tetracycline are taken together, the insoluble complex formed between the antibiotic and Fe^{2+} ions results in neither drug being absorbed.

Increased bioavailability can be achieved by making compounds that are: (i) more stable in gastric acid, (ii) more lipophilic, and therefore more readily able to diffuse across biological membranes, or (iii) by synthesizing compounds that are substrates for endogenous carriers.

Table 15.1 Examples of reduced oral bioavailability or delayed absorption

Mechanism	Example	Notes
Destroyed by gastric acid	Benzyl penicillin (penicillin G) Methicillin	Penicillins with electron-withdrawing side groups (penicillin V) less affected
Digested by proteolytic enzymes	Insulin	
	Oxytocin	
Poorly/erratically absorbed	(+)-Tubocurarine Aminoglycoside antibiotics Pyridostigmine	Give i.v. or via large oral dosage (pyridostigmine)
First-pass metabolism	Glyceryl trinitrate	See Section 15.5.5
Interaction with other drugs/food	Tetracycline antibiotics/divalent metal ions (Fe^{2+}, Ca^{2+}, Mg^{2+})	Absorption of both reduced
Delayed gastric emptying and reduced gut motility	Opiates, anticholinergic drugs	May reduce amount absorbed
P-Glycoprotein efflux	Fexofenadine	

Some compounds are *prodrugs*, that is substances that are not pharmacologically active until they have been metabolized or, in some cases, chemically transformed, to an active moiety. Esterification of ampillicin to talampicillin or pavampicillin increases bioavailability, and reduces the incidence of diarrhoea. Enalapril is a prodrug ester that is hydrolyzed to enalaprilat. Levodopa is actively transported from the lumen of the GI tract, and into the brain, where it is rapidly decarboxylated to dopamine. Clorazepate dipotassium spontaneously decarboxylates in gastric acid to give nordazepam.

15.2.1.1 P-Glycoprotein

P-Glycoprotein (PGP) (1280 amino acids) is an efflux 'pump' that transports substrates from the intracellular to the extracellular side of the cell membrane. It was first recognized in tumour cells that were resistant to a number of cytotoxic agents as a result of overexpression of what was then referred to as the multidrug resistance (*MDR1*) gene. PGP is classified as being a member of the adenosine triphosphate (ATP)-binding cassette family of transport proteins (ABC transporters). The energy required for PGP to function derives from binding of ATP. The human gene encoding PGP is now known as the *ABCB1* gene.

The location of PGP in non-cancerous tissue suggests that it has evolved to transport potentially toxic substances out of cells (Fromm, 2004). It is highly expressed in the apical membrane of enterocytes lining the GI tract, renal proximal tubular cells, the canalicular membrane of hepatocytes and other important blood–tissue barriers such as those of the brain, testes and placenta. Many of the observations that were once ascribed to enzyme induction or inhibition (i.e. to changes in drug metabolizing enzyme activity) may in fact be due to changes in PGP activity. Inhibitors of PGP are being refined to see if they can be used to overcome the problem of MDR in cancer therapy. Quinidine is often used if a PGP inhibitor is required in experimental work.

15.2.1.2 *Presystemic metabolism*

Even when a drug is transferred rapidly across the gastric mucosa there is no guarantee that it will reach the systemic circulation. The mesenteric capillary network that carries anything that has been absorbed to the liver via the hepatic portal vein surrounds most of the GI tract, apart from the buccal cavity and the lower part of the rectum. Drugs, such as glyceryl trinitrate (trinitrin, GTN) are totally metabolized as they pass through the liver and, because the metabolites have very reduced activity compared with the patent drug, GTN is usually administered via alternative routes. The phenomenon is frequently referred to as first-pass metabolism. However, the term presystemic metabolism covers all eventualities, including metabolism by the gastric mucosa and metabolism of β-adrenoreceptor agonists by the lungs. There is some evidence that large doses of some drugs (e.g. propranolol) saturate first-pass metabolism and so the oral bioavailability may increase with increasing doses.

Presystemic metabolism is important for the activation of some prodrugs. The now withdrawn nonsedating antihistamine, terfenadine, provides one such example. Terfenadine undergoes extensive presystemic metabolism to the antihistamine fexofenadine and was not normally measurable in plasma after a single oral dose. However, when the enzyme responsible (CYP3A4, Section 15.5.1.1) was inhibited the parent drug was absorbed and caused fatal ventricular arrhythmias in susceptible patients. Fexofenadine has been marketed as an alternative non-sedating antihistamine, and although its metabolism in man is negligible, it is also a substrate for PGP. Inhibitors of PGP can increase the absorption of fexofenadine by up to threefold with a corresponding increase in antihistaminic effect. However, no adverse effects have been reported as a result of this interaction (Shon *et al.*, 2005).

15.2.2 *Intravenous injection*

Drugs may either be injected as a single (bolus) dose or infused over minutes, hours or even days. An i.v. bolus should be injected cautiously and with careful observation of response. If the injection is controlled, it should be possible to stop the injection if an adverse response is observed. The first major organ the drug will reach is usually the heart, hence the rate of injection should not be so fast that a concentrated bolus is delivered. If the rate of injection has to be more carefully controlled (e.g. lidocaine to control ventricular arrhythmias) or for prolonged action (e.g. morphine to alleviate terminal pain) then the drug may be infused via an i.v. cannula.

Other drugs that may be given i.v. include *N*-acetylcysteine, the antidote to paracetamol poisoning, which has low oral bioavailability, general anaesthetics (propofol, thiopental), other drugs used in surgery such as skeletal muscle relaxants (e.g. atracurium), and the anticholinesterase neostigmine, which may be used to reverse the effect of atracurium. The fact that these latter drugs are quaternary ammonium compounds and are extremely poorly absorbed from the GI tract is not an issue when they are given i.v.

The disadvantages of i.v. injection are chiefly that this route of administration requires sterile preparations and equipment, and a high degree of skill and training to ensure safety (Box 15.1). Also, it is not easy to perform self-injection. On occasions it may be difficult to locate a suitable vein and some formulations may cause considerable damage to surrounding tissue if the entire volume of the injection is not confined to the vein. An example is thiopental, which is formulated as the sodium salt – the solution is very alkaline, and thus the injection has to be given with extreme care.

Box 15.1 Advantages/disadvantages of intravenous dosage

Advantages

- All the dose enters the systemic circulation – no absorption problems
- Almost instantaneous peak plasma concentration after a bolus injection – rapid onset of effects
- Plasma drug concentrations can be controlled carefully

Disadvantages

- Requires care and skill in establishing i.v. access – may not be possible if the patient has badly damaged veins/very low blood pressure
- Inadvertent injection of air may cause pulmonary embolism
- Requires stable solution of drug in sterile, i.v. compatible medium
- Requires special equipment (syringe or infusion apparatus)
- Injection/infusion may damage vein wall if injection incautious
- Extravasation (injection near to the vein or leakage of the i.v. injection through the vein wall into adjacent tissue) may cause localized tissue damage

15.2.3 Intramuscular and subcutaneous injection

Intramuscular (i.m.) and subcutaneous (s.c.) injection avoid the problems of oral administration but, unlike i.v. injection, the drug must be absorbed from the injection site. This may, or may not be an advantage. Sterile preparations and equipment are required, as for i.v. injection, but less skill is required and patients can be trained to inject themselves. There is no immediate peak plasma concentration, and the onset of action may be quicker or slower than after oral dosage. In some preparations, diazepam, for example, the drug may precipitate in muscle, and absorption may be slow and unpredictable. The onset of action of diazepam is quicker when it is given orally than if injected i.m. However, slow release from muscle is exploited in the case of sustained-release (depot) injections of drugs such as procaine penicillin G and fluphenazine decanoate (Section 16.5.1). There are several preparations available that release insulin at differing rates when injected s.c.

15.2.4 Sublingual and rectal administration

Drugs absorbed from the buccal cavity and lower rectum do not enter the hepatic portal circulation and so avoid first-pass metabolism. Sublingual (under the tongue) administration is particularly useful for GTN, which is rapidly absorbed from the buccal cavity to give relief from the pain of angina. Because GTN is volatile, tablets have a limited shelf-life and are being increasingly replaced by aerosols that are used to spray a metered dose of the drug into the mouth. Buprenorphine, a potent opioid partial agonist, is available as a sustained-release lozenge administered sublingually.

Rectal administration is used for a number of drugs that have systemic effects. The route may be used to administer antiemetics (prochlorperazine) when the oral route is unlikely to be suitable. Diazepam suppositories are available for use in epileptic infants as attempts to establish i.v. access in a convulsing child may be dangerous. The rectal route is also used for compounds that cause gastric irritation and to give a prolonged duration of action, as is the case with aminophylline (a combination of theophylline and ethylenediamine), which can be given rectally to asthmatic children to ease their breathing, particularly at night.

The disadvantages of using suppositories, apart from patient acceptability considerations, are that prolonged use may lead to irritation and inflammation, and secondly that the patient may need to defaecate shortly after insertion.

15.2.5 Intranasal administration

Although drugs may be applied to the nasal passages for a local effect, lipophilic drugs are rapidly absorbed into the systemic circulation. The structure of the nose, with its rich blood supply, highly fenestrated capillaries and an epithelium layer with gaps around the goblet cells (Mygind and Dahl, 1998), allows absorption of molecules that cannot be given orally, such as peptide hormones including insulin, calcitonin and desmopressin (Türker *et al.*, 2004).

With regard to drugs of abuse, snuffs (not necessarily tobacco-based) have probably been used for thousands of years. Cocaine is rapidly absorbed into the systemic circulation from the nasal cavities after insufflation of crystals. Some of the dose may be directly transferred to the brain (Chow *et al.*, 1999).

15.2.6 Transdermal administration

Transdermal drug administration for absorption *across* the skin rather than topical administration for local effect is primarily limited to potent, lipophilic drugs. Skin is a major barrier for many drugs and even very lipophilic drugs are slowly absorbed. Consequently, transdermal administration is a useful way of increasing the duration of action of these drugs. GTN is available as ointments to be rubbed onto the skin, or as plasters to be stuck on. Some plasters contain a rate-limiting membrane so that the rate of absorption is more predicable. Other drugs available as sticking plasters include hyoscine (motion sickness), oestrogens (hormone replacement therapy, HRT), nicotine (tobacco withdrawal) and buprenorphine (analgesia).

15.2.7 Inhalation

The large surface area of the lungs leads to rapid absorption of inhaled drugs and hence rapid onset of effect. Gaseous/volatile general anaesthetics are given this way. The lungs are also the major route by which these drugs are eliminated. Thus, the correct level of anaesthesia can be maintained by adjusting the partial pressure of the anaesthetic in the apparatus used to administer the drug.

Bronchodilators, such the β_2-adrenoceptor agonists (salbutamol, terbutaline), muscarinic antagonists (ipratropium) and steroids (beclomethasone), are given by inhalation. The advantages are that, because the drug is being administered to the site of action, the dose is less than that required if it were given orally, and unwanted systemic effects are reduced. Sodium cromoglicate is inhaled as a dry powder via a spin-hailer, not to ensure local effect on the bronchi, but because it is very poorly absorbed if given orally.

15.2.8 Other routes of administration

There are other sites at which drugs may be administered, sometimes for local effects and sometimes for systemic effects. Intraperitoneal (i.p.) injection provides a convenient way of dosing laboratory animals; many drugs are rapidly absorbed and injecting into the peritoneal cavity reduces the risk of damaging a major organ. Intravaginal applications of fungicidal creams are usually for a local effect on *Candida* infections (thrush), whereas prostaglandin pessaries (preparations designed to release drug after insertion into the vagina) are used to induce labour. Systemic effects can be caused by a drug applied to have a local action. Timolol eye-drops, applied to the cornea for the treatment of glaucoma, have been known to cause bradycardia for example.

15.3 Absorption

The transfer of a drug or other xenobiotic from its site of administration to the bloodstream is termed absorption. Most drugs are absorbed by passive diffusion, although there are a few examples of alternative mechanisms. Filtration and passive and facilitated diffusion are 'passive' processes; no energy is expended, and net movement is down a concentration gradient. Facilitated diffusion is typified by the co-transport of glucose and sodium. Like active transport, it is carrier-mediated and therefore saturable. There are examples of drugs being transported by facilitated diffusion, such as the absorption of vitamin B_{12} from the GI tract. However, active transport, in which compounds are transported against a concentration gradient with the expenditure of energy, is important for some drugs. Presumably, the carrier (transport) proteins evolved to transport important nutrients and other naturally occurring substances, but exogenous compounds with similar structures, such as thiouracil and levodopa, are also transported. Important points about carrier-mediated transport are: (i) it can become saturated at high substrate concentrations and (ii) it can be inhibited by other molecules. In some situations, drug overdose for example, this may have major implications for the interpretation of kinetic parameters and analytical results, and hence, treatment.

Pinocytosis is the process in which microscopic invaginations of cell wall engulf drops of surrounding fluid along with any solutes contained therein. It is probably unimportant for small molecules ($M_r < 5000$), but larger molecules such as botulinum toxin are transported into mucosal cells by pinocytosis, from where they drain into the lymphatic system.

15.3.1 Passive diffusion

Passive diffusion is the most important mechanism by which xenobiotics are absorbed. The rate of diffusion across a membrane is determined by Fick's Law:

$$rate = D \cdot R \cdot A \frac{\Delta C}{\Delta x} \tag{15.1}$$

where D is the diffusion coefficient (number of moles of drug that diffuse across unit area in unit time when $\Delta C = 1$), R is the partition coefficient between the membrane and aqueous phase, A is the area of the membrane, Δx is the thickness of the membrane and ΔC is the concentration gradient across the membrane. Simplification gives:

$$rate = P \cdot A \frac{\Delta C}{\Delta x} \tag{15.2}$$

where P is the permeability coefficient, which is characteristic of the membrane and the molecule in question. Thus, the surface area over which the drug is being transferred, the nature of the membrane and the physicochemical properties of the substance affect the rate of diffusion. The specialized epithelial cells of the small intestine have rows of microvilli that vastly increase the effective surface area for absorption, the brush-like appearance giving the term 'brush border'. A drug must have some water solubility to be able to approach the cell membrane and compounds with very poor water solubility may not be suitable for use as drugs. Small molecules diffuse more rapidly than larger ones. In aqueous solution, the rate of diffusion of a molecule is inversely proportional to the square root of the relative molecular mass (M_r) up to $M_r \approx 1000$, after which the rate is inversely proportional to $M_r^{1/3}$. However, the most important properties are the partition coefficient and degree of ionization. Ionized molecules do not readily dissolve in the lipid bilayer of cell membranes and are generally not absorbed unless there are specific transport mechanisms.

15.3.1.1 Partition coefficient

To be absorbed by passive diffusion, substances have to be soluble in the lipid bilayers of the mucosal cells, and xenobiotics with high organic solvent/water partition coefficients rapidly dissolve in the lipid membranes of cells. Such substances are described as lipophilic (lipid-loving) and the term is, in many ways, more useful than partition coefficient when describing the nature of a drug. Partition coefficients, for example between octanol and water, can be measured, but lipophilicity cannot, as it is only a concept. The characteristics of a lipophilic drug are that it is: (i) rapidly absorbed, (ii) rapidly and widely distributed in body tissues and (iii) extensively metabolized before it is excreted. Aberrations in partitioning behaviour, such as ion-pair extraction, often seen with partitioning into halogenated solvents (Section 3.2.5.2), can give a misleading impression of the lipophilic nature of a drug. Generally, partition coefficients (and lipophilicity) increase with increasing hydrocarbon chain length, numbers of aromatic moieties and so on, and decrease with increasing numbers of polar groups, such as hydroxyl and carbonyl.

15.3.1.2 Ionization

Ionization markedly reduces lipophilicity (cf. pH partitioning, Section 3.2.5.1) and can have a major effect on absorption. The degree of ionization of a weak electrolyte is a function of the pH of the environment in which it is dissolved and the pK_a of the ionizable moiety, and can be calculated from the Henderson–Hasselbalch equation:

$$\mathrm{p}K_a = \mathrm{pH} + \log\frac{[\mathrm{AH}]}{[\mathrm{A}^-]} \tag{15.3}$$

Aspirin (p$K_a \approx 3.4$) is largely unionized (99.01 %) at gastric pH and can be absorbed from the stomach. Thus at equilibrium, for every ionized aspirin molecule in gastric acid there will be 100 unionized molecules. As it is the unionized aspirin that equilibrates across the membrane, there will be the same number (100) of unionized molecules in plasma. However, in plasma (99.99 % ionized) there will be 1,000,000 ionized molecules for every 100 unionized aspirin molecules (Figure 15.2). Counting up the total amounts in each fluid gives a ratio of 1,000,100:101 = 9902:1 in favour of plasma. Plasma protein binding and bulk flow of aspirin in the bloodstream help to maintain the concentration gradient so that even more aspirin is absorbed. This is known as the pH-partition hypothesis. Despite the favourable pH of the stomach and the less favourable pH of the small intestine, approximately 75 % of an oral dose of aspirin is absorbed from the

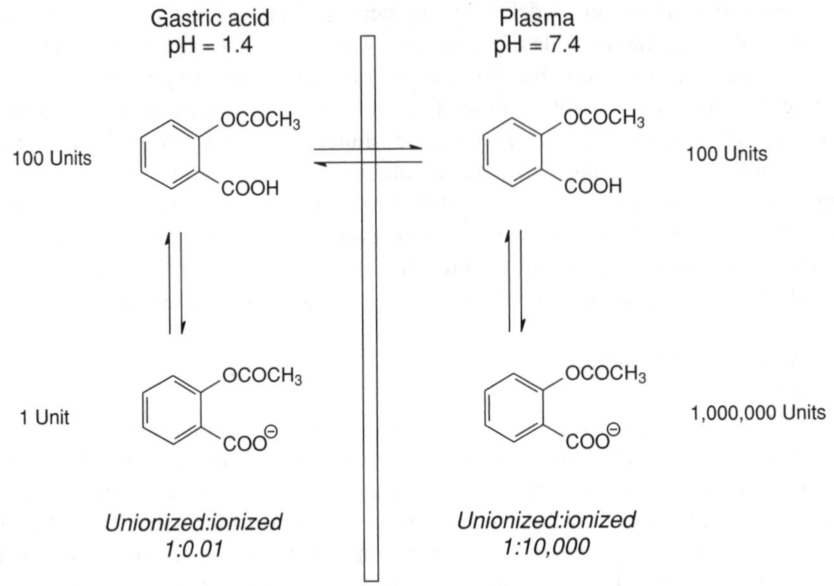

Gastric acid pH = 1.4 | Plasma pH = 7.4

100 Units — [aspirin structure: $OCOCH_3$, COOH] ⇌ [aspirin structure: $OCOCH_3$, COOH] — 100 Units

1 Unit — [$OCOCH_3$, $COO^{\ominus}$] | [$OCOCH_3$, $COO^{\ominus}$] — 1,000,000 Units

Unionized:ionized 1:0.01 | *Unionized:ionized 1:10,000*

Figure 15.2 Distribution of aspirin ($pK_a = 3.4$) between gastric fluid and plasma water.

small intestine, illustrating the importance of the large surface area of this part of the bowel and the fact that the stomach wall is relatively thick.

The converse is true for bases. Amfetamine (pK_a 9.8) cannot be absorbed from the stomach because it is almost totally ionized at pH 1.4 ($\alpha = 4 \times 10^{-9}$). On the other hand, basic drugs may diffuse from plasma into the stomach (a phenomenon sometimes referred to as *'ion-trapping'*). Therefore, the presence of a basic drug in stomach contents does not necessarily indicate that the drug has been taken orally.

Unless there are specialized carrier mechanisms, very polar/ionized drugs would not be expected to be well absorbed from the GI tract. Quaternary ammonium compounds are poorly and erratically absorbed if taken orally. Pyridostigmine when used orally to treat myasthenia gravis is typically given in 60 mg doses, but when used i.v. to reverse the effects of competitive neuromuscular antagonists, 0.5–1 mg may be sufficient. (+)-Tubocurarine, a purified alkaloid from curare used as an arrowhead poison, contains two charged nitrogen moieties and is poorly absorbed from the GI tract, hence poisoned prey can be eaten safely. Aminoglycoside antibiotics are very polar and are poorly absorbed across the gastric mucosa. They are usually given by injection, although oral preparations are available for treating GI infections.

15.3.2 Carrier-mediated absorption

It has long been known that transport systems exist for the absorption of essential amino acids, pyrimidines and related compounds, and that structurally related xenobiotics may utilize these carriers (Schanker and Tocco, 1960). Widely distributed transport systems are the organic anion-transporting proteins (OATPs) of which there are several human isoforms (DuBuske, 2005). These are found in brain, liver, kidney, small intestine and many other tissues. They are important in the transfer of conjugates of glucuronic acid, sulfate and glutathione across cell membranes. OATPs in the small intestine may be important for the absorption of xenobiotics, for example rat OATP3 transports fexofenadine from the intestinal lumen. Organic cation transporters (OCTs) transport small cations such as tetramethylammonium, choline, histamine and noradrenaline.

A notable exception to the generalization that quaternary ammonium compounds are poorly absorbed is, of course, paraquat, which may be transported across biological membranes by several carrier-mediated mechanisms (Table 15.2). Bennett *et al.* (1976) observed dose-dependant GI absorption of paraquat. In dogs dosed at 0.12 mg kg^{-1} 46–66 % of the dose was absorbed, but only 25–28 % at 5 mg kg^{-1}. Previous studies in the rat had used doses up to three orders of magnitude higher, and so the effect had not been recognized. Paraquat uptake by rat intestinal loops is saturable, and is blocked by tetramethylammonium and choline salts, but not by putrescine (Nagao *et al.*, 1993).

Table 15.2 Putative transport systems for paraquat

Putative substrate	Location	Inhibitor	Reference
Choline	Intestine	Tetramethylammonium	Nagao *et al.* (1993)
Putrescine	Lungs	Putrescine	Smith (1988), Smith *et al.* (1990)
	Kidney (basolateral)	Quinine	Chan *et al.* (1996)
Neutral amino acids	Brain	L-Valine	Shimizu *et al.* (2001)

15.3.3 Absorption from muscle and subcutaneous tissue

In peripheral blood capillaries the endothelial cells are not tightly packed and the gaps (fenestrations) between the cells allow free passage of small molecules ($M_r < 5000$). Thus, polar, ionized drugs that are poorly absorbed across the tightly packed epithelial cells of the GI tract are absorbed into the systemic circulation following i.m. injection. Absorption from muscle (and subcutaneous tissue) is normally perfusion limited, in other words the greater the blood flow to the injection site, the faster the rate of absorption. Absorption is quicker from muscles with a greater blood flow, and tends to be slowest following s.c. injection. Massaging and warming the injection site may increase the rate of absorption. The rate of absorption can be deliberately reduced by co-administration of vasoconstrictors, for example the use of adrenaline with local anaesthetics to increase the duration of effect and increase the amount of anaesthetic that can be given safely. Absorption is likely to be reduced in shock (circulatory failure).

15.4 Distribution

Once absorbed into the blood stream, drugs must be distributed to their site(s) of action. The mechanisms by which this occurs are similar to those described above for absorption, with passive diffusion and filtration being the most important. Initially the net movement will be from blood to tissues, but as the blood concentration falls, the net movement will be from tissues to blood. This phenomenon has been termed 'random walk'. Thus, when injected into the bloodstream small molecules rapidly equilibrate between plasma and interstitial fluid.

The specialized capillaries supplying blood to the brain have membranes consisting of tightly packed endothelial cells and so only lipophilic compounds that can diffuse across the cell walls, or those molecules that are substrates for the active transport systems that exist, presumably, to transfer endogenous materials into the brain, can normally enter the CNS. The observation that certain dyes did not enter the brain led to the concept of the blood–brain barrier (BBB). Ionized molecules such as neostigmine and pyridostigmine, developed for the treatment of myasthenia gravis, are excluded from the CNS. In meningitis, the BBB is disrupted and substances, including penicillins, which do not normally enter the brain, may do so.

Small molecules such as amoxicillin and (+)-tubocurarine that do not bind to plasma protein and do not penetrate cells are distributed in extracellular fluid. Molecules that enter cells, but do not bind to plasma protein or cell constituents, such as ethanol, are uniformly distributed in body water. However, many xenobiotics are concentrated (sequestered) in some tissues rather than others. An indicator of the extent of sequestration is the apparent volume of distribution (Section 16.2.2.1). The greater the proportion of drug in tissues the smaller will be the proportion, and hence for a given amount in the body, the concentration, in the plasma. The mechanisms by which substances are sequestered in tissues include ion-trapping, binding to macromolecules, distribution in lipids and transporter systems.

15.4.1 Ion-trapping

Differences in pH may result in intracellular (pH 6.8) concentrations of basic drugs being higher than the concentrations in plasma water (pH 7.4) due to ion-trapping, in a manner analogous to the sequestration of bases in gastric acid discussed above. Under normal physiological conditions one would expect the concentrations of acidic substances such as salicylate (Figure 15.3) to be

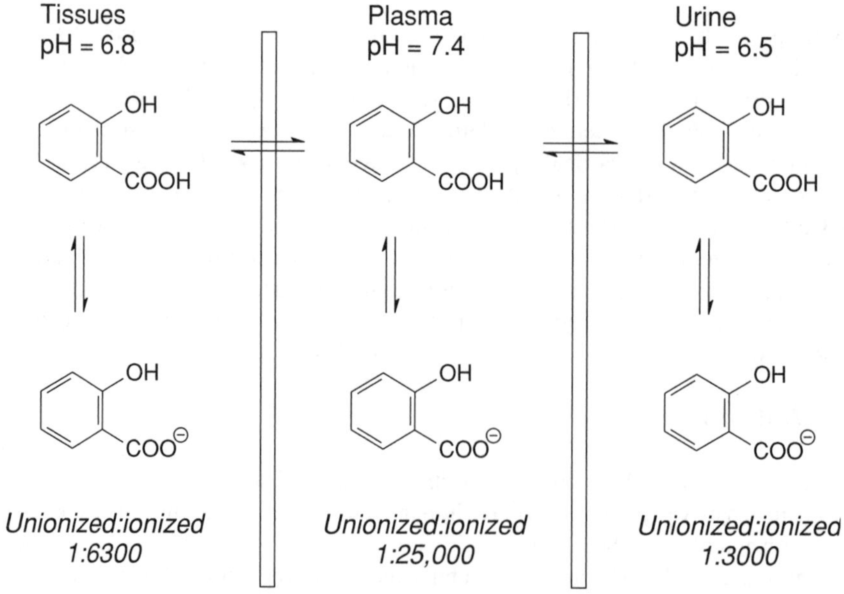

Tissues
pH = 6.8

Plasma
pH = 7.4

Urine
pH = 6.5

OH
COOH

OH
$COO^{\ominus}$

Unionized:ionized 1:6300

Unionized:ionized 1:25,000

Unionized:ionized 1:3000

Figure 15.3 The effect of pH on the distribution of salicylate ($pK_a = 3.0$) between plasma water, intracellular fluid (tissues) and urine. The pH values represent normal physiological conditions.

higher in plasma water than those in intracellular fluid. However, in salicylate overdose a resulting acidosis can shift the equilibrium from plasma to tissue (Section 15.6.1).

15.4.2 Binding to macromolecules

Quinacrine and chloroquine are very highly localized in cells as a result of their interaction with DNA. Basic drugs accumulate in melanin-containing tissues including hair, which in turn can be used to assess chronic exposure to such compounds (Section 17.5.3). High erythrocyte concentrations of chlortalidone are due to binding to carbonic anhydrase. Digoxin binds to sodium/potassium-ATPase and so tends to concentrate not only in cardiac tissue, but also in skeletal muscle. That the quaternary ammonium compounds neostigmine and pyridostigmine have apparent volumes of distribution greater than extracellular fluid volume is explained by the fact they bind to acetylcholinesterase on the surface of erythrocytes. Similarly, lead and other heavy metals such as mercury bind to thiol groups on erythrocytes, hence the need measure whole blood and not plasma concentrations of these analytes. Binding to thiol moieties at the active sites of enzymes is thought to be important in the toxicity of lead, for example (Amici *et al.*, 1994). Carbon monoxide binds to haemoglobin and other cytochromes, whilst cyanide ion binds to cytochromes containing iron(III) such as cytochrome c, thereby inhibiting cellular respiration. Cyanide also binds to methaemoglobin, haemoglobin in which iron(II) has been oxidized to iron(III), and this is the rationale for administering nitrites, which increase the proportion of methaemoglobin present in blood, in the treatment of cyanide poisoning.

15.4.2.1 Plasma protein binding

Small molecules may bind to plasma protein. Acidic drugs are often bound to albumin, and bases to albumin and also to AGP. Binding to plasma proteins is an important mechanism by which molecules are transported in the blood. Lipophilic molecules tend be bound extensively and the concentrations in plasma (i.e. bound + non-bound) can exceed the aqueous solubilities of such compounds. Protein binding is normally reversible and can be described in terms of the Law of Mass Action. For a single class of binding sites, the number moles of drug bound (D_b) to the total concentration of protein (P_t) is given by the Scatchard equation:

$$\frac{D_b}{P_t} = \frac{nKD_f}{1 + KD_f} \tag{15.4}$$

where K is the equilibrium binding constant, n is the number of binding sites per molecule of protein, and D_f is the concentration of free (non-bound) drug. Because there is very little protein in cerebrospinal fluid (CSF), drug concentrations in CSF are often very close to the non-bound plasma concentration. A similar argument applies to saliva for weakly acidic compounds such as phenytoin, primidone, ethosuximide and carbamazepine (Soldin, 1999).

Binding to plasma proteins reduces the apparent volume of distribution by 'holding' the drug in plasma, in other words it reduces the concentration of drug that is free to diffuse into tissues. Thus, protein binding will reduce the activity of a drug if it reduces the amount available to reach its site(s) of action. More importantly, drug activity, and, possibly toxicity, may increase if plasma protein binding is reduced. This may occur in some disease states that result in reduced plasma protein concentrations. Displacement of one drug by another from plasma protein binding sites is also a potential mechanism of drug–drug interactions.

Many *in vitro* studies have demonstrated displacement of one drug by another, but *in vivo* the situation is more complex. The 'total' concentration of a displaced drug in plasma will be reduced as some of the liberated drug diffuses into tissues as new equilibria are established. The increased concentration of nonbound drug may lead to greater, possibly toxic, effects. Hence, measurement of the 'total' (bound + nonbound) concentration of a drug in plasma may be misleading in certain circumstances. When phenytoin was displaced by salicylate, for example, the percentage nonbound increased from 7.14 to 10.66 %, and this was accompanied by a significant decrease in total serum phenytoin concentration from 13.5 to 10.3 mg L^{-1}. The salivary phenytoin concentration rose from 0.97 to 1.13 mg L^{-1} (Leonard *et al.*, 1981).

Whether a plasma protein displacement interaction is clinically important depends on a number of factors. The displacing agents will usually attain plasma concentrations approaching that of the binding protein and show concentration-dependent binding. Such agents include phenylbutazone, salicylate and valproate. However, for a displaced drug with a large apparent volume of distribution, the amount displaced will represent a small proportion of the dose so the increase in tissue concentration is unlikely to be significant.

15.4.3 Distribution in lipid

Lipophilic substances concentrate in lipid membranes and fat cells. For some drugs, such as thiopental, distribution into lipid is a major determinant of its duration of action (Section 16.7). Generally, lipophilic drugs have high apparent volumes of distribution, but a high value does not always equate with CNS activity. Amiodarone (V 62 L kg^{-1}) is mainly found in fat, lung and liver, with little being present in brain.

15.4.4 Active transport

Active transport systems can maintain higher concentrations of a drug inside a cell against a prevailing concentration gradient. Amfetamine enters noradrenergic nerve terminals via the amine (uptake 1) transporter. Levodopa (a prodrug of dopamine) is actively transported into the brain. Active transport of paraquat into and out of the lungs is an important factor in its pulmonary toxicity. It is concentrated in lung by a diamine transport system with a high affinity for putrescine, but its efflux is dependent on PGP (Section 15.3.2). Induction of PGP by dexamethasone reduced the amount of paraquat in rat lung by approximately 50 % (Dinis-Oliveira *et al.*, 2006). Entry of paraquat into the brain is mediated by a system that can be blocked by L-valine, a high affinity substrate for the neutral amino acid transporter, although this cannot be exploited in the treatment of paraquat poisoning.

15.5 Metabolism

The kidney usually rapidly excretes polar, hydrophilic drugs and other xenobiotics. However, lipophilic drugs are usually poorly excreted because they tend to be highly protein bound and filtered poorly at the glomerulus, and even if filtered, are readily reabsorbed by the renal tubule. Therefore, most drugs have to be metabolized to more polar metabolites before they can be excreted. Often drug metabolism reduces the biological activity of the drug. On the other hand: (i) many drugs have active metabolites that contribute to the overall pharmacological profile,

(ii) some drugs are not active until they have been metabolized and (iii) some metabolites are highly toxic, as in the case of the postulated intermediary paracetamol metabolite *N*-acetyl-*p*-benzoquinoneimine (NAPQI).

The liver is the most important site of xenobiotic metabolism, but most tissues are capable of metabolizing drugs, and hydrolysis by plasma esterases may also be important. Metabolism by intestinal flora and mucosa (and excretion by PGP) is important in influencing the oral bioavailability of many drugs, as discussed above.

It is convenient to consider xenobiotic metabolism under two headings, phase 1 and phase 2. Phase 1 reactions involve chemical modification of the molecule by oxidation, reduction or hydrolysis, whilst phase 2 reactions are conjugation reactions, in which a second, hydrophilic molecule such as D-glucuronic acid is added to the molecule. Phase 2 reactions cannot occur unless a suitable reactive group, such as hydroxyl or primary amine, is present and because phase 1 reactions often introduce such a group, the process has been termed 'functionalization'. Consequently, phase 1 reactions often, but not exclusively, precede phase 2 reactions. For example, diazepam undergoes a phase 1 conversion to temazepam followed by conjugation with glucuronic acid, whereas temazepam and oxazepam can be conjugated directly (Section 15.5.3, Figure 15.11).

15.5.1 Phase 1 metabolism

Phase 1 reactions can be subdivided into the types of products that are formed: aromatic hydroxylation, *N*-oxidation, deamination and so on. Of major importance in the oxidation of xenobiotics is the MFO system (Section 15.1.1). These membrane-bound enzymes are found in the smooth endoplasmic reticulum (SER) of cells, and may have evolved for steroid (terpene) and lipid metabolism. When liver cells are homogenized the SER forms small vesicles known as microsomes, which can be collected as a pellet after high-speed centrifugation of homogenate (100,000 *g*, 60 min), usually after cell debris has been removed (9000 *g*, 10–20 min). Reactions catalyzed by these enzymes are often referred to as microsomal reactions. Enzymes such as ADH which remain in the 100,000 *g* supernatant layer, are referred to as soluble or cytosolic enzymes.

Microsomal oxidations involve a relatively complex chain of redox reactions and require: (i) NADPH, (ii) a flavoprotein (NADPH-cytochrome P450 reductase) or cytochrome b_5 (iii) a cytochrome P450 (CYP) enzyme, a haem-containing protein and (iv) molecular oxygen (Figure 15.4).

Thus, xenobiotic (RH) combines with the CYP, and the Fe^{3+} in the complex is reduced to Fe^{2+} by acquiring an electron from NADPH-P450 reductase. The reduced complex combines with oxygen $[(RH)Fe^{2+}O_2)]$ and combination with a proton and a further electron (from NADPH-cytochrome P450 reductase or cytochrome b_5) produces a RH-Fe^{2+}OOH complex. The addition of a proton liberates water and a ferric oxene complex $[RH(FeO)^{3+}]$, which extracts H from the drug with the formation of a pair of free radicals. Finally, the oxidized drug is released from the complex with the regeneration of the P450 enzyme. The last stage of the reaction involves free radicals. Treatment of the compound under study with Fenton's reagent (a source of hydroxyl radicals) can sometimes give the same products.

15.5.1.1 The cytochrome P450 family

Cytochrome P450 was so called because in its reduced form it appeared pink and reacted with carbon monoxide to produce a characteristic absorption spectrum with a peak at 450 nm. Treating animals with certain agents induced the synthesis of a cytochrome with a maximum closer to 448 nm. Today, we know that these cytochromes are a superfamily of related, but distinct, enzymes.

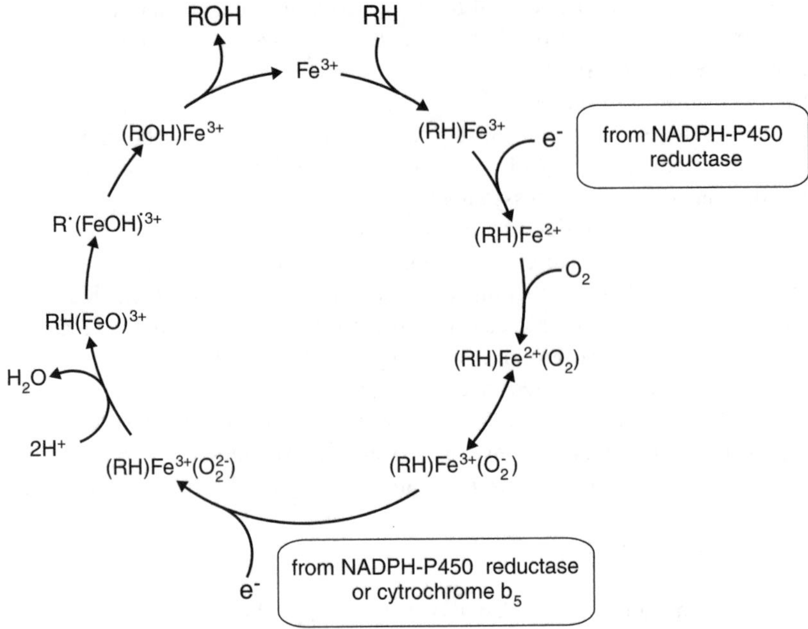

Figure 15.4 A Schematic representation of the mono-oxygenase cycle showing the oxidation states of the haem iron and incorporation of molecular oxygen.

They are classified on the basis of cDNA cloning according to similarities in amino acid sequence. A family contains genes that have at least a 40 % sequence homology. The three major drug metabolizing families are CYP1, CPY2 and CPY3. Members of a subfamily (denoted by a letter) must have at least 55 % identity. The final number denotes the individual gene products (Table 15.3). There are some 50 CPY genes important in man.

CYP3A4 is the most prevalent CYP in the body and has many substrates. It constitutes approximately 30 % of the total CYP content of the liver and 70 % of the total CYP in the gastrointestinal mucosa. It is found in tissues that express large amounts of PGP, with which it appears to work in concert, xenobiotics that are not PGP substrates being metabolized by CYP3A4. CYP3A5 may be more abundant than was originally thought, particularly in Africans. It is the most abundant CYP3A form in the kidney. CYP3A7 is a foetal form of the enzyme that is believed to be important in the metabolism of endogenous steroids and *trans*-retinoic acid. It rapidly declines during the first week of life and is rarely expressed in adults.

Although CYP2D6 represents only 2 % of hepatic CYP, it metabolizes a large number of substrates, including many antipsychotic and antidepressant drugs, β-blockers and the antihypertensive, debrisoquine. It is also responsible for activating the prodrug, tramadol and for the *O*-demethylation of codeine to morphine (Figure 15.5).

15.5.1.2 Other phase 1 oxidases

Flavin-containing monooxygenases (FMOs) are microsomal enzymes that catalyze the NADPH-dependent oxidation of a large number of sulfur-, selenium- and nitrogen-containing compounds,

Table 15.3 Some drug metabolising isoforms of cytochrome P 450

Isoform	Substrates	Inhibitors	Inducers
CYP1A2	Caffeine, clozapine, theophylline	Cimetidine, ciprofloxacin	Phenobarbital, phenytoin, polycyclic, hydrocarbons, rifampicin
CYP2A6	Nicotine		Ritonavir
CYP2B6	Cyclophosphamide[a]	Orphenadrine	Cyclophosphamide, phenobarbital
CYP2C9	Losartan[a], phenytoin, *S*-warfarin	Amiodarone, ketoconazole, sulfaphenazole	Phenobarbital, rifampicin
CYP2C19	Diazepam, omeprazole	Cimetidine, omeprazole, sulfaphenazole	Phenobarbital, rifampicin
CYP2D6	Codeine[a], debrisoquine, haloperidol, oxycodone[a], phenothiazines, SSRIs, tramadol[a]	Cimetidine, quinidine, phenothiazines, SSRIs	None known
CYP2E1	Dichloromethane, enflurane, ethanol, halothane, paracetamol	Disulfiram, ethanol (acute), miconazole	Ethanol (chronic), isoniazid, rifampicin
CYP3A4	Clozapine, nifedipine	Calcium channel blockers, cimetidine, erythomycin, ketoconazole, SSRIs	Carbamazepine, phenobarbital, rifampicin
CYP3A5	Caffeine, midazolam		Dexamethasone

[a] pro-drug

such as the *N*-oxidation of tertiary amines and stereospecific oxidation of sulfides. They are not cytochromes. A genetic failure to express one isoform (FMO3) results in trimethylaminurea, also known as ‘fish-odour syndrome’. The inability to metabolize trimethylamine leads to large quantities being excreted in the urine and sweat, a very distressing condition for the sufferers, who smell of putrefying fish.

Monoamine oxidase (MAO), which is bound to the surface of mitochondria, is located in aminergic nerve terminals, liver and intestinal mucosa. As well as deaminating endogenous neurotransmitters (noradrenaline, dopamine and serotonin), chemically similar molecules such as tyramine are also substrates. Amfetamine, having an α-methyl substituent, is a poor substrate for this enzyme, however.

Other important phase 1 enzymes include xanthine oxidase that metabolizes 6-mercaptopurine (Figure 15.6) and ADH, the mammalian form of which oxidizes several alcohols as well as ethanol, including methanol, ethylene glycol and 2,2,2-trichloroethanol, the active metabolite of chloral hydrate.

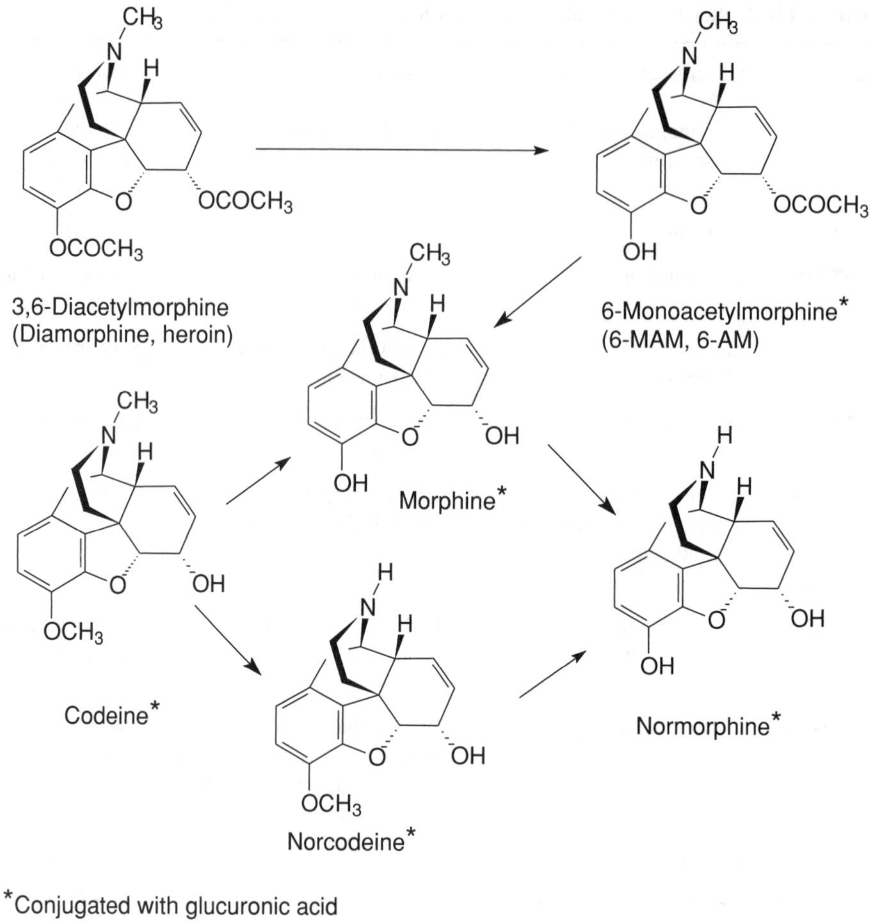

3,6-Diacetylmorphine (Diamorphine, heroin)

6-Monoacetylmorphine* (6-MAM, 6-AM)

Morphine*

Codeine*

Normorphine*

Norcodeine*

*Conjugated with glucuronic acid

Figure 15.5 Metabolic pathways of diamorphine, morphine and codeine.

15.5.1.3 Microsomal reductions

Aromatic nitro- and azo-compounds are reduced by systems that require NADPH and are stimulated by flavins. The nitro-moieties in chloramphenicol and nitrazepam are reduced to primary aromatic amines, which may then be acetylated (Section 15.3.4.3). Azo-reductase reduces and cleaves the double bond in azo compounds with the formation of two aromatic primary amines. The azo dye prontosil is converted to triaminobenzene and sulfanilamide.

Under anaerobic conditions, enzymes that require NADPH and oxygen reductively dehalogenate halothane and methoxyflurane. These compounds also undergo oxidative dehalogenation.

15.5.1.4 Hydrolysis

Esterases and amidases hydrolyze substrates to reveal reactive groups that may then undergo phase 2 conjugation. Esters are sometimes used as prodrugs (Section 15.2.1) to increase bioavailability

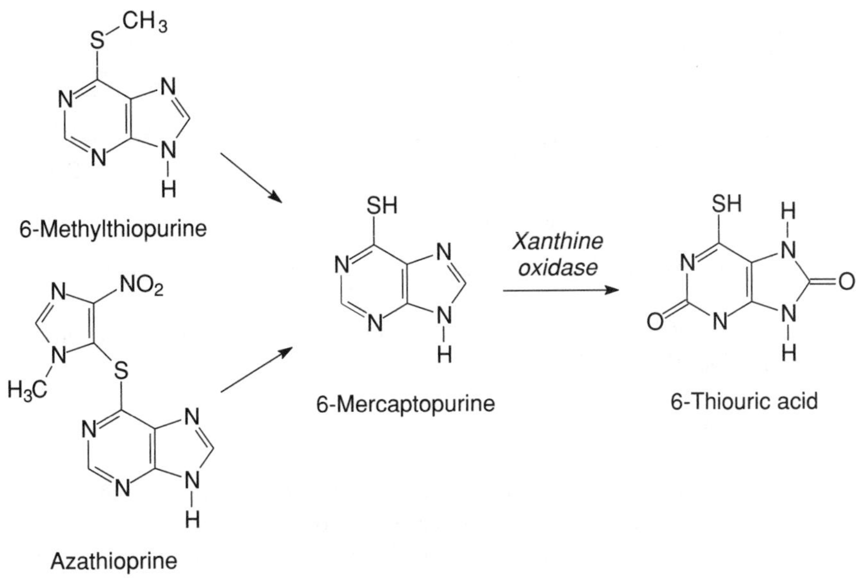

Figure 15.6 Role of xanthine oxidase in the metabolism of purines.

and/or reduce GI irritation. Plasma contains a number of esterases, including butyrylcholinesterase (also known as pseudocholinesterase or plasma cholinesterase), which rapidly hydrolyzes cocaine, procaine, suxamethonium and physostigmine. Acetylcholinesterase is a membrane-bound enzyme located primarily at cholinergic synapses and on erythrocytes.

15.5.2 Phase 2 reactions

These reactions include glucuronidation, sulfation, N-acetylation and N- and O-methylation, as well as reaction with reduced glutathione (GSH) and with amino acids such as glycine.

15.5.2.1 D-Glucuronidation

This is an important reaction as D-glucuronides are usually very polar, water soluble, inactive metabolites that are rapidly cleared by the kidney or via the bile. The increase in size (glucuronidation increases M_r by 207) and their acidic nature also make glucuronides good substrates for active efflux into bile. Morphine may be unique in having a glucuronide metabolite (morphine-6-glucuronide) that is at least as active as an analgesic as the parent molecule. This metabolite accumulates in the plasma of renal failure patients given diamorphine or morphine, and is responsible for the apparent paradox that such patients may need decreasing doses of drug to maintain analgesia (Osborne *et al.*, 1986).

Glucuronidations require a donor molecule, uridine diphosphate glucuronic acid (UDPGA), which is synthesized in the cytoplasm from a combination of α-D-glucose-1-phosphate and uridine triphosphate, followed by oxidation to the acid. Ether O-glucuronides are formed with phenols and alcohols, whereas ester glucuronides are formed with carboxylic acids. N- and S-Glucuronides (formed from amines and thiols, respectively) are less stable than O-glucuronides.

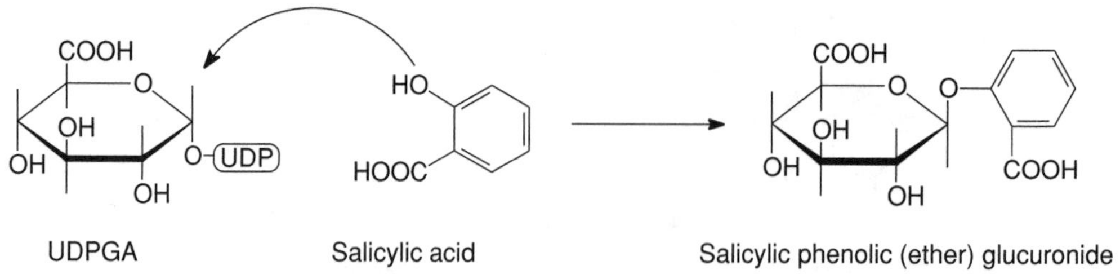

Figure 15.7 Conjugation of salicylic acid to produce the β-D-glucuronide.

D-Glucuronyl transferases are found in the microsomal fraction and the reaction always produces β-D-glucuronides (Figure 15.7).

15.5.2.2 O-sulfation and N-acetylation

Ethereal sulfate conjugation may be catalyzed by a number of cytoplasmic sulfotransferases depending on the substrate. The sulfate donor is 3′-phosphoadenosine-5′-phosphosulfate (PAPS). Sulfates are normally very water soluble and readily excreted. Sulfation is important in the excretion of paracetamol, for example.

Aromatic primary amines and hydrazines are *N*-acetylated by a soluble enzyme, *N*-acetyltransferase 2 (NAT-2) and acetyl-CoA. Genetic polymorphism in the expression of NAT-2 can be a major determinant of the plasma concentrations of drugs acetylated by this enzyme (Section 15.4.2). Drugs which are acetylated include isoniazid, several sulfonamide antimicrobials, hydralazine, phenelzine, dapsone and procainamide (Figure 15.8). Primary amine metabolites formed by reduction of nitro moieties such as nitrazepam and chloramphenicol may also be acetylated. Generally, the metabolites are less pharmacologically active than the parent drug, although *N*-acetylprocainamide (acecainide, NAPA) has similar activity to procainamide (Bagwell *et al.*, 1976). Unlike the products of glucuronidation and sulfation, acetyl metabolites do not show increased water solubility. Indeed, the opposite might be expected. The acetyl metabolites

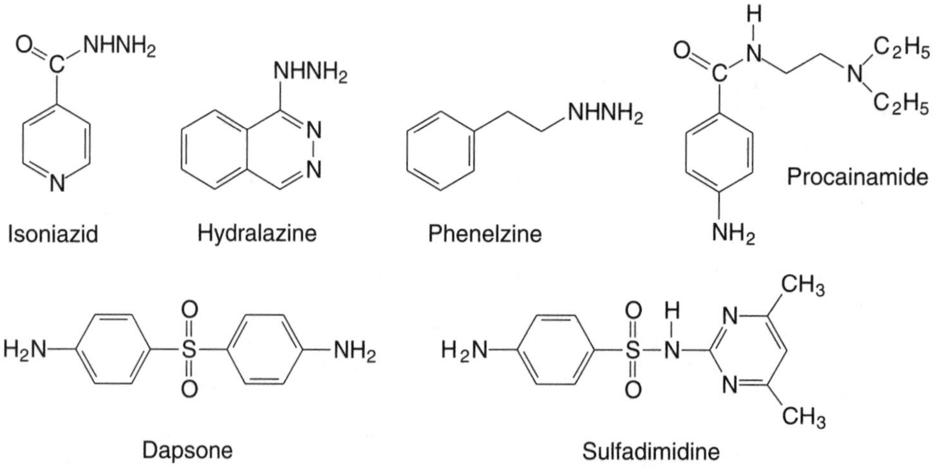

Figure 15.8 Some drugs metabolized by *N*-acetylation.

of older sulfonamide antimicrobials were implicated in causing crystalluria. Recently introduced sulfonamides are more water soluble and the decrease in pK_a on acetylation increases the degree of ionization at physiological pH values, which appears to compensate for the increase in log P.

15.5.2.3 O-, N- and S-methylation

The methyl donor for these reactions is *S*-adenosylmethionine (SAM). Several methyl transferases catalyze the methylation of endogenous substances. Catechol *O*-methyl transferase (COMT) methylates endogenous catechols (noradrenaline, dopamine, adrenaline) and some drugs such as isoprenaline to give 3-methoxy metabolites. Phenolic metabolites may also be methylated, for example 7-hydroxychlorpromazine is metabolized to 7-methoxychlorpromazine. A small amount of morphine may be *O*-methylated to codeine, but this is unlikely to be clinically important.

The biosynthesis of adrenaline from noradrenaline is catalyzed by phenylethanolamine *N*-methyl transferase. Other *N*-methylations include that of histamine to 1-methylhistamine and the methylation of acetylserotonin to give melatonin. Examples of xenobiotics that undergo *S*-methylation include 2-mercaptoethanol and thiouracil.

Microbiological methylations of heavy metals such as mercury are of toxicological significance: mercury salts are toxic to the kidney, whereas methylmercury is neurotoxic and more difficult to remove from the body by use of chelating agents. Non-volatile inorganic compounds of sulfur, selenium and tellurium are metabolized to volatile dimethyl metabolites that are excreted via the lungs.

15.5.2.4 Conjugation with glutathione

The thiol group in reduced glutathione (GSH) usually acts as a nucleophile and may react chemically or enzymatically via glutathione-*S*-transferase (GST), actually a family of enzymes, found in the soluble fraction of cells. Substrates include aromatic halogen and nitro compounds, and reactive phase 1 metabolites such as epoxides. Thus, GSH conjugation normally has a protective role by removing potentially toxic metabolites, as with paracetamol. At therapeutic doses (500–1000 mg) most of a dose of paracetamol is conjugated with glucuronic acid or sulfate (Figure 15.9). A small proportion of the dose is oxidized by CYP2E1 to a reactive intermediate, *N*-acetyl-*p*-aminobenzoquinoneimine (NAPQI), which reacts with GSH. This conjugate undergoes further metabolism via glutathionases (sequential metabolism by γ-glutamyltransferase, cysteinylglycinase and *N*-acetyltransferase) to give the mercapturate (the *N*-acetylcysteine conjugate of paracetamol), which is excreted in urine. However, larger doses (>10 g = 20 tablets, perhaps less in susceptible subjects) may saturate the sulfate and glucuronide pathways and the increased amount of NAPQI produced exceeds the capacity for GSH detoxification. In the absence of antidotal therapy, the intermediate reacts with macromolecules in the hepatocytes and in other ways, leading in some cases to liver failure, which may prove fatal.

15.5.2.5 Amino acid conjugation

Conjugation with amino acids is typified by the formation of salicyluric acid from salicylic acid and glycine (Figure 15.10). Reaction with coenzyme A produces salicyloyl-CoA, the acyl donor for glycine. In man over 80 % of a 250 mg dose of aspirin can be eliminated as salicyluric acid in the urine. However, as this pathway is saturable, the proportion falls with increasing dose (70 % for a 1000 mg dose, 60 % for a 2000 mg dose).

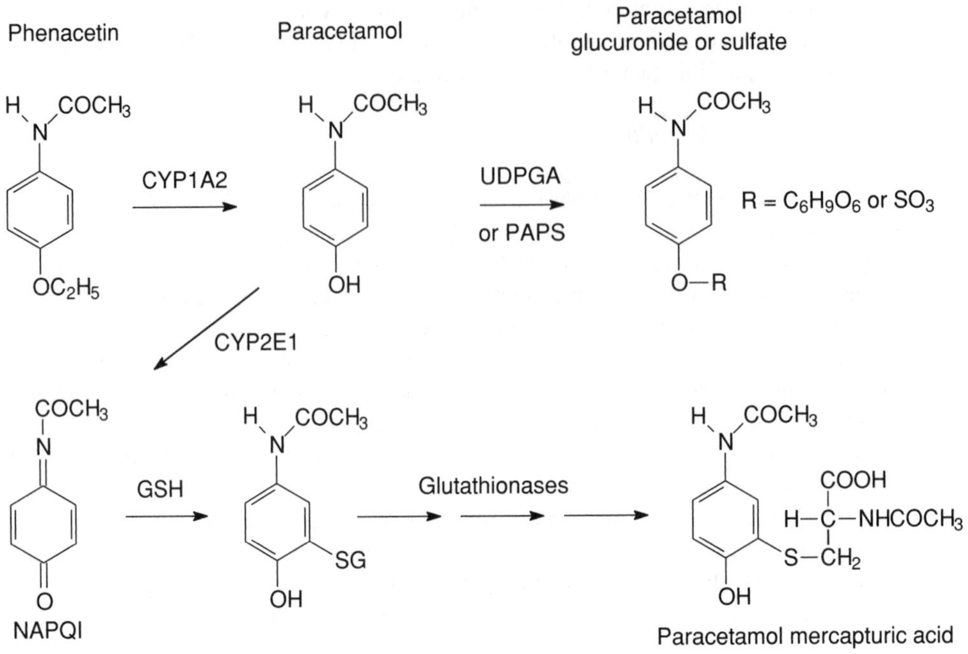

Figure 15.9 Metabolic pathways of phenacetin and paracetamol.

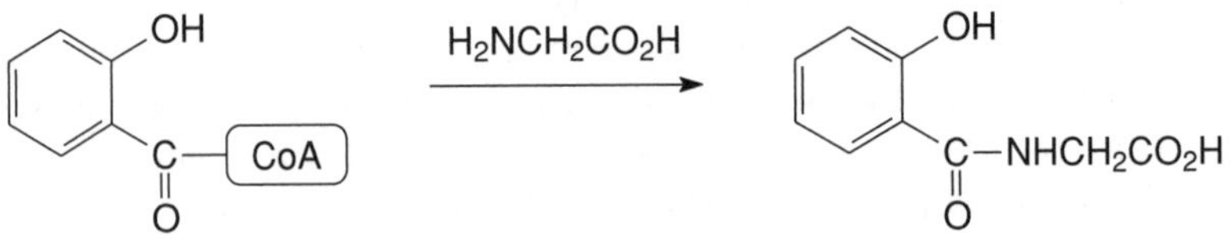

Figure 15.10 Conjugation of salicylic acid with glycine.

15.5.3 Metabolic reactions of analytical or toxicological importance

15.5.3.1 Oxidative dealkylation

N-Dealkylation frequently leads to the formation of pharmacologically active metabolites, which may have different pharmacodynamic or pharmacokinetic profiles to the parent compound. Secondary and tertiary amines are oxidatively dealkylated, the alkyl group being oxidized to an aldehyde. Thus, in addition to the dealkylated product demethylation gives formaldehyde, whilst de-ethylation produces acetaldehyde, both of which are incorporated into intermediary metabolism. Tertiary amines may be di-dealkylated, for example imipramine is demethylated to desipramine and further demethylation gives desmethyldesipramine. Amitriptyline and morphine are *N*-demethylated to nortriptyline and normorphine, respectively. Metabolism of diazepam and temazepam gives nordazepam and oxazepam, respectively (Figure 15.11). In man, the plasma half-life of nordazepam is over twice that of diazepam.

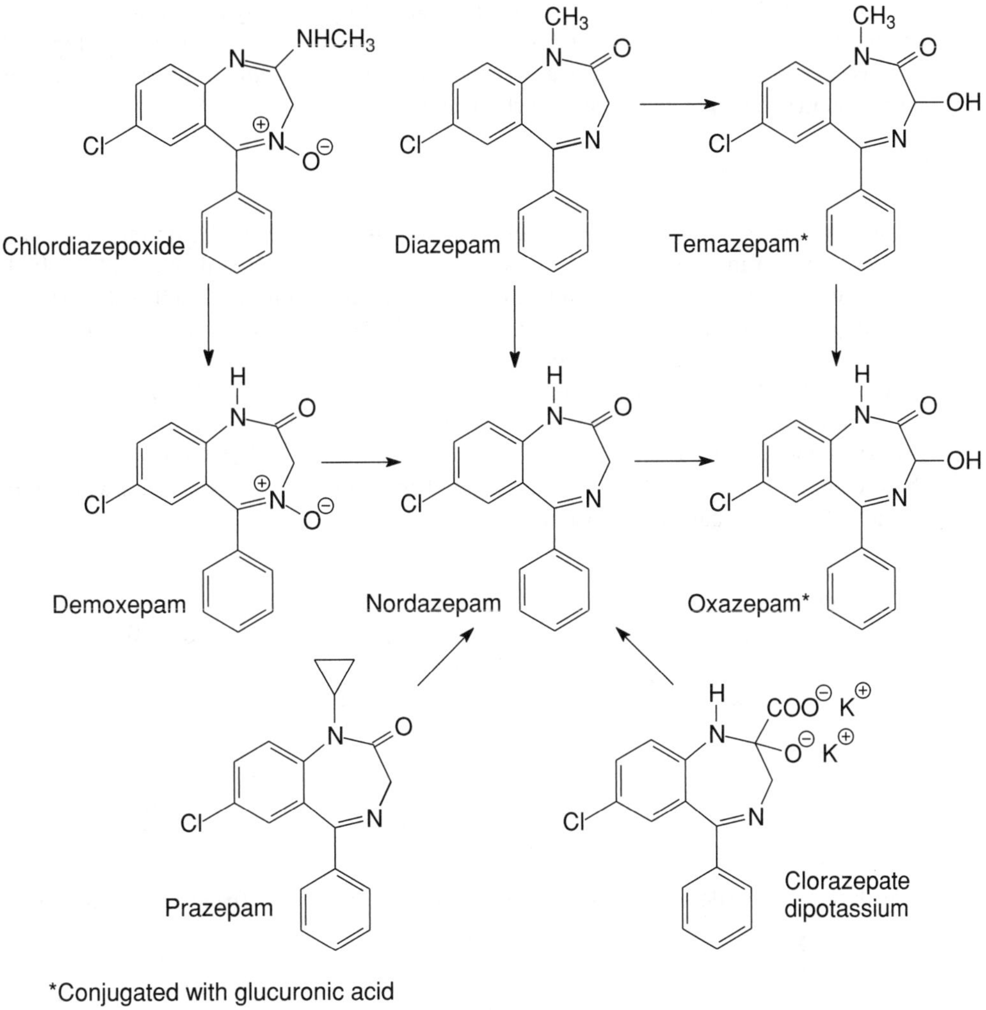

Figure 15.11 Metabolic pathways of some 7-chloro-1,4-benzodiazepines.

O-Dealkylation is exemplified by the demethylation of codeine to morphine (Figure 15.5) and of phenacetin to paracetamol (Figure 15.9). Methylmercaptopurine is *S*-demethylated to mercaptopurine (Figure 15.6).

15.5.3.2 Hydroxylation

Hydroxylation is a common metabolic reaction important in 'functionalizing' many xenobiotics prior to phase 2 conjugation. 2-Substituted phenothiazines such as chlorpromazine are hydroxylated chiefly in the 7-position, whereas in phenothiazines without such a substituent 3-hydroxylated products are formed. Because of the different convention in numbering, tricyclic antidepressants such as imipramine are hydroxylated in the 2-position. 7-Hydroxychlorpromazine has been implicated in the retinal toxicity sometimes found in patients taking chlorpromazine.

Imipramine also undergoes aliphatic hydroxylation to give 10-hydroxyimipramine. Diazepam and nordazepam are 3-hydroxylated to temazepam and oxazepam, respectively (Figure 15.11). Because the 3-carbon is asymmetrically substituted, the glucuronide conjugates are diastereoisomeric and may be separated by achiral chromatography.

15.5.3.3 S- and N-oxidation

Sulfides may be oxidized to sulfoxides and, in some cases, further oxidized to sulfones. Oxidation of the ring sulfur in phenothiazines gives relatively inactive 5-sulfoxide metabolites. In the case of thioridazine, the side-chain 2-thiomethyl sulfur is oxidized to the 2-sulfoxide, mesoridazine, and further oxidized to the 2-sulfone, sulforidazine. Both these compounds are pharmacologically active and have been marketed as drugs.

Asymmetrically substituted sulfides will give chiral sulfoxides. *S*-Oxidation of methionine, for example, gives diastereoisomeric methionine sulfoxides. Thus, 5-sulfoxides of 2-substituted phenothiazines are enantiomeric, and will not be separable by achiral chromatography. However, the situation is different if the molecule has another chiral centre, as is the case with thioridazine (Figure 15.12), and the diastereoisomers may be separable.

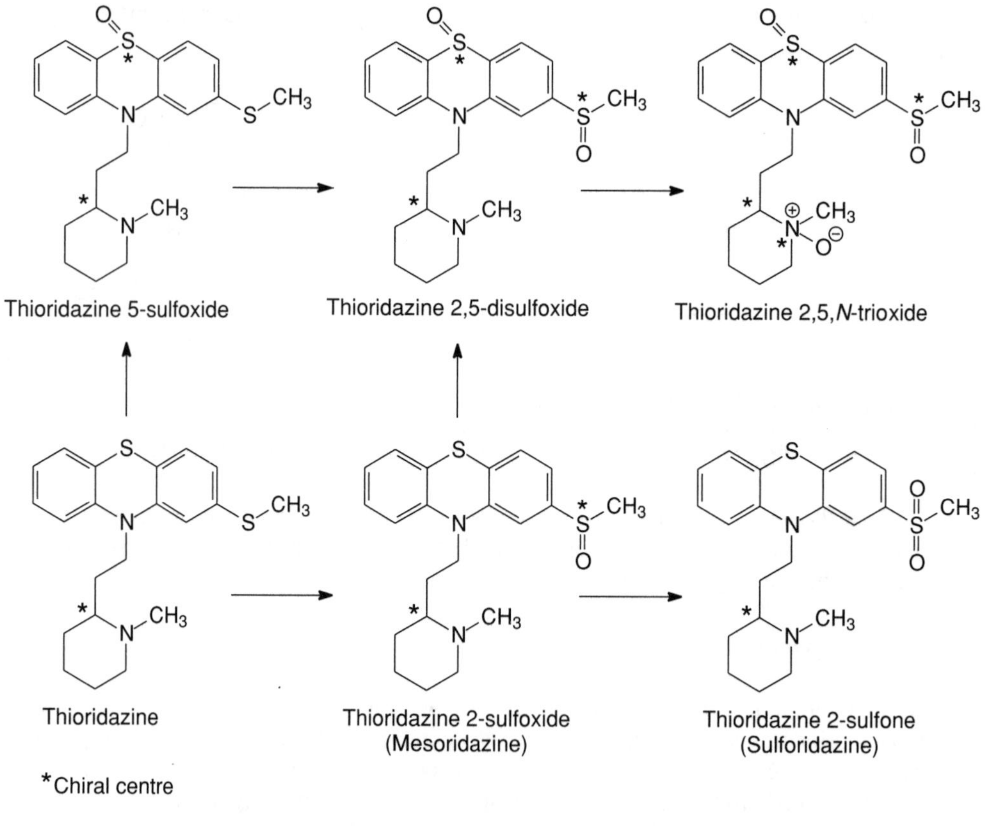

Thioridazine 5-sulfoxide

Thioridazine 2,5-disulfoxide

Thioridazine 2,5,*N*-trioxide

Thioridazine

Thioridazine 2-sulfoxide (Mesoridazine)

Thioridazine 2-sulfone (Sulforidazine)

*Chiral centre

Figure 15.12 Some oxidation pathways of thioridazine.

Most tertiary amines, whether aliphatic or alicyclic, form *N*-oxides. The reaction may be catalyzed by FMO (see above) or, depending on the substrate, by one of the drug metabolizing cytochromes. *N*-Oxidation of clozapine, for example, is catalyzed by CYP3A4. The piperazine side chain of fluphenazine can be oxidized on one or both of the nitrogen atoms.

N-Oxides are labile and relatively easily reduced to the parent amine, for example by antioxidants (ascorbic acid or metabisulfite added to stabilize samples), exposure to alkaline conditions, or simply on storage (Section 2.4). Reduction may be accompanied by dealkylation, which should be borne in mind when samples that might contain *N*-oxides are to be analyzed. On heating, *N*-oxides undergo Cope elimination, for example chlorpromazine *N*-oxide decomposes to an allyl derivative on GC. Asymmetric tertiary amines are chiral, but because of rapid molecular vibrations (inversion) they are not separable. However, *N*-oxidation slows the rate of inversion so that may be possible to separate *N*-oxide enantiomers. Because the thioridazine 5-sulphoxide, thioridazine 2-sulfoxide and the *N*-oxide are all chiral, thioridazine 2,5,*N*-trioxide can be any one of 16 isomers, including eight diastereoisomers that may be separable by achiral chromatography (Figure 15.13). Clearly situations such as this can make metabolite identification difficult.

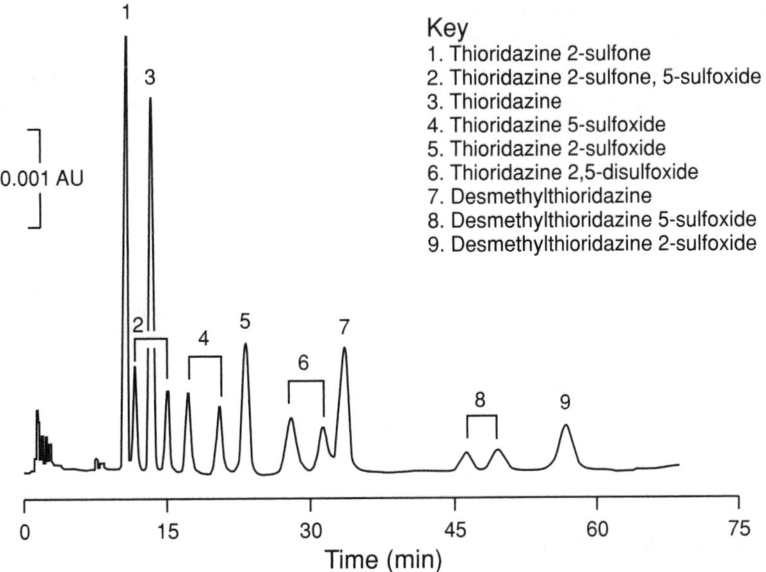

Figure 15.13 HPLC of thioridazine and some of its metabolites. Reproduced from Watkins, G.M., Whelpton, R., Buckley, D.G. and Curry, S.H. (1986) Chromatographic separation of thioridazine sulphoxide and N-oxide diastereoisomers: identification as metabolites in the rat. J Pharm Pharmacol, **38**, 506–9, with permission from RPS Publishing.

15.5.3.4 Oxidative dehalogenation

Halothane undergoes oxidative dehalogenation to trifluoroacetyl chloride and hydrogen bromide. The acid chloride may be hydrolyzed by water to trifluoroacetic acid, or may acylate neighbouring proteins to produce antigens (Figure 15.14). The trifluoroacetylated proteins are found chiefly

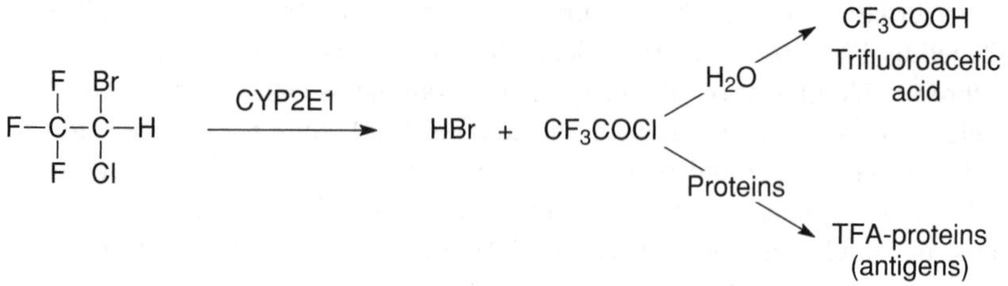

$$F_2ClC\text{-}CHBrF \xrightarrow{CYP2E1} HBr + CF_3COCl$$

$CF_3COCl \xrightarrow{H_2O} CF_3COOH$ (Trifluoroacetic acid)

$CF_3COCl \xrightarrow{Proteins}$ TFA-proteins (antigens)

Figure 15.14 Oxidative debromination of halothane.

in the endoplasmic reticulum, but some are found in the cell membrane, which is why repeat exposure to halothane or to a structurally related compound can produce a potentially fatal immune reaction.

The metabolism of dichloromethane in mammals is dose dependent and proceeds via two major pathways. The predominant pathway at low doses in man involves mainly CYP2E1-dependent oxidation to carbon monoxide and carbon dioxide, probably via an unstable intermediate, formyl chloride. Alternatively, cytosolic GST-dependent metabolism leads to the formation of carbon dioxide, presumably following the formation of a GSH conjugate (Figure 15.15). Carboxyhaemoglobin derived from dichloromethane has a longer half-life in blood than if the carbon monoxide were derived from an exogenous source, presumably reflecting continuing carbon monoxide production *in vivo*. Other dihalomethanes (dibromomethane, bromochloromethane) also give rise to carbon monoxide *in vivo*.

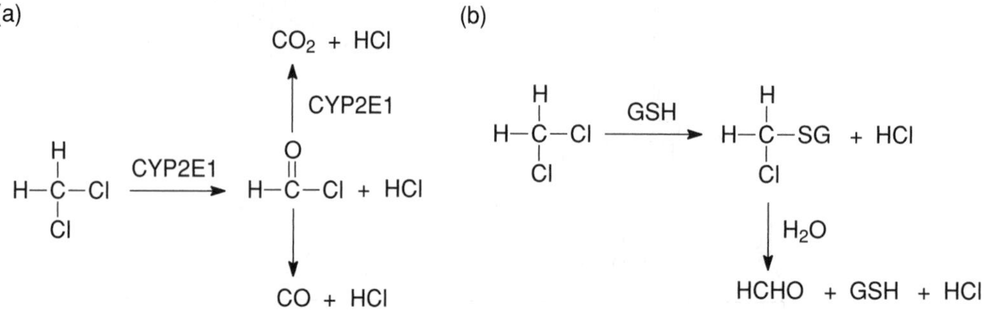

(a)

$$CH_2Cl_2 \xrightarrow{CYP2E1} HC(=O)Cl + HCl$$

$HC(=O)Cl \xrightarrow{CYP2E1} CO_2 + HCl$

$HC(=O)Cl \rightarrow CO + HCl$

(b)

$$CH_2Cl_2 \xrightarrow{GSH} H_2C(Cl)\text{-}SG + HCl$$

$H_2C(Cl)\text{-}SG \xrightarrow{H_2O} HCHO + GSH + HCl$

Figure 15.15 Pathways of metabolism of dichloromethane: (a) microsomal (b) cytosolic.

The CYP pathway has high affinity for dichloromethane, but is saturable at inhaled concentrations of about 1800 mg m^{-3} (500 ppm), whereas the cytosolic pathway has low affinity, but shows no signs of saturation at concentrations up to 36,000 mg m^{-3} (10,000 ppm) in air. Human and rat microsomal liver fractions show similar rates of dichloromethane metabolism via the GST pathway *in vitro*, but mouse preparations have a 10-fold higher activity, which is thought to explain the higher carcinogenic potential of dichloromethane in mice as opposed to man.

As regards aromatic oxidative dechlorination, metabolism of clozapine gives 8-hydroxydechloroclozapine (Dain *et al*., 1997), in which the chlorine atom is replaced by a hydroxyl moiety, but dechlorination of chlorpromazine gives promazine. It is important not to conclude that promazine has been administered to patients taking chlorpromazine simply because promazine has been detected in a biological sample.

15.5.3.5 Desulfuration

The thiobarbiturate thiopental is converted to pentobarbital. Desulfuration of malathion and parathion to organophosphates increases the toxicity of these insecticides. In man, malathion is chiefly hydrolyzed to the mono- and di-acid metabolites, whereas in insects it is converted to malaoxon (Figure 15.16).

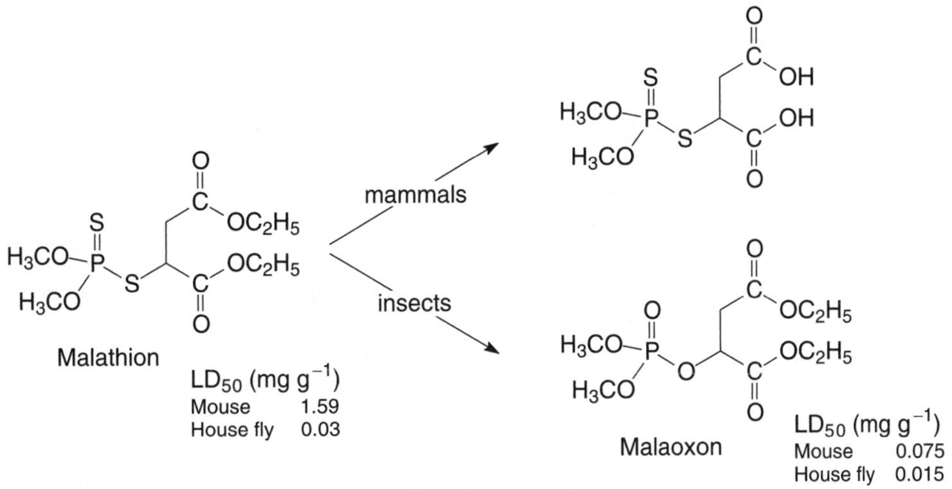

Malathion

LD_{50} (mg g^{-1})	
Mouse	1.59
House fly	0.03

mammals

insects

Malaoxon

LD_{50} (mg g^{-1})	
Mouse	0.075
House fly	0.015

Figure 15.16 Comparative metabolic pathways of malathion in mammals and in insects.

15.5.3.6 Trans-sulfuration and trans-esterification

Oxygen–sulfur exchange is a mechanism by which small quantities of cyanide may be detoxified. The sulfur donor is thiosulfate, and the reaction is catalyzed by the widely-distributed enzyme rhodanese:

$$CN^- + S_2O_3^{2-} \rightarrow CNS^- + SO_4^{2-}$$

This is the rationale for giving thiosulfate to treat cyanide poisoning. The reverse reaction is catalyzed by thiocyanate oxidase, an enzyme found in erythrocytes.

When ethanol is co-ingested with methylphenidate, a small quality of ethylphenidate may be detected (Figure 15.17). This unusual drug interaction occurs with cocaine to give the ethyl homologue, cocaethylene and detection of this metabolite may be used confirm co-ingestion of ethanol (Section 17.6.4.6). Although trans-esterification may occur when esters are dissolved in alcohols, *in vivo* these reactions are thought to be catalyzed by carboxyesterases.

15.6 Excretion

Major excretory organs are the liver, the lungs and the kidneys. The liver is not only responsible for eliminating xenobiotics by metabolism, but also for the excretion of drugs and their metabolites via bile; those drugs and metabolites that are not reabsorbed from the GI tract are excreted in the

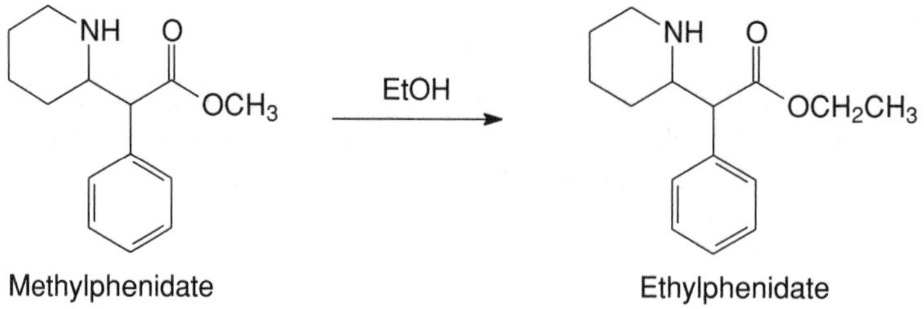

Methylphenidate —EtOH→ Ethylphenidate

Figure 15.17 Trans-esterification of methylphenidate in the presence of ethanol.

faeces. Excretion via fluids other than urine and bile may be insignificant in terms of the amount of drug excreted, but could have important consequences – excretion of drugs via breast milk, for example, may transfer significant quantities of a drug such as lithium to a breast-fed infant (Moretti *et al.*, 2003). The lungs are important for the excretion of gaseous anaesthetics and some other volatile compound and metabolites such as dimethyl selenide (Section 15.5.2.3). The proportion of ethanol excreted by the lungs is small, but this route is exploited when performing breath alcohol measurements. Similarly, amniotic fluid, sweat, saliva, hair and nail, although usually of minor importance in terms of the total amount of drug excreted, are sometimes important samples for detecting exposure to xenobiotics such as drugs of abuse and heavy metals.

15.6.1 The kidney

The functioning unit is the nephron. Plasma is filtered at the glomerulus. The ultrafiltrate is similar in composition to plasma water as the plasma proteins are, for the most part, too large to pass through the basement membrane. As the filtrate passes down the nephron most of the sodium and water, and many other constituents, including many drugs, are reabsorbed. What remains is urine. Drugs may enter tubular fluid in the ultrafiltrate or by active secretion from the cells of the proximal convoluted tubule (PCT).

The more a drug is ionized in tubular fluid the less it will be reabsorbed by passive diffusion. The renal clearances of weak acids and bases may be markedly affected by changes in the pH of urine. Aspirin is hydrolyzed to salicylic acid (pK_a 3.0) and metabolized to other polar metabolites *in vivo*. The pH of urine varies, but 6.3–6.6 is a reasonable estimate under normal physiological conditions. The percentage of salicylic acid ionized at pH 6.3 is 99.90 %, whereas at pH 7.3 the percentage ionized is 99.99 %. This may seem to be a very small proportionate change, but the proportion of nonionized salicylate, the species which diffuses through the tubular cells has decreased some 10-fold (0.1–0.01 %). The pH of urine can be easily raised to over pH 8 by giving sodium bicarbonate i.v. and such treatment may be valuable in treating severe poisoning with acids such as salicylates and chlorophenoxy herbicides. Alkalinization is important because it corrects the metabolic acidosis and reduces the amount in tissues, as well as increasing the renal clearance.

Increasing urine flow by administering large volumes of fluid and possibly a diuretic with the aim of enhancing drug elimination is dangerous as there is the risk of causing electrolyte imbalance and pulmonary oedema due to fluid overload. It is also largely futile because to have any hope of efficacy the drug in question would have had to be largely eliminated via the kidney and it would have had to be possible to affect the proportion of drug that was reabsorbed to an appreciable degree by increasing the diuresis. For example, if 50 % of a drug is normally passively reabsorbed from tubular fluid then the maximum effect that intervention can have is to reduce the reabsorption to zero, that is increase the renal clearance by a factor of 2. In the case of salicyclate and other strong acids, simple adjustment of urinary pH is sufficient to increase renal excretion without the need to enhance the diuresis.

Weak bases (pK_a 6–12) such as amfetamine are more readily excreted in acidic urine. Making the urine acidic, for example with ammonium chloride, increases the ionization of such bases and hence increases their renal clearance. However, this is of no relevance in treating acute poisoning with these compounds as they are largely eliminated by metabolism.

15.6.1.1 Tubular secretion

The transporters of the PCT may be in the basolateral or apical membranes. OATPs and OCTs are found in the basolateral membrane. Drugs and metabolites that are actively secreted by these transport systems include penicillins, most diuretics, and glucuronide and glycine metabolites. Probenecid competes for the carrier for organic acids and blocks the secretion of penicillins. It was used to increase the duration of action of some penicillins because, without it, penicillins have short elimination half-lives (0.5–1 h). Probenecid is not now used for this purpose as slow-release microcrystalline salts (e.g. procaine penicillin G, benzathine penicillin G) are available for i.m. injection. Probenecid also blocks uric acid reabsorption and may be used to treat gout.

The basolateral uptake of paraquat into proximal renal epithelial cells can be blocked by quinine, but not by putresecine, suggesting that the diamine transport is not involved. A second active-transport system, possibly PGP, appears to be responsible for transfer across the luminal membrane of the cell (Chan *et al.*, 1996).

15.6.1.2 Excretion of metabolites

Polar molecules such as pyridostigmine that are filtered will not be reabsorbed. Lipophilic drugs tend to be protein bound and any filtered drug is reabsorbed, hence in general such compounds must be metabolized before elimination. More polar metabolites are not reabsorbed and therefore these, rather than the parent compound, are found in the urine. For example, the urine of a patient taking diazepam may contain only small amounts of the parent drug, but oxazepam is found in larger amounts and oxazepam glucuronide in even larger amounts. Hence detection of metabolites in urine (e.g. benzoylecgonine from cocaine) is often a useful indicator of exposure to the parent compound.

15.6.2 Biliary excretion

Drugs may enter bile by passive diffusion, but there are thought to be at least three active-transport mechanisms, including cationic and anionic transporters, and PGP. There appears to be a size cut-off, with substrates having to have M_r values >325. Polar, acidic molecules such as glucuronides are usually good substrates for active secretion. Some compounds do not need to be conjugated to be actively transported, for example the cardiac glycoside ouabain, $M_r = 728.8$, can have a bile:plasma ratio as high as 500.

15.6.2.1 Enterohepatic recirculation

Drugs such as phenolphthalein that form glucuronides may be subject to enterohepatic recirculation (Figure 15.18). Absorbed drug is conjugated with glucuronic acid and the glucuronide is transported actively into bile. Once in the intestine, β-glucuronidases from the gut flora hydrolyze

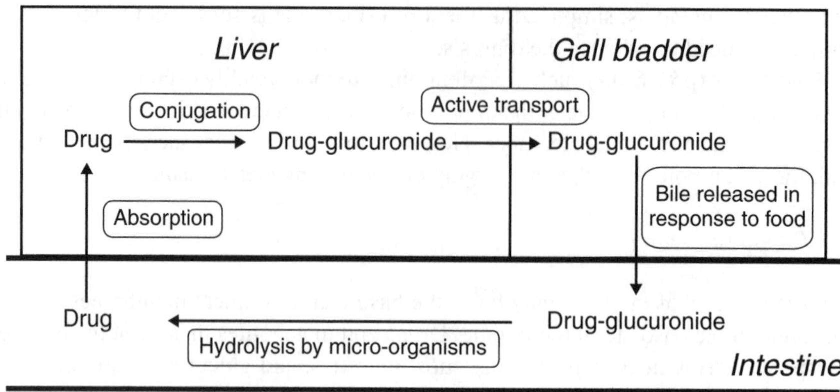

Figure 15.18 Schematic representation of enterohepatic recirculation.

the glucuronide and the parent drug is reabsorbed. Because bile is released in response to (fatty) food, increases in plasma drug concentrations may be observed after meals.

15.7 Summary

An understanding of the factors affecting the disposition and fate of xenobiotics after different routes of administration is important to understanding clinical effect. Although metabolic and excretory processes play a vital role in limiting the duration of drug action and in removing drugs and other potentially harmful substances from the body, biotransformation in some cases produces metabolites that are considerably more toxic and/or have a longer duration of action in the body than the parent substance. Knowledge of xenobiotic disposition and metabolism informs not only the choice of body fluid or other sample for analysis, but also the target analyte and the interpretation of the results obtained. The effects of time of sampling in relation to exposure, and of disease, previous exposure, age, genetics and other factors on the interpretation of analytical results are discussed in the next two chapters.

References

Amici, A., Emanuelli, M., Ferretti, E., Raffaelli, N., Ruggieri, S. and Magni, G. (1994) Homogeneous pyrimidine nucleotidase from human erythrocytes: enzymic and molecular properties. *Biochem J*, **304**, 987–92.

Bagwell, E.E., Walle, T., Drayer, D.E., Reidenbert, M.M. and Pruett, J.K. (1976) Correlation of the electrophysiological and antiarrhythmic properties of the *N*-acetyl metabolite of procainamide with plasma and tissue drug concentrations in the dog. *J Pharmacol Exp Ther*, **197**, 38–48.

Barnes, M.M., James, S.P. and Wood, P.B. (1959) The formation of mercapturic acids. 1. Formation of mercapturic acid and the levels of glutathione in tissues. *Biochem J*, **71**, 680–90.

Bennett, P.N., Davies, D.S. and Hawkesworth, G.M. (1976) *In vivo* absorption studies with paraquat and diquat in the dog. *Br J Pharmacol*, **58**, 284P.

Caldwell, J. (2006) Drug metabolism and pharmacogenetics: the British contribution to fields of international significance. *Br J Pharmacol*, **147** (Suppl 1), S89–99.

Chan, B.S., Lazzaro, V.A., Seale, J.P. and Duggin, G.G. (1996) Transport of paraquat in a renal epithelial cell line LLC-PK1. *J Pharmacol Exp Ther*, **279**, 625–32.

Chow, H.-H.S., Chen, Z. and Matsuura, G.T. (1999) Direct transport of cocaine from the nasal cavity to the brain following intranasal cocaine administration in rats. *J Pharm Sci*, **88**, 754–8.

Colebrooke, L. and Kenny, M. (1936) Treatment of human puerperal infections, and of experimental infections in mice, with prontosil. *Lancet*, **i**, 1276–86.

Dain, J.G., Nicoletti, J. and Ballard, F. (1997) Biotransformation of clozapine in humans. *Drug Metab Dispos*, **25**, 603–9.

Dinis-Oliveira, R.J., Duarte, J.A., Remiao, F., Sanchez-Navarro, A., Bastos, M.L. and Carvalho, F. (2006) Single high dose dexamethasone treatment decreases the pathological score and increases the survival rate of paraquat-intoxicated rats. *Toxicology*, **227**, 73–85.

DuBuske, L.M. (2005) The role of P-glycoprotein and organic anion-transporting polypeptides in drug interactions. *Drug Saf*, **28**, 789–801.

Fromm, M.F. (2004) Importance of P-glycoprotein at blood-tissue barriers. *Trends Pharmacol Sci*, **25**, 423–9.

Leonard, R.F., Knott, P.J., Rankin, G.O., Robinson, D.S. and Melnick, D.E. (1981) Phenytoin-salicylate interaction. *Clin Pharmacol Ther*, **29**, 56–60.

Mason, H.S. (1957) Mechanisms of oxygen metabolism. *Adv Enzymol Relat Subj Biochem*, **19**, 79–233.

Moretti, M.E., Koren, G., Verjee, Z. and Ito, S. (2003) Monitoring lithium in breast milk: an individualized approach for breast-feeding mothers. *Ther Drug Monit*, **25**, 364–6.

Mygind, N. and Dahl, R. (1998) Anatomy, physiology and function of the nasal cavities in health and disease. *Adv Drug Deliv Rev*, **29**, 3–12.

Nagao, M., Saitoh, H., Zhang, W.D., Iseki, K., Yamada, Y., Takatori, T. and Miyazaki, K. (1993) Transport characteristics of paraquat across rat intestinal brush-border membrane. *Arch Toxicol*, **67**, 262–7.

Nebert, D.W., Adesnik, M., Coon, M.J., Estabrook, R.W., Gonzalez, F.J., Guengerich, F.P., Gunsalus, I.C., Johnson, E.F., Kemper, B., Levin, W., Phillips, I.R., Sato, R. and Waterman, M.R. (1987) The P450 gene superfamily: recommended nomenclature. *DNA*, **6**, 1–11.

Osborne, R.J., Joel, S.P. and Slevin, M.L. (1986) Morphine intoxication in renal failure: the role of morphine-6-glucuronide. *Br Med J*, **292**, 1548–9.

Schanker, L.S. and Tocco, D.J. (1960) Active transport of some pyrimidines across the rat intestinal epithelium. *J Pharmacol Exp Ther*, **128**, 115–21.

Shimizu, K., Ohtaki, K., Matsubara, K., Aoyama, K., Uezono, T., Saito, O., Suno, M., Ogawa, K., Hayase, N., Kimura, K. and Shiono, H. (2001) Carrier-mediated processes in blood–brain barrier penetration and neural uptake of paraquat. *Brain Res*, **906**, 135–42.

Shon, J.H., Yoon, Y.R., Hong, W.S., Nguyen, P.M., Lee, S.S., Choi, Y.G., Cha, I.J. and Shin, J.G. (2005) Effect of itraconazole on the pharmacokinetics and pharmacodynamics of fexofenadine in relation to the MDR1 genetic polymorphism. *Clin Pharmacol Ther*, **78**, 191–201.

Smith, L.L. (1988) The toxicity of paraquat. *Adverse Drug React Acute Poisoning Rev*, **7**, 1–17.

Smith, L.L., Lewis, C.P., Wyatt, I. and Cohen, G.M. (1990) The importance of epithelial uptake systems in lung toxicity. *Environ Health Perspect*, **85**, 25–30.

Soldin, S.J. (1999) Free drug measurements. When and why? An overview. *Arch Pathol Lab Med*, **123**, 822–3.

Türker, S., Onur, E. and Özer, Y. (2004) Nasal route and drug delivery systems. *Pharm World Sci*, **26**, 137–42.

16 Pharmacokinetics

16.1 Introduction

Pharmacokinetics (sometimes abbreviated to PK) is the study of the rates of the processes involved in the absorption, distribution, metabolism and elimination of drugs and other agents. By subjecting the results of observations such as the change in plasma concentration of a drug as a function of time to mathematical analysis, pharmacokinetic parameters such as plasma half-life ($t_{0.5}$) and apparent volume of distribution (V) can be calculated. Having derived appropriate pharmacokinetic parameters for a given drug it may then be possible to predict future dose requirements, the effects of changing the dose or the frequency of dosing on plasma drug concentrations, and also the effects of changes in metabolism or the co-administration of other drugs on these parameters. In forensic toxicology, knowledge of pharmacokinetic parameters may allow the time or size of a dose to be estimated. However, there are often several caveats that must be applied to pharmacokinetic modelling in such circumstances. This chapter provides most of what is likely to be required by way of pharmacokinetic calculations. Additional material can be found in standard texts (Gibaldi and Perrier, 1982; Rowland and Tozer, 1995).

Deriving mathematical equations to describe the changes in the concentration of a drug or other xenobiotic that occur with time in different parts of the body (the *time course* of the drug in the body) is known as mathematical modelling. The commonest form of modelling is to treat the body as if were one or more volumes or *compartments*. When a drug enters a compartment it is assumed that it is distributed instantly and uniformly throughout the compartment. In the *single-compartment* model, the body is treated as if it were one homogeneous solution of the drug. The equations used to describe the time course of a drug are relatively simple and many fundamental concepts of pharmacokinetics can be understood using a single-compartment model (Sections 16.3–6). However, it is often necessary to use more complex models (the *two-compartment* and *three-compartment* models – Section 16.7). The available data are rarely good enough to justify using more than three compartments. Finally, the derivation of basic pharmacokinetic parameters without having defined a model is discussed in Section 16.8.

16.1.1 Historical development

Friedrich Hartmut Dost (1910–1985) introduced the term 'pharmacokinetics' (Dost, 1953), but before this date groups were clearly conducting experiments and performing calculations that would now be referred to as pharmacokinetics. The work of Henry Bence Jones on the absorption and distribution of lithium and of quinine in the mid 1860s was cited in Section 4.1.1. Erik M.P. Widmark (1889–1945) pioneered the study of ethanol pharmacokinetics, and was the first to describe a one-compartment model (Widmark and Tandberg, 1924). Teorell (1937a, 1937b) introduced the concept of the two-compartment model. Prior to that, Gold and DeGraff (1929) had shown that urinary concentration data could be used to define pharmacokinetics when plasma measurements were not available.

16.1.2 Symbols and conventions

By convention, italics are used for variables such as time, t, or concentration, C. When labels are constants, for example to denote a concentration at $t = 0$, C_0, they are not italicized. The concentration at any time, t, should be written C_t and some authors add a further label to denote the medium to which the concentration refers, for example the concentration in plasma at time t might be written:

$$C_{\mathrm{p}_t}$$

However, in order to make subsequent equations easier to read concentrations at time t are written as C.

Increasingly, λ is used to denote rate constants, with suitable labels being added as appropriate. Thus, a general equation to describe the concentration of drug being eliminated from a multicompartment model can be written:

$$C = \sum_{i=1}^{n} A_i \exp(-\lambda_i t) \tag{16.1}$$

where n is the number of compartments. Again as this notation is difficult to read, the older notation in which A_1, A_2, λ_1 and λ_2 are represented by A, B, α and β, respectively, is used here. The terms α- and β-phase, to represent the steeper and shallower parts of the plasma concentration–time curve in a two-compartment model, respectively, are moreover widely used. The rate constant of the terminal elimination phase is here denoted by λ_z and k is used for other rate constants. Finally, half-life is denoted by $t_{0.5}$ rather than $t_{1/2}$.

16.2 Fundamental concepts

The equations used to describe the time course of substances in the body are mainly those used in chemical kinetics. At therapeutic concentrations most compounds exhibit first-order elimination, although the elimination of some analytes, notably high concentrations of ethanol, can be described using zero-order equations (Section 16.2.1.2). The kinetics of other analytes, phenytoin, for example, can only be adequately described using the Michaelis–Menten (M–M) equation (Section 16.6). Many drugs that exhibit first-order elimination kinetics at therapeutic doses require the use of M–M kinetics to describe their time course following overdosage. The plasma half-life is a convenient and easily understood way of describing the kinetics of a substance, but it is important to realise that plasma half-life is controlled by clearance and the apparent volume of distribution (Section 16.2.2).

16.2.1 Rates, rate constants and reaction order

It is important to distinguish between rate and rate constant. A general equation relating rate ($-\mathrm{d}C/\mathrm{d}t$), rate constant (k) and concentration (C) is:

$$-\frac{\mathrm{d}C}{\mathrm{d}t} = kC^n \tag{16.2}$$

where n is known as the order of the reaction. In chemical kinetics n would be measured experimentally and is often close to an integer, 0 or 1, and so reactions are referred to as zero-or first-order, respectively.

16.2.1.1 *First-order elimination*

For a first-order reaction, substituting $n = 1$ in Equation (16.2) gives:

$$-\frac{dC}{dt} = kC \tag{16.3}$$

hence the rate of the reaction is directly proportional to the concentration (amount) of substance present. As the reaction proceeds and the concentration of substance falls, the rate of the reaction decreases. This is exponential decay, analogous to radioactive decay where the probability of disintegration is proportional to the number of unstable nuclei present. The first-order rate constant has units of reciprocal time (e.g. h^{-1}). Integrating Equation (16.3) gives:

$$C = C_0 \exp(-kt) \tag{16.4}$$

which is the equation of a curve that asymptotes to zero from the initial concentration, C_0 [Figure 16.1(a)]. Taking natural logarithms of Equation (16.4):

$$\ln C = \ln C_0 - kt \tag{16.5}$$

gives the equation of a straight line of slope, $-k$ [Figure 16.1(b)]. If common logarithms are used (log C versus t) the slope is $-k/2.303$. Another way of presenting the data is to plot C on a logarithmic scale. This approach was often used when computers were not readily available, but a common mistake was to label the slope $-k$. The slope is the same as that of the exponential plot and decreases with time, but the half-life can be read easily from the semilogarithmic plot [Figure 16.1(c)]. The half-life ($t_{0.5}$) is the time for the initial concentration (C_0) to fall to $C_0/2$,

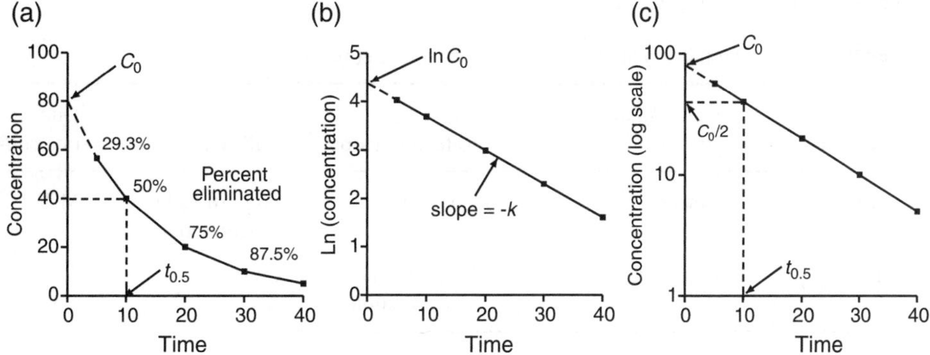

Figure 16.1 First-order elimination curves: (a) C versus t, (b) ln C versus t and (c) C versus t using a semilogarithmic scale.

and substitution in Equation (16.5) gives:

$$t_{0.5} = \frac{\ln 2}{k} = \frac{0.693}{k} \tag{16.6}$$

as ln 2 = 0.693. This important relationship where $t_{0.5}$ is constant (independent of the initial concentration) and inversely proportional to k, is unique to first-order reactions. Because $t_{0.5}$ is constant, 50 % of the compound is eliminated in $1 \times t_{0.5}$, 75 % in $2 \times t_{0.5}$ and so on. Thus, when 5 half-lives have elapsed less than 5 % of the analyte remains, and after 7 half-lives, less than 1 % remains.

16.2.1.2 *Zero-order elimination*

For a zero-order reaction, $n = 0$, and:

$$-\frac{\mathrm{d}C}{\mathrm{d}t} = kC^0 = k \tag{16.7}$$

Thus, a zero-order reaction proceeds at a constant rate, and the zero-order rate constant has units of rate (e.g. g L^{-1} h^{-1}). Integrating Equation (16.7):

$$C = C_0 - kt \tag{16.8}$$

gives the equation of a straight line of slope, $-k$, when concentration is plotted against time. The half-life can be obtained as before, substituting $t = t_{0.5}$ and $C = C_0$ gives:

$$t_{0.5} = \frac{C_0}{2k} \tag{16.9}$$

Thus, the zero-order half-life is inversely proportional to k, but $t_{0.5}$ is also directly proportional to the initial concentration (Table 16.1). In other words, the greater the amount of drug present initially, the longer the time taken to reduce the amount present by 50 %. The term 'dose-dependent half-life' has been applied to this situation.

Table 16.1 Comparison of zero-order and first-order elimination

Reaction order	Concentration versus time plot	Rate of reaction	Half-life	Dimensions of rate constant
Zero	Linear	Constant	Proportional to concentration	M T^{-1}
First	Exponential	Proportional to concentration	Constant	T^{-1}

16.2.2 *Dependence of half-life on volume of distribution and clearance*

The elimination half-life is dependent on two fundamental parameters: apparent volume of distribution (V) and plasma clearance (Cl). Changes in $t_{0.5}$ may be a result of changes in one or both of these parameters. Increasing the apparent volume of distribution increases $t_{0.5}$, while increasing Cl decreases $t_{0.5}$.

16.2.2.1 *Apparent volume of distribution*

The apparent volume of distribution (V) is defined as the volume of fluid that the amount (X) of a substance in the body would have to be dissolved in to give the same concentration as the plasma concentration (C) of the substance at the time in question:

$$V = \frac{X}{C} \tag{16.10}$$

Apparent volume of distribution has also been described as a 'constant of proportionality' that allows one to calculate the amount of drug in the body from the plasma concentration by re-arrangement of Equation (16.10). It is normally measured in an experiment in which a dose (X_0) of drug is injected i.v. and timed blood samples are taken. C_0 can be obtained from extrapolation to $t = 0$ (Figure 16.1):

$$V = \frac{X_0}{C_0} \tag{16.11}$$

Suitable markers, such as the dye Evans' Blue, which binds so avidly to plasma albumin it is restricted to plasma, inulin, which cannot penetrate cells, and isotopically labelled water, can be used to measure anatomical volumes (Table 16.2). Some substances are confined to these volumes, but many have values of apparent volume of distribution much larger than total body water because they are extensively distributed in tissues.

Table 16.2 Some examples of apparent volumes of distribution

Substance	V (L kg^{-1} body weight)	V in 70 kg subject (L)
Evans' Blue[a]	*0.05*	*3.5*
Heparin	0.06	4.2
Amoxicillin	0.2	14
(+)-Tubocurarine	0.2	14
Inulin[b]	*0.21*	*14.5*
Phenazone	0.6	42
Ethanol	0.65	45
Deuterium oxide (^2H_2O)[c]	*0.55–0.7*	*38–50*
Digoxin	5	350
Chlorpromazine	20	1400
Quinacrine	500	35 000

Anatomical volumes: [a]Plasma [b]Extracellular fluid [c]Total body water

16.2.2.2 *Organ clearance*

Clearance has long been used by physiologists to quantify the ability of an organ to extract a substance from the plasma flowing through it. Renal clearance can be calculated using the formula:

$$Cl_{\text{ren}} = \frac{U}{P} \cdot \textit{urine flow rate} \tag{16.12}$$

where U and P are the concentrations of substance in urine and plasma, respectively. The urine flow rate is usually measured in mL min^{-1}.

In pharmacokinetic terms, organ clearance is the volume of plasma from which a drug is completely removed per unit time as it flows through that organ. Hence clearance has units of flow (usually mL min^{-1} or L h^{-1}) and is *not* synonymous with the term 'elimination'. As an example, if the concentration of drug entering the liver is C_{in} and that leaving via the hepatic vein is C_{out}, the proportion of drug that has been removed is the extraction ratio, E:

$$E = \frac{C_{\text{in}} - C_{\text{out}}}{C_{\text{in}}} \tag{16.13}$$

The organ clearance, Cl_{org}, is the extraction ratio multiplied by the plasma flow-rate, Q:

$$Cl_{\text{org}} = EQ \tag{16.14}$$

Thus, if the hepatic plasma flow was 800 mL min^{-1} and $E = 0.7$ (i.e. 70 % of the drug is removed from the plasma as it flows through the liver) the hepatic clearance would be $800 \times 0.7 = 560$ mL min^{-1}.

16.2.2.3 Whole body clearance

Whole body (or plasma) clearance (*Cl*) is the sum of the individual organ clearances. Thus, for a drug that is eliminated via the liver and the kidneys:

$$Cl = Cl_{\text{hep}} + Cl_{\text{ren}} \tag{16.15}$$

Whole body clearance can be calculated from plasma concentration–time data even though it may not be possible to define all the individual organ clearances that contribute. Thus, in Figure 16.2 the oval represents all the organs eliminating the particular substance and hence the flow is the total plasma flow, Q, to those organs. The amount of drug, X, will be the plasma concentration, C, multiplied by the apparent volume of distribution. Some drug is removed by the organs and the plasma returns to the systemic circulation. Elimination is first-order, and so the rate of elimination

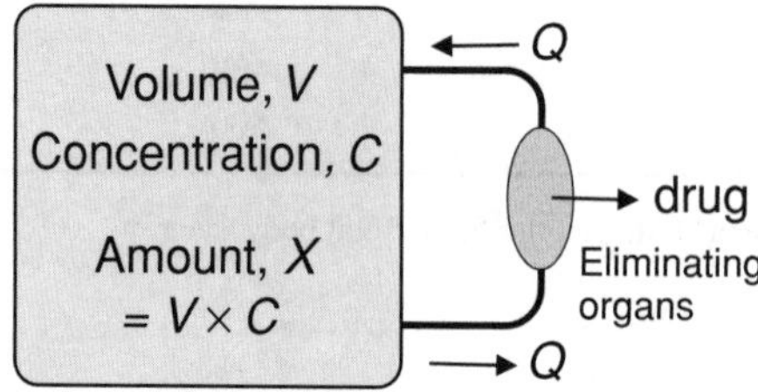

Figure 16.2 Representation of whole body clearance.

of drug is:

$$-\frac{\mathrm{d}X}{\mathrm{d}t} = kX = kVC \tag{16.16}$$

However, the rate of elimination can be written in terms of clearance:

$$-\frac{\mathrm{d}X}{\mathrm{d}t} = C \cdot Cl \tag{16.17}$$

Thus, from Equations (16.16) and (16.17), combining and rearranging gives:

$$Cl = V \cdot k \tag{16.18}$$

Experimentally, k can be obtained from the slope of a plot of $\ln C$ versus t (Figure 16.1) and V from Equation (16.11) and so Cl can be calculated from Equation (16.18). Furthermore, $k = 0.693/t_{0.5}$, so substituting for k in Equation (16.18) gives:

$$t_{0.5} = \frac{0.693 \cdot V}{Cl} \tag{16.19}$$

Clearance is a measure of how well the eliminating organs can metabolize or excrete a substance. Enzyme induction (Section 17.4) may increase Cl, as may manipulation of urine pH to increase the excretion of a weak electrolyte. Enzyme inhibition may reduce Cl. Liver or kidney disease may reduce Cl, but there may be accompanying changes in V, so predicting the effect on $t_{0.5}$ is not straightforward.

Generally lipophilic drugs acting in the CNS have relatively large values of V and correspondingly long plasma half-lives. Chlorpromazine is metabolized extensively by the liver ($Cl_{\mathrm{hep}} = 600$ mL min^{-1}), but with $V = 20$ L kg^{-1} has a plasma half-life of approximately 27 h. Penicillin G on the other hand, with a renal clearance of 320 mL min^{-1}, has a plasma half-life of only 0.5 h because V is only 0.2 L kg^{-1}.

16.3 Absorption and elimination

16.3.1 First-order absorption

Other than following i.v. or intra-arterial (i.a.) injection, administered drug has to be absorbed, and so the plasma concentration–time curve must have a rising phase. The kinetics of absorption after i.m. injection might be expected to be first-order, in other words the greater the amount of drug at the injection site, the faster the rate of absorption. Absorption from the GI tract may be more complex, but frequently first-order absorption is a reasonable approximation. The equation for the plasma concentration as function of time in a single compartment model with simultaneous first-order input and output is:

$$C = F \cdot \frac{Dose}{V} \cdot \frac{k_a}{k_a - k}[\exp(-kt) - \exp(-k_a t)] \tag{16.20}$$

where k_a is the first-order rate constant of absorption and F is the fraction of the dose that reaches the systemic circulation. The concentration is maximal (C_{max}) when the rate of absorption equals the rate of elimination, after which elimination dominates.

When $k_a > k$ (as is often the case, or at least assumed to be the case) the term $\exp(-k_a t)$ approaches zero as t increases faster than $\exp(-kt)$ approaches zero. Consequently, the equation approximates to a single exponential at later times, as can be seen if $\ln C$ is plotted against t (Figure 16.3). Such plots can be solved graphically by drawing a line through the terminal points to estimate k from the slope of the line and the intercept (Y). This allows the contribution of the term $Y.\exp(-kt)$ to be calculated at earlier time points. Subtraction of these values from the experimental points gives estimates of $Y.\exp(-k_a t)$ for the corresponding values of t. A plot of the logarithm of these should give a straight line of slope $-k_a$. This graphical method of deriving the parameters, known as the method of residuals, has been largely superseded by iterative least squares fitting of the data to the model (curved line).

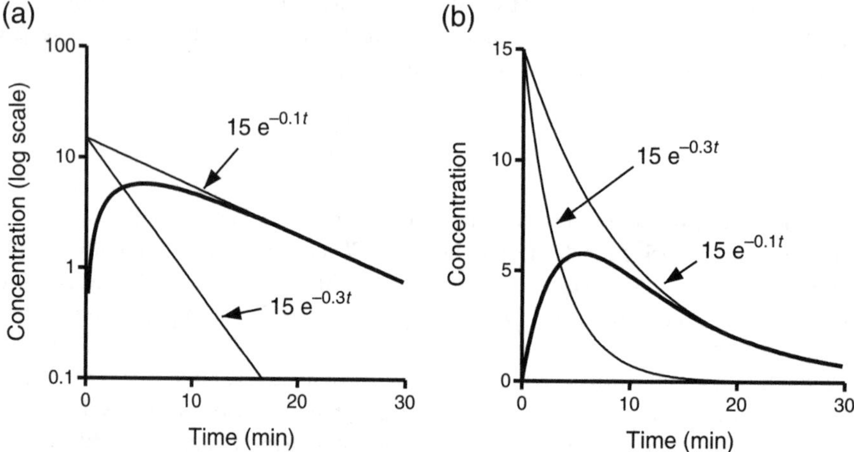

Figure 16.3 Concentration–time curves showing first-order input into a single compartment model with (a) logarithmic y-axis and (b) linear y-axis. Model based on Equation (16.20) with y-intercept of 15, $k_a = 0.3$ and $k = 0.1$.

16.3.2 Bioavailability

In Equation (16.20) the term F is included because in many cases not all the dose administered reaches the systemic circulation. F is sometimes confused with bioavailability. The FDA definition of bioavailability is: 'The rate and extent to which the therapeutic moiety is absorbed and becomes available to the site of drug action'. For some xenobiotics, bioavailability is complex, particularly for prodrugs, and so F is calculated instead, by dividing the area under the plasma concentration–time curve (AUC) after a test dose given, for example, by mouth by the AUC obtained after giving an equal sized i.v. dose:

$$F = \frac{AUC_{po}}{AUC_{iv}} \tag{16.21}$$

It is very important to consider the effect of F when estimating expected plasma concentrations. Even if a literature value of F is known, the extent of absorption may be altered in overdose or by the presence of other drugs. Measurement of AUC using the trapezoidal rule is explained below (Section 16.8).

16.3.3 Maximum concentration

$C_{\max}$ is sometimes taken as the maximum concentration in the data set. However, it can be calculated. For a single compartment model, the time the maximum concentration occurs, $t_{\max}$:

$$t_{\max} = \frac{1}{k_a - k} \ln \frac{k_a}{k} \tag{16.22}$$

and

$$C_{\max} = \frac{F \cdot Dose}{V} \exp(-k \cdot t_{\max}) \tag{16.23}$$

Note that $t_{\max}$ is dose independent, but $C_{\max}$ is directly proportional to the dose; this is an important feature of first-order pharmacokinetics.

16.4 Drug accumulation

A single dose of aspirin may alleviate a headache, but treatment of rheumatoid arthritis requires continual dosage. Because subsequent doses of drug are often given before all of a previous dose has been eliminated, the amount of drug in the body will increase, but provided the drug is eliminated according to first-order kinetics, will not increase indefinitely. This is most easily understood by considering a constant rate infusion.

16.4.1 Intravenous infusion

When a drug is infused at a constant rate, k_0, the plasma concentration will increase as the infusion progresses, but as the plasma concentration increases, the rate of elimination also increases [Equation (16.16)] until the rate of elimination equals the infusion rate. When this steady-state is reached the plasma concentration will be constant, C_{ss}:

$$k_0 = X_{ss} \cdot k = V C_{ss} k \tag{16.24}$$

This must always be the case whilst the elimination kinetics are first-order. The concentration during the rising phase is given by:

$$C = C_{ss}[1 - \exp(-kt)] \tag{16.25}$$

Equation (16.25) represents an exponential curve, which starts at zero and asymptotes to C_{ss}. It is, in essence, a decay curve that has been flipped over. Thus, as the decay curve goes from C_0 to $C_0/2$ in $1 \times t_{0.5}$ the infusion curve goes from 0 to $C_{ss}/2$ in $1 \times t_{0.5}$, that is 50 %

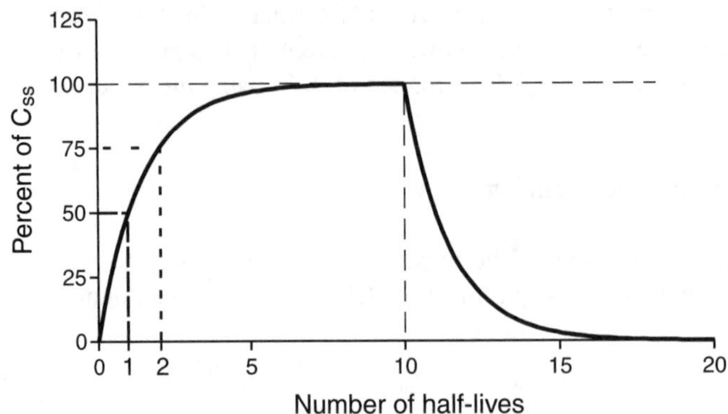

Figure 16.4 Constant rate infusion into a single compartment model. The infusion was stopped after 10 plasma half-lives.

of C_{ss} in one half-life, 75 % in 2 and 87.5 % of the steady-state value in three half-lives. This is most easily seen from a plot of time (in half-lives) versus % of steady-state concentration (Figure 16.4). Because the rate of attainment of steady-state conditions is a function of plasma $t_{0.5}$, a drug with short a half-life reaches steady-state before a drug with a longer half-life. The plasma concentration will >99 % C_{ss} within seven elimination half-lives.

16.4.1.1 *Loading doses*

To reduce the time required to reach steady-state conditions, a loading dose (LD) may be given. If the dose is chosen to give C_{ss} instantaneously (by injecting a bolus LD) at the same time as the infusion is started, the rate of elimination will equal the infusion rate at $t = 0$. The amount of drug (LD) that needs to be injected to give instantaneous C_{ss} is:

$$\mathrm{LD} = V \cdot C_{ss} \tag{16.26}$$

but by rearranging Equation (16.24):

$$V \cdot C_{ss} = \frac{k_0}{k} \tag{16.27}$$

hence:

$$C_{ss} = k_0/Cl \tag{16.28}$$

In summary, C_{ss} is directly proportional to the dose (infusion rate) and is independent of the size of LD.

16.4.2 *Multiple dosage*

A drug given as equal sized doses at equal intervals will produce a plasma concentration-time plot similar to one of those illustrated in Figure 16.5, depending on the plasma $t_{0.5}$ of the drug. The

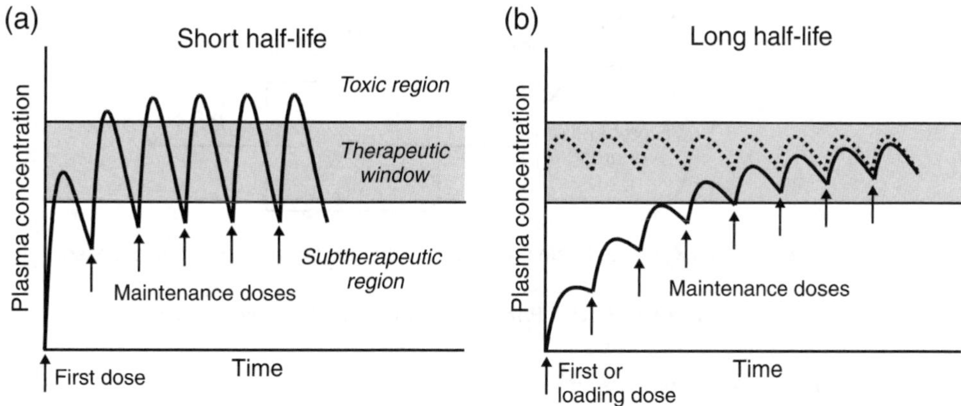

Figure 16.5 Plasma concentration–time plots following repeated doses at equal intervals for a drug with (a) a short half-life and (b) a long half-life. The effect of a loading dose (broken line) is shown in (b).

mean plasma concentrations will asymptote to a steady-state value, in the same way as during a constant rate infusion, but now the concentration will fluctuate between doses. The fluctuations will be greater if the drug has a shorter plasma $t_{0.5}$ because a greater proportion of the dose will be eliminated before the next dose. If such drugs have a small 'therapeutic window', that is a narrow range associated with optimal effect, it may prove difficult to maintain the concentration in the required range. Morphine, for example, may cause respiratory depression at the peak concentration, but patients may experience pain before the next dose.

For a drug with a long half-life, it may take several doses before the plasma concentrations are stabilized within the target range [Figure 16.5(b)]. This delay can be prevented by administration of a suitable LD as discussed above. The average concentration at steady-state is:

$$C_{\text{ss,av}} = F\frac{Dose}{Cl \cdot \tau} \tag{16.29}$$

where τ is the dosing interval. This is analogous to Equation (16.28).

16.5 Sustained-release preparations

Sustained-release (SR) preparations are designed to deliver drug at a constant rate over a prolonged period thereby simplifying life for the patient and hopefully improving the efficacy of therapy. The term controlled release (CR) or modified release (MR) may be applied to SR preparations, but also include preparations such as enteric coated aspirin, which is formulated to disintegrate in the bowel rather than the stomach in an attempt to reduce the extent of gastric bleeding caused by the drug. By making the absorption rate constant (k_a) smaller than the elimination rate constant (k) one can prolong the duration of action. As with any sequential reaction, the rate constant of the slowest step determines the overall rate, and under these conditions, k_a becomes rate limiting (Figure 16.6).

After acute overdosage, prolonged absorption due to either a drug delaying its own absorption, or the presence of other drug(s) delaying absorption, can produce what appears to be a greatly enhanced plasma $t_{0.5}$.

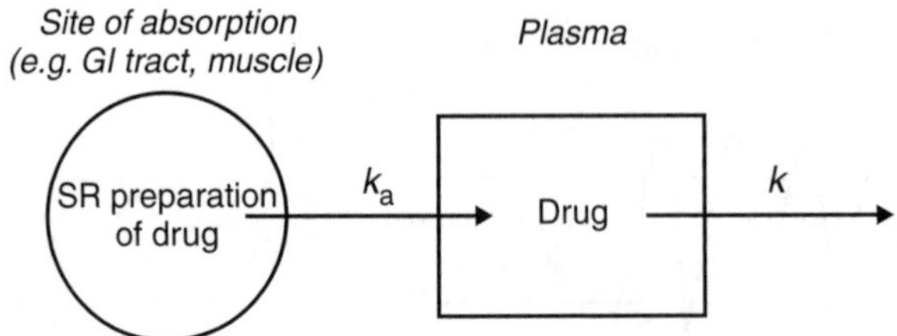

Figure 16.6 Principle of sustained-release preparations.

16.5.1 *Intramuscular depot injection*

Long-acting depot injections illustrate the principal of SR preparations. When fluphenazine decanoate, a very lipophilic ester of fluphenazine, is injected i.m. in oil it is slowly hydrolyzed to fluphenazine, which is released into plasma. Doses may be given 2–4 weekly, which is useful when adherence to oral medication is an issue. Drugs given in this way must be potent, as otherwise the doses required would be too large to be injected.

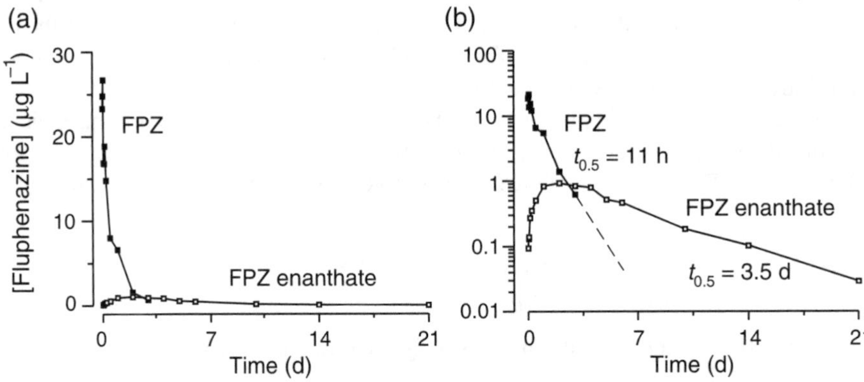

Figure 16.7 Plasma fluphenazine (FPZ) concentrations following i.m. injection of (a) hydrochloride salt and (b) enanthate ester (after Curry *et al.*, 1979).

The way in which slow release is rate limiting can be seen from the kinetics of fluphenazine. Following injection of the nonesterified parent drug, the plasma half-life was 11 h, but when the enanthate (heptanoate) was injected, the terminal half-life of fluphenazine was 3.5 d (Figure 16.7). Resolution of the curve after enanthate injection into its exponential components gave $t_{0.5}$ values of 12 h and 3.5 d, respectively. Thus, $k = 0.6\ \text{h}^{-1}$ and $k_a = 0.008\ \text{h}^{-1}$. Clearly the slow release of ester from the injection site makes it *appear* that the elimination $t_{0.5}$ is 3.5 d, when in fact it is the absorption $t_{0.5}$, that is 3.5 d. That is, the rate of absorption is rate limiting and the rate constants k_a and k *appear* to have been exchanged – this is sometimes referred to 'flip-flop' kinetics, and is the principle behind all SR preparations.

16.5.2 Other sustained-release preparations

SR formulations are available for most routes of drug administration, including oral, subcutaneous (s.c.) and transdermal. Oral SR preparations make use of different particle sizes, of wax matrixes, or with tablets made of layers of material, so that different rates of dissolution give prolonged drug release. Several formulations of insulin are available, including soluble insulin and several crystalline forms that release insulin at different rates.

Transdermal delivery of drugs for systemic effect is a relatively new phenomenon, although absorption of toxic solvents though the skin has been recognized for many years. Glyceryl trinitrate (GTN) is readily absorbed though the skin and may be applied as an ointment rubbed onto an area of skin or as a 'sticking plaster' patch. In some preparations the plaster incorporates a membrane to control the rate of GTN release. Hyoscine, nicotine, buprenorphine and some steroid hormones may be given this way.

16.6 Non-linear pharmacokinetics

The models described above are referred to as linear models, and although initially the need to use exponential functions may seem complicated, first-order elimination results in simple relationships, which make dosing and interpretation of analytical results relatively simple. Clearance, $t_{0.5}$, t_{max} and time to reach C_{ss} are constant, and AUC, C_{max} and C_{ss} are directly proportional to dose. However, there are situations when such models are inadequate. This might be expected to be the case for drugs that are eliminated by metabolism when the M–M equation should apply:

$$\text{rate} = \frac{V_{max} C}{K_m + C} \tag{16.30}$$

However, because most such drugs at therapeutic concentrations are eliminated according to first-order kinetics, it seems that in most cases the amount of drug-metabolizing enzyme present is in great excess compared to the 'effective' concentration of drug (i.e. the concentration of drug at the site of metabolism). Under these conditions, $K_m \gg C$ and denominator $(K_m + C)$ in Equation (16.30) approximates to K_m (C making a negligible contribution to the sum) so:

$$\text{rate} \approx \frac{V_{max}}{K_m} C \tag{16.31}$$

which is a first-order equation, and $k = V_{max}/K_m$. Thus, even for drugs that are extensively metabolized, the elimination kinetics will be first-order provided that drug-metabolizing enzyme activity is greatly in excess of the amount of drug present. In other words, the amount of active enzyme present is, in effect, constant and therefore does not influence the kinetics. This is likely to be the case with a drug with a high apparent volume of distribution because the plasma concentration will be relatively low, hence the concentration of drug being delivered to the liver or other site of metabolism will also be low. Indeed, the apparent first-order rate constant may reflect the rate of return from tissues to the liver rather than any intrinsic ability of the liver to metabolize the drug. Moreover, drugs that exhibit first-order elimination when used at normal therapeutic doses are more likely to be marketed.

If the drug concentration is high compared with drug-metabolizing enzyme capacity, $C \gg K_m$ and $(K_m + C) \rightarrow C$, hence:

$$\text{rate} \approx V_{\text{max}} \tag{16.32}$$

This is a zero-order equation because the reaction rate is constant. The enzyme is saturated with substrate and the reaction is at its maximal rate. For first-order reactions, steady-state concentrations are proportional to dose, but as one moves from first- to zero-order the concentration rises disproportionately (Figure 16.8).

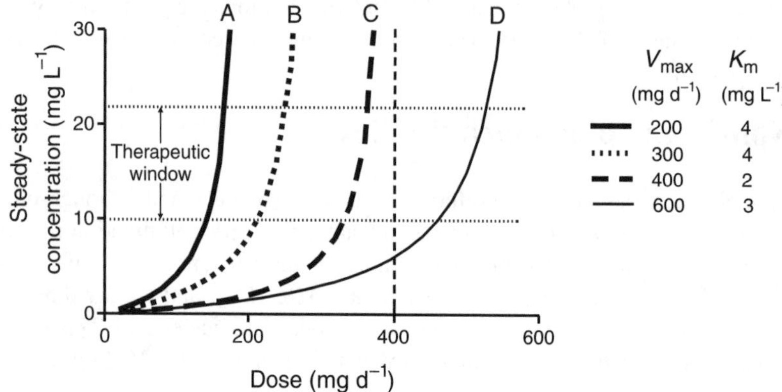

Figure 16.8 Simulation of phenytoin pharmacokinetics in four subjects (A-D). The vertical broken line represents the V_{max} value in subject C.

Drugs whose pharmacokinetics can only be adequately described by M–M kinetics include phenytoin (Richens and Dunlop, 1975), ethanol and (at higher doses) salicylate. Adjusting the dose of a drug such as phenytoin to ensure that plasma concentrations remain in the therapeutic window is complicated by the fact that there are large individual variations in K_m and V_{max} (Figure 16.8). Although 'population' values of K_m and V_{max} could be used to calculate doses to obtain a required steady-state concentration, it is clearly better to use individual values. Because there is a need to solve equations for two unknown values, steady-state concentration data for two doses are required.

If the daily dosing rate is R, then the M–M equation can be written thus:

$$\text{R} = \frac{V_{\text{max}} \cdot C_{\text{ss}}}{K_m + C_{\text{ss}}} \tag{16.33}$$

Rearrangement gives:

$$R = V_{\text{max}} - \frac{R}{C_{\text{ss}}} K_m \tag{16.34}$$

which is the equation of a straight line, of slope $-K_m$ and y-intercept, V_{max}. Thus, values for V_{max} and K_m can be obtained graphically (Figure 16.9).

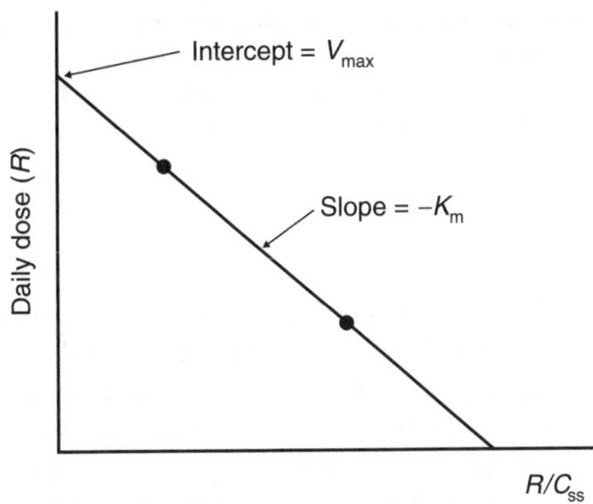

Figure 16.9 Graphical solution for V_{max} and K_m.

Once V_{max} and K_m have been estimated, then C_{ss} for a particular dose can be obtained by rearranging Equation (16.34):

$$C_{ss} = \frac{K_m R}{V_{max} - R} \tag{16.35}$$

Because of the disproportionate increase in plasma concentration with dose of drugs like phenytoin, the dose has to be carefully adjusted, and although a daily dose might be typically 300–500 mg, small tablet sizes are available so that the dose can be adjusted appropriately. Similarly, anything that changes the 'effective' dose, for example changes in bioavailability, or enzyme induction or inhibition (Section 17.4), is likely to have a big effect on the plasma concentration and hence pharmacological action. For example, an individual with a V_{max} of 300 mg d^{-1} and K_m of 4 mg L^{-1}, taking 300 mg d^{-1} of a preparation with $F = 0.8$, would be stabilized with a mean C_{ss} of 16 mg L^{-1}. However, if a preparation with $F = 0.9$ were prescribed, the mean C_{ss} would be 36 mg L^{-1}. Note how Equation (16.35) approaches infinity as R approaches V_{max} and so it becomes more difficult to control the plasma concentration. Furthermore, the time it takes phenytoin concentrations to reach steady-state values becomes progressively longer as the dose is increased.

As C_{ss} is proportional to K_m [Equation (16.35)] changes in this constant, as a result of competitive enzyme inhibition by a second xenobiotic, for example, will lead to proportionate changes in C_{ss}. The example of phenytoin illustrates not only the complexity of prescribing a drug that exhibits M–M kinetics, but also the difficulty in interpreting plasma concentration data if an overdose of drug or other xenobiotic has saturated either metabolism, or carrier-mediated transport.

16.6.1 Ethanol

Although, the kinetics of ethanol at blood concentrations above approximately 0.02 g L^{-1} have been referred to as being zero-order, the kinetics of ethanol are best described by the M–M equation. Substituting numbers into this latter equation clearly demonstrate that there cannot be an abrupt switch from first-order to zero-order kinetics as the ethanol concentration increases. The

Table 16.3 Rate of ethanol metabolism as a function of concentration

C (g L^{-1})	0.01	0.02	0.05	0.1	0.2	0.04	0.8	1.6	4.0	8.0
Rate (g h^{-1})	0.73	1.33	2.67	4.0	5.33	6.4	7.11	7.54	7.8	7.9

$K_m = 0.1$ g L^{-1}
$V_{max} = 8$ g h^{-1}

rate of ethanol metabolism calculated using population values of $K_m = 0.1$ g L^{-1} and $V_{max} = 8$ g h^{-1} is shown in Table 16.3. If first-order kinetics applied, then the values calculated below 0.2 g L^{-1} would be directly proportional to the concentration. Clearly they are not. Similarly, if the kinetics were zero-order above 0.2 g L^{-1}, then the rate would be constant. Again this is not the case, although the rate does asymptote to V_{max} at very high concentrations. This being said, at the concentrations encountered in clinical and forensic situations 'zero-order' is a reasonable description of ethanol kinetics.

In a theoretical decay curve for ethanol [Figure 16.10(a)], the initial part of the curve is almost linear, reflecting approximately zero-order elimination, but the line becomes progressively more curved. Plotting the data with a logarithmic y-axis [Figure 16.10(b)] gives the typical shape expected for M–M elimination. At high concentrations the elimination rate approaches V_{max}, but the *proportionate* change is small, hence the shallow slope of the log-transformed curve. At later times (lower plasma concentrations) the rate of elimination is much lower, but the proportionate change is greater and will become constant when the elimination is first-order.

In the UK, Eire, Canada and in most US states the legal BAC limit for driving a motor vehicle is 0.80 g L^{-1} (80 mg per 100 mL, 0.080 % w/v). The limit in Sweden and Norway is 0.2 g L^{-1}, and in most other European countries and in Australia it is 0.5 g L^{-1}. 'Backtracking' or back-calculation of blood ethanol values to some prior time point is fraught with difficulties, including the assumption that zero-order kinetics apply, the likelihood of individual variation in the rate of ethanol metabolism, including whether the subject is naïve or a heavy drinker (Paton, 2005) and most importantly, the effect of continued ethanol absorption from the GI tract (Jackson *et al.*,

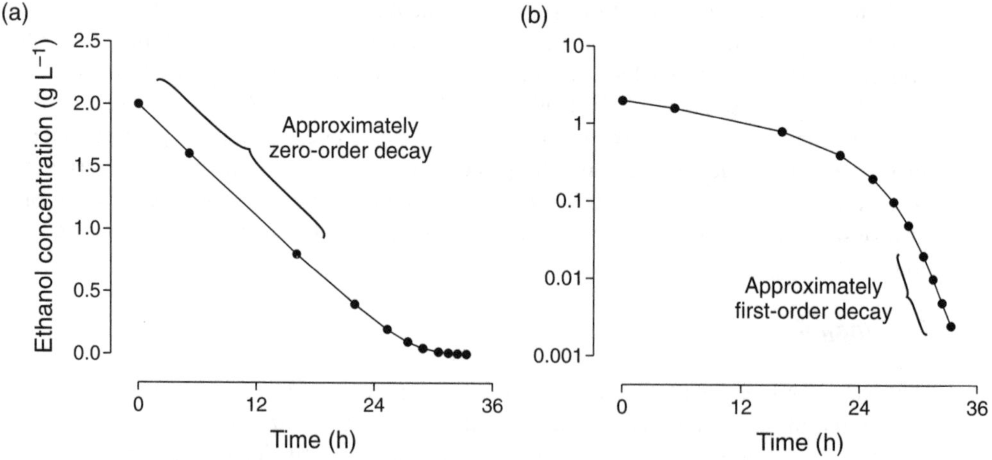

Figure 16.10 Simulated decay curve for ethanol ($C_0 = 2$ g L^{-1}, $V_{max} = 8$ g h^{-1}, $K_m = 0.1$ g L^{-1}).

1991). Absorption from this latter site may be influenced by gastric contents (Sedman *et al.*, 1976).

16.7 Multicompartment models

A large number of basic pharmacokinetic concepts can be understood using a simple one-compartment model. However, there are situations when this model is inadequate. If following an i.v. bolus injection, for example, the plasma concentration decay curve can be best described by the sum of *two* exponential terms:

$$C = A\exp(-\alpha t) + B\exp(-\beta t) \tag{16.36}$$

then the situation may be referred to as a two-compartment model (Figure 16.11).

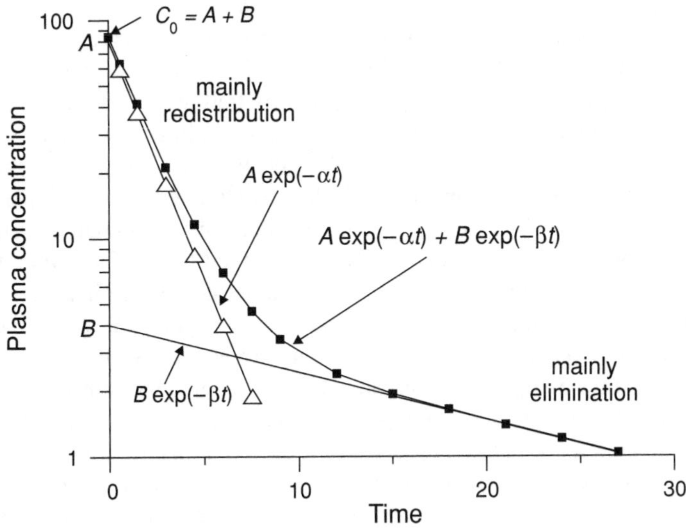

Figure 16.11 Decay from a two-compartment kinetic model: the sum of two exponential terms.

This model now requires the concept of a second apparent volume of distribution, V_2, into which the drug moves more slowly (determined by k_{12}) than the rapid ('instantaneous') distribution of drug to the tissues of the central compartment (V_1) (Figure 16.12). Distribution to the second, peripheral compartment explains the initial steep slope of the plasma concentration–time curve. Later, after equilibration between compartments, the slope of the decay curve is shallower and the main mechanism of decay is elimination from the body (Figure 16.11). If three exponential terms are required to fit the data then a three-compartment model would be required, in other words the number of compartments is given by the number of exponential terms required to define the decay curve.

The central compartment *always* includes plasma and often includes well-perfused tissues, such as liver, heart and lung. Tissues of the peripheral compartment(s) may be skeletal muscle and adipose tissue. The brain may be in either compartment depending on the drug. Alternative

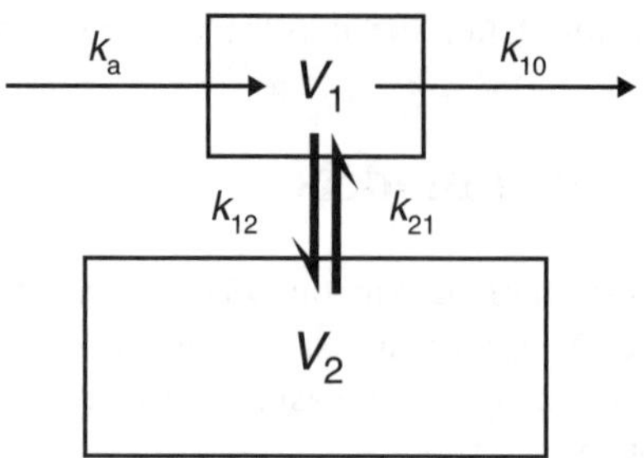

Figure 16.12 Schematic representation of a two-compartment pharmacokinetic model.

models in which drug is eliminated from the peripheral compartment, or indeed from both compartments, are possible. However, in many cases the liver and/or the kidneys are the main organs of elimination, and these tend to be in the central compartment because they have a high blood flow and consequently rapidly equilibrate with plasma.

Usually, the parameters describing multicompartment models are derived from plasma concentration data. However, for thiopental (thiopentone) there are data for tissue concentrations. In this study in the dog (Brodie et al., 1952), the log(concentration) plots for liver and plasma declined in parallel indicating that the liver:plasma concentration ratio was constant. Thus, liver and plasma were in the central compartment. The actual concentrations do not have to be identical for tissues to be part of the same compartment, but the kinetic parameters do. Thiopental concentrations in skeletal muscle rose initially, peaked and then declined in parallel with the plasma concentrations, whereas the concentrations in adipose tissue continued to rise for several hours post-injection, indicating that the model for thiopental required at least two peripheral compartments.

The plasma concentration–time curve following i.v. injection of thiopental in man can be resolved into three exponential phases (Brodie and Mark, 1950), which clearly indicates that there are at least two peripheral compartments in addition to the central compartment. The initial, very short duration of action when thiopental is used as an i.v. general anaesthetic is explained by its almost immediate uptake into brain (in the central compartment), followed by rapid redistribution to peripheral compartments, hence recovery from anaesthesia occurs during the 'α-phase' of the plasma concentration–time curve. With larger or repeat doses, the plasma concentrations rise to the point where recovery from anaesthesia occurs during the 'β-phase' and so the duration of action is markedly increased. During prolonged thiopental administration, for example in the treatment of status epilepticus, the terminal plasma $t_{0.5}$ is several days (Russo and Bressolle, 1998), which may have profound implications in the differential diagnosis of brainstem death.

Compartmental modelling is simply a way of interpreting complex plasma concentration-time curves. The crucial difference between a single compartment and a two-compartment model is that in the former, plasma and tissues equilibrate so rapidly that a distribution phase cannot be discerned on a log plasma concentration–time plot. To some extent this will depend on the quality of the data collected and the route of drug administration. An i.v. bolus injection and early blood sampling may reveal a distribution phase, whereas an oral dose and/or not taking early samples may not (Section 16.8.2).

It is often not possible to define exactly which tissues comprise which compartments. However, plasma is always part of the central compartment as discussed above, and usually kinetic parameters are derived from plasma concentration–time data. It is important to remember this, as concentration–time plots rarely show the time course of the substance in the peripheral compartment. For small, non-protein bound molecules, it is to be expected that ECF will be part of the central compartment as the fenestrations in the capillary walls will provide little in the way of a barrier to the movement of such molecules. Lipophilic molecules will rapidly cross cell walls and the rate of equilibration is largely controlled by the blood flow to the tissues, as is the case with thiopental. If the equilibration occurs within the first few circulations of the blood, it will not be possible to monitor a plasma-tissue distribution phase and such tissues will be classed as being in the central compartment. Thus, for a lipophilic substance, distribution might be predicted as: central compartment: plasma and well-perfused tissues; peripheral compartment: tissues, which for their size, receive less of the cardiac output (muscle, fat, bone, etc.) (Figure 16.13).

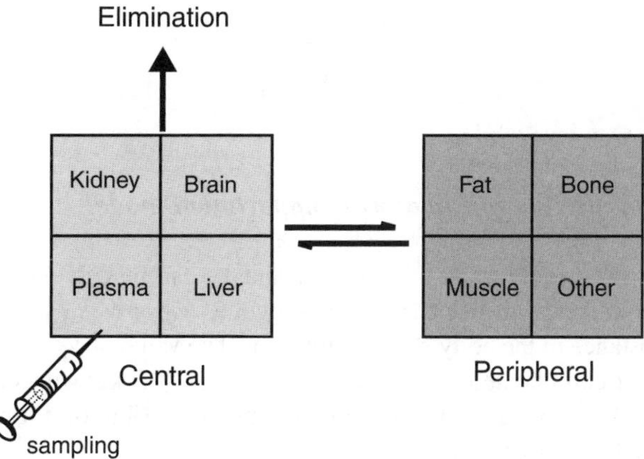

Figure 16.13 Representation of tissues that might constitute part of the central and peripheral compartments.

16.7.1 Calculation of rate constants

The rate constants α and β, should not be confused with k_{21} or k_{10}. The method of residuals may be used to fit complex decay curves graphically, and although iterative curve fitting is now possible with personal computers and is the way in which such values are normally derived, an understanding of the principle of the graphical approach is important. As $-\alpha$ is the slope of the steeper, initial phase, $\alpha > \beta$ and the term $A\exp(-\alpha t)$ approaches zero faster than $B\exp(-\beta t)$. Therefore at later times, the contribution from the first exponential term is negligible and Equation (16.36) approximates to:

$$C \cong B\exp(-\beta t) \tag{16.37}$$

Provided the plasma concentration–time curve is monitored for long enough, the terminal portion of the ln C *versus* t curve will be a straight line from which B and β can be estimated (Figure 16.11). Values of $B\exp(-\beta t)$ can be calculated for earlier time points, that is when $A\exp(-\alpha t)$ is making a significant contribution to the plasma concentration. Therefore, if the $B\exp(-\beta t)$ values are subtracted from the experimental values at those times this will give estimates of $A\exp(-\alpha t)$ which are referred to as residuals. A semilogarithmic plot of the residuals (Δ) should give a straight line of slope $-\alpha$. Thus, values for A and B, and for α and β, can be obtained and the microconstants calculated (Box 16.1). When fitting the data iteratively with a computer it is still important that the terminal phase approximates to a single exponential. Ideally, the data should be collected over at least three half-lives.

Box 16.1 Equations for calculating microconstants for a two compartment model

$$k_{21} = \frac{A\beta + B\alpha}{A + B} \tag{16.38}$$

$$k_{10} = \frac{\alpha\beta}{k_{21}} \tag{16.39}$$

$$k_{12} = \alpha + \beta - k_{10} - k_{21} \tag{16.40}$$

16.7.2 Volumes of distribution in a two-compartment model

Although it is reasonable to assume that V_1 and V_2 can be calculated, for a two-compartment model there is in fact a further parameter, known as V_β or V_{area}, which is used to calculate the amount of substance in the body post-equilibrium. The value of V_β changes with changing values of *Cl*. To further complicate matters, the sum of $V_1 + V_2$ is known as the volume of distribution at steady-state, V_{ss}. Note that the apparent volume of distribution for a two-compartment model is *not* calculated from *Dose*/*B*.

V_1 can be calculated in the same way as the apparent volume of distribution for a single compartment model. At $t = 0$, following an i.v. bolus injection all the drug is in the central compartment, but none in the peripheral compartment ($C_0 = A + B$), so:

$$V_1 = \frac{Dose}{A + B} \tag{16.41}$$

When the concentration (C) in the peripheral compartment is the same as that in the plasma, there is no net flow and the forward and backward rates are the same:

$$V_1 \cdot C \cdot k_{12} = V_2 \cdot C \cdot k_{21} \tag{16.42}$$

Thus, V_2 is given by:

$$V_2 = V_1 \frac{k_{12}}{k_{21}} \tag{16.43}$$

The distribution of the drug throughout the body is the sum of the volumes and, as it is derived under steady-state conditions, it is referred to as V_{ss}:

$$V_{ss} = V_1 + V_2 \tag{16.44}$$

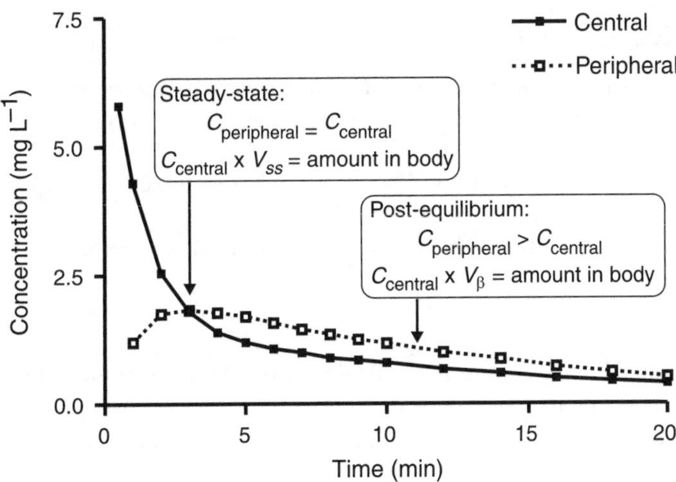

Figure 16.14 Central and peripheral concentration in a two-compartment model.

In a two-compartment model there is a delay in equilibration of the concentrations in the tissues of the central compartment and those of the peripheral compartment. Initially, the concentrations in V_2 rise and then decline (Figure 16.14). When the concentrations in each compartment are equal there is no net flow of drug and, the amount of drug in the body is given by the plasma concentration (C_{central}) multiplied by V_{ss}. However, post-equilibrium the concentration in the peripheral compartment is higher than that in the central compartment, as drug has to return to the central compartment to be eliminated. Thus, multiplying plasma concentration by V_{ss} would give an underestimate of the amount of drug in the body, and the third volume is derived, V_{β}:

$$V_{\beta} = \frac{Dose}{AUC \cdot \beta} \tag{16.45}$$

It is often unclear whether literature values refer to V_{ss} or V_{β}. The situation is further complicated by the fact that V_{β} changes if clearance changes. As it takes a finite time for drug to return from the peripheral compartment, increasing the rate of elimination from the central compartment means that return from the peripheral compartment cannot keep pace with elimination. In the peripheral compartments the concentrations will also fall, but not to the same extent as the decline in the plasma concentration. Hence the *differences* in concentrations between the two compartments will be greater, that is the terminal phases of the curves in Figure 16.14 will be further apart. Thus, a larger value of V_{β} will be required to compensate for this. If clearance is reduced, then V_{β} will be reduced. If there were no elimination, the concentrations in each compartment would equalize and V_{β} would equal V_{ss}.

16.8 Model-independent pharmacokinetic parameters

If the compartmental model approach seems inappropriate, some pharmacokinetic parameters can usually be derived from *AUC* and the terminal plasma $t_{0.5}$ without defining a model. *AUC* is usually obtained using the trapezoidal method (Box 16.2, Figure 16.15).

Box 16.2 Trapezoidal method to obtain the area under the curve

- Treat the plasma concentration–time data as a series of trapeziums
  - The total area will be the sum of all the trapeziums to the last time point, t_n:

$$AUC_{(0-t_n)} = (C_1 + C_2) \cdot (t_2 - t_1)/2 + \cdots + (C_{n-1} + C_n) \cdot (t_n - t_{n-1})/2$$

  - Plus the final portion, which is extrapolated from t_n to $t = \infty$

$$AUC_{(t_n - \infty)} = C_n/\lambda_z$$

- The terminal rate constant (λ_z) is obtained from the slope of the terminal portion of the ln C versus t plot

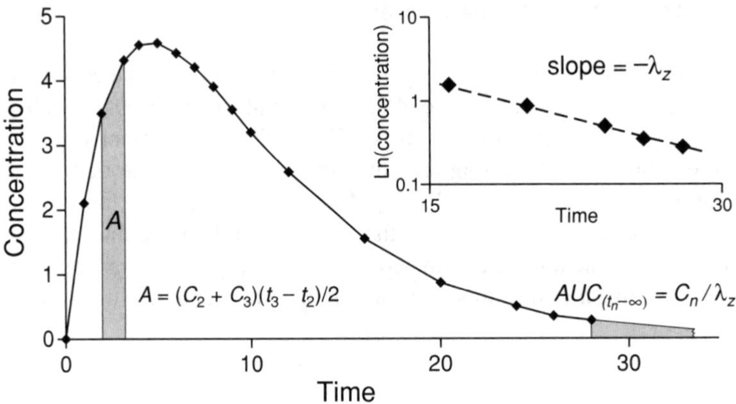

Figure 16.15 The trapezoidal method of measuring *AUC* (Inset: measurement of terminal elimination rate constant, λ_z).

16.8.1 Apparent volume of distribution

The formula for *AUC* can be obtained by integrating the appropriate concentration–time equation. For a single compartment model this gives:

$$AUC = \frac{C_0}{k} \tag{16.46}$$

Combining Equations (16.11) and (16.46), gives:

$$V = \frac{Dose}{AUC \cdot k} \tag{16.47}$$

for a one-compartment model. When the number of compartments is unknown:

$$V_{\text{area}} = \frac{Dose}{AUC \cdot \lambda_z} \tag{16.48}$$

where λ_z is the rate constant of the final or terminal exponential phase.

16.8.2 Clearance

From Equations (16.18) and (16.47), it follows that for an i.v. injection:

$$Cl = \frac{Dose}{AUC} \tag{16.49}$$

and for an oral dose:

$$Cl = \frac{F \cdot Dose}{AUC} \tag{16.50}$$

16.8.3 Model-independent approach

Without plasma concentration–time data for an i.v. dose it may not be possible to define a suitable compartmental model. This often happens following oral or i.m. doses if $k_a < k_{21}$ (Figure 16.16) when the steeper 'α-phase' cannot be distinguished. However, using the trapezoidal rule to measure the *AUC* and λ_z allows values of Cl/F and V_{area}/F to be calculated from Equations (16.48) and (16.50). From the data of Figure 16.16, the *AUC* values for the two cases were in good agreement, 68 and 68.5 mg h L^{-1}, respectively. Note that it is not possible to measure *F* without an i.v. dose.

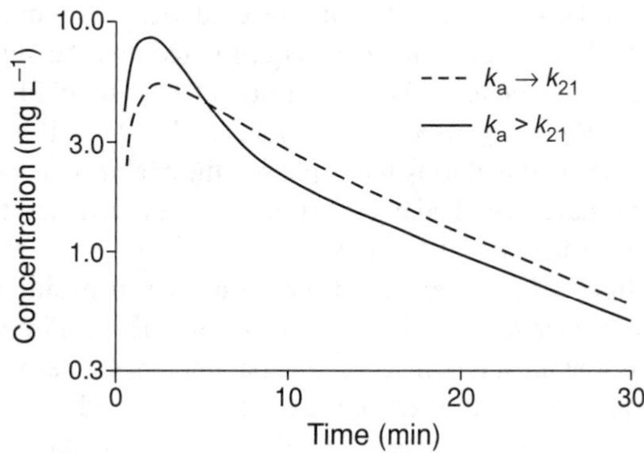

Figure 16.16 Simulated curves for first-order absorption into a two-compartment model using two values of k_a.

16.9 Pharmacokinetics and the interpretation of results

16.9.1 Back-calculation of dose or time of dose

Assuming that a drug or other xenobiotic, or a metabolite, has been detected in a biological sample, questions that may be posed are:

(i) How much substance was administered?
(ii) When was the substance administered?
(iii) Did the substance cause or contribute to clinically apparent toxicity or behavioural disturbance, or death?

Always given that many other factors have to be taken into account in attempting to interpret analytical results in blood samples taken postmortem (Section 17.6), providing estimates in an attempt to answer the first two questions relies on being able to apply pharmacokinetic equations with appropriate caveats as to the inherent uncertainties in such matters. Answering the third, although requiring knowledge of the pharmacodynamics and toxicology of the substance, will also be influenced, in part, by knowledge of its pharmacokinetics.

To help decide whether a given plasma concentration indicates that an overdose has been administered, the result can be compared with published plasma concentration data. In attempting to calculate either time of dosing or dose, the two most useful parameters are apparent volume of distribution and *Cl*. If these parameters are known then the elimination rate constant and half-life can be calculated from Equations (16.18) and (16.19). Some indication of F is vital for interpretation of results after ingestion. If a typical range of plasma concentrations is not known, a mean value for the steady-state plasma concentration can be calculated from the dose rate [Equation (16.29)].

If whole blood has been analyzed, then for a drug that does not partition into erythrocytes, the plasma concentration may be calculated by assuming a haematocrit of 0.45–0.55. However, there will be a large discrepancy between blood and plasma concentrations for drugs and other poisons that are sequestered in erythrocytes. If the sample is taken some time (more than 2–4 h for most drugs) after the substance was administered then it is probably safe to assume that drug concentrations in body compartments have equilibrated and the simpler equations used for one compartment models may be used. Whenever possible V_β should be used rather than V_{ss} (e.g. in a study of diazepam V_{ss} was $\leq$50 % of V_β; Klotz *et al.*, 1976). If blood or plasma samples were collected pre-equilibrium (i.e. during the α-phase), the analyte concentration may be considerably higher than it would have been had equilibrium been attained, and the calculated dose may be a gross overestimate, particularly if $V_1 \ll V_2$.

The potential effects of age, sex and disease on drug disposition and consequently on pharmacokinetics are considered in Section 17.3. Probably the most important caveat when there has been an overdose is that most pharmacokinetic parameters are derived from healthy volunteers or patients given therapeutic or sub-therapeutic doses. Continued absorption and non-linear kinetics following overdose may make the use of published pharmacokinetic parameters unreliable. For drugs that normally have a high oral bioavailability ($F > 0.9$) then this is unlikely to change after overdosage, but for drugs with low oral bioavailability, saturation of first-pass metabolism and efflux proteins such as PGP could markedly increase the proportion of the dose entering the systemic circulation. There may also be saturation of excretory mechanisms (hepatic metabolism, renal excretion) or inhibition by co-ingested compounds.

16.9.1.1 *How much substance was administered?*

If the sample was collected during life and the time of sampling and of dosage are known, then it may be possible to calculate the dose. For an i.v. bolus, the concentration at $t = 0$ can be obtained from rearranging Equation (16.4):

$$C_0 = C \cdot \exp(kt) \tag{16.51}$$

where t is the time since the dose was taken. An alternative approach is to calculate the number of plasma half-lives that have elapsed, N, because:

$$C_0 = C \cdot 2^N = C \cdot 2^{(t/t_{0.5})} \tag{16.52}$$

Equation (16.52) can be derived from Equation (16.51) by substituting $k = \ln 2/t_{0.5}$. N need not be an integer. The published volume of distribution is used to convert C_0 to dose. The calculation for an oral dose is more complex, requiring rearrangement of Equation (16.20). However, provided that absorption is essentially complete and $k_a \gg k$, then $(k_a - k) \to k_a$, so simplification and rearrangement of Equation (16.20) gives:

$$C_0 = \frac{C \exp(kt)}{F} \tag{16.53}$$

In other words, either Equation (16.51) or Equation (16.52) may be used provided allowance is made for the fraction of the dose that is absorbed.

It may be possible to take serial blood samples to obtain a value of $t_{0.5}$ and hence k for the patient, rather than using a population value.

16.9.1.2 *When was the substance administered?*

Provided that the plasma half-life is known and there is a reasonable estimate of the amount of substance taken, then Equation (16.5) can be rearranged to calculate t:

$$t = \frac{(\ln C_0 - \ln C)}{k} \tag{16.54}$$

As $C_0 = Dose/V$ and $k = 0.693/t_{0.5}$, this gives:

$$t = \frac{1}{k} \ln\left(\frac{Dose}{VC}\right) = \frac{t_{0.5}}{0.693} \ln\left(\frac{Dose}{VC}\right) \tag{16.55}$$

This is the equation for an i.v. bolus injection. For an oral dose the equation is:

$$t = \frac{1}{k} \ln\left(\frac{F k_a\, Dose}{VC(k_a - k)}\right) \tag{16.56}$$

However, if $k_a \gg k$, Equation (16.56) can be simplified as before:

$$t = \frac{1}{k} \ln \left(\frac{F Dose}{V C} \right) \tag{16.57}$$

As explained above these equations will only give approximate answers. Although more rigorous calculations would use equations applicable to multicompartment models: (i) large individual variation in the parameters, (ii) uncertainties about the size or time of dose and (iii) difficulty in finding published values for α, k_a, V_1 and so on, mean that the conclusions are unlikely to be any more reliable than those obtained using the simpler approach.

16.9.1.3 *Practical examples*

A question might be: Is a plasma concentration of 0.12 mg L^{-1} diazepam consistent with ingestion of a single 10 mg tablet 45 min before sampling? Using a value of apparent volume of distribution of 1 L kg^{-1}, and assuming a body weight of 70 kg then: $0.12 \times 1 \times 70 = 8.4$ mg. Assuming the time to peak plasma concentration is about 1 h and F is close to 1, then this is consistent with a single tablet having been taken at the specified time.

In TDM the analyte is measured, usually in plasma or serum, and the result compared with target concentrations. If appropriate, the dose may be adjusted until the measured concentration falls within the therapeutic or target range. The process may be assisted by the use of appropriate pharmacokinetic calculations, as described for phenytoin (Section 16.6). To give a further example, from Equation (16.29) the mean C_{ss} for 5 mg diazepam given twice daily assuming $t_{0.5} = 40$ h and body weight = 70 kg is:

$$Cl = (0.693 \times V)/40 = 0.0173 \text{ L h}^{-1} \text{ kg}^{-1}$$
$$C_{ss} = (F \times 5)/(0.0173 \times 12 \times 70) = 0.34 \text{ mg L}^{-1}$$

The plasma half-life of nordazepam is about twice that of diazepam in man and the volume of distribution is similar hence the C_{ss} is approximately twice that of diazepam. If dosage ceases, nordazepam will still be detectable in plasma for some time after diazepam has been cleared from the systemic circulation. If, on the other hand, the plasma diazepam concentration is higher than that of the metabolite, this suggests that sampling has occurred soon after diazepam dosage. Such calculations cannot be used with postmortem data, however, because blood concentrations of lipophilic compounds are likely to rise after death, especially if exposure has been chronic (Section 17.6).

16.9.1.4 *Calculation of time of cannabis exposure*

The time since marijuana was smoked or ingested can be important when assessing whether a subject's actions are likely to have been affected by exposure to the drug. Equations for calculating the elapsed time (T) since marijuana smoking have been derived empirically (Huestis *et al.*, 1992; Huestis *et al.*, 2006), based on either Δ^9-tetrahydrocannabinol (THC) plasma concentrations:

$$\log T = -0.698 \log[\text{THC}] - 0.687 \tag{16.58}$$

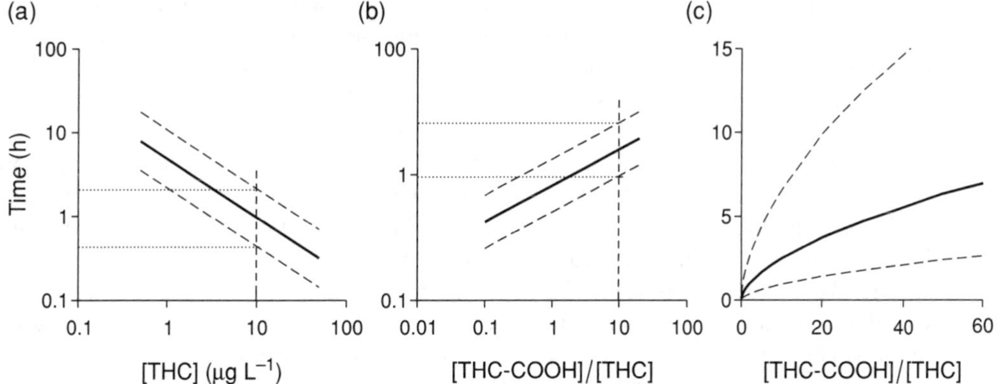

Figure 16.17 **Models for predicting time of cannabis exposure (a) from THC concentration and (b) from [THC-COOH]/[THC] ratios, using logarithmic scales, (c) data for [THC-COOH]/[THC] ratios plotted on linear scales** (Bold lines calculated according to Equations 16.58 and 16.59, 95 % CI values (dashed lines) calculated according to Equations 16.60 and 16.61).

or the ratio of the concentrations of inactive metabolite 11-nor-9-carboxy-Δ^9-tetrahydrocannabinol (THC-COOH) to THC in plasma:

$$\log T = 0.576 \log \left(\frac{[\text{THC-COOH}]}{[\text{THC}]} \right) - 0.176 \tag{16.59}$$

The equations were derived for plasma concentration data from six subjects who smoked either a 1.75 %, or a 3.55 % THC-containing cigarette. The accuracy of the models was tested by comparison with the results from nine clinical studies, either absolute time error (predicted elapsed time − actual time elapsed) or comparison of the actual time with the 95 % confidence intervals (CI) for the predicated time of exposure.[1] For Equation (16.58):

$$\log \text{CI} = \log T \pm 1.975 \sqrt{0.030 \left(1.006 + \frac{(\log[\text{THC}] - 0.996)^2}{89.937} \right)} \tag{16.60}$$

while for Equation (16.59):

$$\log \text{CI} = \log T \pm 1.975 \sqrt{0.045 \left(1.006 + \frac{(\log[\text{THC-COOH}]/[\text{THC}] - 0.283)^2}{123.420} \right)} \tag{16.61}$$

From Equations (16.58)–(16.61) it is possible to draw the time concentration relationships (Figure 16.17).

---

[1] The equations for CI in Huestis *et al.* (1992) are incorrect and also contain typographical errors. The correct equations are given in Huestis *et al.* (2006).

The accuracy of Equation (16.58) was shown to be 92.9 and 100 % (CI analysis) in infrequent and frequent smokers, respectively. The corresponding mean absolute errors were 0.45 and 0.17 h, respectively. The accuracy of Equation (16.59) was less in both groups of smokers, 89.7 and 71.4 %, respectively (CI analysis), but was deemed 100 % accurate in a small group of subjects who took 20 mg THC orally. Equation (16.58) was only 38.5 % accurate in the oral group (absolute mean time error = 2.27 h). However, the 95 % CI values range from approximately −50 % to +100 % of the value predicted by Equation (16.58). For example, the predicted time for a plasma THC concentration of 10 $\mu g\ L^{-1}$ is 0.97 h (CI 0.44–2.15 h) [Figure 16.17(a)]. The confidence intervals are slightly larger for Equation (16.59), for example the predicted value for a THC-COOH:THC ratio of 10 is 2.5 h (CI 0.95–6.22 h) [Figure 16.17(b)]. The relative sizes of the CI values are more easily appreciated from a linear plot of the data. [Figure 16.17(c)]. More recently, Huestis *et al.* (2006) have applied Equations (16.58) and (16.59) to studies with single and multiple oral doses of cannabis in the form of either Hemp oil or dronabinol (synthetic THC). As with the oral study noted above, use of THC-COOH:THC ratios was considered to be more accurate.

The use of models of this kind has been criticized, particularly when THC-COOH:THC ratios are employed because residual THC-COOH concentrations from previous cannabis use might influence the outcome (Bogusz, 1993). More fundamentally, the models are based on studies in which the doses were controlled carefully. No allowance for the amount of cannabis consumed is made in Equation (16.58) [cf. Equation (16.57)]. This may explain the greater accuracy of Equation (16.59) when predicating elapsed time following an oral dose, when variability in bioavailability (4–12 %; McGilveray, 2005) may change the effective dose. Furthermore, six is a small number of subjects from which to derive population pharmacokinetic parameters.

Giroud *et al.* (2001) measured the plasma:blood ratios of THC and THC-COOH with a view to investigating whether blood rather than plasma could be used in the Huestis models. Mean ratios of about 1.6 were obtained for both compounds in volunteer samples. Postmortem blood was centrifuged to separate what the authors referred to as 'serum'. Mean serum:blood ratios were about 2.4, with more scatter, particularly for THC-COOH. The authors concluded that blood from living subjects could be used with caution, but did not apply the models to their data.

16.9.2 Toxicokinetics

An understanding of xenobiotic disposition and pharmacokinetics may also be valuable in clinical toxicology. Chlorophenoxy acid herbicides such as 2,4-dichlorophenoxyacetic acid (2,4-D) are largely eliminated unchanged by the kidney via OATPs and possibly PGP – dogs are deficient in OATPs and these herbicides are highly toxic in these animals (oral LD_{50} dog and rat, 100 and 700 mg kg^{-1}, respectively). At high plasma concentrations, the OATPs and perhaps PGP, of the human renal proximal convoluted tubule may be saturated. Unless ionization of the acid is promoted to increase renal clearance by giving sodium bicarbonate, prolonged coma with the attendant risks of pulmonary complications, rhabdomyolyis and renal failure may result.

Plasma 2,4-D, urinary 2,4-D excretion and urine pH during treatment of a patient who had ingested 2,4-D are shown in Figure 16.18. The initial decline in plasma 2,4-D concentration is zero-order. With an initial concentration of ~700 mg L^{-1} and $V_{max} = 5$ mg L^{-1} h^{-1}, calculated

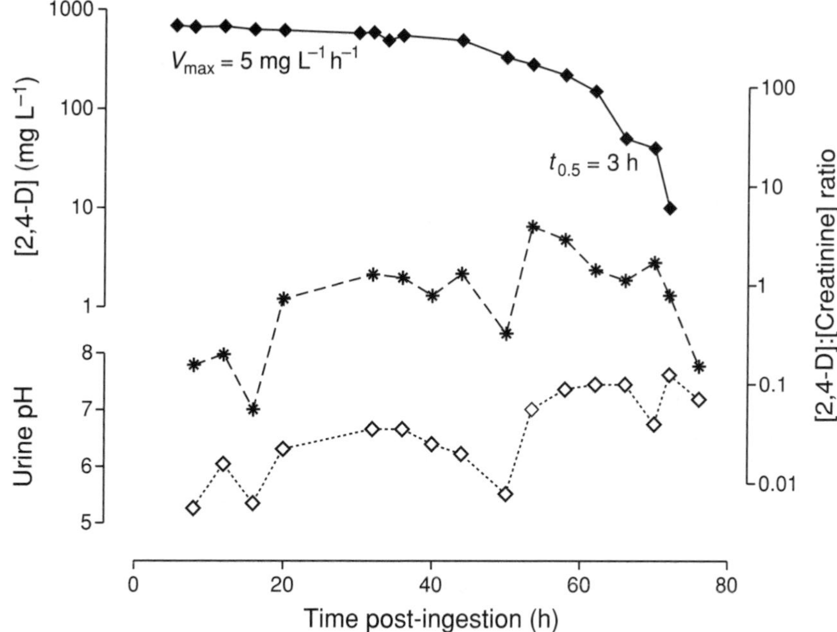

Figure 16.18 Plasma 2,4-D concentrations (◆), urinary excretion (∗) and urine pH (◇) in a patient poisoned with 2,4-D and treated with i.v. sodium bicarbonate.

using the first eight plasma points, $t_{0.5}$ is about 70 h. When the plasma concentration had fallen by two orders of magnitude, the data show first-order kinetics with $t_{0.5}$ about 3 h. Without further data it is impossible to be sure that the first-order plasma $t_{0.5}$ of 2,4-D was not even shorter (2,4-D was not detected in a 76 hour sample). However, assuming $t_{0.5} = 3$ h, then the rate of change in plasma concentration at 7 mg L^{-1} = 0.693/3 × 7 = 1.6 mg L^{-1} h^{-1}. Hence the rate of change in plasma concentration would have been 160 mg L^{-1} h^{-1} at 700 mg L^{-1} had there been no saturation (i.e. had elimination been first-order throughout).

However, in this case the situation has been complicated by the effect of therapy. 2,4-D has a low pK_a (2.6), hence 99.8 % will be ionized at pH 5.3 and 99.9996 % at pH 8.0. Cautious treatment with sodium bicarbonate markedly reduced the hydrogen ion concentration in the urine, and this was reflected in an increased urinary 2,4-D excretion rate, as indicated by the 2,4-D:creatinine ratio (Figure 16.18). The correlation between pH and urinary excretion rate can be seen clearly. At later time points the excretion rate is reduced somewhat because it is also a function of the plasma concentration. Sodium bicarbonate is important in treating any acidosis, and in reducing the amount of herbicide in tissues, because the increased degree of ionization shifts equilibrium towards plasma, as with salicylate (Figure 15.3). In effect, this *reduces* the apparent volume of distribution, thereby introducing a second mechanism by which $t_{0.5}$ is reduced. This effect is more apparent in data from a patient who had ingested a mixture containing 2,4-D and 2,4,5-trichlorophenoxyacetic acid (2,4,5-T; pK_a 2.8) (Figure 16.19).

A plasma paracetamol measurement performed four hours or more after ingestion to allow for absorption of a single overdose of the drug is used to guide treatment. If the concentration

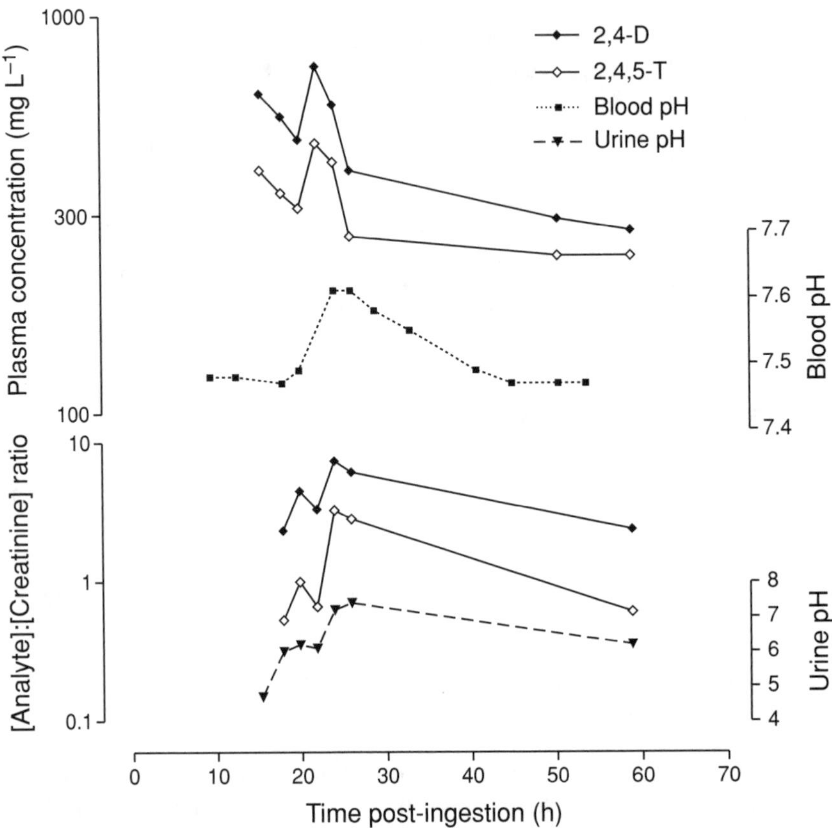

Figure 16.19 Self-poisoning with mixture of 2,4-D and 2,4,5-T: effect of i.v. sodium bicarbonate on blood and urine pH, and on plasma concentrations and renal excretion of 2,4-D and 2,4,5-T.

is above a line joining 200 mg L^{-1} at four hours and 50 mg L^{-1} at 12 h (100 and 25 mg L^{-1}, respectively, for 'high-risk' patients, i.e. patients assumed to have low hepatic GSH or thought to be able to produce the toxic paracetamol metabolite more quickly), antidotal therapy [*N*-acetylcysteine (NAC) or methionine] is indicated (Figure 16.20). However, these lines are quite literally guidelines. Paracetamol is poorly absorbed from the stomach and indeed paracetamol absorption can be used to assess gastric emptying (Tarling *et al.*, 1997). Thus, delayed gastric emptying, due for example to coingestion of drugs such as opioids (Tighe and Walter, 1994; Spiller, 2001) or tricyclic antidepressants, or continued paracetamol release from SR preparations (Vassallo *et al.*, 1996), may result in the plasma paracetamol concentration being below the selected treatment line at four hours, but above it at later times (Figure 16.20). Because the 'guideline' of the nomogram was followed, two patients (Figure 16.20) were not given antidotal treatment – one did not develop features of hepatotoxicity (Spiller, 2001), but the other developed fulminant hepatic failure and died some 24 h after an initial admission (Flanagan, unpublished). There was no evidence that this latter patient had ever taken a paracetamol overdose before.

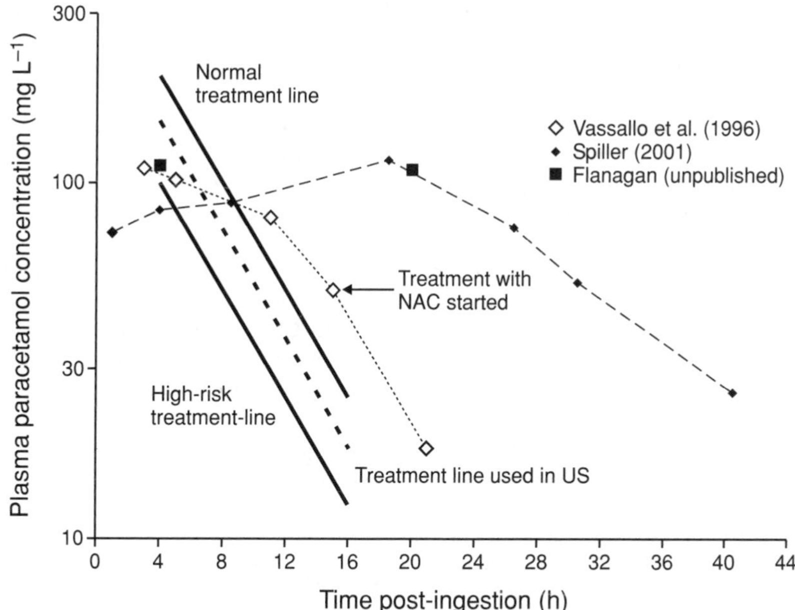

Figure 16.20 Plasma paracetamol concentrations after paracetamol self-poisoning.

16.10 Summary

Pharmacokinetics plays an important part in drug development by helping to establish suitable dosage regimens. For some drugs, notably anticonvulsants, cardiac glycosides and lithium, TDM may be used to adjust the dose to individual need. In the absence of other information, the apparent volume of distribution and dose may be used to predict plasma concentrations, so that, for example, assay calibrators may be prepared over an appropriate concentration range. The predicative value of pharmacokinetics (estimating the size or time of dose) is more limited in analytical toxicology because of the caveats discussed above. However, knowledge of how clearance and distribution determine the time course of a substance in the body is essential to understanding the influence of genetics, disease, age, sex and other parameters on pharmacokinetics and clinical effect, as discussed in the next chapter.

References

Bogusz, M. (1993) Concerning blood cannabinoids and the effect of residual THCCOOH on calculated exposure time. *J Anal Toxicol*, **17**, 313–6.

Brodie, B.B., Bernstein, E. and Mark, L.C. (1952) The role of body fat in limiting the duration of action of thiopental. *J Pharmacol Exp Ther*, **105**, 421–6.

Brodie, B.B. and Mark, L.C. (1950) The fate of thiopental in man and a method for its estimation in biological material. *J Pharmacol Exp Ther*, **98**, 85–96.

Curry, S.H., Whelpton, R., de Schepper, P.J., Vranckx, S., Schiff, A.A. (1979) Kinetics of fluphenazine after fluphenazine dihydrochloride, enanthate and decanoate administration to man. *Br J Clin Pharmacol*, **7**, 325–31.

Dost, F.H. (1953) *Der Blutspiegel—Kinetik der Konzentrationsabläufe in der Kreislaufflüssigkeit*, Thieme, Leipzig.

Gibaldi, M. and Perrier, D. (1982) Pharmacokinetics, in *Drugs and the Pharmaceutical Sciences*, Vol. 15 2nd edn. (series ed. J. Swarbrick) Marcel Dekker, New York.

Giroud, C., Ménétrey, A., Augsburger, M., Buclin, T., Sanchez-Mazas, P., Mangin, P. (2001) Δ^9-THC, 11-OH-Δ^9-THC and Δ^9-THCCOOH plasma or serum to whole blood concentrations distribution ratios in blood samples taken from living and dead people. *Forensic Sci Int*, **123**, 159–64.

Gold, H. and DeGraff, A.C. (1929) Studies on digitalis in ambulatory cardiac patients. II. The elimination of digitalis in man. *J Clin Invest*, **6**, 613–26

Huestis, M.A., Henningfield, J.E. and Cone, E.J. (1992) Blood cannabinoids. II. Models for the prediction of time of marijuana exposure from plasma concentrations of delta 9-tetrahydrocannabinol (THC) and 11-nor-9-carboxy-delta 9-tetrahydrocannabinol (THCCOOH). *J Anal Toxicol*, **16**, 283–90.

Huestis, M.A., Elsohly, M., Nebro, W., Barnes, A., Gustafson, R.A., Smith, M.L. (2006) Estimating time of last oral ingestion of cannabis from plasma THC and THCCOOH concentrations. *Ther Drug Monit*, **28**, 540–4.

Jackson, P.R., Tucker, G.T. and Woods, H.F. (1991) Backtracking booze with Bayes—the retrospective interpretation of blood alcohol data. *Br J Clin Pharmacol*, **31**, 55–63.

Klotz, U., Antonin, K.H. and Bieck, P.R. (1976) Comparison of the pharmacokinetics of diazepam after single and subchronic doses. *Eur J Clin Pharmacol*, **10**, 121–6.

McGilveray, I.J. (2005) Pharmacokinetics of cannabinoids. *Pain Res Manag*, **10**, 15A–22A.

Paton, A. (2005) Alcohol in the body. *Br Med J*, **330**, 85–7.

Richens, A. and Dunlop, A. (1975) Serum-phenytoin levels in management of epilepsy. *Lancet*, **2**, 247–8.

Russo, H. and Bressolle, F. (1998) Pharmacodynamics and pharmacokinetics of thiopental. *Clin Pharmacokinet*, **35**, 95–134.

Rowland, M. and Tozer, T.N. (1995) *Clinical Pharmacokinetics: Concepts and Applications*, 3rd edn, Williams & Wilkins, Baltimore.

Sedman, A.J., Wilkinson, P.K., Sakmar, E., Weidler, D.J., Wagner, J.G. (1976) Food effects on absorption and metabolism of alcohol. *J Stud Alcohol*, **37**, 1197–214.

Spiller, H.A. (2001) Persistently elevated acetaminophen concentrations for two days after an initial four-hour non-toxic concentration. *Vet Hum Toxicol*, **43**, 218–9.

Tarling, M.M., Toner, C.C., Withington, P.S., Baxter, M.K., Whelpton, R., Goldhill, D.R. (1997) A model of gastric emptying using paracetamol absorption in intensive care patients. *Intensive Care Med*, **23**, 256–60.

Tighe, T.V. and Walter, F.G. (1994) Delayed toxic acetaminophen level after initial four hour nontoxic level. *J Toxicol Clin Toxicol*, **32**, 431–4.

Teorell, T. (1937a) Kinetics of distribution of substances administered to the body. I: The extravascular modes of administration. *Arch Intern Pharmacodyn*, **57**, 205–25.

Teorell, T. (1937b) Kinetics of distribution of substances administered to the body. II: The intravascular modes of administration. *Arch Intern Pharmacodyn*, **57**, 226–40.

Vassallo, S., Khan, A.N. and Howland, M.A. (1996) Use of the Rumack-Matthew nomogram in cases of extended-release acetaminophen toxicity. *Ann Intern Med*, **125**, 940.

Widmark, E.M.P. and Tandberg, J. (1924) Uber die Bedinungen fiir die Akkumulation indifferenter narkotiken: theoretische Bereckerunger. *Biochem Z*, **147**, 358–69.

17 Clinical Interpretation of Analytical Results

17.1 Introduction

Clinical interpretation of analytical results is a complex area. The aim of the analysis is to help understand a clinical or forensic scenario, or to provide evidence for the courts. Detailed knowledge not only of the limitations of the analytical method(s) used, but also of the clinical pharmacology, toxicology and pharmacokinetics of the compound(s) involved is important. Patients often respond differently to a given dose of a given compound, especially as regards behavioural effects. Further complicating factors may include the role of pharmacologically active/toxic metabolites (Fraser *et al.*, 2002), and possible toxic effects of drugs on the liver (Buratti and Lavine, 2002). Some additional considerations are summarized in Table 17.1.

The aim of this concluding chapter is to give some information to help in the interpretation of results as regards: (i) therapeutics (Sections 17.2–17.4), the investigation of acute poisoning (Section 17.5) and finally (iii) postmortem toxicology (Section 17.6). Clearly there is potential for overlap amongst all of these areas.

17.2 Pharmacogenetics

Genetic differences in drug response may arise from differences in pharmacodynamics (changes in receptor response) or in drug disposition. Most of the recorded polymorphisms relate to differences in the expression of drug metabolizing enzymes and transporters, and hence drug disposition. When genetic differences in drug disposition are due to a single gene mutation and the incidence is relatively common (>1 %), then such differences may be detected in population studies as bi- or tri-modal frequency distributions (Figure 17.1). Phenotyping may be performed using a suitable test drug or nowadays genotyping may be performed directly. However, the degree of the expression, that is the phenotype, may be more useful than genotype. If a compound is pharmacologically active, subjects who lack 'normal' drug metabolizing enzyme capacity may be more likely to suffer dose-related adverse effects, whereas those with multiple copies of the intact gene may show no response unless the dose is increased. The converse may be true for prodrugs. Thiopurine methyltransferase (TPMT) phenotyping is used to guide treatment with azathioprine to avoid life-threatening agranulocytosis (Lennard and Lilleyman, 1996). However, should a patient have received blood products this may skew the results of phenotyping. Genotyping would detect two cell populations and enable a potentially serious mismatch to be avoided.

Table 17.1 Some factors that may affect interpretation of toxicology results

Factor	Comment
Acidosis/alkalosis	Influence volume of distribution of water soluble ionizable poisons if pK_a is within two pH units of the blood pH
Age	Very young and elderly have lower metabolic capacity, elderly have lower hepatic blood flow, children have low volume of distribution
Body mass	Related to age; males have different body composition to females
Burns	State of hydration; metabolic response to injury
Concomitant drug therapy	Long term and recent: effect on absorption, protein binding, distribution and/or clearance
Disease	Liver or renal disease may reduce metabolic capacity, plasma protein binding may be altered; decreased renal clearance affects renally excreted drugs (Section 17.3.2)
Duration/intensity of exposure	Possible effect on tolerance, body burden, induction or inhibition of metabolism (Section 17.4)
Ethanol consumption	Short- and long-term effects on clearance, potentiation of effect and so on (Lane *et al.*, 1985; Tanaka, 2002, 2003)
Formulation	Sustained release, racemate and so on. If pesticide, vehicle may be more toxic than active ingredient
Genetics/ idiosyncrasy	Acetylator status, CYP polymorphisms and so on (Section 17.2)
Haemolysis	Altered plasma:red cell distribution
Hobbies	Access/exposure to unusual poisons
Infection	Increased cellular permeability, changes in clearance mechanisms (Section 17.3.2)
More than one poison present	Potentiation of effect, altered disposition/clearance (Koski *et al.*, 2003)
Nutrition	Possible change in protein binding and xenobiotic disposition
Occupation	Access/exposure to poisons in the workplace
Pregnancy	May alter drug disposition (Section 17.3.3)
Route of exposure	Many compounds higher acute toxicity if given i.v. or by inhalation rather than orally
Sex	Males have greater body mass, but lower proportion of fat than females: affects disposition of some xenobiotics, notably ethanol (Section 17.3.3)
Shock	Absorption of orally administered drugs may be reduced
Site of sampling	Especially important if patient undergoing an infusion (Chiou, 1989a; 1989b), and in postmortem work (Section 17.6)
Surgery/trauma	Absorption of orally administered drugs may be reduced; state of hydration; metabolic response to injury

Table 17.1 *(Continued)*

Factor	Comment
Time of sampling relative to exposure and/or death	The longer the time between exposure and sampling: (i) the more difficult poison detection (except with hair/nail) in clinical samples, (ii) the greater the potential for change in postmortem work (Section 17.6)
Treatment	May get displacement from binding sites including receptors; F_{ab} antibody fragments and chelating agents will increase total plasma concentration temporarily
Tolerance	Previous exposure may have produced pharmacological tolerance or cross-tolerance; induction of metabolism

17.2.1 Acetylator status

17.2.1.1 Isoniazid

The plasma concentrations of isoniazid were shown to be bimodally distributed when 483 subjects were given identical doses (Evans, et al. 1960). The drug is *N*-acetylated and those with plasma isoniazid concentrations <2 mg L^{-1} were defined as fast acetylators. The rapid gene (R) is autosomal dominant, and so only homozygotes (rr) are slow acetylators. The distribution of fast to slow acetylators is approximately 50:50 in caucasians and black Americans, whereas Inuits and Japanese are fast acetylators (95 %), but some Mediterranean Jews are mainly (80 %) slow acetylators.

Fast acetylators may require higher doses of isoniazid for therapeutic efficacy. Slow acetylators may develop peripheral neuropathy due to depletion of pyridoxal by imine formation with isoniazid. Vitamin B_6 supplements may be given as this has no effect on the antibiotic activity. Isoniazid-related hepatotoxicity was first described in Japan and is thought to be due to *N*-acetylhydrazine released from the acetyl metabolite, which, of course, is in much higher concentrations in fast acetylators. The antidepressant iproniazid (*N*-isopropylisoniazid) shows similar hepatotoxicity.

Phenelzine, hydralazine, procainamide and dapsone are amongst the drugs that are *N*-acetylated, although the distribution of fast and slow acetylators may not completely overlap with that of isoniazid.

17.2.1.2 Sulfonamides

Some sulfonamides show the same polymorphic distribution pattern as isoniazid, and sulfadimidine has been used to assess acetylator status (Figure 17.1). The Bratton–Marshall colorimetric method (Section 4.3.3.4) is relatively simple, although it can only measure sulfonamides possessing a primary amine moiety hence *N*-acetyl metabolites have to be hydrolyzed to the parent compound prior to the analysis. Using HPLC, both drug and metabolite can be measured directly. The percentage of *N*-acetylsulfadimidine in plasma or urine is measured 6–8 h after a test dose (0.5–1 g). The cut-off is taken as 40 % acetylsulfadimidine in plasma or 70 % in urine.

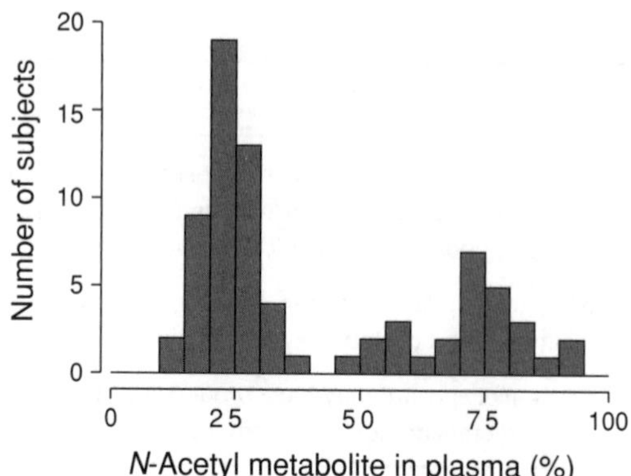

Figure 17.1 Frequency distribution of plasma *N*-acetylsulfadimidine 6 h after a test dose of sulfadimidine.

17.2.2 Cytochrome P450 polymorphisms

Several important polymorphisms in drug metabolizing enzymes, chiefly CYP2D6, CYP2C9 and CYP2C19, have been identified. By convention, genes (alleles) coding for specific enzymes are italicized, for example the *CYP2D6* allele codes for CYP2D6. Poor metabolizers (PMs) are homozygous with two copies of the defective gene and lack a functional enzyme, whereas extensive metabolizers (EMs) are homozygous for the 'normal' enzymes. Intermediary metabolizers (IMs) are heterozygous with one defective and one normal allele, and ultrarapid metabolizers (UMs) have several copies of the 'normal' gene.

17.2.2.1 CYP2D6 polymorphism

Some 5–10 % of Caucasians have no CYP2D6 and are poor metabolizers of several drugs, including the obsolete antihypertensive drug debrisoquine, one of the first compounds for which this polymorphism was observed. To ascertain CYP2D6 status, the concentrations of debrisoquine and 4-hydroxydebrisoquine were measured in a urine sample collected over 8 hours after giving a test dose of the drug:

$$\text{metabolic ratio} = \frac{\%\ \text{dose as debrisoquine}}{\%\ \text{dose as 4-hydroxymetabolite}}$$

PMs are defined as those with a log(metabolic ratio) >1.1, which equates to a ratio of $10^{1.1} \approx 12.6$. CYP2D6 status varies widely between different ethnic groups.

Several drugs exhibit CYP2D6 polymorphism, notably metroprolol, propranolol, dextromethorphan and several centrally acting drugs, including some tricyclic antidressants, SSRIs and antipsychotics (Caraco, 2004). CYP2D6 polymorphisms are responsible, in part, for the wide variations in clinical response to drugs such as nortriptyline, for which the doses required to achieve the same plasma concentrations can vary 10–20 fold between individuals. UMs may never receive sufficient drug and in the past have been dismissed as 'non-responders' or thought to be non-adherent (Ingelman-Sundberg, 2004).

There is evidence that codeine and some other analgesics (e.g. tramadol) are activated by CYP2D6 and hence should be considered as prodrugs. Codeine toxicity has been attributed to excessive morphine production in CYP2D6 UMs (Gasche *et al.*, 2004; Koren *et al.*, 2006), whilst it has been suggested that PMs do not benefit from the analgesic effect of codeine as they do not produce morphine. Inhibiting CYP2D6 with quinidine reduced the abuse potential of codeine (Kathiramalainathan *et al.*, 2000), further suggesting that morphine has a role. However, in many studies pain was induced experimentally and there were relatively few subjects. An alternative view, that is that analgesia is due to codeine-6-glucuronide, has been proposed (Vree *et al.*, 2000). Because: (i) CYP2D6 is expressed in brain, (ii) both morphine and codeine have active metabolites and (iii) morphine is a PGP substrate, it is difficult to come to a definitive conclusion. Tamoxifen is another prodrug which is reliant on CYP2D6 activation; PMs do not respond as well to treatment with this drug as do EMs.

17.2.2.2 CYP2C9 and CYP2C19 polymorphisms

It has been estimated that CYP2C9 catalyzes approximately 10 % of P450-mediated drug metabolism. Substrates of this enzyme include phenytoin, tolbutamide, valproate and warfarin. Adverse drug reactions to phenytoin have been ascribed to patients having defective *CYP2C9* alleles. Substrates for CYP2C19 include carbamazepine and the proton pump inhibitor omeprazole. The degree to which the pH of gastric contents was increased and the success of ulcer treatment with omeprazole was highly dependant on phenotype, with 100 % success in PMs, but only 25 % in EMs. It has been suggested that CYP2C19 genotyping is cost effective in predicting response to omeprazole and amoxicillin in the treatment of *Helicobacter pylori* infection and peptic ulcer (Furuta *et al.*, 1998).

17.2.2.3 Other CYP polymorphisms

Cyclophosphamide is a prodrug with complicated activation and inactivation pathways, some of which are nonenzymatic. The first step, oxidation to 4-hydroxycyclophosphamide, is catalyzed by CYP2B6 and individuals with a particular allele (*CYP2B6*6*) metabolize the drug at a faster rate. Polymorphism in nicotine oxidation has been attributed to an inactive variant allele of *CYP2A6*. An alternative explanation is that in some Asian subjects a *CYP2A6* gene may be deleted and may be responsible for reduced nicotine metabolism. Some individuals may have multiple copies of *CYP2A6* (Rao *et al.*, 2000).

As CYP3A4/5 has been estimated to metabolize some 50 % of commonly used drugs and has an important role in first-pass metabolism, polymorphisms could have major affects on bioavailability, and hence pharmacological/toxicological activity. However, because of the similar substrate specificities of CYP3A4/5 and PGP it is not always possible to ascertain whether individual differences in bioavailability are due to polymorphisms in the enzymes or the transporter. For example, Yates *et al.* (2003) showed that increased oral clearance of ciclosporin was due to increased expression of PGP, but could not independently assess whether any pharmacokinetic parameter correlated with CYP3A5 expression.

17.2.3 Atypical cholinesterase

Several genotypes of plasma pseudocholinesterase are known. The frequency of the usual form is 94 %. An atypical form, which results in a prolonged duration of action of suxamethonium, is widely distributed (1:2800). The duration of action of suxamethonium is normally about 6 minutes, but those with the atypical enzyme may be paralyzed for over 2 h and will require respiratory support until the drug is eliminated by other routes. Individuals at risk can be identified from their family history and the susceptibility of their cholinesterase to the local anaesthetic dibucaine, which does not inhibit the atypical form as much as it inhibits the normal form. Individuals in the UK identified as having the atypical form of the enzyme have for many years carried cards identifying their status. As with other significant genetic polymorphisms, the status of family members should be investigated if these are not already known.

17.2.4 Glucose-6-phosphate dehydrogenase (G6PD)

The gene that codes for an abnormal form of G6PD is carried on the X-chromosome, and is prominent in Sephardic Jews and Africans. Over 80 variants of G6PD have been identified. The reaction catalyzed is the oxidation of glucose-6-phosphate to 6-phosphogluconate with concomitant reduction of NADP to NADPH. This in turn reduces GSSG to GSH. GSH is important in maintaining haemoglobin in its reduced (Fe^{2+}) form. Decreased G6PD activity ($<$30 % of normal) leads to increased methaemoglobinaemia and an increased risk of haemolysis on exposure to oxidants. Several drugs including aspirin, primaquine and some sulfonamides, can cause haemolysis in G6PD-deficient patients.

17.2.5 Alcohol dehydrogenase and aldehyde dehydrogenase

Alcohol dehydrogenase (ADH) is a dimeric enzyme made up of six separate subunits, encoded by three genes, ADH_1, ADH_2 and ADH_3. Many combinations of isoenzymes exist, leading to different rates of metabolism amongst white, black African and Asian populations.

Aldehyde dehydrogenase (ALDH) is a mitochondrial enzyme. Some Asians have ALDH different from that of Caucasians and about 50 % of Asians (principally Chinese) have inactive ALDH, leading to flushing and other unpleasant effects when these individuals consume ethanol. These effects are similar to those seen with disulfiram.

17.3 Effects of age, sex and disease on drug disposition

17.3.1 Age

In neonates and a relatively large proportion of the elderly, gastric pH values are high (achlorhydria) and the bioavailability of acid labile drugs such as penicillin G may be much higher than expected. Some transport processes may not be fully developed in the young, for example riboflavin absorption is slow in the first 5–6 days of life. It should also be remembered that increases in the oral bioavailability of drugs that undergo significant first-pass metabolism may be due to reduced GI or hepatic metabolism.

In neonates the proportion of body water, particularly extracellular water, is high and the proportion of fat low. These factors will influence the distribution of both lipophilic and hydrophilic drugs. Neonates have lower plasma pH values than adults and this will affect the distribution of weak electrolytes such as salicylate. The proportion of body fat is higher in the elderly, and changes in tissue perfusion, possibly caused by arteriosclerosis and a reduction in cardiac output, may be responsible for changes in tissue distribution. Plasma protein binding may be less in both neonates and the elderly, and this may lead to a fall in *total* blood concentrations, but to an increase in the non-bound fraction in plasma, and hence increased activity (Section 15.4.2.1).

In neonates, several drug metabolizing enzyme systems are not fully developed, including glucuronidation, acetylation and plasma esterase activity. High plasma concentrations of bilirubin in neonates may lead to jaundice with the risk of developing kernicterus and subsequent brain damage. Morphine has a long half-life in neonates and is not normally used in obstetrics. Chloramphenicol (metabolized by acetylation) is safe provided the dose is appropriate otherwise cardiovascular collapse ('grey-baby' syndrome) may ensue. The half-life of diazepam is longer in neonates, particularly premature neonates, than children, because of reduced drug metabolic capacity. In adults the terminal half-life (20–90 h) increases linearly with age (20–80 yr) (Klotz *et al.*, 1975).

17.3.1.1 Effect of age on renal function

After the neonatal period, renal blood flow and glomerular filtration rate (GFR) decrease with age and this affects the half-lives of drugs cleared primarily by the kidneys. In healthy young men, GFR is 125 mL min^{-1}, but is less in neonates (20 mL m^{-2} min^{-1}) and the elderly. By age 50 yr GFR is reduced by 25 % (50 % by age 75). Normal creatinine clearance as a function of age (yr) and weight (kg) can be calculated from Equation (17.1):

$$Cl_{\text{Creatinine}} = (140 - \text{Age})\frac{\text{Weight}}{\text{Factor}} \tag{17.1}$$

The 'factor' is to account for differences between men (70) and women (85). Creatinine clearance gives some indication of renal function as the elimination rate constant (k) of a drug excreted via the kidneys is proportional to the creatinine clearance. However, the use of serum creatinine in this way is complicated by the fact that serum creatinine is not solely dependant on GFR and that creatinine values are method dependent.

More recently the modified diet in renal disease (MDRD) formula has been introduced to calculate estimated GFR (eGFR). Although said to be suitable in patients with moderate to advanced kidney disease and diabetic neuropathy (GFR <60 mL min^{-1} per 1.73 m^2 body surface area), it was not recommended for healthy subjects with higher GFR values (Poggio *et al.*, 2005). The calculation is also inaccurate, for example, in those aged <18 years and in pregnancy, and for GFRs >90 mL min^{-1}.

17.3.2 Disease

Disease can cause both pharmacokinetic and pharmacodynamic changes in the response to drugs and poisons. The major systems where disease might be expected to affect pharmacokinetic parameters are the GI tract, the liver and the kidney. However, the cardiovascular system should not be overlooked.

Achlorhydria, arising from pernicious anaemia or gastric carcinoma, reduces the absorption of some acidic drugs such as aspirin, salicylamide and cephalexin. Gastric stasis and pyloric stenosis delay the absorption of paracetamol, and indeed this drug has been used as a marker of gastric emptying. Coeliac disease delays the absorption of some drugs (e.g. rifampicin), but increases the absorption of others such as clindamycin, erythromycin and sulfamethoxazole.

Crohn's disease has been shown to decrease the absorption of trimethoprim, but increase the absorption of sulfamethoxazole.

Although it may seem reasonable to assume that diseases of the liver would reduce the metabolism of drugs, increase elimination half-lives and lead to elevated plasma concentrations, the situation is much more complex. Even for drugs that are extensively metabolized, it should be remembered: (i) that there may be extrahepatic metabolism and (ii) that for many drugs there appears to be excess metabolic capacity under normal conditions, such that liver failure may have to be extreme before an effect is apparent. Furthermore, the elimination half-life is a function of the volume of distribution and clearance (Chapter 16), hence changes in plasma protein concentration in liver disease may change drug distribution and hence elimination.

Care should be taken (both clinically and when interpreting results) if renal function is impaired. The elimination of drugs that are normally cleared unchanged by the kidney will be most affected if the urine flow is markedly reduced. If the drug is normally metabolized prior to excretion, the metabolites may accumulate as with morphine-6-glucuronide (Section 15.5.2.1).

Changes in the concentration of binding proteins can markedly influence the pharmacokinetics and pharmacodynamics of highly bound drugs. In otherwise healthy subjects plasma total (free and protein bound) quinine concentrations of 10–15 mg L^{-1} may be associated with serious toxicity, but in acute malaria, plasma AAG is increased and total quinine concentrations in this same range are attained during effective treatment without manifestation of toxicity (Krishna and White, 1996). Protease inhibitors used in treating acquired immunodeficiency syndrome (AIDS) patients are also strongly bound to AAG, hence renewed interest in this area (Israili and Dayton, 2001).

17.3.3 Sex

Although a genetic phenomenon, the effects of sex on drug disposition are normally considered separately from pharmacogenetics *per se*. Men are generally heavier and have larger muscle mass and organ blood flow than women, factors that can influence drug disposition. Women have a higher percentage of body fat than men and so the volume of distribution of lipophilic drugs such as diazepam and trazodone may be higher in women. On the other hand, the volume of distribution of ethanol is lower in women. GFR is directly proportional to body weight, and so differences in renal clearance may be attributable to differences in average weight. Men have higher serum creatinine concentrations (larger muscle mass) and so if serum creatinine is used to estimate GFR, a correction must be applied [as with Equation (17.1)].

Although women have not been studied as frequently as men, there is evidence to suggest that women have higher CYP3A4 activity than men, but that glucuronyltransferase activity is greater in men than women. These findings may explain the sex-related differences that have been reported in the clearance of some drugs (Table 17.2). The situation is further complicated by possible differences in hepatic PGP expression (Cummins *et al.*, 2002). It has been suggested that women have reduced PGP activity and so drugs remain in hepatocytes for longer, allowing for more metabolism. This being said, erythromycin is one of the few compounds for which studies have shown consistently higher clearance in women compared to men.

There is some evidence that apart from CYP3A4 (which might be induced by high concentrations of progesterone) men have higher activities of other drug-metabolizing cytochromes than women. It has been suggested that theophylline, caffeine and thiothixene metabolism is reduced in women. Women attain higher plasma clozapine concentrations than men on a given clozapine dose (Rostami-Hodjegan *et al.*, 2004).

Table 17.2 Reported sex-related differences in drug clearance

Greater in Women	Greater in Men	No difference
Diazepam	Digoxin	Amatadine
Erythromycin	Fluorouracil	Lidocaine
Midazolam	Oxazepam	Nitrazepam
Prednisolone	Paracetamol	
Theophylline	Salicylic acid	
Verapamil	Temazepam	

The menstrual cycle, pregnancy, menopause, hormone replacement therapy and the use of oral contraceptives may all influence xenobiotic disposition. Drug concentrations may fluctuate with the menstrual cycle, for example the metabolic clearance of methaqualone is approximately two-fold higher at mid-cycle. Similarly, the elimination half-life of paracetamol is shorter mid-cycle. GI transit time varies during the cycle, elevated plasma progesterone concentrations being associated with relaxation of smooth muscle and reduced transit times. Absorption of alcohol and salicylates may be slowed mid-cycle.

Although drugs should be avoided during pregnancy if possible, some drugs cannot be withdrawn. Plasma concentrations of phenytoin, phenobarbital and carbamazepine may be lowered in pregnancy unless dosage is adjusted. The clearance of penicillin antibiotics is increased in pregnancy. Conversely, the plasma half-life of caffeine is increased. The influence of oral contraceptives is really a sex-specific drug interaction (Harris *et al.*, 1995). Failure of oral contraception due to concomitant administration of antibiotics may be due in part to enzyme induction (rifampicin) and in part to destruction of GI micro-organisms leading to reduced enterohepatic recirculation (Section 15.6.2.1) and consequently reduced plasma concentrations. Ethynyl-containing contraceptives inactivate cytochrome P450 enzymes and can reduce the hepatic clearance of a number of other drugs, including ciclosporin and theophylline.

17.4 Enzyme induction and inhibition

17.4.1 Enzyme induction

Exposure of xenobiotic-metabolizing enzymes to certain substances, usually lipophilic compounds that are slowly metabolized, can lead to induction of synthesis of these same enzymes. The increased enzyme activity may lead not only to a reduction in the plasma concentration of the inducing agent, but also those of other substrates of the same enzyme(s). This is an important mechanism underlying some drug–drug interactions. Phenobarbital induces a number of drug metabolizing cytochromes (Table 15.3) and also increases glucuronyl transferase activity as evidenced by increased bilirubin conjugation. Rifampicin (rifampin) is a potent enzyme inducer and several drug interactions have been described including pregnancies following failure of oral contraception as discussed above.

Induction of CYP2E1 by ethanol may explain why chronic alcohol consumption is associated with increased paracetamol toxicity after overdosage as the production of the reactive intermediate,

NAPQI, is catalyzed by this enzyme (Figure 15.9), but reduced hepatic GSH in chronic alcoholics is a further factor here. Acute alcohol ingestion, on the other hand, may be protective, presumably because of competition for the enzyme.

Drug-metabolizing enzyme activity returns to pretreatment levels, usually over a few days, when the inducing agent is withdrawn. Consequently, dosage of other drugs metabolized by the affected enzyme(s) may need to be reduced. Clozapine is metabolized by several enzymes including CYP1A2, which is induced by cigarette smoke and by cannabis smoke. On cessation of smoking, clozapine concentrations may rise if dosage is not adjusted, and serious adverse effects have been reported (McCarthy, 1994).

17.4.2 Enzyme inhibition

The mechanism of action of some drugs is by enzyme inhibition and so interactions with drugs that are metabolized by the same enzyme should be predictable. Allopurinol is a xanthine oxidase inhibitor used to treat gout and so it is to be expected that it will potentiate the action of mercaptopurine and azathioprine (Section 15.5.1.2). Similarly, suxamethonium is hydrolyzed by plasma cholinesterase and interaction with anticholinesterases is predictable. However, some inhibitors may not be specific for the target enzyme. For example, disulfiram, used to inhibit aldehyde dehydrogenase (ALDH) in aversion therapy for alcoholism, also inhibits CYP2C9 leading to interactions with other drugs, including warfarin. The antimicrobial metronidazole has been claimed to produce a disulfiram-like effect with alcohol, but this may not be due to elevated plasma acetaldehyde (Visapaa *et al.*, 2002). However, metronidazole inhibits CYP2C9.

Monoamine oxidase inhibitors (MAOIs) such as tranylcypromine and phenelzine that are used to treat depression potentiate the effects of indirectly-acting sympathomimetic drugs such as ephedrine and phenylpropanolamine. The interaction of MAOIs with tyramine, found in cheese and other fermented foods, is potentially life threatening. Tyramine normally undergoes first-pass metabolism by intestinal and hepatic MAO. However, when these enzymes are inhibited, the tyramine is absorbed and displaces vesicular noradrenaline, the concentrations of which are elevated by the MAOI. This can lead to a 'hypertensive crisis' with the subsequent risk of stroke. Although this has become known as the 'cheese reaction', patients often report that ingestion of even a small quantity of many foods produces throbbing headaches.

Other interactions have been observed. Several drugs, including erythromycin, ketoconazole and also some components of grapefruit juice, inhibit CYP3A4. This enzyme acts in partnership with PGP (Section 15.5.1.1) to reduce the bioavailability of many xenobiotics, hence reduced activity may result in potentially dangerous interactions with a large number of drugs (Bailey *et al.*, 1998). The H_2-receptor blocker cimetidine inhibits several drug-metabolizing cytochromes (Table 15.3). The antiarrhythmic drug amiodarone increases plasma concentrations of digoxin, a known PGP substrate, but *in vivo* inhibition of liver OATP2 (Section 15.3.2) may also be important in this interaction.

17.5 Investigation of acute poisoning

Detection of a drug or other poison is sometimes sufficient to prove exposure, although there are always quantitative considerations (LoD) to any qualitative analysis. At the other extreme the interpretation of individual measurements may be simplified by regulation. In the UK and in many other countries whole blood ethanol measurements in regard to driving a motor or other vehicle and blood lead measurements performed to assess occupational exposure, provide examples. With drugs 'therapeutic', 'normal', 'normally expected' or 'target' plasma or serum concentrations for parent drugs and metabolites often provide a basis for the interpretation of quantitative measurements as far as assessing the magnitude of exposure is concerned. However, in all cases the interpretation of results in individual patients, which is often the responsibility of the laboratory or analyst providing the report, at least in the first instance, is absolutely dependent on the context in which the results are to be viewed. Reliable interpretation of analytical findings can be made only by comparing the results with information on analyte concentrations, clinical response and outcome, and the circumstances under which exposure occurred in other patients.

The investigation of poisoning with psychoactive and other drugs may be complicated by: (i) the development of tolerance to the toxicity of drugs taken chronically, (ii) lack of information on the blood concentrations of drugs, and in some cases metabolites attained during chronic treatment, and (iii) that some drugs have idiosyncratic toxicity on the heart and/or brain. In addition, people given psychoactive drugs may be at higher risk of self-harm, especially if they have become poorly- or non-adherent to treatment, and also the underlying illness may carry a higher risk of sudden death. Such considerations also apply in the case of illicit drug use, and there are the added complications of method of administration and purity of the preparation(s). Ethanol is often a complicating factor.

Some information that should be available before attempting to interpret the results of toxicological investigations is listed in Table 17.3. Clearly relevant circumstances such as the time since

Table 17.3 Information important when interpreting the results of toxicological investigations

For each sample:	Name of patient or other identification
	Nature of sample collected
	Time/date of sample
	Sampling site (if appropriate)
	Sample container used
	Sample preservation/transport/storage
	Name of person collecting sample
For each analytical report:	Time/date of analysis
	Nature of sample analyzed (plasma, whole blood, etc.)
	Unambiguous nomenclature
	Units used to report result(s)
	Name and accreditation status of laboratory and analyst performing the analysis and reporting result(s)
	Other poisons or groups of poisons looked for in the specimen or a related specimen
	Analytical method(s) used
	Reference compound(s) used (source, purity, storage conditions, expiry date)

ingestion/exposure and the number of different poisons involved are important factors. The recent medical or occupational history, age and state of health of the deceased, amongst other factors, may all have a bearing on the interpretation of analytical results.

Proper interpretation is often difficult in the absence of adequate background information, or where specimen collection has been inadequate. It is particularly difficult to comment on the significance of quantitative measurements carried out on a single blood specimen collected from an unspecified site in a cadaver (Section 17.6). Interpretation of findings can also be difficult: (i) if a prescribed drug is involved and (ii) in drug abusers because the likely degree of tolerance to the toxic effects of a drug is unclear. In the case of some compounds encountered in the workplace current biological monitoring guidance values are available (Health and Safety Executive, 2005). After acute exposure, higher concentrations of particular poisons may sometimes be attained with little evidence of toxicity. On the other hand, deliberate abuse of solvents may result in death despite the blood concentration being little more than that which may encountered after legitimate workplace exposure (Flanagan *et al.*, 1997).

17.5.1 Selectivity and reliability of analytical methods

A factor that must not be neglected is the selectivity and reliability of the analytical method. The results of immunoassays especially require confirmation using a physical method if the data are to have any evidential value (Drummer and Gerostamoulos, 2002). Possible shortcomings in methodology as well as in sample collection and so on, must be remembered when studying (older) published reports or compilations of analytical data to help guide interpretation, especially as regards rarely encountered poisons or unusual situations.

There is much interest in the role of chirality in drug action. However, although it has been estimated that some 25 % of drugs and pesticides are marketed as racemates, there is, in general, little need for the enantiomers to be measured separately as far as analytical toxicology is concerned (Williams and Wainer, 2002).

Analyte stability is of far greater import. Some compounds that are unstable in whole blood or plasma are listed in Chapter 2 (Table 2.8). Herbal medicines are a further area of concern. Traditional medicines have always been very widely used in developing countries and are becoming more widely used elsewhere. Toxicity related to use of traditional medicines is being recognized increasingly (Bogusz *et al.*, 2002). Users of these preparations in Western countries are also at risk as herbal remedies, ethnic medicines and other such preparations may contain undeclared pharmaceuticals or inorganic poisons such as toxic metal compounds (Corns, 2003).

17.5.2 Route and duration of exposure and mechanism of toxicity

It is important to bear in mind the time course of an episode of poisoning or suspected poisoning when providing interpretation. For example, results obtained from samples taken before absorption and/or distribution of an ingested drug are complete may be misleading as in the case of paracetamol and with sustained release preparations such as those containing theophylline (Bernstein *et al.*, 1992). With poisons that can give rise to delayed toxicity such as ethylene glycol, methanol, paracetamol and paraquat, interpretation may be unreliable without some knowledge of the time of ingestion. Examples of compounds giving rise to delayed or irreversible toxicity are listed in Table 17.4. With paraquat, plasma measurements are at present only valuable in assessing the prognosis (Proudfoot *et al.*, 1979). In other cases of poisoning, continued metabolism after

Table 17.4 Some examples of delayed/irreversible toxicity associated with particular poisons

Compound/Group of compounds	Examples	Manifestation of toxicity[a]
Acetylcholinesterase inhibitors	OP insecticides, nerve agents	Respiratory failure
α-Amanitin		Hepatorenal toxicity
Amfetamine, metamfetamine		Hepatic toxicity
Anticoagulants	Warfarin and other coumarins and indanediones	Impaired heamostais
Antineoplastic agents	Cyclophosphamide	
Aspirin		Hepatic toxicity
Bromomethane		Pulmonary oedema
Carbon tetrachloride		Hepatorenal toxicity
Chloroform		Hepatorenal toxicity
Chloroquine		Retinal toxicity
1,2-Dichloropropane		Hepatorenal toxicity
Ethylene glycol		Renal and CNS toxicity
Heavy metals	Cadmium, lead, mercury, thallium	
Herbal medications	Germander, greater celandine, kava, skullcap, valerian	Hepatotoxicity
Hexane and 2-hexanone		Peripheral neuropathy
Iron salts		Hepatorenal toxicity
Methadone		Respiratory depression
Methanol		Retinal and CNS toxicity
MDMA ('ecstasy')		Hepatic toxicity
Monoamine oxidase inhibitors (MAOIs)		Convulsions, cardiac arrest, cerebrovascular accident
Paracetamol		Hepatic and sometimes renal toxicity
Paraquat		Lung toxicity
Pemoline		Hepatic toxicity
Phosgene		Pulmonary oedema
Quinine		Retinal toxicity
Sustained release preparations	Diltiazem, lithium, theophylline, morphine	

[a] Only examples in some cases

a period of hypoxia or anoxia, for example, may mean that the actual poison has been largely eliminated from the body before a sample is obtained.

Detection of tablet residues in a sample of stomach contents may point to ingestion. In postmortem work, measurement of the amount of a poison in a sample of stomach contents may be helpful in establishing acute ingestion of an 'overdose' amount. Injection may be indicated by the presence of needle marks on the body. Injection, especially intrathecal (spinal) and i.v. injection, or inhalation, may be associated with serious toxicity even if blood concentrations are low, as the poison may have been transported rapidly to the brain. As an example, postmortem blood metamfetamine concentrations can range up to 0.8 mg L^{-1} or so, whilst a blood metamfetamine concentration of 9.5 mg L^{-1} has been reported some 1 h postingestion in a patient who survived with supportive therapy (Logan *et al.*, 1996). After i.m. or i.p. injection transport to the brain is slower and appreciable quantities of compound may remain at the injection site.

Sometimes measurement of metabolites can provide valuable information not only as to exposure to a given compound, but also the route of administration, time of sampling relative to the dose, and whether a large single dose has been administered. For example diazepam is demethylated to nordazepam or hydroxylated to temazepam, oxazepam being formed from these products (Figure 15.11). The urinary excretion of nordazepam and temazepam peaks between 6–12 and 12–24 h after a dose of diazepam, respectively, whereas the excretion rate of oxazepam is maximal at 1–3 d. In the first 24 h the urinary oxazepam:nordazepam ratio is ≤ 1.

In man, many anxiolytic and antidepressant drugs such as diazepam, imipramine and amitriptyline are metabolized by *N*-demethylation. The secondary amine metabolites have similar apparent volumes of distribution, but lower clearances, than the parent drugs. Thus, these metabolites accumulate on chronic dosing to concentrations approaching or exceeding those of the drug administered. Thus drug:metabolite ratios for these compounds during chronic therapy are normally less than 1. Ratios >2 usually indicate that a large single dose has been administered recently. As an example, in 101 diazepam overdoses, the average diazepam:nordazepam ratio was 3 (Jatlow *et al.*, 1979).

17.5.3 Hair analysis

Hair analysis can sometimes help distinguish between episodic or continuous exposure. As head hair grows at approximately 1 cm month^{-1}, segmental analysis can give information on exposure weeks to months prior to collection, whereas blood and urine specimens will only provide evidence of recent exposure (for a few days beforehand at most). Acidic and neutral drugs and metabolites, such as THC and THC-COOH, are, in general, more difficult to detect in hair than basic drugs.

Hair analysis has its drawbacks. Interpretation of the results is based on knowledge: (i) of the anatomy, physiology and biochemistry of growing hair, (ii) of the pharmacokinetics and metabolism of the compound in question, and (iii) of the factors influencing the incorporation, storage and removal of drugs and other poisons from hair. Surface contamination is likely, although deliberate adulteration of the specimen is unlikely. Sample preparation is currently labour intensive and obtaining adequate QA material is difficult except for trace elements and toxic metals. There are differences in hair growth rate depending on anatomical region, age, sex, ethnicity and interindividual variability, and thus interpretation of parent drug and/or metabolite concentrations in hair is not straightforward (Wennig, 2000). An exact value for a dose of a particular poison or an exact time of exposure is not yet possible. Cosmetic treatment, natural and artificial hair

Table 17.5 Some factors that might influence the interpretation of hair analysis results

Factor	Recommendation/Effect on interpretation
Specimen collection (head hair)	Pluck (include root) or tie and cut; label cut end
Hair colour, location on the scalp, diameter, rate of growth	Collect about 200 mg from posterior vertex of scalp, cut as close as possible to skin, and tie together
Storage	Room temperature wrapped in aluminium foil
Surface contamination (smoke, collection procedure, laboratory atmosphere)	Use validated sample pretreatment (washing) method
Cosmetic treatment (shampoos, bleach, dyes, permanent waving, hair relaxers, hair styling products/sprays, hot dryers/curlers)	Many drugs degraded by these or related procedures, but useful qualitative results may still be obtained
Assay reliability	Use GC-MS, LC-MS, or ICP-MS with validated sample preparation procedure. Participate in pre-analytical and analytical EQA schemes
Nature of analyte	Basic drugs better sensitivity than acidic/neutral drugs
Information on 'normal' and 'abnormal'/'toxic' ranges	Base interpretation on comparable methodology
Sex, ethnicity/race, diet, age, geographical location, season	May influence limit of detection

colour, differences in hair structure, and the selectivity of the analytical methods used may all influence the results obtained (Table 17.5).

17.5.4 Sources of further information

Methods for specific analytes and also on interpretation of analytical results can often be found in the literature and in secondary sources such as those listed in Table 17.6 or subsequent editions. Information on dose, structure, pK_a and the occurrence of metabolites can often be obtained from these same sources. Manufacturers of particular compounds may also be able to give details of assay methods, current literature and potential internal standards, although their own recommended methods may become 'fossilized' because of regulatory requirements.

Web sites where further information on aspects of analytical toxicology may be obtained are listed in Table 17.7. Web site addresses in this Table and elsewhere in this Section were correct on 4 August 2007. Assayfinder (http://www.assayfinder.org/) can be used to locate a UK provider of an unusual or rare assay.

The standard National Library of Medicine (NLM) search engine for the Index Medicus (Medline) database of scientific journal citations and abstracts (Pubmed) is at http://www.ncbi.nlm.nih.gov/entrez/query.fcgi. There are many other NLM databases: see http://www.ncbi.nlm.nih.

Table 17.6 Analytical toxicology: basic reference sources

Author/editor	Title/volume/edition	Publication details
Baselt, R.C.	*Disposition of toxic drugs and chemicals in man*. Edition 7	Foster City, CA: Biomedical Publications, 2004
Dart, R.C., ed.	*Ellenhorn's Medical Toxicology: Diagnosis and Treatment of Human Poisoning*. Edition 3	Philadelphia: Lippincott Williams and Wilkins, 2003
Dollery, C., Boobis, A., Rawlins, M., Thomas, S., Wilkins, M., ed.	*Therapeutic Drugs. Volumes I and II.* Edition 2	Edinburgh: Churchill Livingstone, 1999
Drummer, O.	*The Forensic Pharmacology of Drugs of Abuse*	London: Arnold, 2001
Hathaway, G.J., Proctor, N.H., ed.	*Proctor and Hughes' Chemical Hazards of the Workplace*. Edition 4	New York: Wiley, 2004
McEvoy, G.K., ed.	*AHFS Drug Information 2003*	Bethesda, Maryland: American Society of Health-System Pharmacists, 2003
Moffat, A.C., Osselton, M.D., Widdop, B., ed.	*Clarke's Analysis of Drugs and Poisons* Edition 3	London: Pharmaceutical Press, 2004
O'Neil, M.J., ed.	*The Merck Index*. Edition 13	Whitehouse Station, New Jersey: Merck, 2001
Sweetman, S.C., ed.	*Martindale: The Complete Drug Reference*. Edition 34	London: Pharmaceutical Press, 2004

gov/gquery/gquery.fcgi?itool=toolbar; as an example chemical structures, synonyms and so on can be found at http://chem2.sis.nlm.nih.gov/chemidplus/chemidlite.jsp.

A WHO sponsored link: http://www.intox.org/ is a source of toxicological and poisons information. The UK NPIS primary poisons information database (free access to UK National Health Service personnel) is Toxbase (http://www.spib.axl.co.uk/). The Clinical Evidence (BMJ Group) web site (http://www.clinicalevidence.com/ceweb/resources/index.jsp) has a glossary and statistics modules amongst other features. Further glossaries can be found on the Web, for example at http://www.emc.maricopa.edu/faculty/farabee/BIOBK/BioBookglossA.html#axons.

17.6 Postmortem toxicology

The aim of postmortem toxicology is usually to establish the role that drugs or other poisons played in the death or events immediately prior to death. In addition, it may sometimes be useful to attempt to assess, for example, adherence to prescribed medication or occupational exposure. Any death under, or soon after, detention (includes Mental Health Act detention) in England and Wales must be presented at a Coroner's inquest before a jury. Bringing together all the information that may be required to provide proper interpretation of analytical findings in individual

Table 17.7 Sources of information on analytical toxicology and related disciplines

American Academy of Clinical Toxicology (AACT)	http://www.clintox.org/
American Association of Clinical Chemistry, TDM/Clinical Toxicology Division (AACC TDM/CT)	http://www.aacc.org/AACC/members/divisions/tdm/
Association for Clinical Biochemistry (ACB)	http://www.acb.org.uk
European Association of Poisons Centres and Clinical Toxicologists (EAPCCT)	http://www.eapcct.org/
Health and Safety Executive, Health and Safety Laboratory (HSE HSL)	http://www.hsl.gov.uk/
International Association of Therapeutic Drug Monitoring and Clinical Toxicology (IATDM-CT)	http://www.iatdmct.org
International Federation of Clinical Chemistry and Laboratory Medicine (IFCC)	http://www.ifcc.org/
International Programme on Chemical Safety INTOX Programme (IPCS INTOX)	http://www.who.int/ipcs/poisons/en/
London Toxicology Group (LTG)	http://www.ltg.uk.net/
National Institute for Occupational Safety and Health (NIOSH)	http://www.cdc.gov/niosh/about.html
Société Française de Toxicologie Analytique (SFTA)	http://www.sfta.org/
Society of Forensic Toxicologists (SOFT)	http://www.soft-tox.org/
The International Association of Forensic Toxicologists (TIAFT)[a]	http://www.tiaft.org/

[a] A web database to assist in the interpretation of analytical results is available to TIAFT members.

cases is not easy, especially as investigations tend to be compartmentalized: (i) circumstances: coroner's officer/emergency services/accident and emergency physicians, (ii) background medical: GP/consultant, (iii) postmortem examination: pathologist and (iv) toxicology: analyst. The importance of individual items may only become clear if and when all the evidence has been assembled.

Analyses postmortem must take into consideration the circumstances under which death occurred, the age of the deceased and the possible presence of other drugs or alcohol (Leikin and Watson, 2003). Further potential variables are: (i) the suitability of the analytical investigation(s) employed, (ii) the stability of the analyte(s), (iii) the nature of the specimen(s) sent for analysis and (iv) the fact that changes may occur postmortem in the composition of body fluids especially blood (Skopp, 2004).

17.6.1 Choice of sample and sample collection site

Blood obtained postmortem is a very variable sample. A degree of haemolysis is usual and for this reason whole blood (ideally obtained before opening the body, by needle aspiration from a femoral or other peripheral vein after ligation proximal to the collection site) is normally analyzed directly. However, sedimentation of cells and/or clot formation may have occurred before collection of the sample. Nevertheless, whole (i.e. unseparated) blood is commonly used in postmortem toxicology because: (i) it is relatively easy to collect, (ii) it is relatively homogeneous making it easier to dispense in the laboratory and (iii) there are often data on the plasma or serum (or sometimes even whole blood) concentrations of many analytes attained during normal therapy and after acute exposure to provide at least some basis for the interpretation of results. If the patient was admitted to hospital prior to death, for example, there may be ante- or peri-mortem samples available for analysis. A compilation of postmortem femoral blood concentrations observed in poisoning fatalities is also available to assist in the interpretation of results (Druid and Holmgren, 1997), but simply relying on figures from other cases does not always recognize the possible magnitude of the changes that may take place after death.

Although tissue analysis has become unfashionable to an extent, in part because of a desire to reduce costs, and also because of the comparative lack of information on tissue concentrations attained during life, the possibility that measurement of tissue drug and metabolite concentrations could aid in the interpretation of particularly difficult or important cases should not be neglected. Apple (1989) drew attention to the use of liver tissue as well as postmortem blood analyses in helping differentiate acute tricyclic antidepressant overdosage from chronic exposure to these drugs, the relatively high liver drug:metabolite ratio after acute overdosage providing additional interpretative information. Analysis of other tissue, such as hair, nail, bone marrow or bone may be the only option if all soft tissues have decomposed, but confirmation of exposure may be all that can be hoped for (McIntyre *et al.*, 2000).

Clearly where: (i) the development of tolerance to the acute toxicity of a drug and (ii) the possibility of postmortem changes in blood analyte concentrations may make interpretation of the results difficult, the availability of additional information such as the results of tissue analyses could provide further evidence of the nature and magnitude of exposure. However, it should be remembered that site-to-site variation in postmortem drug concentrations has been reported from different skeletal muscles. Different drug concentrations have also been found in samples collected from different regions of liver and lung (Pounder *et al.*, 1996b). Hair or nail analysis does, of course, provide a further option if noncompliance with therapy or illicit drug use in the days or weeks before death is a subject of debate (Section 17.5.3).

Measurement of electrolytes such as sodium and potassium has no value postmortem. The presence of acetone in blood and urine may be an indication of ketoacidosis or unrecognized diabetes. The measurement of glycated haemoglobin (HbA1c) is a much more reliable guide to undiagnosed diabetes mellitus as this measurement reflects elevated blood glucose concentrations over a period of time before death. Vitreous humour can sometimes be useful for glucose and lactate measurement, as well as for urea and creatinine (Forrest, 1993).

17.6.2 Assay calibration

Many assays intended for TDM or clinical toxicology are aimed at plasma or serum, these two fluids often being used interchangeably, and hence are calibrated using analyte-free plasma or serum from an appropriate source. However, after death, intravascular coagulation, cell sedimentation, haemolysis and a variety of other processes may occur, all of which suggest that well-mixed, haemolyzed whole blood should be analyzed in an attempt to obtain a representative sample of the contents of the blood vessel. Some analysts use haemolyzed whole blood to prepare calibration solutions, but citrated blood is not usually ideal for this purpose (Section 14.1.3).

An alternative approach is to use assays and calibration standards aimed at TDM analyses, but to dilute whole blood samples (after freeze/thawing to ensure haemolysis) $1 + 4$ or $1 + 9$ with the matrix used to prepare the calibration solutions in an attempt to minimize possible matrix effects. Similarly the problems of QC and assay calibration of tissue analyses are often understated. One approach is to use digestion with a proteolytic enzyme such as Subtilisin Carlsberg (Section 3.5), with subsequent dilution with the matrix used to prepare the calibration solutions, again in an attempt to minimize the possible effect of the matrix on analyte recovery.

17.6.3 Interpretation of analytical results

If a patient either under treatment with, or thought to abuse, a drug or drugs dies, the role of the drug(s) in the death may be questioned. Some issues that may be considered include idiosyncrasy, dose-related toxicity, accidental acute or chronic overdosage, and deliberate self-poisoning. In such cases the measurement of the concentrations of the dugs in question in a blood specimen obtained postmortem can be important, but the interpretation of the results obtained may not be straightforward (Drummer, 2004). This is in part because blood concentrations of xenobiotics, notably lipophilic compounds with relatively large volumes of distribution, may increase after death due to diffusion from surrounding tissue or from the GI tract.

If discovery of a body is delayed the extent of decomposition can make not only sample collection, but also the interpretation of quantitative results very difficult. Ensuring that the body is stored at 4 °C prior to the postmortem and that the autopsy is performed as soon as possible after death will minimize the risk of changes in blood analyte concentrations occurring before sampling. Collecting blood by needle aspiration from a peripheral site, ideally after occluding the vein proximally to the site of sampling (Table 17.8), may also help. However, such precautions are not always taken (even proper documentation of the site of blood collection may be lacking), and indeed collection of blood from a central site such as the heart, or even collection of 'cavity blood' (blood remaining in the body cavity when the organs have been removed) may occur. This latter practice, of course, carries the risk of contamination of the 'blood' sample with stomach contents, for example, as well as small pieces of tissue(s).

Many poisons are very potent, that is the blood concentrations associated with severe, possibly fatal, toxicity are very low, typically mg L^{-1} (i.e. parts per million) or even $\mu g\ L^{-1}$ (i.e. parts per thousand million) and thus even trace contamination of a peripheral blood sample can confound the most careful analytical work. In such instances toxicological analysis can often do little more than provide evidence of exposure to a particular substance.

There are no currently accepted biochemical markers that can be used to indicate the magnitude of postmortem changes likely to influence drug redistribution into blood (Langford and Pounder, 1997). Similarly, there are no well-defined experimental models to study postmortem changes in blood concentrations. As there is a need to obtain blood from central sites such as the vena cava *and* from peripheral sites such as the femoral vein at different times after death, small animals, such as rodents and rabbits, are unsuitable. Pigs have some morphological, physiological and metabolic similarities with man and other primates. They are reported to attain maturity as regards hepatic drug-metabolizing enzyme activity by 2 months. However, using young pigs it was only possible to obtain small peripheral blood samples up to 48 h postmortem (Flanagan *et al.*, 2003). A human cadaver model has also been used to study postmortem drug distribution. Diffusion of amitriptyline, lithium and paracetamol from the stomach (acidic or basic conditions) to the base of the left lung and the left lobe of the liver has been demonstrated (Pounder *et al.*, 1996a).

Table 17.8 Factors influencing the likelihood of postmortem change in blood xenobiotic concentrations

Factor	Comment	References
Site of collection	Central sites (heart, vena cava, or 'subclavian' blood) more likely to show changes than peripheral sites (e.g. femoral vein after appropriate isolation). Blood from left ventricle of heart more likely to show change than blood from right ventricle	Dalpe-Scott *et al.* (1995) Moriya and Hashimoto (1999)
Time between death and sample collection	A longer elapsed time gives more potential for changes as tissue pH decreases and autolysis proceeds	
Position body when found	May result in blood draining from central sites to peripheral sites	
Body storage temperature	The higher the temperature, the greater the potential for change	
Method of sampling	Needle aspiration less likely to result in sample contamination with tissue fluid, for example	
Sample preservation	Fluoride needed to help stabilize certain analytes (e.g. ethanol, cocaine, 6-monoacetylmorphine)—does not reverse any pre-collection changes	
Headspace in sample tube	Volatile analytes will equilibrate between sample and headspace; opening the tube when cold (4 °C) will minimize losses	Gill *et al.* (1988)
Volume of blood collected	A larger sample volume less likely to be influenced by localized changes in blood composition	
Nature of xenobiotic	Lipophilic compounds more likely to show increase than lipophobic compounds. See also Table 2.8	
Presence of xenobiotic in the GI tract	Post-mortem diffusion may alter concentrations in adjacent tissues as well as in blood (sample liver from deep inside right lobe as furthest from stomach)	Pounder and Davies (1994); Fuke *et al.*, (1996)

Blood concentrations of some drugs with a relatively small volume of distribution may undergo minimal changes after death. Examples include: lithium (Yonemitsu and Pounder, 1993), trazodone (Martin and Pounder, 1992) and zopiclone (Pounder and Davies, 1994). However, continued absorption from the gastrointestinal tract may occur after death even with these compounds. A further problem is that some analytes, notably ethanol and other volatile compounds, cocaine, cyanide and insulin (Table 2.8), may be lost from, or in the case of ethanol and higher alcohols (propanol, butanol), produced in, postmortem blood. Enzyme activity continues after death, particularly esterase activity and so esters such as heroin are not often detected in postmortem samples. Anaerobic metabolism of 7-nitrobenzodiazepines (nitrazepam, clonazepam, flunitrazepam) produces the corresponding 7-aminocompounds.

The combined effects of postmortem diffusion from tissues and analyte instability can present a confusing picture, as with cocaine (Hearn *et al.*, 1991) and tranylcypromine (Yonemitsu and Pounder, 1993). Both postmortem changes in blood composition and analyte instability may be exacerbated if the body has been kept at a relatively high temperature before sampling. Clearly the longer the duration of treatment with a lipophilic, centrally acting drug and the higher the dose, the greater the potential for postmortem change as the tissue concentrations are likely to be relatively high.

Finally, there is increasing interest in genetic analysis to aid in the interpretation of postmortem data with the aim of detecting those who could have been predisposed to accumulate a potentially toxic concentration of a drug or metabolite (Jannetto *et al.*, 2002; Levo *et al.*, 2003).

17.7 Gazetteer

17.7.1 Antidepressants

Deaths due to TCA poisoning are common and the interpretation of postmortem toxicology poses similar problems to deaths involving antipsychotics and other drugs with relatively large volumes of distribution, and it is likely that a peripheral rather than a cardiac blood sample will more nearly reflect perimortem blood concentrations. Newer antidepressants, notably the selective serotonin reuptake inhibitors (SSRIs), are less toxic than the tricyclics in acute overdose, but nevertheless may be encountered in self-poisoning.

17.7.2 Antiepileptics and antipsychotics

Measurement and interpretation of the results of blood phenothiazine concentrations may be difficult because not only are these drugs, including chlorpromazine, unstable in aqueous solution, but also many metabolites of similarly poor stability and unknown pharmacological activity are often present. Haloperidol measurement is complicated by the presence of its tautomer 'reduced haloperidol' and a further metabolite 4-(4-chlorophenyl)-4-hydroxypiperidine (CPHP), which have to be taken into account when reporting blood haloperidol measurements. With the newer, 'atypical' antipsychotic drugs, analyte stability is generally less of an issue although there are concerns with the sulfur-containing drug olanzapine. However, limited knowledge of the concentrations of many newer drugs and in some cases pharmacologically active metabolites attained on chronic treatment and of the possible impact of factors such as tolerance, adverse reactions to the treatment itself (Burns, 2001), the increased risk of sudden death in epilepsy and in schizophrenia, and the possibility of postmortem changes (Flanagan *et al.*, 2005; Figure 17.2), may make interpretation of postmortem toxicology anything but straightforward.

17.7.3 Carbon monoxide and cyanide

If inhalation of exhaust fumes is suspected, knowing the make, year and model of the vehicle may be helpful, as a catalytic converter, if fitted and properly maintained, should reduce the amount of carbon monoxide produced. Accidental carbon monoxide poisoning from, for example, an improperly ventilated gas fire, often affects more than one person. Children are at especial risk.

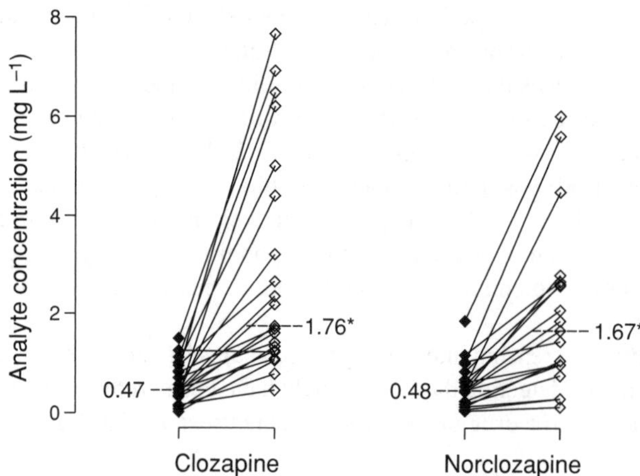

Figure 17.2 Median blood clozapine and norclozapine concentrations before (◆) and after (◇) death in patients dying from causes other than clozapine self-poisoning ($*p < 0.0001$, Wilcox ranked pairs).

Measurement of blood carboxyhaemoglobin saturation (Table 17.9) and cyanide concentration may be helpful when investigating deaths in fires. However, the analysis should be carried out without delay because there is a tendency for the carbon monoxide concentration to decrease after collection, and for the cyanide concentration to increase.

In suspected cyanide poisoning *per se*, specimens of blood from more than one peripheral site, and stomach contents, should be collected. Identifying the source of cyanide exposure is important. If the source is unknown it may be useful to obtain a small specimen of brain (about 20 g) from a site deep within the brain to confirm the presence of cyanide. Blood and tissue specimens are best stored at 4 °C and should be analyzed as soon as possible after collection.

17.7.4 Cannabis

A single dose of cannabis in an inexperienced user, or an overdose in a habitual user, can sometimes induce a variety of unpleasant effects including anxiety, panic, paranoia and feelings of impending

Table 17.9 Blood carboxyhaemoglobin saturation and clinical features of toxicity

Carboxyhaemoglobin saturation (%)	Clinical features
<1	Endogenous carbon monoxide production
3–8	Cigarette smokers
<15	Heavy (30–50 cigarettes/day) smokers
>20	Headache, weakness, dizziness, impaired vision, syncope, nausea, vomiting, diarrhoea (patients with heart disease are at especial risk)
>50	Coma, convulsions, bradycardia, hypotension, respiratory depression, death

doom. The subject may become unusually pallid (a 'whitey'). In addition to difficulty in standing up and staggering, lack of eye coordination may be a further feature. The effects usually persist for only a few hours. Many of the toxic effects of cannabis can be confused with, and compounded by, co-ingestion of alcohol.

There are a number of reports of sudden unexpected death in young people associated with cannabis use alone (Bachs and Morland, 2001; Gupta *et al.*, 2001). A complicating factor is the increasing availability of relatively high strength cannabis ('skunk') in the UK and elsewhere. This has been deliberately cultivated, usually indoors, to contain much larger amounts of THC, the principal active component of cannabis, than field-grown plants. Whereas standard cannabis can be expected to have a THC content of 2–8 % (w/w), 'skunk' may contain over 20 % (w/w) (Drummer, 2004). Clearly the likelihood of accidental overdosage is much higher when 'skunk' cannabis is involved. Back-calculation of time of cannabis use from postmortem blood concentrations of cannabinoids is not recommended (Section 16.9.1.4).

17.7.5 Cardioactive drugs

It has been known for many years that digoxin measurements in heart blood collected postmortem may be misleading when attempting to assess perimortem plasma digoxin concentrations, because of leakage of digoxin from heart muscle after death (Vorpahl and Coe, 1978). Use of peripheral blood is no more reliable. Desethylamiodarone, flecainide and sotalol may also show postmortem changes in blood concentration.

17.7.6 Cocaine

Cocaine is a diester and is hydrolyzed rapidly *in vivo* and *in vitro* by plasma and tissue esterases, and in alkaline solution, to benzoylecgonine (BE) and ecgonine methyl ester (EME) (Toennes and Kauert, 2001; Figure 17.3). Hydrolysis of cocaine can continue after death, even in brain (Moriya and Hashimoto, 1996). Addition of 2 % (w/v) sodium fluoride to a peripheral blood sample and storage at $-20\,^{\circ}C$ reduces plasma esterase activity, thereby increasing the likelihood that the measured blood cocaine concentration reflects the systemic cocaine concentration at the time of sampling. However, if there is significant delay before the postmortem then interpretation of cocaine concentrations can be extremely difficult. It is important to measure benzoylecgonine independently.

Further factors that may be important in the interpretation of results are route of administration, the form of cocaine used (salt or free base), the duration and intensity of prior exposure, and of course the presence of other poisons. Some cocaine is metabolized by cholinesterase, and users may inhibit metabolism by taking cholinesterase inhibiters such as organophosphorus (OP) compounds – this may be a consideration in an otherwise unexplained death in cocaine users. Adulterated cocaine may be yet another factor – diltiazem, phenacetin, and the trazodone and nefazodone metabolite 1-(3-chlorophenyl)piperazine (m-CPP) have been found in illicit cocaine, for example.

Saliva, sweat and hair samples contain relatively high amounts of cocaine, whereas urine contains relatively small amounts of parent drug, but large proportions of BE and EME (Selavka and Rieders, 1995). Cocaine is often taken with alcohol and this can be confirmed by the presence of cocaethylene (a result of transesterification, Section 15.5.3.6) and ecgonine ethyl ester. Cocaethylene has a longer plasma half-life than cocaine and is, if anything, more potent. Anhydroecgonine is a pyrolysis product formed when 'crack' is smoked.

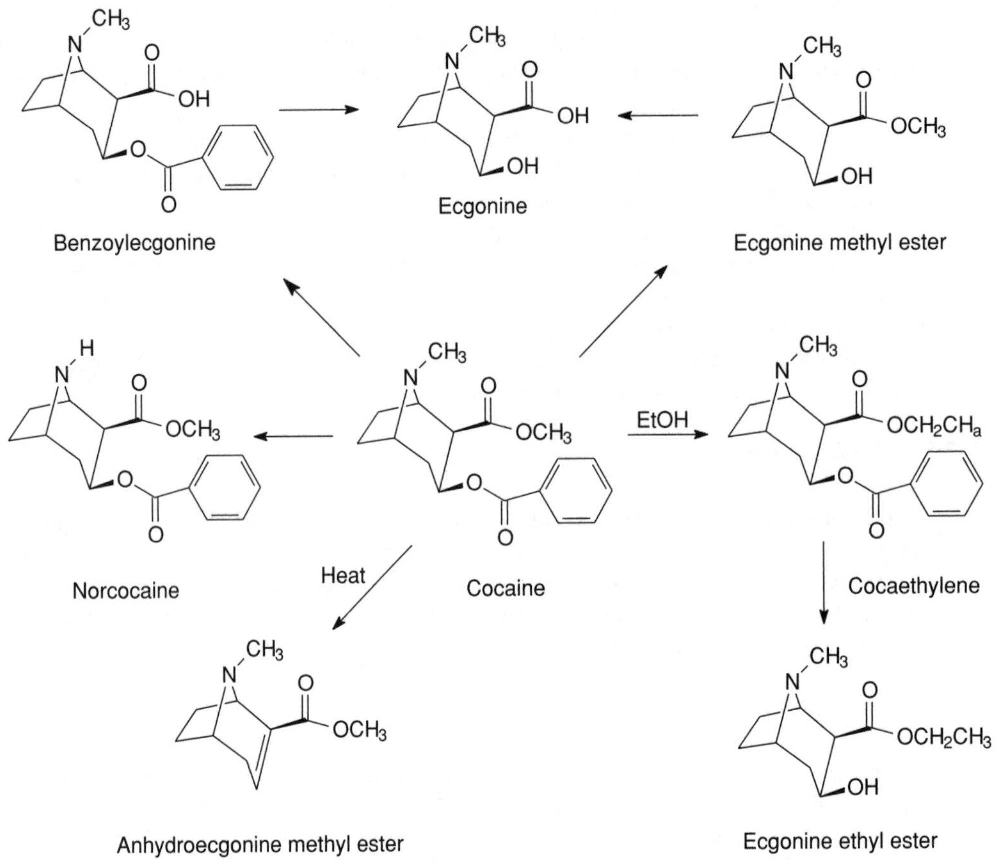

Figure 17.3 Some metabolites and decomposition products of cocaine.

There is no clear site-dependent variation in postmortem blood cocaine, benzoylecgonine or cocaethylene concentrations (Logan *et al.*, 1997), perhaps because of the competing processes of tissue release and degradation of the analytes. In some individuals the acute toxicity of cocaine seems to be dose unrelated (idiosyncratic). It has been suggested that measurement of brain cocaine is important in the investigation of some cocaine-related deaths, such as those preceded by a period of excited delirium (Hernandez *et al.*, 1994). Histological assessment of cardiac muscle to exclude cocaine-induced myocarditis as a cause of death is also important, however. Vitreous humour cocaine concentrations may increase after death (McKinney *et al.*, 1995).

17.7.7 Drug-facilitated sexual assault

'Date-rape drugs' include ethanol, benzodiazepines (such as diazepam, temazepam, clonazepam and oxazepam), cannabis, cocaine, heroin and amfetamines. More recently used drugs include flunitrazepam (Rohypnol, 'roofies'), γ-hydroxybutyrate (γ-hydroxybutyric acid, GHB) and its precursor γ-butyrolactone (GBL), hyoscine (scopolamine) and ketamine ('Special K'). Early sample collection is important. Qualitative identification in urine is often adequate except that with GHB quantitation is needed, as it is an endogenous compound. A urine cut off of 10 mg L^{-1} is recommended to differentiate between endogenous GHB excretion and deliberate administration of GHB or GBL (Kerrigan, 2002).

In blood or plasma, most drugs of abuse can be detected at low mg L^{-1} concentrations for 1–2 days postexposure. In urine, the detection window after a single dose is 1.5–4 days. In chronic users, drugs of abuse can be detected in urine for approximately one week after use, sometimes longer in cocaine and cannabis users. In oral fluid, drugs of abuse can be detected at low mg L^{-1} concentrations for 5–48 h postexposure. The detection 'window' for GHB is much shorter. Analyses for drug metabolites can extend the detection 'window' in some cases. After a single dose of 1 or 2 mg flunitrazepam, for example, sensitive analytical methods can detect 7-aminoflunitrazepam in urine for up to four weeks post-exposure (Verstraete, 2004).

The UK Forensic Science Service has developed an 'early evidence' kit (K106) to facilitate specimen collection in the event of alleged sexual assault (http://www.forensic.gov.uk/forensic_t/inside/news/list_press_release.php?case=12&y, accessed 29 September 2006). Complainants are able to collect urine samples and mouth swabs themselves prior to a full medical examination. Be this as it may, the short 'dwell-time' of some of these drugs linked to the reluctance of victims to come forward immediately, works against the use of urine and oral fluid as media for sample collection. Segmental hair analysis, on the other hand, has been used to detect drug administration several weeks post event (Kintz *et al.*, 2006).

17.7.8 Ethanol (ethyl alcohol, 'alcohol')

Blood ethanol analysis is the commonest request in forensic toxicology, yet the interpretation of results is fraught with problems (Kugelberg and Jones, 2007). Some enzymatic methods for ethanol analysis used by clinical laboratories may give falsely elevated results in critically ill patients or postmortem specimens. Analysis for alcohols should be performed using GC, which provides a reliable ethanol measurement and can be used to quantify acetone, methanol, isopropanol (2-propanol) and other alcohols (Jones and Pounder, 1998). However, ethanol can be lost from or generated in blood specimens if contaminated by bacteria or fungi, or lost by evaporation if the sample tube has a large residual airspace or is not tightly sealed. There is also the possibility of diffusion of ethanol from gastric residue or airways contaminated by vomitus to blood, especially blood in central vessels such as the vena cava.

Where there is putrefaction or extensive injury to the body or there is a delay of several days between death and postmortem, it is advisable to analyze blood taken from a peripheral vessel, urine and vitreous humour specimens (Pounder, 1998). Use of skeletal muscle as an additional specimen for ethanol analysis has also been advocated. In blood, glucose and lactate provide substrates for microbial ethanol production. These compounds are not normally present in vitreous humour or urine, except in the urine of diabetics, and vitreous humour is in any case well protected from bacterial infiltration after death. Thus, analysis of these latter fluids can provide important corroborative information, although this is not always conclusive (Cox *et al.*, 1997). Moreover, Jones and Holmgren (2001) have shown that ethanol distribution ratios (vitreous humour/femoral venous blood) show wide variation in postmortem samples and thus caution is needed when attempting to use vitreous humour to estimate the ethanol concentration in femoral venous blood. Dividing the vitreous humour concentration by two gives a conservative estimate of the femoral blood concentration. Measurement of butanol and other fermentation products, and performing microbiological analysis of the specimen, can also be undertaken to help assess any putrefactive changes. 2-Methyl-2-propanol (*tert*-butanol) has been recommended as an internal standard for GC ethanol assay as it is not likely to be produced by putrefaction (Wigmore, 1993).

Addition of high concentrations of fluoride (at least 0.5 % w/v) to blood obtained during life will inhibit microbial metabolism and prevent any further change in a postmortem blood, urine or vitreous humour sample. Fluoride-containing tubes specifically designed for forensic work are available in the UK from Scenesafe (http://www.scenesafe.co.uk/content.asp?ids=148, accessed 29 September 2006). Available containers include: for urine: 25 mL glass universal; for blood: 25 mL glass universal, 10 mL glass vial with rubber septum closure held in place by white screw cap, 6 mL glass vial with crimped on rubber septum cap. All contain sufficient fluoride to give 0.7–1.0 % (w/v) sodium fluoride when filled. The possible analytical consequences of added fluoride, that is salting out if using HS-GC (Prouty and Anderson, 1987) and inhibition of enzymatic ethanol assays, must be remembered, however.

Ethyl glucuronide is a non-volatile, water-soluble, stable ethanol metabolite that can be detected and measured in body fluids for some time after ethanol is no longer detectable (Wurst *et al.*, 2000). It is also laid down in hair and can be used to monitor abstinence, for example. Ethyl glucuronide has been measured in postmortem fluids and tissues, although there is no clear application of this latter analysis as yet.

17.7.9 Heroin/morphine

The metabolism of heroin is complex and it is important to be clear about what has been measured in a blood, urine or other biological sample as this affects the interpretation of the results. Heroin is metabolized very quickly to 6-monoacetylmorphine (6-MAM; Figure 15.5), which has a plasma half-life of some 5–25 min and is sometimes detectable in biological samples. Detection of 6-MAM provides clear evidence of heroin use, as it is not produced from morphine (Beck and Böttcher, 2006). Morphine, the product of 6-MAM metabolism, has a much longer plasma half-life (2–3 h) than 6-MAM, and is usually the compound that is detected after heroin administration. Morphine is available as a pharmaceutical preparation for oral or parenteral use. Heroin, 6-MAM and morphine have similar toxicity.

Illicit heroin normally contains some 3-acetylcodeine derived from codeine present in opium. 3-Acetylcodeine is metabolized to codeine in the body, but the amounts present in blood after heroin use are usually insignificant as compared to the amounts of 6-MAM and morphine present. Morphine and codeine are metabolized principally by conjugation to give inactive derivatives (morphine-3-glucuronide and codeine-3-glucuronide), which are excreted in urine. A relatively small proportion of the dose of heroin/morphine is also metabolized to morphine-6-glucuronide, which is pharmacologically active, but is only thought to contribute to the toxicity of morphine when the drug is given to patients in renal failure. A small amount of codeine may be converted into morphine in the body (Section 15.5.1.1).

Problems of establishing illicit opiate use are reflected in the number of different approaches that have been advocated. These include the blood morphine:codeine ratio and the presence of additional compounds derived from opium such as: (i) acetylcodeine, (ii) meconin, (iii) noscapine metabolites, (iv) papaverine metabolites and (v) reticuline.

Renal failure apart, the acute toxicity of heroin/morphine is very much dependent on: (i) the dose, (ii) the subject's immediate past exposure to opioid agonists, as people can tolerate exposure to much higher doses if they have been taking such drugs within the last few days, (iii) the method of administration, i.v. dosage being much more dangerous than inhalation of fumes or i.m. injection, (iv) the purity of any mixture injected and (v) in the presence of other centrally acting drugs such as ethanol or methadone. Any interpretation of postmortem morphine data must bear these factors in mind. Because the toxicity of morphine is due to depression of the respiratory centre, a patient maintained via mechanical ventilation may tolerate an otherwise fatal dose. This may give rise to a false suspicion of morphine overdosage if a patient dies subsequently and the immediate cause of death is not clear initially.

Traditionally immunoassays have been available which measure 'total' morphine (i.e. morphine + metabolites, principally the inactive 3-glucuronide). The advantage of this is that, if used together with the 'free' (unconjugated) morphine result, an idea of the magnitude of the heroin/morphine dose can be obtained. If the total morphine is relatively low compared to the free morphine then this might suggest that death ensued quite quickly after heroin/morphine administration, whilst if the total morphine is high and the free morphine low, this might suggest death was not due to morphine administration unless some secondary factor such as hypoxic brain damage had supervened. Of course there are many possible scenarios, but, in general, both free and total morphine should be measured. In addition, the measurements should be performed in blood and in urine, if available, and perhaps in other fluids as the results from these different specimens often provide complementary information to aid clinical interpretation. For example, the presence of low concentrations of free (unconjugated) morphine (<1 mg L^{-1}) in urine obtained postmortem in a suspected heroin-related death strongly suggests that death occurred shortly after a heroin dose. Free morphine concentrations in ventricular postmortem blood are consistently higher than those at peripheral sites, especially at concentrations >0.3 mg L^{-1}, but there appears to be little change with time between death and the blood sample being obtained from either central, or peripheral sites (Logan and Smirnow, 1996).

17.7.10 Hypoglycaemic agents

If the injection of an overdose of insulin or insulin administration to a nondiabetic is suspected, blood must be taken as soon as possible after death and serum separated and deep-frozen for subsequent analysis of insulin and C-peptide (Section 11.7.2). This being said, meaningful interpretation of serum insulin concentrations in insulin-dependant diabetics is often not possible. Blood glucose concentrations generally fall rapidly after death and are an unreliable guide to antemortem glucose.

17.7.11 Methadone

As with heroin and morphine, postmortem blood methadone concentrations may be difficult to interpret especially if there is a history of prior use of methadone or other opioids and/or other drugs are also present (Milroy and Forrest, 2000). Delayed toxicity, especially at night, is a feature of poisoning with this compound, which may further complicate the interpretation of results (Wolff, 2002). Deaths from methadone in children and in subjects previously exposed, but who have lost tolerance to opioids, are not uncommon. Monitoring the main methadone metabolite, EDDP (2-ethylidene-1,5-dimethyl-3,3-diphenylpyrrolidine), in urine not only provides evidence of adherence to methadone treatment, but also guards against non-adherent individuals adding methadone to 'blank' urine in order to obtain continued prescription of the drug with intent to supply to other persons.

17.7.12 Methylenedioxymetamfetamine and related compounds

Deaths related to MDMA ('XTC', 'ecstasy') use are not clearly related to dose and are often associated with use of alcohol and/or other illicit drugs. Deaths attributable to MDA ('adam'), MDEA ('eve') and other related compounds are relatively rare (Schifano *et al.*, 2003). It is important to analyze not only blood, but also urine and stomach contents if available in suspected MDMA-related deaths and to look for other drugs as well as MDMA.

17.7.13 Volatile substance abuse (VSA)

It is especially important to consider all circumstantial and other evidence in cases of possible VSA ('glue sniffing', solvent abuse)-related sudden death, especially if legitimate exposure to solvent vapour was a possibility (Flanagan *et al.*, 1997). Blood should be collected in a glass tube with a closure lined with aluminium foil and with minimal airspace (headspace). Measurement of urinary metabolites of toluene, xylene and trichloroethylene can indicate exposure. Tissue (5 g brain, lung, fat, liver, kidney, spleen) may also be useful, particularly if putrefaction has occurred. Specimens should be transported at 4 or −20 °C. With very volatile substances such as 'butane' from cigarette lighter refills (mixture of propane, butane and isobutane), which are gases at room temperature, identification and hence confirmation of exposure can be performed, but quantification is often not practicable, even in blood, unless special precautions are taken.

17.8 Summary

All the available evidence must be taken into account when investigating any death or other incident where poisoning is suspected. An overall knowledge of the circumstances, time course, clinical/postmortem observations, possible poisons involved and their toxicology and metabolism is paramount. Toxicological analysis can provide objective evidence of exposure and of the magnitude of exposure. Factors which may assist in the interpretation of postmortem blood toxicology results are the possible availability of: (i) antemortem samples (whole blood or plasma/serum) and (ii) routine TDM data. Even if such 'baseline' data are available, comparison with postmortem data must be made in the knowledge of the possible effects of the time elapsed since death, the site of collection of the blood sample and analyte stability, amongst other factors. Furthermore, TDM samples are frequently measured pre-dose ('trough' concentrations) whereas the samples for toxicological analysis are usually taken without regard to the time since the last dose. Measurement of major blood metabolites is important if only to ensure that they have not been quantified together with the parent drug, and the blood drug:metabolite ratio might indicate whether exposure was acute or chronic. Analysis of other fluids and/or tissues may provide further useful information.

References

Apple, F.S. (1989) Postmortem tricyclic antidepressant concentrations: Assessing cause of death using parent drug to metabolite ratio. *J Anal Toxicol*, **13**, 197–8.

Bachs, L. and Morland, H. (2001) Acute cardiovascular fatalities following cannabis use. *Forensic Sci Int*, **124**, 200–3.

Bailey, D.G., Malcolm, J., Arnold, O. and Spence, J.D. (1998) Grapefruit juice-drug interactions. *Br J Clin Pharmacol*, **46**, 101–10.

Beck, O. and Böttcher, M. (2006) Paradoxical results in urine drug testing for 6-acetylmorphine and total opiates: implications for best analytical strategy. *J Analyt Toxicol*, **30**, 73–9.

Bernstein, G., Jehle, D., Bernaski, E. and Braen, G.R. (1992) Failure of gastric emptying and charcoal administration in fatal sustained-release theophylline overdose: pharmacobezoar formation. *Ann Emerg Med*, **21**, 1388–90.

Bogusz, M.J., al Tufail, M. and Hassan, H. (2002) How natural are 'natural herbal remedies'? A Saudi perspective. *Adverse Drug React Toxicol Rev*, **21**, 219–29.

Buratti, S. and Lavine, J.E. (2002) Drugs and the liver: advances in metabolism, toxicity, and therapeutics. *Curr Opin Pediatr*, **14**, 601–7.

Burns, M.J. (2001) The pharmacology and toxicology of atypical antipsychotic agents. *J Toxicol Clin Toxicol*, **39**, 1–14.

Caraco, Y. (2004) Genes and the response to drugs. *N Engl J Med*, **351**, 2867–9.

Chiou, W.L. (1989a) The phenomenon and rationale of marked dependence of drug concentration on blood sampling site. Implications in pharmacokinetics, pharmacodynamics, toxicology and therapeutics (Part I). *Clin Pharmacokinet*, **17**, 175–99.

Chiou, W.L. (1989b) The phenomenon and rationale of marked dependence of drug concentration on blood sampling site. Implications in pharmacokinetics, pharmacodynamics, toxicology and therapeutics (Part II). *Clin Pharmacokinet*, **17**, 275–90.

Corns, C.M. (2003) Herbal remedies and clinical biochemistry. *Ann Clin Biochem*, **40**, 489–507.

Cox, D.E., Sadler, D.W. and Pounder, D.J. (1997) Alcohol estimation at necropsy. *J Clin Pathol*, **50**, 197–201.

Cummins, C.L., Wu, C.Y. and Benet, L.Z. (2002) Sex-related differences in the clearance of cytochrome P450 3A4 substrates may be caused by P-glycoprotein. *Clin Pharmacol Ther*, **72**, 474–89.

Dalpe-Scott, M., Degouffe, M., Garbutt, D. and Drost, M. (1995) A comparison of drug concentrations in postmortem cardiac and peripheral blood in 320 cases. *Can Soc Forensic Sci J*, **28**, 113–21.

Druid, H. and Holmgren, P. (1997) A compilation of fatal and control concentrations of drugs in postmortem femoral blood. *J Forensic Sci*, **42**, 79–87.

Drummer, O.H. (2004) Postmortem toxicology of drugs of abuse. *Forensic Sci Int*, **142**, 101–13.

Drummer, O.H. and Gerostamoulos, J. (2002) Postmortem drug analysis: analytical and toxicological aspects. *Ther Drug Monit*, **24**, 199–209.

Evans, D.A.P., Manley, K.A. and Mc Kusik, K.V. (1960) Genetic control of isoniazid metabolism in man. *Br Med J*, **2**, 485–91.

Flanagan, R.J., Amin, A. and Seinen, W. (2003) Effect of postmortem changes on peripheral and central whole blood and tissue clozapine and norclozapine concentrations in the domestic pig (*Sus scrofa*). *Forensic Sci Int*, **132**, 9–17.

Flanagan, R.J., Spencer, E.P., Morgan, P.E., Barnes, T.R.E. and Dunk, L. (2005) Suspected clozapine poisoning in the UK/Eire, 1992–2003, *Forensic Sci Int*, **155**, 91–9.

Flanagan, R.J., Streete, P.J. and Ramsey, J.D. (1997) *Volatile Substance Abuse—Practical Guidelines for Analytical Investigation of Suspected Cases and Interpretation of Results*. United Nations International Drug Control Programme, Vienna (http://www.undcp.org/odccp/technical_series_1997-01-01_1.html, accessed 17 July 2006).

Forrest, A.R.W. (1993) Obtaining samples at postmortem examination for toxicological and biochemical analysis. *J Clin Pathol*, **46**, 292–6.

Fraser, A.D., Coffin, L. and Worth, D. (2002) Drug and chemical metabolites in clinical toxicology investigations: the importance of ethylene glycol, methanol and cannabinoid metabolite analyses. *Clin Biochem*, **35**, 501–11.

Fuke, C., Berry, C.L. and Pounder, D.J. (1996) Postmortem diffusion of ingested and aspirated paint thinner. *Forensic Sci Int*, **78**, 199–207.

Furuta, T., Ohashi, K., Kamata, T., Takashima, M., Kosuge, K., Kawasaki, T., Hanai, H., Kubota, T., Ishizaki, T. and Kaneko, E. (1998) Effect of genetic differences in omeprazole metabolism on cure rates for Helicobacter pylori infection and peptic ulcer. *Ann Intern Med*, **129**, 1027–30.

Gasche, Y., Daali, Y., Fathi, M., Chiappe, A., Cottini, S., Dayer, P. and Desmeules, J. (2004) Codeine intoxication associated with ultrarapid CYP2D6 metabolism. *N Engl J Med*, **351**, 2827–31 (erratum) *N Engl J Med*, **352**, 638).

Gill, R., Hatchett, S.E., Osselton, M.D., Wilson, H.K. and Ramsey, J.D. (1988) Sample handling and storage for the quantitative analysis of volatile compounds in blood: the determination of toluene by headspace gas chromatography. *J Anal Toxicol*, **12**, 141–6.

Gupta, B.D., Jani, C.B. and Shah, P.H. (2001) Fatal 'Bhang' poisoning. *Med Sci Law*, **41**, 349–52.

Harris, R.Z., Benet, L.Z. and Schwartz, J.B. (1995) Gender effects in pharmacokinetics and pharmacodynamics. *Drugs*, **50**, 222–39.

Health and Safety Executive (2005). *EH40/2005. Occupational Exposure Limits 2005.* HSE Books, Sudbury.

Hearn, W.L., Keran, E.E., Wei, H.A. and Hime, G. (1991) Site-dependent postmortem changes in blood cocaine concentrations. *J Forensic Sci*, **36**, 673–84.

Hernandez, A., Andollo, W. and Hearn, W.L. (1994) Analysis of cocaine and metabolites in brain using solid phase extraction and full-scanning gas chromatography/ion trap mass spectrometry. *Forensic Sci Int*, **65**, 149–56.

Ingelman-Sundberg, M. (2004) Pharmacogenetics of cytochrome P450 and its applications in drug therapy: the past, present and future. *Trends Pharmacol Sci*, **25**, 193–200.

Israili, Z.H. and Dayton, P.G. (2001) Human alpha-1-glycoprotein and its interactions with drugs. *Drug Metab Rev*, **33**, 161–235.

Jannetto, P.J., Wong, S.H., Gock, S.B., Laleli-Sahin, E., Schur, B.C. and Jentzen, J.M. (2002) Pharmacogenomics as molecular autopsy for postmortem forensic toxicology: genotyping cytochrome P450 2D6 for oxycodone cases. *J Anal Toxicol*, **26**, 438–47.

Jatlow, P., Dobular, K. and Bailey, D. (1979) Serum diazepam concentrations in overdose: their significance. *Am J Clin Pathol*, **72**, 571–7.

Jones, A.W. and Holmgren, P. (2001) Uncertainty in estimating blood ethanol concentrations by analysis of vitreous humour. *J Clin Pathol*, **54**, 699–702.

Jones, A.W. and Pounder, D.J. (1998) Measuring blood alcohol concentration for clinical and forensic purposes, in (ed. S. Karch), *Drug Abuse Handbook*, CRC Press, Boca Raton, Florida. pp 327–356.

Kathiramalainathan, K., Kaplan, H.L., Romach, M.K., Busto, U.E., Li, N.Y., Sawe, J., Tyndale, R.F., Sellers, E.M. (2000) Inhibition of cytochrome P450 2D6 modifies codeine abuse liability. *J Clin Psychopharmacol*, **20**, 435–44.

Kerrigan, S. (2002) In vitro production of gamma-hydroxybutyrate in antemortem urine samples. *J Anal Toxicol*, **26**, 571–4.

Kintz, P., Villain, M. and Cirimele, V. (2006) Hair analysis for drug detection. *Ther Drug Monit*, **28**, 442–6.

Klotz, U., Avant, G.R., Hoyumpa, A., Schenker, S. and Wilkinson, G.R. (1975) The effects of age and liver disease on the disposition and elimination of diazepam in adult man. *J Clin Invest*, **55**, 347–59.

Koren, G., Cairns, J., Chitayat, D., Gaedigk, A. and Leeder, S.J. (2006) Pharmacogenetics of morphine poisoning in a breastfed neonate of a codeine-prescribed mother. *Lancet*, **368**, 704.

Koski, A., Ojanperä, I. and Vuori, E. (2003) Interaction of alcohol and drugs in fatal poisonings. *Hum Exp Toxicol*, **22**, 281–7.

Krishna, S., and White, N.J. (1996) Pharmacokinetics of quinine, chloroquine and amodiaquine. Clinical implications. *Clin Pharmacokinet*, **30**, 263–99.

Kugelberg, F.C. and Jones, A.W. (2007) Interpreting results of ethanol analysis in postmortem specimens: A review of the literature. *Forensic Sci Int*, **165**, 10–29.

Lane, E.A., Guthrie, S. and Linnoila, M. (1985) Effects of ethanol on drug and metabolite pharmacokinetics. *Clin Pharmacokinet*, **10**, 228–47.

Langford, A.M. and Pounder, D.J. (1997) Possible markers for postmortem drug redistribution. *J Forensic Sci*, **42**, 88–92.

Leikin, J.B. and Watson, W.A. (2003) Post-mortem toxicology: what the dead can and cannot tell us. *J Toxicol Clin Toxicol*, **41**, 47–56.

Lennard, L. and Lilleyman, J.S. (1996) Individualizing therapy with 6-mercaptopurine and 6-thioguanine related to the thiopurine methyltransferase genetic polymorphism. *Ther Drug Monit*, **18**, 328–34.

Levo, A., Koski, A., Ojanpera, I., Vuori, E. and Sajantila, A. (2003) Post-mortem SNP analysis of CYP2D6 gene reveals correlation between genotype and opioid drug (tramadol) metabolite ratios in blood. *Forensic Sci Int*, **135**, 9–15.

Logan, B.K. and Smirnow, D. (1996) Postmortem distribution and redistribution of morphine in man. *J Forensic Sci*, **41**, 221–9.

Logan, B.K., Smirnow, D. and Gullberg, R.G. (1997) Lack of predictable site-dependent differences and time-dependent changes in postmortem concentrations of cocaine, benzoylecgonine, and cocaethylene in humans. *J Analyt Toxicol*, **21**, 23–31.

Logan, B.K., Weiss, E.L. and Harruff, R.C. (1996) Case report: distribution of methamphetamine in a massive fatal ingestion. *J Forensic Sci*, **41**, 322–3.

Martin, A. and Pounder, D.J. (1992) Post-mortem toxico-kinetics of trazodone. *Forensic Sci Int*, **56**, 201–7.

McCarthy, R.H. (1994) Seizures following smoking cessation in a clozapine responder. *Pharmacopsychiatry*, **27**, 210–1.

McIntyre, L.M., King, C.V., Boratto, M. and Drummer, O.H. (2000) Post-mortem drug analyses in bone and bone marrow. *Ther Drug Monit*, **22**, 79–83.

McKinney, P.E., Phillips, S., Gomez, H.F., Brent, J., MacIntyre, M. and Watson, W.A. (1995) Vitreous humor cocaine and metabolite concentrations: do postmortem specimens reflect blood levels at the time of death? *J Forensic Sci*, **40**, 102–7.

Milroy, C.M. and Forrest, A.R. (2000) Methadone deaths: a toxicological analysis. *J Clin Pathol*, **53**, 277–81.

Moriya, F. and Hashimoto, Y. (1996) Postmortem stability of cocaine and cocaethylene in blood and tissues of humans and rabbits. *J Forensic Sci*, **41**, 612–6.

Moriya, F. and Hashimoto, Y. (1999) Redistribution of basic drugs into cardiac blood from surrounding tissues during early-stages postmortem. *J Forensic Sci*, **44**, 10–6.

Poggio, E.D., Wang, X., Greene, T., Van Lente, F. and Hall, P.M. (2005) Performance of the modification of diet in renal disease and Cockcroft-Gault equations in the estimation of GFR in health and in chronic kidney disease. *J Am Soc Nephrol*, **16**, 459–66.

Pounder, D.J. (1998) Dead sober or dead drunk? *Br Med J*, **316**, 87.

Pounder, D.J., Adams, E., Fuke, C. and Langford, A.M. (1996b) Site to site variability of postmortem drug concentrations in liver and lung. *J Forensic Sci*, **41**, 927–32.

Pounder, D.J. and Davies, J.I. (1994) Zopiclone poisoning: tissue distribution and potential for postmortem diffusion. *Forensic Sci Int*, **65**, 177–83.

Pounder, D.J., Fuke, C., Cox, D.E., Smith, D. and Kuroda, N. (1996a) Postmortem diffusion of drugs from gastric residue: an experimental study. *Am J Forensic Med Pathol*, **17**, 1–7.

Proudfoot, A.T., Stewart, M.J., Levitt, T. and Widdop, B. (1979) Paraquat poisoning: Significance of plasma paraquat concentrations. *Lancet*, **ii**, 330–2.

Prouty, R.W. and Anderson, W.H. (1987) A comparison of postmortem heart blood and femoral blood ethyl alcohol concentrations. *J Anal Toxicol*, **11**, 191–7.

Rao, Y., Hoffmann, E., Zia, M., Bodin, L., Zeman, M., Sellers, E.M. and Tyndale, R.F. (2000) Duplications and defects in the CYP2A6 gene: identification, genotyping, and in vivo effects on smoking. *Mol Pharmacol*, **58**, 747–55.

Rostami-Hodjegan, A., Amin, A.M., Spencer, E.P., Lennard, MS., Tucker, GT. and Flanagan, R.J. (2004) Influence of dose, cigarette smoking, age, sex, and metabolic activity on plasma clozapine concentrations: a predictive model and nomograms to aid clozapine dose adjustment and to assess compliance in individual patients. *J Clin Psychopharmacol*, **24**, 70–8.

Schifano, F., Oyefeso, A., Corkery, J., Cobain, K., Jambert-Gray, R., Martinotti, G. and Ghodse, A.H. (2003) Death rates from ecstasy (MDMA, MDA) and polydrug use in England and Wales 1996–2002. *Hum Psychopharmacol*, **18**, 519–24.

Selavka, C.M. and Rieders, F. (1995) The determination of cocaine in hair: a review. *Forensic Sci Int*, **70**, 155–64.

Skopp, G. (2004) Preanalytic aspects in postmortem toxicology. *Forensic Sci Int*, **142**, 75–100.

Tanaka, E. (2003) Toxicological interactions involving psychiatric drugs and alcohol: an update. *J Clin Pharm Ther*, **28**, 81–95.

Tanaka, E. (2002) Toxicological interactions between alcohol and benzodiazepines. *J Toxicol Clin Toxicol*, **40**, 69–75.

Toennes, S.W. and Kauert, G.F. (2001) Importance of vacutainer selection in forensic toxicological analysis of drugs of abuse. *J Anal Toxicol*, **25**, 339–43.

Verstraete, A.G. (2004) Detection times of drugs of abuse in blood, urine, and oral fluid. *Ther Drug Monit*, **26**, 200–5.

Visapaa, J.P., Tillonen, J.S., Kaihovaara, P.S. and Salaspuro, M.P. (2002) Lack of disulfiram-like reaction with metronidazole and ethanol. *Ann Pharmacother*, **36**, 971–4.

Vorpahl, T.E. and Coe, J.I. (1978) Correlation of antemortem and postmortem digoxin levels. *J Forensic Sci*, **23**, 329–34.

Vree, T.B., van Dongen, R.T. and Koopman-Kimenai, P.M. (2000) Codeine analgesia is due to codeine-6-glucuronide, not morphine. *Int J Clin Pract*, **54**, 395–8.

Wennig, R. (2000) Potential problems with the interpretation of hair analysis results. *Forensic Sci Int*, **107**, 5–12.

Wigmore, J.G. (1993) The distribution of ethanol in postmortem blood samples. *J Forensic Sci*, **38**, 1019–21.

Williams, M.L. and Wainer, I.W. (2002) Role of chiral chromatography in therapeutic drug monitoring and in clinical and forensic toxicology. *Ther Drug Monit*, **24**, 290–6.

Wolff, K. (2002) Characterization of methadone overdose: Clinical considerations and the scientific evidence. *Ther Drug Monit*, **24**, 457–70.

Wurst, F.M., Kempter, C., Metzger, J., Seidl, S. and Alt, A. (2000) Ethyl glucuronide: A marker of recent alcohol consumption with clinical and forensic implications. *Alcohol*, **20**, 111–6.

Yates, C.R., Zhang, W., Song, P., Li, S., Gaber, A.O., Kotb, M., Honaker, M.R. and Alloway, R.R., Meibohm, B. (2003) The effect of CYP3A5 and MDR1 polymorphic expression on cyclosporine oral disposition in renal transplant patients. *J Clin Pharmacol*, **43**, 555–64.

Yonemitsu, K. and Pounder, D.J. (1993) Postmortem changes in blood tranylcypromine concentration: competing redistribution and degradation effects. *Forensic Sci Int*, **59**, 177–84.

Index

B

C

D

E

N

O

P